SCIENCE & TECHNOLOGY FIRSTS

Daniel J. Boorstin, guest foreword

Donna Olendorf, editor

SCIENCE & TECHNOLOGY FIRSTS

LEONARD C. BRUNO

GALE

DETROIT · NEW YORK · TORONTO · LONDON

Science & Technology Firsts

Leonard C. Bruno
Index by Michelle B. Cadoree

Gale Research Staff

Donna Olendorf, *Editor*
Kyung Lim Kalasky, *Production Editor*

Nicole Beatty, Kristine Binkley, *Photo Coordinators*

Erika Berry, Karen Boyden, Christine Jeryan, Jacqueline Longe,
Kimberley A. McGrath, Pamela Proffitt, Bridget Travers
Assisting Editors

Mary Beth Trimper, *Production Director*

Shanna Heilveil, *Production Assistant*

Cynthia Baldwin, *Art Director*

Tracey Rowens, *Cover and Page Designer*

Susan Salas, *Permissions Coordinator*

Jeffrey Muhr, *Editorial Technical Support*

Library of Congress Cataloging-in-Publication Data

Bruno, Leonard C.
    Science & technology firsts / Leonard C. Bruno ; Donna Olendorf,
edtior ; guest foreword by Daniel J. Boorstin.
        p.   cm.
    Includes index.
    ISBN 0-7876-0256-6 (alk. paper)
    1. Technology--History--Miscellanea.   2. Science--History-
-Miscellanea.   I. Olendorf, Donna. II. Title.
T15.B684   1996
609--dc20   96-43595
                                                    CIP

TABLE OF CONTENTS

FOREWORD

We Americans have been understandably obsessed by "firsts" of every kind. Our nation has grown by expanding across a sparsely settled and little-known continent, and our history has been dominated by the story of a frontier. "Getting There First" was the key to success in the race to engross and develop the farming and mineral resources of our New World nation. In my home state of Oklahoma, early settlers were called "Sooners" for having won the race to occupy Indian Territory. The trans-continental race for exploitation and community-building has, of course, been promoted by new technologies—the steamboat and the railroad in the 19th century, and the automobile and the airplane in our century. The pioneers in this technology—John Fitch, Robert Fulton, Henry Ford, and the Wright brothers, among others—have shaped American history.

The desire to protect and encourage innovators was here at the birth of the nation. The federal Constitution (Article I, Section 8) provided the foundation of our patent laws when it gave Congress the power "To Promote the Progress of Science and useful Arts, by securing for limited Times to Authors and Inventors the exclusive Right to their respective Writings and Discoveries." When a Patent Office was established by the Act of 1836 (much modified since), it required that every patent application be examined for the invention's novelty and usefulness. Until 1880 a model was required with every application. By the late 20th century more than 3 million patents had been granted.

The historian in search of "firsts" faces a bewildering variety of challenges. For example, there has long been a striking contrast between the realms of Discoverers and of Inventors. The Discoverer's work—the success of Columbus or Magellan—was often public, praised and celebrated by their nationals. And yet in that great Age of Discovery, because geographic discoveries of the best and shortest routes at sea had great commercial value, they were enshrouded in official secrecy which the reckless sailor would violate at risk of his life. This makes it difficult for the historian who seeks to apportion credit for first discoveries among the Portuguese and the Spanish, and others. Still, the historic reputation of a Discoverer depends on his claim to be "first" in some enterprise. Being second was not good enough.

The great geographic Discoverers, whose works are not within the scope of this book, were often adventurers, who did their work in the open air (cf. *Explorers and Discoverers of the World*). The Inventors, in sharp contrast, were commonly artisans whose habitat was the stuffy workshop. There they forged axles, cut cogs on wheels, polished lenses and mirrors. The great Discoverers needed sponsorship of sovereigns to finance and outfit their expeditions which were inevitably public and invited competition from other nations. The Inventor shaped what he held in his hand. His work, too, was competitive, but until modern times, until the growth of patent and copyright laws, it was less public, often remained anonymous and even secret. The remarkable inventions of Christian Huygens (1629-1695) and Robert Hooke (1635-1703), marking epochs in the history of astronomy and timekeeping, are famous only to the historians of technology. The great works of Inventors seldom made them national heroes. Their innovations were incremental—a new pendulum, a new "grasshopper" escapement—and not designed to strike the lay imagination.

The title of this volume, *Science & Technology Firsts*, bears witness to a modern development, not often enough noted. Until modern times, the worlds of the Discoverer and of the Inventor (with a few remarkable exceptions like Galileo [1564-1642]) were conspicuously separate. But recently, the progress of science has become increasingly a producer and a by-product of technology. Discoverer and Inventor have now been joined in a new unlimited, if involuntary, partnership. They have been assimilated in a single community of questing mankind. The science of astronomy and the technology of the telescope, of biology and the microscope, have advanced in unison. To make sense of my story of the great Discoverers, I needed to chronicle only a few crucial inventions. These included the clock, the compass, the printing press, the telescope, and the microscope. But the most recent advances of science have enlisted varieties of computers and countless inventions in electronics and other realms, that are themselves a spin-off of scientific advances. Our successes in space exploration are brilliant examples of the new collaboration.

The assimilation of advancing knowledge and advancing artisanship—the collaboration of science and technology firsts—has been expressed in the new vocation of "Scientist." What once was the realm of the "Natural Philosopher" (which seemed more akin to philosophy than to physics) has become a new vocation so widespread and so familiar that we have ceased to note its signifi-

cance. Surprisingly, the word "Scientist" is not recorded in English until 1840. The new word signaled a new partnership. "We need very much a name," said the brilliant English philosopher and mathematician William Whewell (1794-1866), "to describe a cultivator of science in general. I should incline to call him a Scientist." About the same time another realm of innovation, the realm of the artisan, was being transformed. And the word "artist" now came into use to describe "one who practices a manual art in which there is much room for display of taste; one who makes his craft a 'fine art'."

Newly accelerated forces in technology and society would soon make the new role of "Scientist"—the seeker of firsts in scientific knowledge—very much like that of the time-honored Discoverer. Since the 19th century, literacy, the diffusion of books, and the rise of periodical and daily journalism have given the day-to-day enterprises of the Scientist prompt—and eagerly received—public notice. And since the Industrial Revolution, the successes of the Inventor, stimulated by the profit-incentives of patent protection and mass-production and marketing, have become increasingly competitive, public, and anything but anonymous. The lengthy litigation over Bell's telephone patent and over Ford's automotive patent made inventors into newsworthy celebrities. And so created a copious new documentary source for the historian of technology. The growth of large corporations with heavy interests in certain inventions, along with patent laws and the efflorescence of the American legal profession, have made "Who was the first inventor?" no mere academic question. The user of this book should be reminded that this survey of "firsts" is also a survey of the limits of our present knowledge. Every assertion of a "first" is a confession that we know no earlier example of the listed scientific concept or invention. This book, therefore, will be an invitation to readers to increase our knowledge by till-now unnoted examples of these same human achievements.

In these and countless other ways science and technology firsts have entered the mainstream of American social and economic history. And the new partnership of the search to know and the passion to innovate have made it more than ever impossible to prophesy impossibilities. While the flood of firsts, described in this book, may have dulled our sensitivity to surprise, we hope that they may help keep our imaginations open to the endless stream of firsts still to come.

—*Daniel J. Boorstin,* Historian and
    the Librarian of Congress Emeritus
    (copyright © 1996, by Daniel J. Boorstin)

"He who thus considers things in their first growth and origin ... will obtain the clearest view of them."
—Aristotle, *Politics*. Benjamin Jowett translation, Franklin Center, PA: Franklin Library, 1977, p. 4.

To this Aristotelian dictum might be added the following: "And he who puts them in chronological order will understand the relation of one to another." Although there is no guarantee that any real insight is gained by knowing things either way, there is in fact much to be said for examining certain areas of human endeavor from a chronological perspective. While chronology alone is certainly not real history, it is in many ways the frame upon which the story of history hangs. Without the ordered, factual progression that chronology provides, history—or organized, related, and interpreted information—would not be possible. Since chronology places events in the sequence in which they occurred, its intrinsic importance might be better appreciated by attempting to write history without the certainty of knowing which event preceded another. Without this time-based foundation, the construction of historical narrative would fall apart, since historical cause and effect would be reduced to speculation and little could be stated with any certainty. Knowledge of the time when something happened and, therefore, of what came first and what came afterwards, is really the skeleton upon which the flesh of history is placed. Neither is fully useful without the other.

Chronology of course necessarily implies a beginning, and the approximately 4,000 entries and 164 photos that comprise this book are the results of a search for the many beginnings of science and technology. Although primacy, or being first in any field or endeavor, is often a relative quality with a decidedly cultural dimension, this aspect is less noticeable in scientific and technical areas. Certain accomplishments in those fields easily lend themselves to absolute categorization—like the first man to walk on the moon. That feat is arguably one of those hard-and-fast "firsts" about which there is little debate. However, science and technology also have their share of less-than-clear-cut "firsts," and the need to add a qualifier to a first is sometimes more the rule than the exception. An example of this might be the first American in space. Although Alan B. Shepard, Jr. was the first American to enter the vacuum of space, his 15-minute *Freedom 7* flight was a ballistic one or simply an up-and-down mission, landing him about 300 miles downrange. John H. Glenn, Jr., however, was the first American to orbit in space (spending nearly four hours in orbit). This may be a minor distinction, but it is nonetheless a significant difference. Because of these important distinctions, this book contains many such qualified firsts.

While much could be written about the seemingly universal importance of being the first of anything (from primogeniture, to being first at the patent office, or the first to publish), often many of these accepted landmarks are really the result of many smaller steps or accomplishments, sometimes by others and sometimes of equal importance. This book includes the well-known firsts in science and technology, but it also attempts to recognize many of those second-level firsts without diminishing the work of the tried-and-true achievers. Those accomplishments which are so well-known as to have attained a sort of mythic status in our culture, folklore, or curriculum need little help in telling their story. It is rather those other firsts which somehow never reached that lofty plane—often remaining anonymous, little-known, or misunderstood—that need some light shed in their direction.

Finally, to get the most out of a book like this one, the reader should be aware of some of the ground rules the author set for himself. Most important is the fact that this is not a comprehensive history of science and technology told in a chronological manner. Rather, it is selected history based on the concept of "firsts." As such, there are naturally many major achievements that are not included here simply because they did not qualify as a first of something. A good example are the various Nobel Prizes. These major international awards are not listed except when there is some aspect of a first involved—such as Marie S. Curie being the first woman or Albert A. Michelson the first American to attain that honor. The reader may also notice that this book is preponderantly heavy on the science side of things as opposed to technology. This was done deliberately because of the availability of so many compilations, chronologies, encyclopedias, and timetables of invention. The author did not want to get bogged down making sure that the first electric coffee pot was included and overlook the discovery and first use of the coffee bean.

Some of the book's 12 chapters also may require explanation as to their contents, while others are exactly what one would expect. The Astronomy chapter is one of the more clear-cut sections, as is Mathematics. However, some astronomical instruments that typically appear in a chapter on astronomy appear rather in a technology-related section. Mathematics naturally includes the intricate variety so typical of that discipline, but its concrete application (such as in computers) will be found elsewhere. Medicine includes surgery but also has some information on a subject like herbals, which is also found in Biology. This is so because the general term biology is considered to include plants (botany) as well as animals (zoology). Earth Sciences goes beyond geology to include earth-based disciplines like oceanography and sometimes activities like mining. Physics attempts to embrace both classical physics—with its many sub-areas like mechanics, optics, and electricity and magnetism—as well as modern or quantum physics. Here, there is some spill-over into astronomy as well. The Computers chapter should more aptly be titled Computing, since it includes far more than hardware and software, encompassing all aspects of calculating and data processing. Transportation is similarly inclusive, embracing humankind moving in any manner on land, air, sea, or space. In this chapter, such transportation-related subjects as bridges and lighthouses can also be found. The Communications chapter surveys its many firsts irrespective of their media, and therefore has information pertinent to the semaphore as well as the telegraph and telephone, and on printing and photography as well as radio, television, and cinema. Chemistry includes alchemy and metallurgy in addition to its many modern sub-disciplines like organic and inorganic chemistry. The Agriculture & Everyday Life chapter attempts to link together the science of agriculture and the art of farming along with what might be called "domestic technology." This broad chapter therefore contains entries on a genetically altered tomato, crop rotation, and on more domestic arts like refrigeration and canning. The final chapter, titled Energy, Power Systems, & Weaponry, discusses energy and power sources from wind and steam through nuclear energy, while also detailing the mechanical advances made in the fields of weapons and warfare.

Another important guideline followed in selecting and composing the entries was that this should be a book consisting not only of major or landmarks firsts, but one that allows for the broadest possible interpretation of that term. Thus, the author often uses certain other terms—such as discovers, invents, founds, establishes, or earliest—as synonyms for first. The entries therefore do not read "first discovers" or "invents the first," but simply says discovers or invents, assuming that the reader can do without the redundancy.

Placement of a particular entry in a particular chapter was sometimes difficult and often arbitrary given the nature of the subject matter. A good example is the 1917 discovery of the new element in uranium, called protactinium, that is written about in the Chemistry chapter. This could just as properly have been placed in the Physics chapter. To locate such a subject, the reader is advised to consult the index. A better example of the blurring of disciplines that sometimes made entry placement difficult is Ernest Rutherford's 1908 Nobel Prize for Chemistry—a chemistry prize given to a physicist for an achievement that ultimately earned him the title "Father of Nuclear Physics."

The exact location of an event in time would seem an easier task, but there, too, reality has a way of making for imprecision. In fact, there were many instances of several equally legitimate dates from which to choose. For a single scientific discovery there could be the date of the actual event itself; the date of its announcement, proof, or verification; or the date of its official publication. Often the author chose the best-known date or that exhibited by the greatest number of secondary sources.

Other details of style or format that will make this book easier to use include the following:

Cross-references are found only within chapters or subjects.

An entry is not duplicated in another chapter despite its relevance to that other subject. Thus if the user cannot find an entry in what appears to be its logical subject area, the index should be consulted.

Life dates (birth and death years) are given only when known completely. No life dates following an individual's name indicate, for older entries, that this information was not obtained; for modern entries, that the individual is probably still living.

Very few, if any, entries relating to voyages of discovery and exploration are included.

Although the research and writing for this book was literally a one-person job, the author could not have devoted his full attention to those tasks without the knowledge that the index was in the best of hands. I am indebted to Ms. Michelle B. Cadoree who spoiled many of her lunches and evenings away from her duties at the Library of Congress making sure that each and every entry would be given its full due in the book's Index. She also was my computer guru, regularly solving any and all technical problems and creating a highly workable and extremely efficient program for me to use. I am in her debt. My editor at Gale Research, Ms. Donna Olendorf, was as understanding and truly helpful as anyone I have worked for, and showed a remarkable ability to be both demanding and supportive at the same time. I would not hesitate a minute to work with such a paragon again. Finally, my wife, Jane, and my children, Nat, Ben, and Nina, allowed this project to come into their lives as well, yet never, ever, made me feel guilty about being less than always available or attentive. Each exhibited an admirable lack of selfishness that I hope to reward by being doubly attentive and available in the future. They have my thanks and love.

—Leonard C. Bruno

PHOTO CREDITS

The photographs and line drawings appearing in this book were obtained from the following sources:

Agriculture & Everyday Life

c. 24,000 B.C. Food is stored by primitive man below ground in pits during winter months in what is now Eastern Europe. This is the earliest known form of cold storage and food preservation.

c. 9000 B.C. First example of a domesticated animal is the dog in Mesopotamia and the Middle East. One of the preconditions of domesticating animals is a settled way of life which, in turn, depends on sufficient food resources. These first tamed dogs probably accompany hunters and guard human settlements, warning the inhabitants of possible danger. Like sheep and goats which later became domesticated, they are probably eaten during the early stages of domestication.

c. 8000 B.C. With the beginning of settled agriculture come the first simple digging and harvesting tools. At this very early point in mankind's agricultural history, hoes or digging sticks are used to break the ground. The Natufians of Palestine possess some form of sickle with which they harvest both wild and sown grain.

c. 8000 B.C. Beer is first brewed in Mesopotamia. The Sumerians reserve as much as 40% of their cereal harvest for brewing this thick drink that has been described as "drinkable bread." Like wine, beer is fermented and undistilled, but unlike wine which is made from basic materials rich in natural sugar, beer is made from materials high in starch content. These starches are then converted to sugar before fermentation can occur. Beer is made from rice in the Orient, from maize in South America, and from other cereal grains in Europe.

c. 6500 B.C. First very primitive plow, called the ard, is used in the Near East. As probably the most important agricultural implement to be developed, this first plow evolves from the simple digging stick and has a handle used by humans to push or pull along the ground. This in turn evolves into the Egyptian crook plow which has stag antlers or a forked branch tied to a pole to break the ground for planting.

(See also c. 1000 B.C.)

c. 6000 B.C. Wine-making first begins in northern Mesopotamia and the Levant (along the eastern Mediterranean shore). Wine is produced in areas conducive to grape cultivation, and it is made from basic materials rich in natural sugar. It is fermented and undistilled.

c. 6000 B.C. Bread-making wheat called "tritium vulgare" is first cultivated in southwestern Asia. This is the most important variety of wheat since it is used for bread making, and it becomes a major source of energy in the human diet.

c. 5600 B.C. Saddle quern is first used for grinding grain. A type of mortar and pestle, it consists of a flat stone bed and a rounded stone that is operated manually against it. By around 1100 B.C., the rotary or true quern appears in the Mediterranean area. It uses a handle that rotates one stone atop a stationary stone. It is the precursor of the heavy querns, later used by Greeks and Romans, that are operated by slaves and donkeys.

c. 4400 B.C. Earliest evidence of the use of a horizontal loom is its depiction on a pottery dish found in Egypt and dated to this time. These first true frame looms are equipped with foot pedals to lift the warp threads, leaving the weaver's hands free to pass and beat the weft thread. This is a much improved version of the earliest peg or ground looms which are primitive devices consisting of two pegs driven in the ground, between which wooden bars are mounted.

c. 2500 B.C. Earliest known wholly glass objects are beads made in Egypt at this time. Early peoples may have discovered natural glass, which is created when lightning strikes sand. The Egyptians make glass beads by heating sand (silica), soda, lime, and other ingredients. Glass proves highly versatile and durable, and plays an important role in human culture.

c. 2500 B.C. Apiculture or beekeeping first begins in Egypt around this time. One of the oldest forms of animal husbandry, it involves the care and manipulation of colonies of honeybees (of the *Apis* genera) so that they will produce and store a quantity of honey above their own requirements.

c. 1000 B.C. Iron-tipped plow first appears. When this strong, sharp plow is drawn by oxen, it proves a major improvement over the older plowshare made of wood or stone. With this

more powerful tool, seed is often sown broadcast in front of the plow, which then covers it. (See also c.6500 B.C.)

c. 800 B.C. First known complete sundial is used by the Egyptians. This early timekeeping device has a pillar or style that casts a shadow across a dial that has markings. The style is aligned parallel to the earth's axis.

600 B.C. Phoenicians first use soap for cleaning cloth. It is made from goat's tallow and wood ashes and is sometimes used as an article of barter. Although they have made soap from animal fat and ashes for many centuries, it was used primarily as a salve or ointment. The early Egyptians also used a soap-like paste to treat skin diseases.

c. 300 B.C. Egyptians are the first to use scissors made of bronze. Although some form of shears or scissors have been in existence for a thousand years, improvements in the forging of metals enable the Egyptians to make excellent scissors using blades connected by a C-shaped spring at the handle ends.

c. 100 B.C. Syrian craftsmen first introduce the blowing iron for glassmaking. This new technique, in which the molten glass is gathered on the end of a hollow pipe and inflated (by blowing) to a bubble, soon leads to the production of glass articles on a larger scale than ever before possible.

c. 25 B.C. Marble quarries at Carrara in central Italy are first opened and developed. This snow-white marble is favored by sculptors. In classical times, this site was called "Luna" after the city nearby. Carrara becomes world-famous for the pureness of its metamorphic marbles.

c. 35 A.D. M. Gabius Apicius, Roman Epicure, writes what many consider to be the oldest cookbook. Titled *De Re Coquinaria*, it contains glimpses of Roman cooking and eating habits as well as recipes.

c. 50 A.D. Large, cylindrical, airtight storage bins for grains, similar to modern silos, first begin to appear in both China and the Middle East.

c. 100 A.D. Glazed windows first appear in Britain after the Romans introduce glass-making techniques. By this time, glassblowing techniques have spread and have made window production almost commonplace.

c. 100 A.D. Candles made of tallow containing a wick are mentioned for the first time. Before this, pottery lamps burned vegetable oil by means of a simple wick. Candles also come to be made of beeswax, and a fiber wick replaces the use of reeds.

c. 100 A.D. First insecticide is believed to be used by the Chinese who discover that dried, powdered chrysanthemum flowers repel and even kill insects.

914 Al-Masudi, Arab historian and geographer, records accurate principles on evaporation and the causes of ocean salinity. He is the first in the East to combine history and scientific geography in a major work. He is known as the "Herodotus of the Arabs."

c. 1000 Avicenna (980-1037), Persian physician also known as Ibn Sina, records the first use of coffee as a beverage. This drink is made by brewing the roasted and ground beans with water. The earliest known and cultivated species is *Coffea arabica*, the coffee shrub of Arabia. Coffee may derive its name from the Arabic "qahwah," but some connect it with the name Kaffa, a province in southwest Ethiopia.

c. 1100 Stained glass windows begin to appear in English and French churches. Formed of pieces of colored glass outlined by lead strips and sometimes assembled into a narrative picture, these spectacularly ornamental windows are the first major advance in glass since the fall of the Roman Empire.

c. 1100 In Italy, wine is first distilled to make brandy. Also made from a fermented fruit mash which is distilled, brandy is usually aged and contains about 50% alcohol by volume. Most wine-producing areas also make brandy, but it is the Dutch who first turn its production into a commercial enterprise in the 16th century.

1185 Manufacture of wool is first mentioned in English literature. Wool is a readily available natural fiber, since sheep have been domesticated for thousands of years. Wool from the animal's secondary follicles produces fleece fibers that are fine, short, and somewhat scaly. These scales allow the fibers to interlock, so that little twisting is needed to make wool into yarn. Since they lock so easily, wool fibers trap a great deal of air between them and act as an insulator. Wool is therefore a desirable material in colder climates.

c. 1200 Buckwheat is first introduced into Europe from Asia. It becomes a staple grain crop for poultry and livestock and is also cooked and served much like rice. It is not suitable for bread making.

1253 Linen is first manufactured in England. This yarn and fabric is made from the flax plant and is stronger and more durable than cotton. It is a lightweight fabric, ideal for a hot, dry climate. It was first produced in Ancient Egypt.

1272 First spinning wheel is believed to be used in Bologna, Italy. This machine that turns fiber into thread or yarn is not an Italian invention but originated in India. It soon comes to replace hand spinning. The important part of the mechanization process is the mounting of the spindle horizontally in bearings so that it can be rotated by a cord encircling a large, hand-driven wheel.

1277 Jean de Joinville (c.1224-1317), French historian of the Second Crusade, is the first to mention velvet in his treatment of Muslim life and customs. This special weave of silk or other fiber, called a pile weave, results in a soft, downy surface that is formed by clipped yarns.

c. 1300 The tulip is first cultivated in Turkey. It is imported as a garden flower from Constantinople to Vienna around 1550 and is enthusiastically adopted by the Dutch. The Netherlands eventually become the world center of tulip breeding.

1305 The "acre" is first used as a measure of land during the reign of Edward I (1239-1307), king of England, and becomes a standard. Probably derived from an old Anglo-Saxon word, it is an area four rods wide by 40 rods long. This total of 160 square rods equals 43,560 square feet. In the metric system, one acre is equivalent to 0.405 hectares (4,047 square meters).

1340 First "worsted" is manufactured at Worsted in Norfolk, England. It is spun wool that is manufactured into cloth. Worsted knitting yarn is usually preferred for hand-knitting and is often used to make sweaters since it is heavy, highly twisted, and soft.

c. 1400 Whiskey is first distilled in Ireland and Scotland from a fermented mash of grain, primarily barley. Its name in Celtic, Irish Gaelic, and Scots Gaelic means "water of life."

c. 1400 Wild coffee plants from Kefa (Kaffa), Ethiopia, are cultivated there and first made into a drink.

1440 Earliest original work in English on gardening is written by Ion Gardener and titled *The Feate of Gardening*. It offers sound, practical advice.

1472 Agricultural writings of the first century A.D. Roman farmer and soldier Lucius Junius Moderatus Columella are first printed in Venice, Italy and titled *De Re Rustica*. Written in 12 books, this work gives a full treatment and description of Roman farming and country life.

Roman farmer plowing his field.

1500 Castrating of horses, or gelding, is first practiced in England during the reign of King Henry VII (1457-1509). This practice makes male animals sufficiently docile for training or work.

1503 Raw sugar is first processed or refined into the granular, white form we consume today. Refining takes advantage of the alkaline stability of sugar and removes a variety of impurities from its raw state by treating it with alkali. After this, syrup preparations are crystallized to form white sugar crystals.

1520   Hops are first cultivated in England. They are used in medicines and to impart a bitter flavor to malt liquors. Hop is the common name for plants of the genus *Humulus* of the family Cannabaceae.

1520   Chocolate and vanilla are first imported to Spain from Mexico. In 1519, the Spanish are served xocoatyl, a bitter cocoa-bean drink sweetened with vanilla, at the court of Montezuma, the ruler of Mexico. They then flavor it with cinnamon, and it remains a Spanish secret for nearly a hundred years. Chocolate houses eventually spread throughout Europe, which also widely embraces vanilla for its many uses in sweets.

1523   Alejandro Geraldine, bishop of Santo Domingo, writes to Europe of turkeys in the new world and is the first to refer to them as fowl. The birds eventually get their name from the mistaken belief that they came to England originally from Turkey.

1524   Soap is first made commercially in London, England. Since English authorities consider soap a luxury, it is heavily taxed until 1853. The modern soapmaking industry begins in 1791 when French chemist Nicholas Leblanc (1742-1806) develops a method of obtaining sodium carbonate from common salt rather than from plant ashes.

1530   Pietro Martire d'Anghiera (1455-1526), Spanish historian, makes the first European reference in print to rubber. In his book on the new world, *De Orbe Nouo*, he mentions "gummi optimum" in his description of the rubber used in an Aztec game.

1530   Matches made of sulphur are first mentioned in England. They are simple wooden sticks dipped in melted sulphur that ignite after contact with a burning object. Friction matches do not appear for a very long time.

(See also 1826)

1547   Catherine de Medici (1519-1589), daughter of Lorenzo of Urbino, arrives in Paris as the bride of Henry II (1519-1559), king of France, and brings high-heeled slippers with her. The high heel then first appears at most of the royal courts.

1585   The potato is taken from South America to Spain. This tuber-bearing species of the genus *Solanum* is a native of the Peruvian-Bolivian Andes. Cooked as a vegetable, it is highly digestible and supplies needed vitamins. Introduced to Europe by Spain, the potato becomes a major crop in continental Europe.

(See also 1719)

1586   John Harrington (1561-1612) of England designs and installs the first water closet in his home near Bath, England. This same year he installs one at Richmond Palace for his godmother, Elizabeth I (1533-1603), Queen of England. His design features an overhead water tank with a valve that releases water on demand. His system does not cope with drainage and venting problems, however, and this early toilet remains smelly and unsanitary.

1587   First beer brewed by Europeans in the New World is made by colonists in the Roanoke Colony of Virginia.

1589   William Lee (c.1550-c.1610), English inventor, invents the first knitting machine. Also called the stocking frame, it forms the basis for all subsequent knitting and lace-making machines. He later improves his loom for making stockings and opens a mill in Nottingham that makes silk stockings.

1600   Olivier de Serres (1539-1619), French agronomist, publishes his *Théâtre d'Agriculture et Mésnage de Champs*. He is also the first to practice systematic crop rotation.

1602   Oats, barley, and wheat are first introduced in America by Bartholomew Gosnold at Massachusetts.

1609   First yield of corn that is produced by

American colonists is harvested by the Jamestown colony in Virginia. The Jamestown settlers learned how to grow corn from the Native Americans two years earlier.

c. 1610 Cornelius Jacobszoon Drebbel (1572-1633), Dutch-English inventor, devises a self-regulating oven (using a thermostat). He uses the same principle to build the first self-regulating incubator for hatching chicken eggs.

1612 Antonio Neri (c.1576-1614) of Italy publishes his *L'Arte Vetraria*, the first systematic account of the treatment and preparation of raw materials to make glass. Fifty years later it is translated into English and becomes very influential.

(See also 1668)

1615 Spanish report from the New World includes the first mention of rubber being used for something other than sports. This report describes how natives collect the milk from cuts made in a tree and brush it on their cloaks. They even make crude footwear by using molds and allowing it to dry.

1616 First known cultivation of tobacco by colonialists occurs in Virginia. It is believed that John Rolfe of Jamestown planted seeds a few years earlier that he may have obtained from the West Indies. Wild tobacco is *Nicotiana rustica* and is grown by Native Americans of eastern North America. Common tobacco is *Nicotiana tabacum* and is native to South America, Mexico, and the West Indies.

1621 Cotton is first planted experimentally in Virginia. As one of the world's most important non-food agricultural commodities, it is one of the first vegetable fibers used for textiles purposes. Materials made of cotton have been found in tombs in India dating to 3000 B.C. The cotton trade is of great importance in the development of the British East India Company, and England sees the southern colonies as an excellent candidate for cotton production.

1623 Flax is first introduced in America and is cultivated solely for its fiber, from which linen and yarn are made. The seeds of the plant are called linseed from which linseed oil is made.

1625 Buckwheat is first cultivated in America on Manhattan Island. Although it is less productive than other grain crops on good soils, it is particularly adapted to arid, hilly land and cool climates. It is also late-maturing and can be grown as a late season crop.

1626 First flour mill in the American colonies is built in New Amsterdam. Such mills to grind grain are necessarily built by a source of moving water which powers the turning stones. When the Dutch are driven from New Amsterdam by the British, it becomes known as New York.

1629 First attempt at raising hops in America is made in New Netherlands (later New York).

1629 First shoemaker in the New World, Thomas Beard of England arrives in Salem, Massachusetts.

1635 Jacques Philippe Cornut (1606-1651), French physician and botanist, publishes his *Canadensium Plantarum*, which is considered to be the first published work on the plants of America.

1640 William Keift, governor of New Netherlands, opens a business that produces what he claims is the first beer in America.

1650 Richard Weston (1591-1652), English agriculturalist, publishes his *Discours of Husbandrie*, which contains the first description of a crop rotation system that uses no fallow break (tilling land without sowing seeds in it for a season).

1653 Bernabe Cobo (1582-1657) of Spain writes his *Historia del Nuevo Mondo* in which he first associates the word "Cachuc," or, in Spanish "caucho," with the liquid resin or rubber that comes from a tree.

1668 First known illustration of a glassmaker's

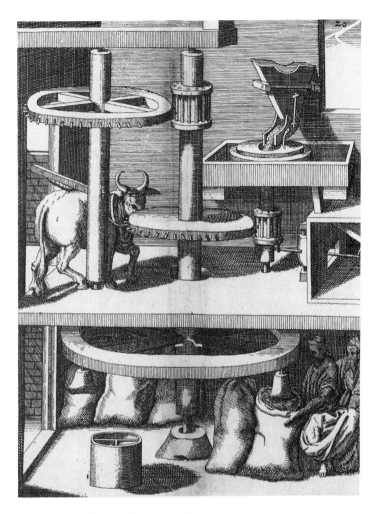

An ox powers this mill that was used to grind grain.

chair—a distinctive combination of seat and workbench—is seen on the frontispiece of the Amsterdam edition of Antonio Neri's (c.1576-1614) classic called *Ars Vitraria*.

(See also 1612)

1670 First cookbook written by a woman is published. Written by Hannah Woolley of England, *The Queenlike Closet; or, Rich Cabinet* sees several editions.

1674 George Ravenscroft (1618-1681), English glassmaker, first produces lead glass, which is also called flint glass. This heavy, blown glass is especially brilliant as well as durable. It is now called lead crystal.

1684 Nicolas van Benschoten of Holland invents the modern thimble. A thimble is a small, bell-shaped implement designed to protect the top of the finger when sewing.

1685 Rice is successfully planted for the first time in America in the Carolinas. This starchy cereal grain is grown on submerged land in the coastal plains, tidal deltas, and river basins of at least semi-temperate regions. The seeds are sown in prepared beds, and when the seedlings are 25-50 days old, they are transplanted to a field or paddy submerged continuously in 2-4 in (5-10 cm) of water.

1688 Dom Pierre Pérignon (1638-1715), French Benedictine monk, invents the sparkling or bubbly wine called champagne. He blends wines, uses sugar to start a second fermentation in the cask, and stores wine in air-tight bottles, conserving the carbon dioxide.

(See also c.1690)

c. 1690 Dom Pierre Pérignon (1638-1715), French Benedictine monk, invents wine bottle corks that are held in place by wires. This solves the problem he encounters when his corked bottles are unable to contain the pressure built up by his newly invented sparkling wine called champagne.

1699 John Woodward (1665-1728), English geologist and physician, is one of the earliest to cultivate plants in water. He is also one of the founders of experimental plant physiology.

c. 1700 Four-course Norfolk system of crop rotation is adopted in Norfolk County, England, and eventually becomes established practice throughout Europe.

1701 Jethro Tull (1674-1741), English agriculturalist and inventor, invents a seed drill that sows seed in neat rows, saving seed and making it easier to keep weeds down. His horse-drawn hoes destroy weeds and keep the soil between the rows in a friable condition.

1701 First printed reference to a contraceptive sheath to be used by men is made in "The Ladies' Visiting Day," a play written by William Burnaby (c.1673-1706) of England.

1712 Whale hunters of New England discover that spermaciti fluid found in a whale's head cavity makes excellent wax candles. This new product creates a flourishing trade.

1719 Potatoes, which originated in South America, are first grown in the American colonies in New Hampshire using stock brought from Ireland.

1719 John and Thomas Lombe, English brothers, introduce the first silk-throwing machinery into England at their silk mill in Derby. This could be considered the first British textile factory.

1720 First tinware made in the American colonies is made at Berlin, Connecticut. Seldom made of pure tin, these common household objects are more often made of tinplate. Tinplate was first made in England where the Allgood brothers developed a process of rolling sheets of iron and dipping them into molten tin.

1723 "Society of Improvers in the Knowledge of Agriculture in Scotland" is organized. It is the first of its kind in the United Kingdom.

1725 Luigi Ferdinando Marsigli (1658-1730), Italian soldier and oceanographer also called Marsili, publishes his *Histoire Physique de la Mer*, the first complete book dealing with the sea. He also is the first to use a naturalist's dredge.

1730 First English cotton mill is built in Gloucester.

1732 Michael Menzies of Scotland invents the first mechanical thresher. It consists of flails attached to a hydraulically operated wheel. It is not completely successful.

MAY 26, 1733 John Kay (1704-c.1764), English engineer, first receives a patent for his flying shuttle. This device is an important step toward automatic weaving and by using it, one weaver can weave fabrics of any width quicker than two weavers can the manual way.

1736 Natural rubber is first scientifically described by Charles Marie de La Condamine (1701-1774), French geographer, and his

French colleague, François Fresneau, following an expedition to South America. They give an accurate account of rubber production in the Amazon region from the condensed juice of the "Hevea" tree.

1740 Benjamin Franklin (1706-1790), American statesman and inventor, invents the Pennsylvania fireplace that comes to be known as the Franklin stove. He improves on an existing fireplace design and adds a flue around which room air can circulate. The flue acts like a radiator and increases heating efficiency. Franklin claims it makes a room twice as warm with one-quarter of the wood.

1740 James Small of Scotland first introduces a plow with a cast-iron moldboard and wrought iron plowshares.

1742 Thomas Boulsover (1706-1788), English inventor, discovers what comes to be known as "Sheffield plate." He finds while working with silver and copper that the two metals can be fused and eventually behave as a single metal. This discovery leads to the economical manufacture of a large number of plated objects, from buttons and snuff boxes to utensils.

1742 First cookbook published in the American colonies is *The Compleat Housewife; or, Accomplished Gentlewoman's Companion*, published at Williamsburg, Virginia. It was first written by a cook, E. Smith, and published in London in 1727. Following this first edition, this popular work goes through a total of 18 editions, one of which is American.

1743 Earliest society in the American colonies to promote a scientific approach to agriculture is the American Philosophical Society organized this year.

1747 Agricultural seeds are first sold commercially in the American colonies.

1748 Ball valve for water supply systems is first introduced by an anonymous inventor. It be-

comes essential for the control of modern plumbing systems.

1750 John Adams Dagyr, a Welsh immigrant to the American colonies, opens the first successful shoe factory in the colonies and uses the division of labor system.

1758 William Cullen of Scotland develops the first known artificial refrigeration device by evaporating ethyl ether in a partial vacuum and obtaining a small amount of ice. In 1777, English physicist Edward Nairne (1726-1806) improves Cullen's refrigeration process and achieves artificial desiccation by using sulfuric acid to absorb water.

1760 First commercial nursery in the American colonies is established by William Prince of Long Island. He becomes a pioneer in this industry.

1761 Robert Henchliffe of England first makes cast steel scissors for domestic use.

1761 First veterinary school is established at Lyons, France.

1764 James Hargreaves, English inventor, builds the spinning jenny, the first practical application of multiple spinning by a machine. His machine allows one person to spin wool, cotton, or flax into a plurality of threads.

1764 First modern greenhouse in the American colonies is built in New York.

1768 James Cook (1728-1779), English navigator, begins the first of three ocean voyages during which he is the first to take the subsurface temperature of the ocean.

1769 Benjamin Franklin (1706-1790), American statesman and scientist, develops the first published chart of the Gulf Stream. He uses temperature measurements and observation of water color to track its course. He suggests that ships going to Europe stay in its current, and that those returning to America avoid it.

1769   Orange is introduced into California, but the first grove is not planted until 1804.

1772   Antoine Laurent Lavoisier (1743-1794), French chemist, makes the first quantitative analysis of seawater.

1779   Tomatoes are first used to make catsup in New Orleans, Louisiana.

1779   Samuel Crompton (1753-1827), English inventor, invents the spinning mule, a machine that can simultaneously draw out and give a final twist to the cotton fibers fed into it. This reproduces mechanically the actions of hand spinning. This device makes possible the large-scale manufacture of high-quality thread and yarn.

1780   Arnoldus Juliaans of the Netherlands submits his thesis, *De Resina Elastica Cajennensi*, to the University of Utrecht. This is the first work devoted exclusively to the study of rubber.

1780   First hat factory in the United States is built in Danbury, Connecticut, by Zadoc Benedict.

1783   First factory for making plows is established in England.

JANUARY 7, 1784   First seed business in the United States is begun by American nurseryman David Landreth (1752-c.1828) in Philadelphia, Pennsylvania.

1784   Benjamin Franklin (1706-1790), American statesman and inventor, is credited with inventing the first bifocal lenses for eyeglasses. He divides his lenses for distant and near vision, with the split part held together by the frame.

c. 1784   Cotton is first exported by the United States.

1785   Robert Ransome (1753-1830), English inventor, first patents a plow with a cast-iron share (the part of a moldboard plow that cuts into the furrow).

Benjamin Franklin wearing his own invention, the bifocal lens.

1785   Edmund Cartwright (1743-1823), English inventor, builds the first successful power loom. His wool-combing machine is the predecessor of the modern power loom.

1785   First organization of American agricultural societies convenes in Philadelphia, Pennsylvania.

1786   Andrew Meikle (1719-1811) of Scotland invents a grain thresher that consists of a rotating drum with four vertical blades and a stationary shield through which the grain is fed. He uses a fan to blow the chaff away from the heavier kernels of grain.

1786 First ice cream company in the United States is started in New York by a man named Hall.

1790 James Rennell (1742-1830), English geographer, begins his scientific study of winds and currents, and makes the first comprehensive study of the Atlantic Ocean currents.

1790 First known patent on caoutchouc or rubber is granted to W. Roberts and W. Dight of England for treating canvas with rubber varnish.

1792 Shoelaces, called shoe strings, first begin to replace buckles in the United States.

1792 Robert Bailey Thomas (1766-1846) of the United States founds the *Old Farmer's Almanac*, which becomes the oldest surviving almanac in America.

APRIL 1793 First cotton gin is invented by Eli Whitney (1765-1825), American mechanical engineer. His gin, or engine, has metal projections or wires that poke through slats, entangle themselves in the cotton fibers, and detach the difficult-to-remove seeds as they pull themselves free. One machine can produce 50 lbs (23 kg) of cleaned cotton per day and greatly stimulates the entire cotton industry in the United States. It makes cotton a big business and enables it to compete with flax.

1793 Samuel Slater (1768-1835), English-American inventor, builds the first successful cotton mill in the United States. After working as an apprentice to Strutt and Arkwright in England and obtaining a thorough knowledge of British textile machines, he is able to duplicate their machines from memory and essentially founds the American cotton textile industry in Pawtucket, Rhode Island.

1795 Nicolas François Appert (1752-1841),

The first cotton gin.

French inventor and chef, discovers how to hermetically seal food and creates what becomes known as canning. He devises a method of putting food in corked glass bottles and immersing them in boiling water. This destroys microorganisms in the food (although bacteriology has not yet discovered this). Appert sets up a bottling plant at Massy, south of Paris, in 1804.

(See also 1812)

1796 First special academy to educate and train farmers is established at Keszthely, Hungary.

JUNE 26, 1797 Charles Newbold of the United States first receives a patent for a single-piece, all-iron plow. Its major drawback is that if any part breaks, the entire plow is ruined.

1798 The McIntosh apple, a winter variety of northern origin, is discovered by John McIntosh of Ontario, Canada.

c. 1800 Wooden screw press is invented in Naples, Italy that makes mass production of pasta possible. Prior to this, pasta is made by hand and is both time-consuming and strenuous.

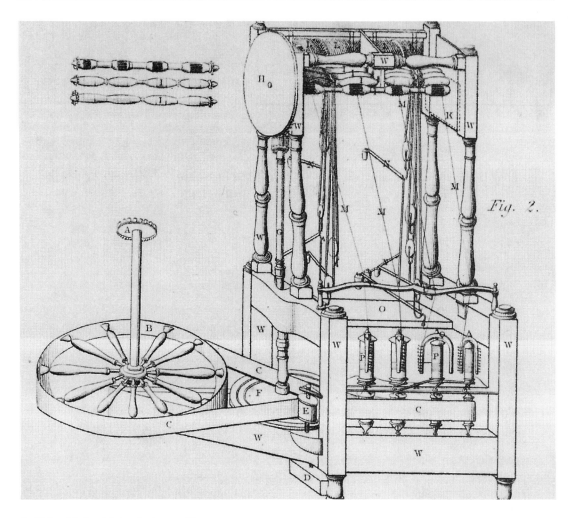

Sir Richard Arkwright's spinning machine.

c. 1801   American legendary figure John Chapman, called "Johnny Appleseed," first begins planting apple seeds throughout Indiana and the adjacent territory. He uses the broadcast or scatter method of seeding, and he lives to see 100,000 acres of apple trees.

1802   David Humphreys, American minister to Spain, first imports Merino sheep into the United States. He sends 100 sheep from Lisbon to Derby, Connecticut. These sheep are descended from a breed developed in Spain and produce a high quality wool.

1802   Sugar beet is first introduced as a field crop into Germany, where it becomes a commercial success.

MAY 1, 1803   First American patent for a machine for cutting grain is issued to Richard French and J. T. Hawkins of New Jersey. Their machine is propelled on three wheels.

1803   American cranberry is first grown at Cape Cod, Massachusetts.

1803   Robert Ransome (1753-1830), English

inventor, invents a plow with a self-sharpening cast-iron share. He later designs a plow with standard parts that can be removed and replaced in the field.

1804 Bananas are first imported into the United States by Captain John N. Chetser of the schooner *Reynard*.

1804 Soybean is first cultivated in the United States. It becomes a major crop after World War II.

1804 First agricultural fair in the United States is held at Washington, D.C.

1805 Antoine Augustin Parmentier (1737-1813), French agriculturalist, produces the first powdered milk.

1805 First herd of corn-fed cattle is driven from the Ohio Valley over the mountains to the markets of the east.

1805 Joseph Marie Jacquard (1770-1840), French inventor, invents the Jacquard loom for figured weaving. This machine uses punched cards that control the weaving of the cloth so that any desired pattern can be obtained automatically. His invention spurs a technological revolution in the textile industry and is the basis of the modern automatic loom. His punched card system is later adopted by the English mathematician Charles Babbage (1792-1871) for the control mechanism of his calculator and by American inventor Herman Hollerith (1860-1929) to feed data into his census machine. Punched cards also become the major means of programming early modern computers.

1806 First agricultural high school is founded with royal support at Moglin, near Berlin, Germany.

1809 Samuel Parker of the United States invents the first leather-splitting machine. Hides can be fed into one end of his machine and emerge at the other end accurately cut into two pieces.

JULY 4, 1810 First farm magazine in the United States, *The Agricultural Museum*, begins publication.

1810 First American-made cigars are made in Connecticut.

1810 Marc Isambard Brunel (1769-1849), French-English engineer, builds the first machine to be used for making shoes. His clamping press is designed to attach soles to the uppers.

1811 Bryan Donkin and John Hall of England buy Peter Durand's patent for canning food and establish the first cannery in England in 1812. They supply canned goods to the Royal Navy and to various Arctic expeditions.

1812 Peter Gaillard of the United States receives the first patent on a mowing machine. Mowing machines are different from reapers (which harvest mature grain crops) because they cut the crop while it is green and must be very sharp.

1812 Nicolas François Appert (1752-1841), French inventor and chef, establishes the first commercial cannery. He also invents the bouillon cube.

1817 First Hereford and Jersey cattle are imported into the United States.

1819 American canning industry first begins in Massachusetts and New York. In Boston, Thomas Underwood packs fruits, pickles, and condiments in bottles, while in New York, Thomas Kensett and Ezra Daggett pack seafood in bottles. In 1839, both factories switch from bottles and jars to tin-plate containers.

1819 First distinctively agricultural journal in the United States, the *American Farmer*, is published in Baltimore, Maryland.

1820 Henry Burden (1791-1871), Scottish-American inventor, produces the first cultivator in the United States. This horse-drawn device stirs the soil around crops to promote growth and destroy weeds.

1821 Thomas Hancock (1786-1865), English inventor, invents a rubber shredder he calls a "masticator." This device works rubber scraps into a shredded mass of rubber that can be formed into blocks or rolled into sheets. Hancock later teams with Scottish chemist Charles Macintosh (1766-1843) and founds the British rubber industry.

JANUARY 1, 1823 First agricultural college in the United States, the Gardiner Lyceum, is established in Maine. Its first president is the American educator Benjamin Hale (1797-1863).

1826 Tea is first sold in individual packets on the Isle of Wight by John Horniman of England.

1826 First friction matches are invented by John Walker of England who does not patent them. Called "lucifers," they are thin wooden sticks coated with sulphide of antimony and chloride of potash and are lit by being drawn between two surfaces of a folded piece of sandpaper. Phosphorous is eventually substituted for antimony. (See also 1855)

1827 Nicolas François Appert (1752-1841), French inventor and chef, first condenses milk to preserve it longer. Seven years later he invents the method of evaporating milk.

1827 First slaughterhouse in Chicago is a log structure built by Archibald Clybourne. This marks the beginning of Chicago's meatpacking industry.

1828 Samuel Lane of the United States obtains the first patent on a combine. A combine harvester "combines" the cutting and threshing functions and eliminates the need to bond the cut stalks. Little is known about Lane's machine. (See also 1836)

1828 First comprehensive work on horticulture

in the United States is William Robert Prince's (1795-1869) *A Treatise on Horticulture*.

1830 First patent for a sewing machine is obtained by Barthélemy Thimonnier (1793-1857), French tailor and inventor. His design uses one thread. He builds 80 machines to manufacture army uniforms, but they are later destroyed by a mob fearing unemployment. (See also 1846)

JULY 1831 Cyrus Hall McCormick (1809-1884), American inventor, gives the first public demonstration of his mechanical reaping machine. Its basic workings, including the gearings, reciprocating knife, projecting teeth, and rotating wheel become the basis for modern harvesting machines. He patents an improved version in 1834 and builds a factory in 1847 that produces 800 reapers a year. By 1858, he is making 4,000 reapers annually.

MAY 19, 1832 Wait Webster of the United States is granted a patent for "attaching India rubber soles to boots and shoes." This is the earliest record of such an idea.

1832 Edmund Ruffin (1794-1865), America's first soil scientist, publishes his *Essays on Calcareous Manures*.

1833 First rubber-manufacturing company in the United States, the Roxbury India Rubber Company, is incorporated.

1833 First washboard is patented in the United States. This simple wooden-framed device has a corrugated rectangular surface that is used for scrubbing clothes clean.

1834 Jacob Perkins (1766-1849), American inventor, obtains the first patent for a refrigerating machine that works by the expansion of volatile fluids (in a closed-cycle system using a compressor). In 1844, American physician John Gorrie first develops a machine that provides both ice and air cooling to his hospital. Compressed air is cooled by circulating water, and as the air re-expands in an engine cylinder, it drops to a low enough temperature to make ice.

1835 Henry Burden (1791-1871), Scottish-American ironmaster and inventor, patents the first horseshoe-making machine. Until now, horseshoes have been shaped by hand at the blacksmith's forge.

1835 Rubber nipples are first introduced for infant nursing bottles.

1836 First successful combine harvester is built in Michigan by Americans Hiram Moore and J. Hascall. This horse-drawn machine is pulled by a team of 20 horses and its cutter bars and thresher are powered by the ground wheels. This means that the combine has to keep moving to perform both functions.

1837 First efficient threshing machine is developed in Massachusetts by the Americans Hiram A. Pitts and John A. Pitts. Threshing is done after harvesting and separates the grain from the chaff. Their machine has an "endless belt" with pins that thresh and clean the grain. It is efficient and inexpensive.

1837 John Deere (1804-1886), American inventor, first develops a steel plow that he fashions from a circular saw blade. It is able to cut through the difficult prairie soils. He also realizes that a successful self-scouring steel moldboard depends upon its shape. His steel plow plays a large role in opening the western states to agriculture. His company becomes a leading maker of farm equipment.

1838 First sugar beet factory in the United States is built in Northampton, Massachusetts.

NOVEMBER 24, 1839 James Nasmyth (1808-1890), Scottish engineer, first demonstrates his steam hammer, a device that allows large materials to be forged with great accuracy. With this huge yet extremely precise hammer, large iron parts needed by steamships and locomotives

can be forged. This device proves to be a key element in England's Industrial Revolution.

1840 Suspenders for trousers are invented by the Frenchmen Rattier and Guibal. They incorporate and take advantage of the stretching ability of rubber.

1841 First major state fair in the United States is held in New York.

1841 Edward Forbes (1815-1854), English naturalist, pioneers the use of a dredge in the scientific study of shallow water. As one of the first to consider the sea as an entity and to divide the ocean into natural zones on a scientific basis, he is regarded—along with American oceanographer Matthew Fontaine Maury (1806-1873)—as the co-founder of the science of oceanography.

(See also 1855)

1842 First shipment of milk by rail in the United States is successfully accomplished.

1842 Candee Rubber Company of Connecticut is the first to make rubber shoes using the new Goodyear process.

1842 John Bennet Lawes (1814-1900), English agriculturalist, patents a process for treating phosphate rock with sulfuric acid to produce superphosphate. He also opens the first fertilizer factory this year, thus beginning the artificial fertilizer industry.

1842 First grain elevator in the United States is built in Buffalo, New York.

1843 John Bennet Lawes (1814-1900), English agriculturalist, and English chemist Joseph Henry Gilbert (1817-1901) found the Rothamsted Experimental Station, which is the oldest agricultural research station in the world.

1843 First soap powder to be sold is made by Babbit's Best Soap of New York.

1846   First practical sewing machine is invented by Elias Howe (1819-1867), American inventor. After working for five years, he devises a lock-stitch machine that improves French tailor and inventor Barthélemy Thimonnier's (1793-1857) 1830 machine and uses two threads, a needle, and a shuttle.

(See also 1851)

1846   Robert Reid of the United States first develops a new variety of corn known as "Reid's Yellow Dent" that eventually comes to dominate the American Corn Belt.

1847   Disc plow is first developed in the United States. Similar to a disc harrow, it can cut through crop stubble and roll over field obstructions in an easier manner than previous plows. However, it is not put to use in America until 1893.

1847   William Samuel Henson (1805-1888), English engineer, invents an early safety razor that uses a guard of comb teeth.

(See also 1895)

1848   First large-scale department store opens in the United States. The Marble Dry Goods Palace in New York occupies an entire city block.

1849   Modern form of the safety pin is invented by Walter Hunt of the United States. He adds the all-important circular twist at the bend that acts as a coil-spring and prevents the pin from slipping.

1849   First guano to be used as fertilizer is exported by Peru.

1850   Levi Strauss (1829-1902), Bavarian-American clothing manufacturer, makes his first pair of jeans. Using an idea offered by Nevada tailor Jacob Davis whom he hires, Strauss begins producing in 1873 blue denim jeans and jackets with small copper rivets to reinforce the seams. They are adopted by gold miners and prospectors who need durable trou-

sers, and eventually become identified with a "Western" look. Today his jeans are still popular and the company has become the world's largest clothing manufacturer.

1850   Abner Cutler of the United States designs the first rolltop desk.

1851   First practical sewing machine for home use is invented by Isaac Merrit Singer (1811-1875), American inventor. He improves Elias Howe's (1819-1867) invention with a rotary hook shuttle and a continuous-feed feature and becomes a millionaire. His convenient machine and his masterful use of merchandising puts the sewing machine into households around the world.

1851   James Harrison, Scottish-Australian inventor, builds the first vapor-compression refrigerating machinery to be used in a brewery.

1851   Linus Yale (1821-1868), American inventor, first designs and patents his "Yale Infallible Bank Lock." His most important invention is the cylinder lock, based on the pin-tumbler mechanism of ancient Egypt. The serrations on the edge of the key raise pin tumblers to exactly the correct height, allowing the cylinder of the lock to revolve and withdraw the bolt. In 1862, he introduces the combination lock.

1851   Joseph Paxton (1801-1865), English landscape gardener and architect, designs the Crystal Palace for the Great Exhibition of 1851 in London. This inspired design using prefabricated sheet glass and cast iron marks the first use of glass as an entirely new material. This stunning and functional building encourages the use of glass in public and domestic building as well as horticultural construction.

1851   James T. King of the United States invents a washing machine that uses a rotating cylinder. It is hand-powered and made for home use.

1852   First livestock are shipped by rail in the

United States, but no provisions are made to care for the animals.

1853   First potato chips are made at a New York hotel by George Crumb of the United States.

1855   Matthew Fontaine Maury (1806-1873), American oceanographer, publishes *The Physical Geography of the Sea*, the first textbook in oceanography. With Edward Forbes (See 1841), the co-founder of oceanography, Maury spends years studying the physical and mechanical aspects of the science of the sea. He also publishes the first chart of the depths of the Atlantic Ocean.

1855   John E. Lundstrom of Sweden patents the first "safety match." These thin wooden sticks use the less dangerous amorphous phosphorous as a coating instead of the volatile antimony. The phosphorous matches are still not entirely safe, despite their name.

1856   First butter factory in the United States is established by W. R. Woodhull at New York.

1856   First commercial ice is produced by Alexander C. Twinning of the United States who uses a vapor-compression machine.

1856   Gail Borden (1801-1874), American surveyor and inventor, first cans his sweetened condensed milk using a heat and vacuum method. In 1858 he establishes the New York Condensed Milk Company.

1857   First toilet paper is made by Joseph Gayetty of the United States. He uses unbleached, pearl-colored manila hemp paper. In the 1880s, Walter James Alcock of England invents the perforated toilet paper roll.

1858   First American patent for a can opener is issued to Ezra J. Warner of Connecticut.

1858   Thread-top mason jars for home canning are invented by John Mason of the United States. They are used mostly by the more

affluent Americans since their blown glass is expensive.

1858   Lyman R. Blake of the United States invents a machine for sewing the soles of shoes. It substitutes thread for the nails or pegs used earlier, and makes shoes lighter and more flexible.

1859   Ferdinand Carré (1824-1900), French inventor, first introduces a refrigeration machine that uses ammonia as a refrigerant and water as the absorbent. This method becomes widely adopted.

1859   Great Atlantic and Pacific Tea Company (A&P) is first established and eventually becomes one of the largest food chains in the United States.

c. 1860   Gang plows first appear in the United States. Built with 12-14 blades, this plow requires teams of horses to maintain the necessary momentum to cut through a wide swath of soil.

1862   First working model of a vacuum milking machine is built by an American named Colvin. The cows' udders are placed in plastic sleeves and a vacuum is created. This draws the milk from the udder into a container.

1864   F. S. Davenport of the United States invents the sulky plow, which offers the farmer a seat to ride on behind his team.

1864   Svend Foyn of Norway invents the harpoon gun and stimulates the beginning of the modern whaling industry.

1865   American shoemakers first produce separate left and right shoes. They will no longer make shoes fabricated on straight lasts, which assumes that all shoes are interchangeable. In 1880, they introduce half-sizes for shoes.

1867   Premixed paints are first offered by the Averill Paint Company of New York.

1867   Lucien Smith of the United States first

takes out a patent on a type of twisted wire that can be used as fencing. His patent does not include a means to manufacture this "barbed wire." William D. Hunt of New York also takes out a patent for a crude type of barbed wire the same year.

(See October 1873)

1867   Abilene, Kansas, becomes the first "cowtown" where cattle are driven from Texas to its railroad and then shipped to eastern markets.

1869   Hippolyte Mege-Mouries (1817-1880), French chemist, first patents his "oleomargarine" and wins a government prize given to the inventor of the best "cheap butter." His product consists of liquid beef tallow, milk, water, and chopped cow's udder churned into a solid form. It is first produced commercially in 1873 as "butterine."

1869   David L. Garver of the United States first patents a spring-toothed harrow. Closely related to a plow, a harrow further loosens the soil and cuts field stubble. This harrow has steel spring teeth that flex when they strike an obstacle.

1870   Rubber condoms are first sold commercially in the United States. In 1930, condoms made of a thin latex are first introduced.

1871   First hot dog is sold as a "dachshund sausage" at Coney Island. It is sold without a bun.

1872   Anton Dohrn (1840-1909), German zoologist, founds the first marine biological center in Naples, Italy, and spurs the creation of oceanographical institutes in other countries.

1872   Aaron Montgomery Ward (1844-1913), American merchant, begins the first mail order firm using handbills sent by mail. Two years later he issues his first catalog.

1872   Luther Burbank (1849-1926), American plant breeder, produces the Burbank potato,

the first of a long series of new or improved varieties of vegetables, fruits, and flowers.

1872   First American patent for a toothpick-making machine is issued to Silas Noble and James P. Cooley of Massachusetts.

1872   First square-bottomed, brown paper bag with longitudinal folds is invented by Luther C. Crowell of the United States. In 1883, Charles Stilwell of the United States receives a patent for a machine that makes a self-opening, pleated, flat-bottomed paper grocery bag.

OCTOBER 1873   Joseph Farwell Glidden of the United States first obtains a patent and builds a machine that is capable of producing barbed wire in large quantities. His major improvement consists of a means of threading in the spurs or holding the barbs in place. Barbed wire transforms the West and is a major factor in America's struggle to settle the plains.

1873   Campbell's soups are first offered for sale in the United States.

1873   Naval oranges that were first introduced into the United States from Brazil in 1870 by William Saunders are planted at Riverside, California, by Jonathan and Eliza C. Tibbetts.

1874   François Royer de la Bastie of France invents tempered glass, which he makes exceptionally strong through a reheating process.

1875   First state agricultural experimental station in the United States is established in Connecticut. Later in the year, another is formed in California.

1876   Karl Paul Gottfried von Lind (1842-1934), German chemist, designs the first successful refrigeration compression system that uses ammonia.

1876   Melville Reuben Bissell (1843-1889), American inventor, patents the first carpet sweeper. His device uses a central bearing brush and is soon made practical by the addition of rubber tires.

1876 British first collect rubber seeds from the wild trees of the Amazon jungles in South America and grow young trees in England, which they then transplant to Ceylon and the Malay Peninsula. These form the basis of a natural rubber plantation industry in Asia that will produce nearly 3 million tons a year.

1877 First cream separator is invented by Carl Gustaf Patrik de Laval (1848-1913), Swedish inventor. Operated by a steam engine, his device centrifugally spins milk and separates out the heavier cream from it.

1877 Eleanor Anne Ormerod (1828-1901), English entomologist, publishes the first of her annual *Notes for Observations of Injurious Insects*, which becomes the bible of agriculturalists worldwide.

1882 The first electric iron is patented in the United States by Henry W. Seely of New York.

1884 Modern, pedestal-shaped toilet with its oval shaped, portrait-frame seat is first exhibited at a Health Exhibition by an Englishman named Jennings.

1884 Contagious epidemic of bovine pleuro-pneumonia of foreign origin leads to the adoption of the first Federal quarantine law by the United States Congress.

1885 First self-service cafeteria, the Exchange Buffet, opens in New York.

1886 Rubber heels for shoes are first sold in the United States in Chicago, Illinois. The traditional leather heel has a hole punched out into which is inserted a round piece of rubber.

1886 American soft drinks Coca-Cola and Dr. Pepper are both first introduced this year.

1887 First machine to successfully produce glass bottles semi-automatically, the Ashland ma-chine begins operations in England. With this machine, two men can produce 200 glass bottles an hour.

(See also 1907)

1888 Meat is shipped in railroad cars cooled by mechanical refrigeration for the first time in the United States.

1888 Marvin Chester Stone of the United States patents the first paper drinking straw.

1891 Whitcomb L. Judson, American inventor, invents the clothing zipper. He calls this first modern zipper a "clasp locker" and exhibits it at Chicago in 1893.

1891 First commercially produced electric oven is sold in the United States by a Minnesota company.

1892 Toothpaste in a tube is first introduced by Washington Sheffield of the United States.

1892 First book matches are patented in the United States by Joshua Pusey of Pennsylvania.

1892 William Painter of the United States invents the pry-off bottle cap. His design, called the "Crown cap," is a cork-lined metal cap with a corrugated edge that is crimped around the bottle lip.

1893 First electric toaster is marketed by the Crompton Company of England. In the United States, General Electric markets its first electric toaster in 1909.

1893 First ready-to-eat breakfast cereal, "Shredded Wheat," is introduced by Henry D. Perky of the United States.

1894 James H. Northrop, English-American inventor, builds the first automatic or power loom that does not need to be refilled manually when the waft runs out. This not only speeds up the weaving process but also eliminates the weavers' inhalation of fabric dust (since they no longer must suck the waft thread through the shuttle).

1895 King Camp Gillette (1855-1932), American inventor, produces the first successful safety razor, which he markets in 1903. In 1955, the first stainless steel razor blades are introduced by the Wilkinson Sword Company of England.

1895 August Campbell of the United States first patents a spindle-type cottonpicker which plucks the cotton from the boll. This is the principle for all modern pickers, and Campbell's patent eventually is purchased by the International Harvester Company in the 1920s.

1898 George Washington Carver (1860-1943), American agricultural chemist, publishes his first agricultural paper. During his long and productive career, he develops hundreds of products from sweet potatoes, peanuts, and soybeans that prove a valuable alternative to cotton and tobacco as staple crops. He also emphasizes crop rotation and diversification.

1898 First ironing board with legs is introduced by the J. R. Clark Company of the United States.

1898 Wired glass is first produced commercially for use as a type of safety glass.

1900 Charles Francis Jenkins (1867-1934), American physicist and inventor, invents the conical paper cup.

1901 First practical suction vacuum cleaner is invented by Herbert Cecil Booth of England.

1905 First business in the United States devoted exclusively to the manufacture of tractors is established by C. W. Hart and C. H. Parr.

1906 Freeze-drying is invented by Jacques Arsène d'Arsonval (1851-1940) of France and his colleague, George Bordas. This food preservation process works on the principle of re-

moving water from food. It is not perfected until after World War II.

(See also 1946)

1907 Alva T. Fisher of the United States designs the first electric washing machine. Manufactured by the Hurley Machine Corporation, it is the first washing machine that does not require an operator to crank a handle to perform the washing action.

1907 Glass bottles are first produced fully automatically by a machine designed by Michael J. Owens of England. This highly efficient and complicated machine produces standardized bottles at the rate of 2,500 an hour.

1907 First powdered soap for home use is called "Persil" and is sold by Henkel & Cie in Germany.

1908 Jacques Edwin Brandenberger, Swiss chemist, first patents cellophane. He develops this thin, flexible plastic film from cellulose acetate. It becomes the first wrapping material that allows the contents of a package to be seen.

1908 First gasoline farm tractor with crawler tracks is manufactured by the Holt Company of California.

1910 First stockings made from synthetic fiber (rayon) are produced in Germany.

1912 First self-service grocery stores are established in California by the Alpha-Beta Company. These eventually evolve into today's supermarkets.

1914 Mary Phelps Jacobs of the United States is the first to patent a design for a woman's brassiere.

1915 First lipstick in a tube is invented by Maurice Levy of the United States.

1915 First university department of Glass Technology is established at the University of Sheffield in England.

1917 First mass-marketed athletic shoes, "Keds," are introduced by the United States Rubber Company.

1917 Clarence Birdseye (1886-1956), American inventor, first develops a process for freezing foods in small packages suitable for retailing. His process is highly efficient, and he founds the General Seafoods Company in 1924.

JUNE 18, 1918 First Federal regulation of livestock market practices is a Presidential proclamation requiring American stockyards and livestock dealers to be licensed.

1919 Earle Dickson, a cotton buyer for the Johnson and Johnson Company, devises a temporary dressing for cuts and wounds. His company then introduces in 1921 the first stick-on bandage called a "Band-Aid."

1922 First electrically heated and electrically regulated incubator used for hatching chicken eggs is patented by Ira M. Petersime of the United States.

1922 First canned baby food is manufactured in the United States by Harold H. Clapp of New York.

1923 First self-winding watch is patented in Switzerland by John Harwood.

1923 First electric shaver is patented by Jacob Schick of the United States.

1925 Arthur A. Williams of the United States introduces the first steel-capped safety shoe.

1926 First pop-up toaster is introduced in the United States by the McGraw Electric Company.

1926 Erik Rotheim of Norway builds the first aerosol container. He discovers that he can produce a fine spray by putting gas or liquid in a container to create internal pressure. In 1949, Robert Abplanalp of the United States develops an inexpensive, efficient valve that makes an aerosol spray can easy to use.

1927 First stainless steel cookware is made in the United States by the Polar Ware Company.

1929 W.F. Gericke of the United States first coins the term "hydroponics" for the growing of plants in water. He later demonstrates the commercial applications of soil-free agriculture.

1930 First stationary electric food mixer, the Mixmaster, is introduced in the United States by the Sunbeam Company.

1930 First sliced and packaged bread, "Wonder Bread" is introduced in the United States.

1930 First insulated window glass, called "Thermopane," is introduced by C.D. Haven of the United States.

1935 Beer is first sold in easy-to-carry cans by the Kreuger Beer Company of New Jersey.

1937 Shopping carts are first made available to food customers at Humpty Dumpty Stores in Oklahoma. They are a basket attached to a folding chair on wheels.

1938 First commercially successful instant coffee, "Nescafe," is introduced in the United States by the Nestle Company.

OCTOBER 16, 1945 Food and Agriculture Organization (FAO) is formally organized as part of the United Nations. It is the oldest permanent specialized agency in the United Nations and its objective is eliminating hunger and improving nutrition.

1945 First microwave oven is patented by Perry LeBaron Spencer of the United States.

(See also 1947)

1946 Earl W. Flosdorff (1904-1958), American bacteriologist, first demonstrates that the process of freeze-drying can be used to preserve coffee, orange juice, and even meat. When food is flash frozen in a vacuum, the water in it sublimates, or changes directly from a liquid into a vapor. Because the water sublimates rather than melts, the food's tissues do not collapse.

1947 First microwave oven to be manufactured is made by the Raytheon Company in the United States. They are not introduced for domestic use until 1965.

1948 Teflon, a new non-stick polymer, is first produced commercially by Dupont of the United States.

1948 Georges de Mestral, Swiss engineer, first develops the idea for the fabric-strip fastener called Velcro. He is inspired when he examines a burr with a microscope and finds it is composed of hundreds of tiny hooks. After eight years, he finally makes two strips of woven nylon, one of hooks and the other of loops. He patents Velcro in 1957, and it eventually becomes a common substitute for zippers, buttons, hooks, and snaps.

1949 First cake mixes are marketed by two United States companies, General Mills and Pillsbury.

1949 Artificial sweetener cyclamate is first marketed as Sucaryl by Abbott Laboratories of the United States. When it is linked to causing bladder cancer in rats, cyclamate is banned in the United States in 1970 and is replaced by other, new artificial sweeteners.

(See also 1981)

1950 Embryos are implanted in cattle for the first time.

1952 First attempt in the United States is made to harvest grapes mechanically.

1954 First frozen TV dinners become available in the United States.

1955 Correcting fluid called "Liquid Paper" is

invented by Bette Nesmith Graham of the United States.

1956 First enclosed shopping mall in the United States, Southdale Center near Minneapolis, Minnesota, opens for business.

1959 First mechanical harvester for cherries is introduced in the United States.

1960 Aluminum cans for soft drinks are first introduced in the United States. They are very light but also strong.

1961 First successful brand of disposable diapers, Pampers, is marketed in the United States.

1962 First sugar-free soft drink to be sold nationally to the general public in the United States is Diet-Rite Cola by Royal Crown Cola. It uses cyclamate as an artificial sweetener.

(See also 1981)

1963 First tab-top or pop-top aluminum can for beverages is introduced by Alcoa in the United States. In 1963, these tabs are designed to be pushed into the can, thus eliminating the disposal and litter problem.

1964 Automated irrigation systems are first field-tested for agricultural use in the United States.

1964 New "miracle" rice, a high-yield dwarf strain, is first introduced experimentally in the Phillipines by the International Rice Research Institute at Los Banos. It is a cross between ordinary *indica* rice and Japan's new *japonica* variety. Although it requires higher-than-usual amounts of fertilizer and insecticides, it contributes to what becomes known as the "Green Revolution."

1967 United States Federal Meat Inspection Act first goes into effect after American consumer advocate Ralph Nader successfully lobbies Congress to strengthen the original 1906 law.

AUGUST 1977   Mechanical lettuce harvester is first field-tested in the United States.

OCTOBER 1, 1980   Member countries of the European Economic Community (EEC) first ban the use of hormones in cattle feed.

1981   G.D. Searle of the United States first introduces its aspartame-based artificial sweetener NutraSweet, which is approximately 200 times as sweet as sugar and has no bitter aftertaste.

1986   United States Department of Agriculture (USDA) approves the release of the first genetically altered virus as well as the first outdoor test of genetically altered plants. The virus is to be used to combat swine herpes, and the plants are high-yield tobacco plants.

1993   First genetically altered food for human consumption, the Flavr Savr tomato awaits approval by the United States Food and Drug Administration (FDA).

1996   First refrigerator to use a combination of propane and butane instead of CFC coolant is produced in Germany. This new cylindrical refrigerator has a more energy-conserving shape than the standard "box." The propane and butane flows in coils that wrap around one-half the appliance's curved walls. It uses about half the energy of a comparable rectangular refrigerator.

Astronomy

c. 10000 B.C.   Earliest representation of stars in the heavens are found carved into rock and are attributed to this time period.

2296 B.C.   Earliest recorded sighting of a comet is made by Chinese astronomers. The Chinese describe astronomy as one of the oldest sciences, and astronomical ideas play an important role in their culture. Astronomy develops independently in China because of the remoteness of the country. Although they cannot predict comets, they do keep very accurate records of cometary sightings. Astronomers are held in high esteem in Chinese society, and they often become officials of the imperial court where their main activity is the study of such cometary phenomena and the prediction of eclipses.

c. 800 B.C.   Babylonians discover that the intervals between lunar eclipses can be calculated. At this point in their history Babylonians have already made astronomy a true science. They have a calendar based on the periodic return of the new moon (months) and also possess some notion of a week. As sophisticated astronomers, they observe celestial phenomena systematically and make predictions based on calculations. The earliest eclipse they observe

scientifically from beginning to end is the eclipse of March 19, 721 B.C.

763 B.C.   Earliest recorded solar eclipse is observed by Babylonian astronomers. Eclipses of the sun are often interpreted by early civilizations as a battle or the devouring of a heavenly body and are usually the cause of much anxiety. Although sometimes frightened by the spectacle, Babylonians study solar eclipses as well as the movements of the sun and the moon and lay the foundation for an eventual ability to predict an eclipse. Such predictive ability also implies that an eclipse is considered to be an expected, regular phenomenon that does not necessarily have ominous implications.

570 B.C.   First notion of spheres is contributed to the study of astronomy by Anaximander (610-c.546 B.C.), Greek philosopher. He recognizes that the heavens revolve around the Pole star and therefore pictures the sky as a complete sphere rather than as an arch over the earth. He also says that space is three-dimensional and recognizes that the earth's surface must be curved to account for the change in the position of the stars as one travels. He is the first Greek to use the sundial (that had been known for centuries in Egypt and Babylonia), and is also the first to attempt to draw a map of the world as he knows it.

530 B.C.   First to recognize that the morning star (Phosphorus) and the evening star (Hesperus) are in fact the same star is Pythagoras (c.582-c.497 B.C.), Greek philosopher. We know it today as the planet Venus (so named by the Romans and also called Aphrodite). He is the first to point out that the sun, moon, and various planets do not move uniformly like the stars, but that each has its own orbital path. He thus is the first to note that the orbit of the moon is not in the plane of the earth's equator but is inclined at an angle to that plane. He also is the first to state and teach that the earth is round. He is probably most influential in turning Greek astronomers toward the usefulness of mathematics in building astronomical models.

432 B.C.   Metonic cycle is first proposed by Greek astronomer Meton. While studying the summer solstice this year, he discovers that in a period of 19 years there are 235 lunar months. Using this, he calculates that if one were to arrange to have 12 years of 12 lunar months and 7 years of 13 lunar months, such a lunar calendar would have months to match the four seasons. This eventually is used as a standard by reference to which the actual calendar is from time to time adjusted, and becomes known as the Metonic cycle. It serves as the basis of all calendars until Roman dictator Julius Caesar (100-44 B.C.) establishes the Julian calendar in 50 B.C.

c. 410 B.C.   First ideas of the motion of the earth, both its axis rotation and its revolution around the sun, are made by Greek philosopher Philolaus. He suggests that the earth is not the center of the universe and that all the planets circle in separate spheres around a central fire of which the sun we see is only a reflection. His theory is important because it contains the first notion that the earth itself moves through space.

370 B.C.   First person to attempt at a mathemati-

cal explanation that reconciles Plato's perfect circles theory with the observable irregularities of planetary movement is Greek astronomer and mathematician Eudoxus (c.408-c.355 B.C.). His famous attempt to "save the appearances" argues that the sphere into which a planet is set has its poles set into another, as does that one, and so on, so that each sphere rotates evenly but irregularly in relation to the others. He is the first Greek to establish the fact that the year is not exactly 365 days long, but is actually six hours longer, and is also the first Greek to attempt a map of the stars, dividing the sky into degrees of latitude and longitude.

352 B.C.   Earliest record of a supernova sighting is made by Chinese observers. These extra-brilliant "novae" or new stars may suddenly flare to hundreds of thousands or even millions of times their former brightness, and then gradually revert to their former state. As their name indicates, they are thought to be new stars (called "guest stars" by the Chinese), but in fact are hot, subdwarf stars that are nearing the end of their life and are casting off great amounts of their mass.

(See also June 1054)

350 B.C.   First to clearly define the idea of the rotation of the earth is Greek astronomer Heracleides (c.388-315 B.C.). He considers heavenly bodies to be solid worlds suspended in the infinite ether, and although he incorrectly places the earth in the center of the universe, he nonetheless is the first to suggest that one heavenly body can revolve around another. This new concept of one body revolving around another (which itself revolves around the center of the universe) eventually shows the way to the heliocentric system. He also explains the daily movement of the fixed stars by the rotation of the earth.

240 B.C.   Chinese astronomers make the first recorded observation of what comes to be known as Halley's Comet. The Chinese chronicles and annals carefully note and record the appearance of this and other celestial events, so that an accurate record exists for Halley's and other regular phenomena. Halley's orbit around the earth takes roughly 76 years, and although there are gaps in the Chinese records after this year, the Chinese leave no gaps from 11 B.C. onward.

165 B.C.   Earliest recorded observation of sunspots is made by Chinese astronomers. Although these dark spots on the sun's surface may sometimes be visible to the naked eye when the sun is seen through clouds, the fact that the Chinese regard them as both different and significant enough to be recorded indicates the high level of their astronomical knowledge. They have no way of knowing that a sunspot is a vortex of gas on the sun's surface, associated with strong local magnetic activity. Although they can appear singly, most appear in pairs or groups. The number of spots reaches a maximum every 11 years on the average as the solar cycle waxes and wanes.

134 B.C.   First accurate, systematic star map is made by Greek astronomer Hipparchus (c.190-c.120 B.C.). Upon observing a star in the constellation of Scorpio for which he has no records, he decides to ascertain and note the exact position of over 1,000 of the brightest stars. He makes his star map by plotting the position of each star according to its latitude (angular distance north or south of the equator) and longitude (angular distance east or west of some arbitrary point). He is also the first to divide stars into classes depending on their brightness. The 20 brightest stars are of "first magnitude," while those of the sixth magnitude are just visible to the naked eye. This system has been essentially kept until today.

c. 150   First example in literature of the notion of a voyage to the moon is written by Lucian of Samosata, Greek satirist. In his moon voyage story titled *Vera Historia*, he tells of a group of sailors caught by a waterspout who find themselves eventually on the moon and in the mid-

dle of a battle between the King of the Moon and the King of the Sun (fighting over Venus). His second book, *Icaromenippus*, owes much to the Daedelus myth, for its hero flies on bird's wings to the moon and other planets.

c. 535   First model of the universe to be based on early Christian concepts of the universe is offered by Cosmas Indicopleustes, Alexandrian geographer and traveler. Although it is purported to be a synthesis between Christian and classical (pagan) thought, his model of the universe is based on Biblical accounts which he uses to support his themes. In his 12-volume *Topographia Christiana*, he strives to remove pagan influence from astronomy, spending one entire book trying to "prove" such notions as the sun being much smaller than the earth. His work is embraced by the Church and serves to stifle scientific astronomy for centuries to come.

900   Albategnius (c.858-929), Arabian astronomer also called Al-Battani, reviews Ptolemy's work and introduces new types of mathematical computations to astronomy, becoming the first to make use of a table of sines for that purpose. As the greatest of the Islamic astronomers, his use of new formulas which employ trigonometric functions is very influential. He determines the time of equinox within an hour or two and further clarifies the angle at which the earth's axis is tipped to its plane of revolution. He also establishes the length of the year, which is used afterwards in the reform of the Julian calendar.

(See also 1537)

c. 945   Dunhuang star map is made in China. It is the first map to employ what becomes later known as a Mercator projection.

c. 1000   First Eastern scholar to study the physical nature and composition of the heavenly bodies is Alhazen (965-1038), Arabian physicist. Like Greek astronomer Ptolemy (c.100-c.170), he assumes that the atmosphere is finite, and estimates its depth to be about 10 mi

(16 km). Unlike Ptolemy, however, his celestial model has 47 spheres all turning in a different way about and within one another. His theories later become widely known in Europe.

JUNE 1054 First recorded, detailed description of the explosion of a supernova is made by Chinese and Japanese astronomers. They report the sudden appearance of an extraordinarily bright star in the constellation of Taurus. This "guest star" shines so brightly for three weeks that it can be seen in daylight. At its peak, it is noted as being two to three times as bright as Venus. It fades slowly and remains visible for nearly two years. Interestingly, there is no record in Europe of anyone taking note of this occurrence.

(See also 352 B.C.)

c. 1075 Franco Liege, superintendent of the French cathedral school of St. Lambert, invents what some claim to be the first useful torquetum. This astronomical observational device evolves into a complicated set of inclined and rotating tables composed of several different instruments that allow the skilled user to calculate the altitude and azimuth of heavenly objects.

c. 1075 First to demonstrate the motion of the solar apogee with reference to the stars is Al-Zarqali (1029-1087), Arabian astronomer in Spain (also called Al-Zarqala or Arzachel). Using his own observations, he deduces a solar apogee (its farthest point from the earth) of 77 degrees, 50 minutes. He also invents a new type of plane astrolabe and publishes a book of astronomical tables called *Toledo Tables*.

c. 1270 Kuo Shou-Ching of China builds the first astronomical device to use an equatorial mounting. It is called a torquetum and is an improvement on the armillary sphere.

1424 Ulugh Beg (1394-1449), Turkish astronomer, publishes his own star map, the first new one since Greek astronomer Hipparchus

(c.190-c.120 B.C.), about 1500 years before. He also builds an astronomical observatory in Samarkand (central Asia) where he had founded a university four years before. During its short life, this observatory is a major center of astronomical research and the last great monument of Muslim astronomy. As the grandson of Timur Lenk (1336-1405), the conqueror of Asia also known as Tamerlane and "The Destroyer," he is murdered by one of his own sons.

1472 First exposition of the Ptolemaic system by a European becomes available as *Sphaera Mundi* is published. Written in the 13th century by the English mathematician and astronomer Johannes de Sacrobosco (c. 1200-1250), also known as John of Holywood, it becomes the standard astronomical textbook of the Middle Ages and continues in use to the end of the 17th century. This extremely popular elementary textbook on spherical astronomy has a very large number of translations and editions.

1472 First to make comets the object of serious, scientific study is Regiomontanus (1436-1476), German astronomer (also known as Johann Muller). Observing a comet this year, he measures its position in the sky. These data prove accurate enough that some 200 years later, English astronomer Edmond Halley (1656-1742) is able to use them to calculate the comet's orbit. Heretofore, it was thought that comets would "announce calamities," and they were regarded with superstitious terror. Regiomontanus also revises Ptolemy's work and founds an observatory at Nuremberg.

1483 *Alfonsine Tables* are first printed in book form by German astronomer Regiomontanus (1436-1476). Until now, these planetary tables commissioned by the Spanish monarch Alfonso X (1221-1284) existed only in manuscript form. Regiomontanus makes important, new observations of the heavens and prepares newly revised tables that bring the old tables up to date. They are considered the best available until the 16th century and become widely used

The torquetum was the first astronomical device to use an equatorial mounting.

by ocean navigators (like Columbus) in the coming age of exploration. Regiomontanus also introduces algebraic and trigonometric methods to Germany via these tables.

1515 First printed edition of Greek astronomer Ptolemy's (c.100-c.170) *Almagest* appears in the form of a Latin translation from the Arabic. The first Latin translation made from a Greek text is published in 1528. The astronomical system offered by the Greek astronomer Claudius Ptolemy (c.100-c.170) placed the earth at the center of the universe. His original work passed

to the Arabs and was translated from Arabic into Latin in 1175. It then comes to dominate European astronomical thinking for the next four centuries.

1538 First important astronomical discovery concerning comets is made by Italian physician, astronomer, and geologist Girolamo Fracastoro (c.1478-1553). He announces in his book, *Homocentrica sive De Stellis Liber*, that the tails of comets are always turned away from the sun. His postulation that the planets revolve in spherical orbits around a central, fixed point paves the way for the Copernican model of the solar system.

(See also 1540)

1539   First to label the stars with letters is Italian physicist Alessandro Piccolomini (1508-1578), who publishes his *Della Sfera del Mondo*. The appendix to this work is titled "Libro Delle Stelle Fisse," and contains a simplified chart of the heavens in which the brightest stars in each constellation are labeled simply A, B, C, and so on. He omits the traditional pictorial representations and the fainter stars, and concentrates on making the bright ones identifiable and easy to find. This method is first adopted by the German astronomer Johann Bayer (1572-1625). In his book, *Uranometria*, he calls the brightest star of each constellation "alpha," the next brightest "beta," and so on. This improves Piccolomini's system and is still used today.

(See also September 1603)

1540   First to publish a diagram showing that a comet's tail always points away from the sun is Petrus Apianus (1495-1552), German astronomer also known as Peter Apian. He depicts this phenomenon in his book, *Astronomicum Caesareum*.

(See also 1538)

1540   First printed announcement of the revolutionary astronomical theory of Copernicus is made by German mathematician Rheticus (1514-1574), also known as Georg Joachim von Lauchen, in his *Narratio Prima*. As an early disciple of Nicolas Copernicus (1473-1543), Rheticus had visited the Polish astronomer in 1539 to learn more of the theory he had only heard about, and it was through his urging and support that the great man finally allowed his work to be published. In this small work by Rheticus, he summarizes the heliocentric theory but deliberately does not name its creator.

(See also 1543)

1543   Heliocentric theory is first publicly stated by Polish astronomer Nicolas Copernicus (1473-

1543) in his book *De Revolutionibus Orbium Coelestium*. In this work, which is little-noticed upon publication, he proposes his sun-centered theory in which the earth is in daily motion about its axis and in yearly motion around a stationary sun. After considerable study Copernicus becomes increasingly dissatisfied with earth-centered ideas, and soon realizes that a sun-centered system explains what is actually observed so well. He waits to publish, however, fearing that such an open declaration against the prevailing earth-centered doctrine might get him into trouble. After his death, the book is indeed banned by the Catholic Church for having removed the earth (and therefore mankind) from the center of the universe. Nonetheless, this work, as much as any single publication, can be said to mark the beginning of the Scientific Revolution.

(See also May 16, 1616)

1551   First published recognition in England of the Copernican heliocentric system is made by English mathematician Robert Recorde (1510-1558). He states his support in his astronomical text *The Castle of Knowledge*, which alludes to Copernican theory and contains an endorsement.

1551   First set of planetary tables based on Copernican theory are calculated by Erasmus Reinhold (1511-1553), German mathematician, who publishes his *Tabulae Prutenicae*. As the only German to publicly endorse this theory, Reinhold uses the calculations of Copernicus to produce a set of tables that are more precise than the *Alfonsine Tables*. From Reinhold's tables, the position of the principal celestial bodies for any epoch can be easily determined. They remain the best available until the "Rudolphine Tables" of German astronomer Johann Kepler (1571-1630), three quarters of a century later. Reinhold's work contributes to the gradual spread of the Copernican system.

1561   First observatory to have a rotating dome is built by Wilhelm IV (1532-1592), Landgrave (Count) of Hesse, Germany. He builds a private astronomical observatory in order to pursue his own interests in astronomy and also befriends the Danish astronomer Tycho Brahe (1546-1601).

(See also 1576)

1569 First authority to publicly declare the need for a radical reform of astronomy is Pierre de la Ramee (1515-1572), French philosopher and logician, also known as Petrus Ramus. He publishes his anti-Aristotelian work, *Scholarum Mathematicarum*, in which he praises and supports the theory of Copernicus. As an opponent of traditional medieval thinking, he is convinced that only mathematics can extricate astronomy from the unproven hypotheses that impede its progress. He admires Copernican theory for having a sound mathematical basis as well as reliable observations, and he sees Copernican doctrine as the basis of a new, liberated, scientific astronomy.

1572 First significant Latin translation of the work of Greek astronomer Aristarchus (c.320-250 B.C.) is published as *De Magnitudinibus, et Distantis Solis, et Lunae*. In this work, the early Greek astronomer calculates the size and distance of the sun and moon (both of which he underestimates). By combining the Pythagorean notion of a moving earth with the suggestion of Heracleides that some planets orbit the sun, he becomes known as the Copernicus of Antiquity.

1576 First complete and original exposition of the Copernican system is offered by English mathematician and astronomer Thomas Digges (c.1543-1595). As the leader of the English Copernicans, he adds an appendix to his father's *Prognostication Everlastinge* in which he translates parts of the *De Revolutionibus Orbium Coelestium*. Titled "A Perfit Description of the Caelestiall Orbes," it describes a physical universe in which the fixed stars are at varying distances in infinite space. It also contains illustrations showing the sun at the center of our solar system.

1576 First complete modern star catalog is assembled by Danish astronomer Tycho Brahe (1546-1601). This work gives the accurate positions of 777 fixed stars and establishes his reputation as the greatest naked-eye astronomer, as well as the founder of modern observational astronomy. His work replaces that of Greek astronomers Hipparchus (c.190-c.120 B.C.) and Ptolemy (c.100-c.170), exceeding them in accuracy, and marks the beginning of a new era of practical astronomy.

1576 First modern astronomical observatory is begun by Danish astronomer Tycho Brahe (1546-1601). This year, Frederick II (1534-1588), king of Denmark and Norway, grants title to an island called Ven to Brahe and also gives him financial support for the observatory and laboratory buildings. Brahe names the observatory Uranaiborg after Urania, the muse of astronomy, and stocks it with the finest astronomical instruments and a printing shop. For 12 years, it is the center of astronomical study and discovery in northern Europe.

1584 Giordano Bruno (1548-1600), Italian philosopher, begins writing on his new view of the universe. Using Copernican ideas, he becomes the first to connect his system with what becomes a truly revolutionary interpretation of the universe. Using the Copernican system as the center of his ideas, he expounds on the infinity of space and the habitability of other worlds, among other things. For these and other astronomically related teachings, he is charged with heresy and arrested by the Inquisition in 1592. He refuses to recant any of his ideas during his years of imprisonment and is burned alive in Rome.

AUGUST 13, 1595 First variable star to be discovered is seen by David Fabricius (1564-1617), German astronomer. On this date he observes a star of the third magnitude, which disappears when he next looks for it on October 1. Assuming it to be a nova, he never learns that it is a star

TYCHONIS BRAHE LIB. I.

1. *In Capite.*
2. *In Pectore.*
3. *In Cingulo.*
4. *Quæ ad Ilia.*
5. *In Poplite.*
6. *In Crure.*
7. *Extrema pedis.*
8. *In Flexura Brachij.*
9. *In Cubito ejusd.*
10. *In altero Cubito.*
11. *In Erectione sedis.*
12. *In medio Cathed.*
13. *In supremit. Sed.*
14. *In extrem. Sella.*
15. *Superior Scabelli.*
16. *Extrema Scabelli.*
17. *Media Scabelli.*
18. *Sequens in scabell.*
19. *Præced. in Tbia.*
20. *Mola Genu.*
21. *In Vmbilico.*
22. *Parvul. ad crines.*
23. *Infer. Arundinis.*
24. *Sequens in arund.*
25. *Tertia Arund.*
26. *Suprema Arund.*

Considered history's greatest naked eye astronomer, Tycho Brahe located a supernova in 1572 in the constellation Cassiopeia, which is popularly known as Cassiopeia's Chair.

which has periodic variations in brightness, later named Omicron Ceti.

(See also 1638)

1596   Variation in the obliquity of the ecliptic is discovered by Danish astronomer Tycho Brahe (1546-1601). The sun's path on the celestial sphere is oblique to the equator (lying partly on one side of it and partly on the other), and the great circle it describes is called the ecliptic. It is

called that because eclipses take place only when the moon is in or near it. The angle at which it cuts the equator is called the obliquity of the ecliptic. Brahe finds that this important angle varies a tiny bit each year.

1596   First important work in astronomy that fully supports the Copernican theory is published by Johann Kepler (1571-1630), German astronomer. Although his *Mysterium Cosmographicum* is the work of a young man trying to make reality fit neatly into a somewhat fantastic theory (that there are five different shapes of solids that fit between the orbits of planets), it contains enough good mathematics and knowledge of astronomy to impress Danish astronomer Tycho Brahe (1546-1601), who gives the young Kepler a position. Kepler is a convinced Copernican from the beginning, and he will go on to lay the foundations of modern astronomy. (See also 1609)

SEPTEMBER 1603   Johann Bayer (1572-1625), German astronomer, publishes *Uranometria*, the first edition of his catalog of the heavens. It is the first catalog to show the entire celestial sphere and describes constellations and locates more stars than even the legendary Danish astronomer Tycho Brahe (1546-1601) had done. More importantly, Bayer lists the stars of each constellation by Greek letters in order of brightness. This systematic naming device exits today.

(See also 1539)

1608   First refracting telescope is credited to Hans Lippershey (c.1570-c.1619), Dutch optician. He discovers that aligning a concave and a convex lens makes distant objects appear closer. A refracting lens bends light rays to a focus near the eye where a second lens magnifies the image.

(See also 1609)

1609   Johann Kepler (1571-1630), German astronomer, publishes his *Astronomia Nova* and announces his discovery that the earth and planets travel about the sun in elliptical orbits. In this work he states his two laws of planetary motion, the first of which states that the orbits of the planets can be drawn as ellipses with the sun always at one of their foci. The second says that a line connecting a planet and the sun will sweep over equal areas in equal times as the planet moves about its orbit. In other words, a planet will move faster the closer it is to the sun, and this can be determined by a rule that can be calculated. Kepler's work puts an end to the perfect circles theory held since the Greeks and becomes the cornerstone of modern astronomy. In 1619 he states his third law in the book *Harmonices Mundi*. There he shows the sun as the controller of planetary movement by stating that the square of the period of revolution of a planet is proportional to the cube of its distance to the sun.

1609   Telescope is first used to study the skies by Galileo Galilei (1564-1642), Italian astronomer and physicist. Having heard in April of this year that the Dutch have invented a telescope, he succeeds in arranging two lenses (one concave and one convex) so that they magnify a distant object. After quickly improving his design, he turns his device toward the sky and sees details that no one had seen before. He later describes the surface of the moon, the phases of Venus, and the four satellites of Jupiter. With Galileo, the age of telescopic astronomy begins.

(See also March 1610)

MARCH 1610   Galileo Galilei (1564-1642), Italian astronomer and physicist, publishes his *Siderius Nuncius*, which gives an account of the first use of the telescope in astronomy and offers details of his discoveries. Like an advertising man rather than a scholar, Galileo writes on his book's title page, "Revealing great, unusual, and remarkable spectacles, opening these to the consideration of every man.... The surface of the Moon, innumerable Fixed Stars, Nebulae, and above all Four Planets swiftly revolving about Jupiter at differing distances and periods...." In fact, Galileo sees all these

things and more, demonstrating conclusively the existence of mountains on the moon, the Milky Way as an endless collection of stars, and Jupiter's moons. Galileo's telescopic discoveries puts into doubt the perfection described by Aristotelian astronomy and gives credence to the new Copernican system.

DECEMBER 15, 1612   First astronomer to mention the spiral or elongated Andromeda Nebula, the most distant object that can be seen without a telescope, is Simon Marius (1570-1624), German astronomer and mathematician also called Mayr. Marius also achieves fame of a sort for naming Jupiter's satellites (Io, Europa, Ganymede, and Callisto), despite the fact that he tries to pass off Galileo's discovery of them as his own.

MAY 16, 1616   Astronomical teachings of Polish astronomer Nicolas Copernicus (1473-1543) are first forbidden by Pope Paul V (1552-1621). He also places the book containing his heretical sun-centered theory, *De Revolutionibus Orbium Coelestium*, on the index of prohibited books.

(See also 1543)

1619   First telescopic observation of a comet is made by Johann Baptist Cysat (c.1586-1657), German astronomer and mathematician. He publishes his findings in *Mathemata Astronomica de Loco, Motu, Magnitudine, et Causis Cometae*, which contains observations he conducted over a period of two months. He agrees with Kepler's description of comets, stating that their motion is along a straight line, and that they come from infinity and disappear into infinity. This concept apparently rules out any notion of a comet's periodic return. He also discovers the Orion Nebula.

NOVEMBER 7, 1631   First planetary transit ever observed is made by Pierre Gassendi (1592-1655), French philosopher and mathematician, who records a transit of Mercury across the face of the sun. A transit describes the passage of a

Galileo and his telescope.

smaller body across the visible disc of a larger one. It occurs within five hours of when German astronomer Johann Kepler (1571-1630) had predicted.

1632 First official astronomical observatory (one associated with a national body) is founded at the University of Leiden in Holland. Built upon a university roof, it has an azimuthal quadrant mounted under a revolving dome.

1634 First piece of authentic science fiction is astronomical in nature. Written by German astronomer Johann Kepler (1571-1630) and titled *Somnium*, this allegorical work is about a man who travels to the moon in a dream. In it, Kepler describes the lunar surface in a realistic fashion. This work also inspires others to consider the same theme.

1638 Johann Phocylides Holwarda (1618-1651), Dutch astronomer, identifies the first variable star, Omicron Ceti. This is the same star that Fabricius saw in August, 1595, and which had disappeared by October of that year. Holwarda observes it appear, disappear, and reappear and notes that it increases and then decreases to invisibility in a period of 11 months.

(See also August 13, 1595).

NOVEMBER 24, 1639 First to observe the transit of Venus across the face of the sun is English astronomer Jeremiah Horrocks (1619-1641), also called Horrox. After correcting Kepler's Rudolphine Tables, he is able to correctly predict the date of this event. He also suggests that observations of such a transit from different places might create a parallax effect that can then be used to calculate the distance of Venus as well as to determine the scale of the solar system. He is eventually proved correct in this as well. As the first astronomer to completely accept Kepler's elliptical orbit ideas, he extends Kepler's work by showing that the moon moves in an elliptical orbit about the earth.

(See also 1662)

1650 First observation of a double or binary star is made by Giovanni Battista Riccioli (1598-1671), Italian astronomer. He uses a telescope and discovers that the middle star of the handle of the Big Dipper (called Mizar) is really two stars very close together. He later studies the moon and names lunar craters in honor of past astronomers. He also is the first to maintain that there is no water on the moon.

MARCH 25, 1655 First satellite of Saturn is discovered by Christiaan Huygens (1629-1695), Dutch physicist and astronomer, who devises an improved telescope. This body proves to be the largest of that planet's satellites and he names it Titan because of its size. He also discovers that Saturn is surrounded by a thin ring that does not touch the planet, and further announces his discovery of a huge cloud of gas and dust, the Orion Nebula. He is also the first to make a specific guess at the distance of the stars (which he highly underestimates).

1659 First to note and record some type of surface marks on the planet Mars is Christiaan Huygens (1629-1695), Dutch physicist and

astronomer. Huygens discovers a new and better way of grinding lenses for telescopes and makes his discovery using his own 23 ft (7 m) long telescope.

1662 Johannes Hevelius (1611-1687), German astronomer also called Hewelcke, edits and publishes *Venus in Sole Visa*. This is a posthumous account of the first observation of the transit of Venus in 1639 written by the English astronomer Jeremiah Horrocks (1619-1641).

(See also November 24, 1639)

1666 Theory of universal gravitation is first conceived by Isaac Newton (1642-1727), English scientist and mathematician. It is during this year he spends at his mother's farm (schools are closed because of the plague) that he supposedly watches an apple fall to the ground and starts to ponder if the same force that pulls the apple downward also holds the moon in its grip. He later refines these early calculations.

(See also 1687)

1666 First observation of the polar caps on Mars is made by Giovanni Domenico Cassini (1625-1712), Italian-French astronomer. In 1672 he again focuses his attention on Mars and uses a colleague's observations of that planet made while in Cayenne (French Guiana) and compares them to his own made from Paris, and is able to calculate the distance of Mars from earth. Using this value, he then makes the first nearly correct determination of the sun-earth distance (off by only 7%). This is the first notion of the actual scale of the solar system

(See also 1672)

1668 First known reflecting telescope is constructed by Isaac Newton (1642-1727), English scientist and mathematician. He realizes that since all wavelengths of light are reflected equally, a curved mirror should not suffer from chromatic aberration (the blurring that occurs with refracting telescopes). He uses a metal alloy mirror in his telescope and obtains a magnification of about 32. (See 1672)

1668 First to give an accurate description of the bands and spots visible on the surface of Jupiter is Italian-French astronomer Giovanni Domenico Cassini (1625-1712). He also issues a table of the motions of Jupiter's moons and establishes that planet's period of rotation. He is the first to study zodiacal light (the sunlight that is reflected from dust particles in interplanetary space).

1669 First observer to notice the changes in brightness (variability) of the star Algol is Italian astronomer Geminiano Montanari (1633-1687). He makes his first recorded observation this year and publishes a report of it in 1672. Algol is one of the most famous variable stars in the sky and a prototype of what becomes known as a binary star.

(See also 1782)

1671 First accurate measurement of the circumference of the earth is made by Jean Picard (1620-1682), French astronomer. This is the first measurement of the earth that is more accurate than that of Greek astronomer Eratosthenes (c. 276-c. 196 B.C.). Picard uses the same principles as Eratosthenes, but substitutes the point of a star for the large body of the sun, and therefore is more accurate. He gives the earth a circumference of 24,876 mi (40,025 km) and a radius of 3,950 mi (6,355 km) (values close to those accepted today). Picard is also the first to use the telescope not only for simple observation but for the accurate measurement of small angles. This innovation shows the way toward combining the telescope with other measuring devices, which will result in the creation of entirely new types of scientific instruments.

(See also 1687)

1671 Second satellite of Saturn is discovered by Italian-French astronomer Giovanni Domenico Cassini (1625-1712). Named Iapetus, after the mythological Titan who was the son of Uranus (heaven) and Gaea (earth), it is the first moon of Saturn to be discovered since Dutch astronomer and physicist Christiaan Huygens (1629-

1695) located what he named Titan in 1655. It is also the third largest of Saturn's 10 satellites and orbits that planet in a nearly circular orbit. It has a diameter of about 895 mi (1,440 km). Cassini uses a telescope that is over 100 ft (30 m) long to make this discovery and other Saturn discoveries. In 1672, he discovers Rhea, the third satellite of Saturn to be found. It is named after a sister of Iapetus.

(See also March 25, 1655)

1672 Parallax of Mars is first obtained by Italian-French astronomer Giovanni Domenico Cassini (1625-1712). He determines the distance of Mars using his observations and those of French astronomer Jean Richer (1630-1696), who takes his readings in French Guiana. Since the relative distances of the sun and planets had been established by the early 17th century, it is only necessary for Cassini to determine any one of those distances accurately to be able to calculate all the rest. Using his value obtained by the parallax of Mars, he calculates that the sun is 87 million miles from the earth—fairly close to today's recognized 93 million miles.

JUNE 22, 1675 Royal Greenwich Observatory is founded by Charles II of Great Britain (1630-1685) to improve knowledge of the positions of the celestial bodies as an aid to navigation. John Flamsteed (1646-1719) becomes the first astronomer royal when the building is completed the next year. It is through Greenwich that the prime meridian passes, from which longitude is measured east and west to all parts of the earth. This is decided in 1884, when Greenwich is chosen at an international conference to be the common prime meridian for the entire world. Using an imaginary line drawn through Greenwich as the line of 0° longitude, the longitudes of all other points are then referenced to it.

NOVEMBER 22, 1675 Speed of light is first calculated by Olaus Roemer (1644-1710), Danish astronomer, who announces his achievement at the Academy of Sciences in Paris. By observing the motions of Jupiter's satellites at different times of the year, he comes to conclude that light must have a finite velocity. Until this, most agreed with Descartes that light had an infinite speed. Roemer then studies the irregularities of Jupiter's satellite as seen from earth and is able to calculate the speed of light at 141,000 mi (227,000 km) per second. He was not far off from the correct value of 186,282 mi (299,792 km) per second. He does the actual measurement by timing an eclipse of one of Jupiter's moons when the earth is closest to Jupiter and comparing it to when the earth is the farthest away.

1675 Giovanni Domenico Cassini (1625-1712), Italian-French astronomer, first determines that the ring around the planet Saturn is actually a double one. He notes that a dark line seems to separate the ring into a narrow, bright outer ring and a broad, less bright inner ring. Many think the ring is a single object with a dark line running around it, but Cassini proves correct. The dark line becomes known as "Cassini's division" or "Cassini's gap."

1676 First observatory in the southern hemisphere is established on the island of St. Helena in the South Atlantic by Edmond Halley (1656-1742), English astronomer. Although only 20 years old and working under poor meteorological conditions, Halley determines the coordinates of 360 southern stars and observes a transit of Mercury across the sun.

(See also 1678)

1678 First catalog of telescopically obtained coordinates of stars in the southern hemisphere is published by Edmond Halley (1656-1742), English astronomer. Even before the young Halley begins his observation of comets, his *Catalogus Stellarum Australium* gains him fame as the "southern Tycho."

(See also 1676)

1679 Jean Richer (1630-1696), French astrono-

mer, discovers that gravity increases with latitude. Sent to Cayenne in French Guiana by Italian-French astronomer Giovanni Domenico Cassini (1625-1712) to make observations of Mars, he finds that a pendulum of a given length swings more slowly in Cayenne than at Paris (losing two and one-half minutes a day in Cayenne). He concludes that if the rate of beat of a pendulum varies with the size of the force of gravity acting upon it, then the force of gravity must be weaker in Cayenne because that spot is farther from the center of the earth than is Paris. This proves valuable to Isaac Newton (1642-1727), English scientist and mathematician, and Christiaan Huygens (1629-1695), Dutch physicist and astronomer, who use it to prove that the shape of the earth is an oblate spheroid.

1680 First to attempt to measure the quantity of heat received by the earth from the sun is Isaac Newton (1642-1727), English scientist and mathematician. He estimates that it is "about 2,000 times that of red-hot iron." In fact, transforming the amount of energy radiated by the sun into an effective temperature received by the earth is not possible until the law that connects the temperature of a surface with the amount of heat radiated per second (the Stefan-Boltzmann law) is established in 1880. The sun itself is a star made of intensely hot gas. At its surface the temperature is of the order of 6,000° Centigrade. Near its core, the temperature rises to an incredible 14,000,000° Centigrade. It does not burn in the conventional sense, for its energy is being created by nuclear reactions, and as it grows older it loses mass. Although the sun is at least 5,000 million years old, it will go on shining for at least another 5,000 million years.

1684 Giovanni Domenico Cassini (1625-1712), Italian-French astronomer, discovers two more satellites of the planet Saturn. He names them Dione and Tethys. Dione has a diameter of 696 mi (1,120 km) and a non-uniform surface brightness; Tethys has a diameter of 650 mi

(1,046 km). In Greek mythology, both are Titans and sisters of Iapetus.

1686 Edmond Halley (1656-1742), English astronomer, is acknowledged with publishing the first meteorological chart. It shows the distribution of prevailing winds over the oceans.

1686 Gottfried Kirch (1639-1710), German astronomer, discovers the variability of the star Zeta Cygni. The observed light of a variable star changes noticeably. Its cause may be an eclipsing body or some intrinsic property of the star, and its variations can be periodic, semiregular, or completely irregular. Kirch is also the first director of the Berlin Observatory and one of the earliest systematic observers of comets.

1687 First transit instrument using a telescope is built by Danish astronomer Olaus Roemer (1644-1710), who sets up an observatory in his home. Called a meridian telescope, it has a telescope mounted so that it can rotate around the axis but remain always on the meridian (north-south) plane. With the telescope mounted this way, he can determine the longitude and altitude (the angular distance of a body above the observer's horizon) of a celestial body. He later adds illuminated cross-hairs and a clock that strikes the seconds. Later astronomers improve his mounting so that it can be fully rotated. This mount comes to be known as an equatorial mount.

1687 Law of universal gravitation is first offered by English scientist and mathematician Isaac Newton (1642-1727). This year he publishes his *Philosophiae Naturalis Principia Mathematica* in which he states the three laws of motion that lay the groundwork for his law of universal gravitation. These laws state: (1) that a body remains at rest unless it is compelled to change by a force impressed upon it; (2) that the change of motion (the change of velocity times the mass of the body) is proportional to the forces impressed; and (3) that to every action there is an equal and opposite reaction. Publi-

cation of this work culminates a creative process that began over 20 years before when Newton was a student, and which concludes with his law of universal gravitation. This principle states that the physical laws of the heavens and those of the earth are one and the same, with both planets and apples being subject to the same natural forces. With this, and for the first time, a single mathematical law can explain movement both in the heavens or on earth. Newton shows the entire cosmos to be unified by knowable laws and predictable phenomena, and culminates the Scientific Revolution that began with Polish astronomer Nicolas Copernicus (1473-1543).

1690 Johannes Hevelius (1611-1687), German astronomer also called Hewelcke, has the third and final part of his great work on astronomy published posthumously as *Prodromus Astronomiae*. It is the first work in astronomy to use coordinates with respect to the equator instead of the ecliptic (the orbit of the earth around the sun). This celestial atlas also introduces a number of new constellations made up of small stars located in what had been thought to be blank spaces. It is the last great work of naked-eye astronomy.

1702 First textbook in astronomy to be based on Newtonian principles is published by Scottish mathematician and astronomer David Gregory (1659-1708). In his textbook on gravitational principles titled *Astronomiae Physicae et Geometricae Elementa*, he tries to remodel astronomy so as to conform with Newtonian physical theory.

1704 First observer to notice the changes in brightness (variability) of the star R Hydrae is Italian astronomer Giacomo Filippo Maraldi (1665-1729). Located in Hydra, or the Watersnake, which is the largest constellation in the sky, it has at times naked-eye visibility. Maraldi is the nephew of Italian-French astronomer Giovanni Domenico Cassini (1625-1712).

1705 First scientific prediction of a comet's reappearance is made. Edmond Halley (1656-1742), English astronomer, first predicts that the comet he observed in 1682 would return again in roughly 75 years from that date. He argues that comets are subject to the same gravitational laws as planets and that the comets recorded in 1456, 1531, and 1607 were in fact the same comet, but one which was in a closed but very elongated orbit about the sun and which was visible only when it came close to the earth (every 75 or 76 years). He is proven correct when the comet returns in 1758, making this one of the great achievements in theoretical astronomy.

(See also December 1758)

1715 A total eclipse is observed and recorded for the first time by a large number of coordinated observers throughout Europe. This is due largely to the efforts of English astronomer Edmond Halley (1656-1742), who organizes this international scientific effort.

1718 Edmond Halley (1656-1742), English astronomer, first demonstrates that stars actually move independently in space. Publishing his findings in an article in *Philosophical Transactions*, he points out that at least three stars, Sirius, Procyon, and Arcturus, had changed their positions markedly since Greek times, and that even a change since the time of Danish astronomer Tycho Brahe (1546-1601) was perceptible. He proposes from this that stars have proper motions of their own which are apparent to us only after long periods of time because of the vast distances between the stars and earth. His proof that stars actually change their positions shatters the traditional belief in the heavens filled with fixed stars.

1725 First great star map of the telescopic age is published posthumously by John Flamsteed (1646-1719), English astronomer and first astronomer royal at the Greenwich Observatory. His massive star catalog, *Historia Coelestis Britannica*, is three times as large as Danish astronomer Tycho Brahe's (1546-1601) work,

and the individual stars listed in it are located with six times greater precision. As a perfectionist who wanted his observations to be absolutely correct, Flamsteed was so slow to publish that the Royal Society published them without his permission. Furious at this rash and premature deed, Flamsteed bought up and destroyed every copy he could find. His complete work is published only after his death and helps set a standard of excellence for Greenwich as an observational observatory.

1729 First observational proof of Galileo's argument that the earth is moving is given by English astronomer James Bradley (1693-1762), who discovers the phenomenon he calls the "aberration of light." This theory assumes that light does not travel instantaneously but has a finite or given velocity. It also says that the earth itself is moving around the sun at a given orbital speed. Together, these ideas explain the apparent displacement of a star toward the direction in which the earth is moving. Bradley's discovery confirms not only Galileo's theory but that of Danish astronomer Olaus Roemer (1644-1710) as well. He argued in 1675 that light has a finite velocity.

1729 First use of the term aberration is made by Eustachio Manfredi (1674-1739), Italian mathematician and astronomer. He uses the term in his *De Annuis Inerrantium Stellarum Aberrationibus* in which he announces his discovery of the annual aberration of fixed stars. In astronomy aberration means the small, apparent displacement of a star from its true position in the sky. Bradley correctly explains this phenomenon this same year as being caused by the earth in its orbit.

1731 First sextant-type instrument able to measure the position of the heavenly bodies from the deck of a ship is invented by English instrument maker John Hadley (1682-1744). Called the reflecting quadrant and later the reflecting sextant, this device is able to bring by reflection the images of two celestial bodies (or of the

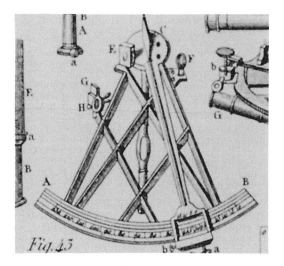

Hadley's sextant.

horizon) into exact coincidence as seen in the telescope. It becomes indispensable to every pilot since with this device the position of a ship on the high seas can be determined by a hand-held instrument, even on the deck of a rolling ship. It is also claimed that the American optician Thomas Godfrey (1704-1749) simultaneously invents the same device, but it remains unknown in Europe.

APRIL 19, 1739   First scientific solar observations made in America are conducted by John Winthrop (1714-1779), American mathematician and astronomer, in the colony of Massachusetts. Considered the first astronomer in America, Winthrop records his solar findings in *Notes on Sunspots* published this year. He later publishes his lectures on the return of Halley's comet and studies the transit of Mercury in 1740. He also directs the first Harvard-sponsored astronomical expedition to St. John's, Newfoundland to study the 1761 transit.

1740   First to try to determine the magnitude of the stars by measuring the intensity of their light using a device other than the human eye is Anders Celsius (1701-1744), Swedish astronomer. Using colored glass as a primitive pho-

tometry instrument, he makes a series of observations of certain stars and is able to measure their intensity (brightness). He also studies the aurora borealis and is the first to associate it with changes in the earth's magnetic field.

1744   First analytical work on planetary motion is published by Leonhard Euler (1707-1783), Swiss mathematician. In his *Theoria Motuum Planetarum et Cometarum* he calculates planetary orbits and applies his mathematics to astronomy, making him a precursor of both Italian-French astronomer and mathematician Joseph Louis Lagrange (1736-1813) and French astronomer and mathematician Pierre Simon Laplace (1749-1827). This text actually begins the study of celestial mechanics. Euler is also the first to note that as a consequence of the law of universal gravitation, a planet does not describe an ellipse around the sun, but both sun and planet describe ellipses around their common center of mass. The same is true for a planet and its satellites.

1747   First approximate solution to the "three-body problem" (concerning the interaction between three different masses in space) is offered by French mathematician Alexis Claude Clairaut (1713-1765). His calculations are contained in his *Théorie de la Lune* published this year. While Newton's laws of universal gravitation could be applied easily to two bodies in space, the problem of computing the motion of three bodies moving under the action of their mutual gravitational attractions has no exact analytical solution. The motion of the three bodies can be computed numerically, however, to any required level of accuracy.

1748   James Bradley (1693-1762), English astronomer, publishes an article in which he announces his discovery of the nutation (small periodic shifts) of the earth's axis. He attributes this to the changes in direction of the gravitational pull of the moon. In order to detect nutation, Bradley must be able to determine differences as small as two seconds of arc.

Although Bradley actually discovered this in 1729, he carefully studies star positions for 19 years and waits until he has gathered sufficient evidence to prove it.

(See also 1749)

1749 First proof of the moon's role in the earth's nutation is offered by French mathematician Jean le Rond D'Alembert (1717-1783). This year he publishes his mathematical demonstration proving that the earth's nutation (the small, slow variation or "nodding" in the earth's axis) is due to the influence of the moon. For the first time, these regular changes in the earth's orientation are explained mathematically.

(See also 1748)

1750 First modern cosmological theory is suggested by English astronomer Thomas Wright (1711-1786). He offers a God-centered model of the universe which includes a perceptive way of picturing the Milky Way. He reasons that the Milky Way is a group of stars as they appear in a flattened system. His original idea says that the stars in the Milky Way form a single system shaped like a lens, thick at the center and thin at the edges, and he then boldly concludes that the stars do not form a spherical system but rather a somewhat flattened one. This is the first time that stars are seen as existing in a flattened, rotating galaxy. He states his new theory in *An Original Theory of the Universe*, published this year.

1755 First truly scientific hypothesis on the formation of the solar system is proposed by Immanuel Kant (1724-1804), German philosopher. He publishes his physical view of the universe in his *General History of Nature and Theory of the Heavens*. It contains three important astronomic "anticipations" or ideas that are later fleshed out and proven by others. He describes the nebular hypothesis; he suggests that the Milky Way is a lens-shaped collection of stars and that other such "island universes" exist; and he suggests that tidal friction slows the rotation of the earth. Altogether, Kant not

only anticipates the existence of galaxies but also hints at the infinity of the universe.

1757 First reasonably accurate figures for the mass of Venus and the moon is obtained by Alexis Claude Clairaut (1713-1765), French mathematician. Although his estimate for Venus is somewhat too small (two-thirds the size of earth) and too large for the moon (one sixty-seventh), they are the closest anyone has yet to offer.

DECEMBER 1758 Halley's Comet is sighted and makes its first predicted return past earth. The reappearing comet is first seen during this month by the German amateur astronomer Johann Georg Palitzsch, and passes its perihelion on March 13, 1759, remaining visible until June 1759. It proves English astronomer Edmond Halley (1656-1742) correct in his 1705 prediction and is an unequivocal confirmation of the Newtonian system.

1761 First astronomer to take seriously English astronomer Edmond Halley's (1656-1742) suggestion of using a transit observation to determine the scale of distances in the solar system is French astronomer Joseph Nicolas Delisle (1688-1768). He applies the method during this year by organizing a worldwide study of the phenomenon.

1761 First to infer that Venus has an atmosphere is Mikhail Vasilievich Lomonosov (1711-1765), Russian chemist and writer. Considered to be the first great Russian astronomer, he observes what he believes to be a dense atmosphere on Venus during its transit across the sun this year. This discovery remains unknown in the West for a century and a half, as does nearly all of Lomonosov's work.

1765 First to make time measurements that are accurate to the nearest tenth of a second is Nevil Maskelyne (1732-1811), English astronomer and the fifth astronomer royal. During his career, he produces lunar tables as well

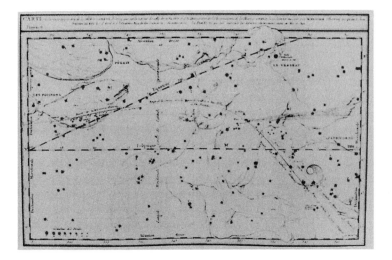

The route of Halley's comet on January 21, 1759, is indicated on this celestial chart by small circles on the arched solid line that bisects the map.

as the *Nautical Almanac*, which remains a useful navigational aid for over a century.

(See also 1767)

1767 First astronomical ephemeris, *The British Nautical Almanac and Astronomical Ephemeris for the Meridian of the Royal Observatory at Greenwich*, is published and includes tables for the moon's positions at midnight and noon. Such tables are used by navigators to determine the position of a ship at sea. This requires the solution of what is known as the "navigational" or "astronomical triangle," the three vertices of which represent, respectively, the position of the ship, the geographical position of the stars (or sun), and the earth's north or south pole.

1770 First short-term comet to have its orbit calculated comes within 1.5 million miles of earth. Although discovered by French astronomer Charles Joseph Messier (1730-1817), it is Swedish astronomer Anders Johan Lexell (1740-1784) who works out its orbit and determines that its period of revolution is five and a half years. It is named after Lexell who demonstrates that the comet's orbit was altered drastically by its close encounter with Jupiter in 1767.

He also shows that Jupiter's massive gravity changed the comet's orbit from a much larger ellipse to its smaller one. This explains why the comet was never seen before this date and was never seen again after it.

1771 First systematic listing of nebulae is published by French astronomer Charles Joseph Messier (1730-1817). As a comet hunter, he is regularly distracted or fooled by fuzzy nebulae (named after the Latin for "cloud"), which are permanent objects in the sky, usually made up of dust or gas. He makes a compilation of them so as to not be fooled a second time and publishes what comes to be known as Messier's Catalog. By 1781, his catalog contains over 100 such objects. His compilation proves highly useful to later astronomers who check each new "discovery" against Messier's catalog before announcing it.

1774 First female astronomer of note, Caroline Lucretia Herschel (1750-1848), builds the best reflector telescope to this time with her brother William Herschel (1738-1822), German-English astronomer. Together they grind the finest lenses in Europe and use them to search the

The armillary, a navigational instrument made in Italy during the 17th century.

skies. Caroline later discovers eight comets and produces a revised edition of Flamsteed's catalog in 1798. She is awarded the Gold Medal of the Royal Astronomical Society in 1828.

MARCH 13, 1781   First new planet to be discovered in historic times is Uranus, discovered by German-English astronomer William Herschel (1738-1822). While using his 7 ft (2 m) reflector telescope searching for double stars, he notices an object that experience tells him is "curious" and looks for it again in a few days.

Finding that it has moved, he initially assumes it to be a comet, but after constant examination (its disc appears to have a sharp edge) and calculations of its orbit, he eventually declares it is indeed the first new planet to be found since recorded history. It is named Uranus after the father of Saturn, the planet beyond which it lies.

1782   First to show that the well-known variations in the brightness of the star Algol change according to a regular pattern is John Goodricke (1764-1786), Dutch-English astronomer. He also makes the bold but correct suggestion that the light change in this second-magnitude star is caused by an invisible companion star that periodically eclipses Algol. His accomplishments are remarkable given that he is only 18 when he makes this observation and deaf mute from birth.

1783   William Herschel (1738-1822), German-English astronomer, discovers that the sun and the rest of the solar system are moving through space relative to the stars. He states that it is moving toward a point in the constellation of Hercules.

1784   First realistic estimate of the distance between the earth and the stars is made by John Michell (1724-1793), English geologist and astronomer. His suggestion that the stars are light-years away indicates a modern sense of the scale of the universe. He also reasons that if light is a particle (as Newton suggests), then the force of gravity from a sufficiently large body could prevent light from moving outward. The idea that gravity could affect light the same way as it does any material object is scoffed at by his contemporaries and relegated to obscurity. More than a century later, however, this concept proves essential to the theories of German-Swiss-American physicist Albert Einstein (1879-1955). It is also from Michell's idea that the concept of the black hole originates.

1785   First systematic attempt to determine the

shape of the Milky Way galaxy is made by German-English astronomer William Herschel (1738-1822). This year he publishes *On the Construction of the Heavens* in which he offers his conclusions about the shape of the universe. He views the visible universe as representing a gigantic collection of stars arranged roughly in the shape of a grindstone with our own sun somewhere near the center. He also believes the Milky Way and the earth to be only one small part of a huge universe.

AUGUST 28, 1789 William Herschel (1738-1822), German-English astronomer, discovers two new satellites of Saturn. After building a telescope with a focal length of 40 ft (12 m) and a 48-in (122 cm) reflector (mirror), he makes his discovery on the first night he uses it. He also devotes much observation time to a study of Saturn's rings.

1794 Ernst Florens Friedrich Chladni (1756-1827), German physicist, is one of the first scientists to write that meteors fall to earth from the heavens. While most peasants believe this, few scientists of this time agree. Chladni is convinced that these iron stones are the debris of exploded planets, but most scientists feel that meteorites are a "superstition unworthy of these enlightened times."

(See also 1803)

1796 Nebular hypothesis is first elaborated by French astronomer and mathematician Pierre Simon Laplace (1749-1827) in a note at the end of his *Exposition du Systeme du Monde*. This idea speculates that the sun originated as a great cloud of gas or nebula that was in rotation, and that by centrifugal force, the core of the nebula became the sun and the rim of gases, the planets. A similar idea had been suggested in 1755 by German philosopher Immanuel Kant (1724-1804), who suggested that the planets condensed out of diffuse "primordial" matter.

1800 Infrared radiation is first detected from the sun by German-English astronomer Wil-

liam Herschel (1738-1822). He tests various portions of the sun's spectrum with a thermometer to see if the different colors vary in temperature, and finds that the highest is in a spot beyond the red spectrum where there is no color at all. He concludes only that sunlight contains invisible light beyond the red. Although he can detect this form of light, it cannot be seen, and it is not adequately explained for another half-century when the infrared is discovered.

JANUARY 1, 1801 First asteroid to be discovered is found by Giuseppe Piazzi (1746-1826), Italian astronomer. Since this new object changes its position over a period of several days, he initially believes it to be a planet located between Mars and Jupiter. During the next few years, he realizes that what he is seeing is really an object that is neither a star nor a planet. What Piazzi discovers is in fact an entire zone of little planets or planetoids now called by the popular name "asteroids," meaning starlike. He names the first and largest of these ever to be seen Ceres, after the Roman goddess associated with Sicily. When the 1,000th asteroid is discovered in 1923, it is named Piazzia in his honor.

1802 First to observe what becomes known as "Fraunhofer lines" in the solar spectrum is William Hyde Wollaston (1766-1828), English chemist and physicist. While studying the spectrum, he notes dark lines in the solar spectrum, but does not investigate them, assuming they are simply the natural boundaries between the various colors of the spectrum. It is not until Fraunhofer that the significance of these lines are realized.

(See also 1814)

MARCH 28, 1802 First to suggest that asteroids or planetoids are fragments of an exploded planet is German astronomer Heinrich Wilhelm Matthaus Olbers (1758-1840), who discovers the asteroid he names Pallas on this date. This is the second asteroid ever discovered, and

Olbers finds another in 1804 that he names Vesta. Pallas is also the second largest known asteroid or planetoid (minor planet) to be found.

1803   Jean Baptiste Biot (1774-1862), French physicist, first convinces a skeptical scientific community that meteorites do fall to earth from the heavens. Commissioned by the Institut National, he travels to L'Aigle, France, where meteorites supposedly fell during the day on April 26 of this year. Biot collects evidence and testimony concerning meteorites and makes a very convincing case. His report settles the matter.

1814   Dark lines in the solar spectrum are first studied at length by German physicist and optician Joseph von Fraunhofer (1787-1826). While using a new prism, he discovers hundreds of dark lines in the solar spectrum and maps them in detail. He finds that reflected light from the moon and other planets contains the same absorption lines, but that light from other stars does not. Later called Fraunhofer lines, these are now known to be caused by elements absorbing certain wavelengths of light. These elements are in the solar atmosphere and partially block the passage of light, so that when it reaches the earth, the sun's light is missing a number of wavelengths. These missing wavelengths are characterized by dark lines in the spectrum. Fraunhofer lines lead to the development of astronomical spectroscopy, which becomes a crucial factor in the development of astrophysics.

1818   First highly accurate star positions are obtained by German astronomer Friedrich Wilhelm Bessel (1784-1846), who publishes a new star catalog titled *Fundamenta Astronomiae*, which contains the positions of 50,000 stars. He also establishes a theory of astronomical measurement and introduces many improvements to astronomical calculations.

(See also 1838)

1818   Jean Louis Pons (1761-1831), French

astronomer, discovers a comet that proves to have the shortest period (smallest orbit) of any discovered before or since. Its period turns out to be only 3.3 years, and its course is calculated in 1819 by German astronomer Johann Franz Encke (1791-1865), after whom it is eventually named. Pons began his career as a janitor in the Marseilles observatory in France and ended it as director of an observatory in Florence, Italy. During that time, he discovers 37 comets and becomes known as the comet-chaser.

1820   Astronomical Society of London is first formed. It later becomes the Royal Astronomical Society.

1821   Fearon Fallows (1789-1831), English astronomer and Mathematical Fellow of St. John's College, Cambridge, becomes the first Director of the Royal Observatory, Cape of Good Hope, South Africa. Fallows later publishes a catalog of southern stars.

1825   Actinometer is invented by English astronomer John Frederick William Herschel (1792-1871). His father, German-English astronomer William Herschel (1738-1822), first noted that the heating effect of the sun's rays was greatest beyond the red end of the spectrum, and John Herschel becomes a pioneer in the chemical analysis of the solar spectrum. His new device is able to measure at any instant the direct heating power of solar radiation.

FEBRUARY 27, 1826   Wilhelm von Biela (1782-1856), Austrian astronomer, discovers a comet with a seven-year period that becomes known as "Biela's comet." When it comes around for its 1846 pass, however, he discovers it has split in two. By 1872, it had become a small crowd of meteors. This is the first time astronomers are able to watch a member of the solar system die before their eyes. It also offers evidence that there is a connection between comets and meteors.

1827   First calculation of the orbit of a binary star is made by French physicist Felix Savary

(1797-1841). Using Newton's law of universal gravitation, he is the first to apply gravitational principles to the computation of the orbits of binary stars. He calculates that the orbit for Ursae Majoris is an eccentric eclipse, thus showing that binary (double) stars obey the movements predicted by Newton's law.

1830 First systematic surface chart and complete map of the planet Mars is drawn by German astronomers Wilhelm Beer (1797-1850) and Johann Heinrich von Madler (1794-1874). It establishes the point of departure for latitude and longitude on the planet that is still in use today. It also shows no canals. Mars comes closer to the earth this year than it does at any time in the 19th century, so the astronomers take advantage of it. They also compare this map to another they draw in 1832 and are able to establish Mars' period of rotation as 24 hours, 37 minutes, and 23.7 seconds.

1831 Heinrich Samuel Schwabe (1789-1875), German astronomer, draws a picture of Jupiter on which the "great red spot" is shown for the first time. Despite the changeable character of Jupiter's cloud patterns, the Red Spot has been a permanent feature since the first telescope was directed at it. Images taken by the United States spacecraft *Voyager* during July 1979 reveal that it is a high-pressure storm large enough to engulf two earths.

1834 First scientifically designed chart of the moon using a proper trigonometrical survey is issued by German astronomers Wilhelm Beer (1797-1850) and Johann Heinrich von Madler (1794-1874). Using a telescope with a 4 in (10 cm) aperture, they indicate the positions of 919 formations and the heights of 1,095 mountains. They view the moon as an airless, lifeless, and changeless place.

MAY 15, 1836 First to describe the eclipse phenomenon known as "Bailey's Beads" is English astronomer Francis Bailey (1774-1844). He describes an effect he observes during an

eclipse in which just before the last sliver of sunlight disappears behind the moon, it appears to break into bits and pieces ("a row of lucid points, like a string of bright beads"). They are caused by the mountains of the moon breaking up the very thin ring of the sun, leaving isolated rays of sunlight shining through the lunar valleys. The same phenomenon occurs on earth at sunrise when sunlight first breaks through between mountains.

APRIL 25, 1837 Johann Franz Encke (1791-1865), German astronomer, first sees a small gap within the outermost ring of Saturn. It soon becomes known as "Encke's division." He places it at one third of the width of the outer ring, counting from its inner edge.

1838 First determination of the parallax of a star is achieved by Friedrich Wilhelm Bessel (1784-1846), German astronomer. After careful observations with an instrument of his own design called a heliometer, he determines that the star 61 Cygni is some 6 light-years away (35,000,000,000,000 mi). This is the first calculation of the distance of a star as well as the introduction of the term "light-year" (based on the fact that the velocity of light is 186,282 mi [299,728 km] a second). Bessel's achievement is, like Bradley's aberration of light (see also 1729), visible evidence that the earth itself moves through space. It also shifts astronomers' attention away from the solar system to the outer universe of stars.

1839 Harvard College Observatory is founded, becoming the first official observatory in the United States. With William Crunch Bond (1789-1859) as its first director, it obtains a 15 in (38 cm) refractor telescope in 1847.

MARCH 23, 1840 First photograph of the moon is taken by English-American chemist John William Draper (1811-1882). Using the newly invented Daguerreotype process, he attaches a telescope to his camera and exposes the plate for a full 20 minutes.

JULY 8, 1842 First attempt to photograph a total solar eclipse is made by the Austrian astronomer Magic. He only obtains a photograph of the partial eclipse.

(See also July 28, 1851)

1843 Heinrich Samuel Schwabe (1789-1875), German astronomer, discovers that sunspots increase and decrease according to a regular cycle. He begins daily observations of the sun in 1826, searching for planetary transits, but soon shifts his attention to recording sunspots. For 17 years he keeps careful records, sketching sunspots for every possible sunny day. By this year he is able to announce the sunspot pattern he perceives in his years of recorded data, and states that sunspots wax and wane according to a 10-year cycle. Although they actually have an 11-year cycle, this discovery is considered the founding mark for modern solar studies. Sunspots eventually are seen as influencing the earth in many ways.

1844 First to begin a detailed study of variable stars is Friedrich Wilhelm August Argelander (1799-1875), German astronomer. He introduces the modern system of naming these stars whose brightness changes over time, using letter prefixes beginning with the letter R for "rot" (red in German). He chooses this letter since so many variable stars are red. His methods of estimating their changing brightness establishes this subject as a new branch of astronomy.

APRIL 1845 First to detect independent galaxies is Irish astronomer William Parsons, Earl of Rosse (1800-1867). Using a 72 in (183 cm) telescope of his own construction, he becomes the first to detect the spiral shapes of cloudy objects millions of light-years away that are much later recognized as independent galaxies (like our own Milky Way). By 1850, he has discovered 14 of them. His huge telescope is named "Leviathan" and is the largest telescope of its time. It is so large it must be supported by two large stone walls and operated by machinery.

SEPTEMBER 23, 1846 New planet Neptune is observed for the first time by German astronomer Johann Gottfried Galle (1812-1910). Its existence is actually first predicted by French astronomer Urbain Jean Joseph Leverrier (1811-1877), who writes to his colleague Galle and asks him to look in a certain spot in the sky. Leverrier suspects that a new planet may exist as he tries to account for the slight orbital deviations of the planet Uranus. He discovers this new planet solely by pure calculation and then tells Galle where he should look to find it. This is perhaps the most dramatic demonstration of the validity of Newtonian theory. Leverrier names it Neptune after the Greek god of the ocean (supposedly because the new planet has a green color). Although English astronomer John Couch Adams (1819-1892) makes the same calculations months before Leverrier, he is unable to persuade his superiors to make the necessary observations, and his potential discovery is thwarted.

OCTOBER 10, 1846 William Lassell (1799-1880), English astronomer, discovers the first of Neptune's two satellites. It is named Triton after the mythical son of Neptune. Larger than the earth's moon, it moves around Neptune in the reverse direction compared to the satellites of other planets, and also rotates in the opposite direction in which Neptune itself spins on its own axis.

OCTOBER 1, 1847 Maria Mitchell (1818-1889), American astronomer, discovers a comet which brings her to the attention of the scientific world. Her discovery earns her a gold medal from the king of Denmark, and in 1848, she is elected the first female member of the American Academy of the Arts and Sciences. She is recognized as the first significant female astronomer in the United States.

(See also 1865)

1847 First to measure the brightness of stars with real precision is English astronomer John Frederick William Herschel (1792-1871), son

Professor of Astronomy Maria Mitchell.

of William Herschel (1738-1822), German-English astronomer. He achieves this while creating a map for the southern hemisphere as his father had done for the northern. This year he publishes his great star map of the southern hemisphere. As the results of 14 years of work, it contains the discovery of 2,306 nebulae and 3,347 double stars. It also contains his cyclonic theory of sunspots.

SEPTEMBER 19, 1848 William Lassell (1799-1880), English astronomer, discovers an eighth satellite of Saturn, later named Hyperion. It is also discovered simultaneously across the Atlantic by the father-son team of American astronomers William Crunch Bond (1789-1859), and George Phillips Bond (1825-1865). In Greek mythology, Hyperion is one of the Titans and the son of Uranus (heaven) and Geae (earth).

1849 First precision measurement of the velocity of light at the earth's surface is obtained by Armand Hippolyte Fizeau (1819-1896), French physicist. He devises his own experiment using a terrestrial and not an astronomical method, and obtains a value for the speed of light that is within 5% of the true figure. He sets up a light source and spinning gear on the peak of a hill, arranged so that the light would shine through the gear's teeth and be reflected back from a mirror on another hill 5 mi (8 km) away. When Fizeau spins the gear very fast, light passes through the gap in the gear's teeth and is reflected back, reentering through the next gap. By using a timer, he is able to determine the amount of time it takes light to travel the 10 mi (16 km). He arrives at a figure of 195,615 mi (314,744 km) per second, slightly faster than the modern figure of 186,000 mi (299,274 km) per second.

1849 First American astronomical periodical, the *Astronomical Journal*, is founded by Benjamin Apthorp Gould (1824-1896), American astronomer. Its publication continues until it is halted by the American Civil War in 1861.

JULY 17, 1850 First photograph of a star is made by William Crunch Bond (1789-1859), American astronomer, who photographs the bright star Vega. He obtains a Daguerreotype after a few minutes' exposure. This marks the beginning of stellar photography. Bond also photographs the moon the following year, taking the first really good photograph, and the image becomes a sensation at the Great Exhibition at the Crystal Palace in London.

JULY 28, 1851 First successful photograph of a total eclipse is taken by German astronomer Berkowski. His photograph shows the sun's corona (the sun's outer atmosphere) and its prominences (clouds of incandescent ionized gas extending sometimes hundreds of thousands of kilometers above the sun's atmosphere).

OCTOBER 24, 1851 William Lassell (1799-1880), English astronomer, discovers two satellites of Uranus. Their names, Ariel and Umbriel, are suggested by English astronomer John Frederick William Herschel (1792-1871), son of the discoverer of Uranus, German-English astronomer William Herschel (1738-1822).

1851 Variation in the earth's magnetic field is

first found by Scottish-German astronomer Johann von Lamont (1805-1879). He obtains this data after taking daily readings from all over Europe for 10 years. In 1862 he publishes his determination that the intensity of the earth's magnetic field rises and falls in a 10-year period that matches the recently discovered sunspot cycle. The actual relation between the two goes unknown for 50 years until the discovery of subatomic charged particles.

APRIL 27, 1857 First to photograph a double star showing its two components is George Phillips Bond (1825-1865), American astronomer. The son of Harvard University astronomer William Crunch Bond (1789-1859), he photographs the double star Mizar. He also shows that estimates of stellar magnitude can be made from photographs of stars.

1857 First photograph of the moon that is sharp enough to be magnified 20 times is taken by Warren de la Rue (1815-1889), British astronomer. He also photographs the sun and stars. This achievement makes astronomers aware of the potential that photography can offer in terms of seeing and documenting more than the real-time observer can see.

JUNE 2, 1858 Giovanni Battista Donati (1826-1873), Italian astronomer, discovers a comet that becomes known as Donati's comet. Later this year it develops a long, curved tail and becomes the first comet to be photographed (by Usherwood). This comet is fairly well celebrated because of the beauty of its curved tail and the brightness of its head.

(See also 1864)

1858 Lewis Morris Rutherfurd (1816-1892), American astronomer, develops the first telescope adapted solely for use with a camera. He devises what is really a camera with a telescope serving as the lens, and later proceeds to advance the method of measuring stellar positions on photographs.

OCTOBER 27, 1859 New technique of recording spectroscopy is first reported by German physicist Gustav Robert Kirchhoff (1824-1887) and German chemist Robert Wilhelm Bunsen (1811-1899). A spectroscope separates the spectral components of a source and presents them so they can be examined visually. Kirchhoff and Bunsen use it to study the chemical composition of the sun.

(See also December 15, 1859)

DECEMBER 15, 1859 Modern astrophysics first begins as German physicist Gustav Robert Kirchhoff (1824-1887) and German chemist Robert William Bunsen (1811-1899) use their newly developed spectroscopic technique and discover part of the chemical composition of the sun. Their experiments show that when light passes through a gas, the wavelengths that are absorbed are the same as those which the gas would emit when incandescent. From this Kirchhoff's law is established, which describes quantitatively the relationship between emission and absorption (stating that the ratio between the emission and absorption powers for rays of the same wave length is constant for all bodies at the same temperature). Thus the same characteristics obtained from the emission spectra of the various laboratory substances are found in the celestial spectra, which are mostly absorption spectra. Kirchhoff then concludes that other terrestrial elements must be present in the sun in greater or lesser degree. In 1861 he draws a precise map of the solar spectrum which the Berlin Academy publishes the next year. He uses three different shades to indicate the different intensities of the lines. His work constitutes a major advance in knowledge of the physical composition of the sun and stars.

1859 Richard Christopher Carrington (1826-1875), English astronomer, observes a starlike point of light that bursts from the surface of the sun. This is the first recorded observation of a solar flare (although it is not recognized as such by Carrington who believes it to be a meteor falling into the sun).

1859 Friedrich Wilhelm August Argelander (1799-1875), German astronomer, publishes the first volume of his giant work *Bonner Durchmusterung*. This four-volume work (1859-1862) locates the positions of 457,848 stars and contains the last star maps to be compiled without photographs. It is reprinted as late as 1950.

1860 Johann Karl Friedrich Zollner (1834-1882), German astronomer, invents the polarizing photometer. Until this device, measuring the brightness of a star was unsystematic and full of errors since there were no standards. Zollner, however, devises a photometer in which the brightness of an artificial star is made equal to that of a real star. The artificial star is formed by a pinhole in front of a flame, and it is diminished at an exactly known rate by an interposed polarizing apparatus consisting of two Nicol prisms. With his instrument, astronomers can measure the intensity or brightness, as well as other properties, of visible light, including infrared and ultraviolet radiation.

1861 Nicolas Camille Flammarion (1842-1925), French astronomer, publishes *La Pluralite des Mondes Habités* at the age of 19. This is the first of some 50 works he writes, popularizing astronomy.

1862 Discovery of hydrogen in the sun's atmosphere is first announced by Swedish physicist Anders Jonas ngstrom (1814-1874). He makes this discovery by applying the newly discovered technique of spectroscopy to study the solar spectrum. He continues and produces in 1868 a map of the normal solar spectrum that remains authoritative for a long time.

AUGUST 1863 Astronomische Gesellschaft, an international astronomical society, is founded by 26 astronomers who meet in Heidelberg, Germany. In 1869, this society produces the first great precision star catalog.

1863 Friedrich Wilhelm August Argelander (1799-1875), German astronomer, founds the Astronomische Gesellschaft, the first large international organization of astronomers.

1863 *Astronomical Register* is founded and is the first English periodical devoted exclusively to astronomy.

1863 First to announce that stars are composed of the same elements as those found on earth is English astronomer William Huggins (1824-1910). While analyzing stars spectroscopically, he notices that the lines of a star's spectrum match those of the spectral lines of oxygen on earth. He then states that there is oxygen in that star. This conclusion puts a final end to the 2,000-year-old idea, begun by Greek philosopher Aristotle (384-322 B.C.), that the heavens are made of unique matter not found on earth.

1864 First to observe the spectrum of a comet is Italian astronomer Giovanni Battista Donati (1826-1873). He studies the spectrum of a comet as it nears the sun and discovers that while it is far away it glows from reflected sunlight. As it gets closer, however, it heats and begins to glow with a radical change in its spectra. The results of this first spectroscopic examination of a comet indicates that the tails of comets consist of gases and, overall, marks the beginning of an understanding of the structure of a comet.

1864 First to classify solar prominences (gas clouds) as quiescent or eruptive and to describe solar spicules (fast-moving, short-lived jets of gases) is Italian astronomer Pietro Angelo Secchi (1818-1878). He also becomes one of the first to systematically adapt the new science of spectroscopy to astronomy. Over the next four years he makes the first spectroscopic survey of the heavens and studies the spectra of 4,000 stars. He discovers that stellar spectra are different from each other, showing for the first time that there is a difference among stars that is more than just position, brightness, and color. He introduces the classification of stars into four classes based on color and characteristics of their spectra. His work lays the foundations for modern spectral classification.

JANUARY 16, 1865   First coherent theory on the constitution of the sun is offered by French astronomer Herve Auguste Etienne Albans Faye (1814-1902). He states that the sun can be considered as a vast heat-radiating machine. He further elaborates that it is a gaseous body whose radiation is due to the transport of heat upwards by convection currents with cooler matter descending; that its photosphere is a condensation at the outer limits; and that sunspots are breaks in photospheric clouds. Only his first statement proves correct.

1865   Maria Mitchell (1818-1889), American astronomer, is appointed first professor of astronomy and director of the observatory at the newly founded women's college in New York, Vassar College. During her career, she works on the *Nautical Almanac* and for the United States Coast and Geodetic Survey, tours European observatories, and becomes a leading advocate for women's rights, founding the Association for the Advancement of Women. After her death, the Maria Mitchell Observatory is founded in 1902 on her native Nantucket Island, off the coast of Massachusetts.

(See also October 1, 1847)

1866   First to examine a nova and a comet spectroscopically is William Huggins (1824-1910), English astronomer. He shows that a nova is enveloped by a shell of hydrogen gas at a temperature higher than that of the star's surface. He also shows that comets emit light from a luminescent carbon gas.

1866   First to study sunspots spectroscopically is English astronomer Joseph Norman Lockyer (1836-1920). He also discovers that solar prominences are upheavals in a layer around the sun (which he names the chromosphere). He shows in 1868 that these mountains of flaming

gas can be observed spectroscopically without the aid of an eclipse.

1866   First to suggest that falling stars are of cometary origin is Italian astronomer Giovanni Virginio Schiaparelli (1835-1910). He compares the orbits of the annual Persied showers of meteors that fall every August with the orbits of known comets, and finds that they are nearly identical with the comets 1866 I and 1862 III, respectively. He then claims that meteor swarms are debris left along the trail of comets on their journey around the sun.

1867   First to study the spectrum of the aurora borealis is Swedish physicist Anders Jonas Ångstrom (1814-1874). He is also the first to detect and measure the characteristic bright line in its yellow-green region. In his spectroscopic studies, Ångstrom measures the wavelengths in units equal to a ten billionth of a meter and arrives at a value that is used to measure wavelengths in light. This useful unit, equal to one ten-billionth of a millimeter in length, is now known as the Ångstrom unit.

1867   French astronomers Charles Joseph Etienne Wolf (1827-1918) and Georges Antoine Pons Rayet (1839-1906) discover gaseous stars whose steady brightness and high temperatures make them an entirely new class of stars. Spectroscopic analysis of these special stars indicates their spectra contain entirely bright instead of dark lines, with strong lines of helium and hydrogen. They come to be known as Wolf-Rayet stars.

1867   First system of names for the topography of the planet Mars is offered by English astronomer Richard Anthony Proctor (1837-1888). His system, which gives mostly English astronomer's names to the continents, seas, bays, and straits of Mars, is later superseded by that of Italian astronomer Giovanni Virginio Schiaparelli (1835-1910), who uses more objective names. Unlike Schiaparelli, however, Proctor sees none of the "canals" that Schiparelli does. He is also the first to suggest that the moon's craters resulted from meteoric bombardment rather than volcanic action.

(See also 1873)

1868   Helium is first found to exist in the sun by Joseph Norman Lockyer (1836-1920), English astronomer. He studies a newly noticed line in the spectrum of the sun and concludes that it belongs to a yet unknown element that he names helium after the Greek word for sun, "helios." This strange spectral line is first seen by the French astronomer Pierre Jules César Janssen (1824-1907), who passes on his discovery to Lockyer, the solar spectra expert. Helium is not discovered on earth for another 40 years.

1869   First to mathematically investigate the sun as a gaseous body is American astrophysicist Jonathan Homer Lane (1819-1880). His development of the first star model (with the sun and stars as gaseous spheres) marks the birth of theoretical astrophysics. Lane's pioneer work determines the tremendous gravitational pressures on the gases in the sun's interior, and he calculates the temperatures and densities necessary to provide the expansive forces balancing these pressures. This work demonstrates the interrelationships of pressure, temperature, and density inside the sun and is fundamental to the emergence of modern theories of stellar evolution. His solar studies also result in Lane's law, which states that as a gaseous body contracts (by cooling or by any other means), the contraction generates heat.

1869   First to photograph the spectrum of the sun's corona is Charles Augustus Young (1834-1908), American astronomer. He carefully observes a solar eclipse (and another in 1870) and discovers the "reversing layer" of the sun. These are seen as dark lines in the spectrum that gleam brightly at the moment of totality. He is also the first to prove the gaseous nature of the corona by observing in its spectrum the green emission line, the wavelength of which he determines.

1871   Società degli Spettroscopisti Italiana, the first institution of its kind, is founded by Italian astronomers Pietro Angelo Secchi (1818-1878) and Pietro Tacchini (1838-1905). Its purpose is to coordinate the research on solar phenomena and celestial physics done by observatories around the world. It becomes the present-day Società Astronomica Italiana.

1871   First spectroscopic proof of the rotation of the sun is obtained by German astronomer Hermann Carl Vogel (1842-1907). He achieves this by comparing the spectra of the two opposite edges, east and west, and thus confirms the existence of the Doppler effect. Using this method, he also is able to make a good measurement of the sun's period of rotation.

1872   Spectrum of a star is photographed for the first time by American astronomer Henry Draper (1837-1882). He grinds his own mirror, builds a 28-in (71 cm) reflector telescope, and succeeds in photographing the spectrum of the star Vega. He achieves this by placing a quartz prism before the focus (since quartz, unlike glass, does not absorb the ultraviolet rays). He later studies the spectrum of the Orion Nebula and shows it to be a cloud of dust and gas lit by starlight. He is the son of English-American chemist John William Draper (1811-1882), who first photographed the moon in 1840.

(See also September 30, 1880)

1873   First rigorous mathematical analysis of Laplace's nebular hypothesis is made by Edouard Albert Roche (1820-1883), French astronomer. This theory, offered by French astronomer and mathematician Pierre Simon Laplace (1749-1827), states that the solar system was formed from nebulous material in space. Roche also suggests important additions that make the hypothesis more self-consistent.

(See also 1796)

1873   Richard Anthony Proctor (1837-1888), English astronomer, is the first to suggest that lunar craters are created by meteoric bombardment. He suggests in his book *The Moon* that the moon's surface had to be in a semimolten condition for this process to work.

DECEMBER 8, 1874 Photographic techniques are used during a transit of Venus by astronomers for the first time. During this first transit of Venus to occur since 1769, various expeditions are sent around the world to attempt to discover how long it takes for that planet to cross the face of the sun. The English and American contingents use photographic telescopes so that the position of Venus upon the sun could be found by measuring the plates afterwards. Altogether, the results obtained from all the observers (including non-photographic) are disappointing since they all obtain slightly different values. This underscores the difficulty of measuring positions of an object against a luminous background like the sun.

1874 William de Wiveleslie Abney (1843-1920), English astronomer, invents a dry photographic emulsion that is superior to the old wet method. He uses this method to photograph a transit of Venus across the Sun.

(See also 1877)

1875 William Huggins (1824-1910), English astronomer, discovers that the cumulative effect of long photographic exposures of the sky can make objects visible that are normally too faint to be seen by the naked eye. As one of the first to experiment with photography as an adjunct of astronomy, he also devises methods of photographing spectra and makes the first use of dry plate photography in 1876.

1876 Asaph Hall (1829-1907), American astronomer, discovers a white spot on Saturn's surface that he uses to discover Saturn's period of rotation (ten and three quarter hours).

(See also August 11, 1877)

AUGUST 11, 1877 First satellite of Mars is discovered by American astronomer Asaph Hall (1829-1907). Using a 26-in (66 cm) refractor

telescope (then the largest in the world) at the Washington Naval Observatory, he locates a tiny, moving object near Mars. He confirms its existence the next time the weather permits him to observe (August 16), and discovers yet another Mars satellite on August 17. Both prove to be very small, fast-spinning bodies, the larger of the two being about 15 mi (24 km) in diameter and the smaller about 7.5 mi (12 km). He names them Deimos ("Terror") and Phobos ("Fear") after the two sons of the Greek war god Ares.

1879 Josef Stefan (1835-1893), Austrian physicist, publishes the law that bears his name, showing that total radiation increases as the fourth power of temperature. Stefan's law is one of the first important steps towards an understanding of blackbody radiation (a blackbody being a theoretical object that absorbs all radiation that falls upon it), and it states that the radiant energy of a blackbody is proportional to the fourth power of its temperature. It is from this understanding that the quantum idea of radiation eventually springs. Further, since the sun's total radiation is known, it is possible—using his law—to calculate for the first time the sun's total radiance, which is about 6,000° centigrade.

SEPTEMBER 30, 1880 First to photograph a nebula is American astronomer Henry Draper (1837-1882). He obtains a photo of the Orion Nebula, the gas cloud in the constellation Orion that is visible to the unaided eye and is illuminated by faint, young stars.

(See also 1883)

1881 Samuel Pierpont Langley (1834-1906), American astronomer, invents a bolometer, an instrument for accurately measuring tiny quantities of heat. This delicate device measures the heat a telescope receives from a celestial body by its effect on the balance of an electrical circuit. It can tell differences to a hundred thousandth of a degree by way of the size of the tiny electric currents that are created by that heat in a blackened platinum wire. He uses this to measure solar radiation and extends knowledge of the solar spectrum into the far infrared for the first time. This instrument founds spectrophotometry.

1881 First to discover a comet by using a photographic method is Edward Emerson Barnard (1857-1923), American astronomer. Barnard is a pioneer of celestial photography and the leading observational astronomer of his time.

1883 First really successful direct photographs of a nebula (Orion) are taken by Andrew Ainslie Common (1841-1903), English astronomer. He goes on to take some of the finest astronomical photographs of his time.

1884 First catalog of photometric star magnitudes is produced by Edward Charles Pickering (1846-1919), American astronomer. He publishes his *Harvard Photometry*, which is a determination of the magnitude of all the brightest stars, and includes over 4,000 stars. He is able to compile this data because of his newly invented meridian photometer. With this device, which he invents in 1879, he is able to compare the brightness of any star to that of a standard reference star (Polaris).

1887 William de Wiveslie Abney (1843-1920), English chemist, invents a red-sensitive photographic emulsion that makes it possible for the first time to photograph the solar spectrum in the infrared. This then makes possible the study of how sunlight is altered as it passes through the atmosphere (since some of the infrared is absorbed by the air).

1888 James Edward Keeler (1857-1900), American astronomer, studies the rings of Saturn and observes a second major gap in its ring system. Further studies will allow him to prove his theory. This discovery becomes known as the "Keeler gap" as opposed to the first "Cassini gap."

(See also April 1895)

1888 Lick Observatory on Mt. Hamilton in northern California is founded. Named after its donor, the American philanthropist James Lick (1796-1876), the observatory has a 26-in (66 cm) refractor telescope and becomes the first of the great California observatories as well as the first mountaintop observatory (elevation 4,200 ft [1,280 m]). It is eventually turned over to the University of California.

1889 First photograph of the Milky Way is taken by Edward Emerson Barnard (1857-1923), American astronomer. Taken with large-aperture lenses, his photos reveal much new detail that could never have been obtained by the eye alone. These and later photographs reveal that the bright clouds and streams of the Milky Way are made up of hundreds of thousands of very faint star, from the 13th, 14th, and 15th magnitude downwards. They will also show the Milky Way's dark features that appear empty.

1890 Spectroscopic binaries are discovered by German astronomer Hermann Carl Vogel (1842-1907). These are two stars that are so close together that no telescope is able to view them separately. Steady spectral analysis makes them appear as one star that is alternately advancing and receding from the viewer. Although they appear visually to be a single object, their separate identity can be seen clearly with a spectroscope. Spectroscopic binaries later prove to be plentiful in the universe.

1890 Spectroheliograph is first developed by American astronomer George Ellery Hale (1868-1938). This device presents a monochromatic image of all or part of the solar disk and its prominences and makes it possible to photograph the light of a small band of wavelengths of the sun. With this instrument, Hale is able to photograph wavelengths and to distinguish elements. At about this same time French astrophysicist Henri Deslandres (1853-1948) independently builds a similar instrument for the same purpose. The spectroheliograph

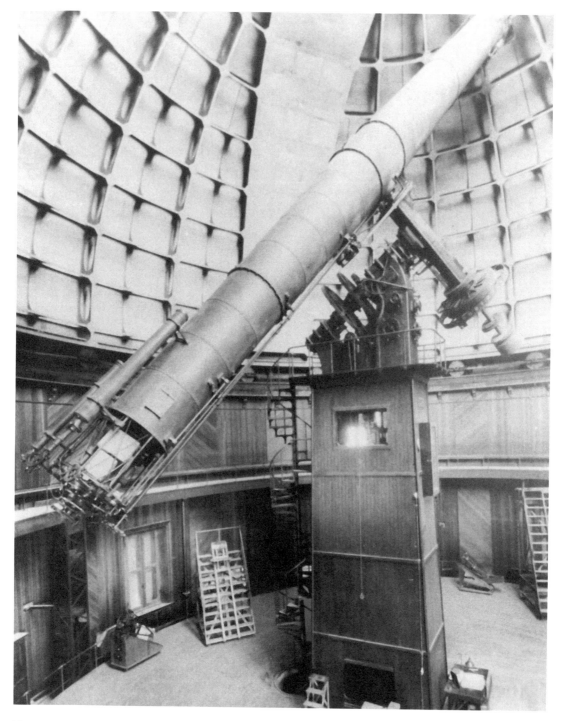

The telescope at Lick Observatory in California.

California's Lick Observatory in 1907.

proves invaluable to the study of the chemistry of the sun's outermost layer.

(See also 1908)

DECEMBER 22, 1891 First discovery of an asteroid from photographs is made by German astronomer Maximilian Franz Joseph Wolf (1863-1932). He adapts photography to the study of asteroids and demonstrates that stars appear as points in photographs while asteroids show up as short streaks. During his lifetime he discovers over 500 asteroids in this manner.

1891 First to measure the rotation of the sun and show that it does not rotate uniformly is Swedish astronomer Nails Christopher Diner (1839-1914). Employing the Doppler effect, he finds that its equatorial region has a rotational period of about 25 days, but its poles are about 38 days. This strange phenomenon still has not been satisfactorily explained.

SEPTEMBER 9, 1892 Edward Emerson Barnard (1857-1923), American astronomer, discovers a fifth satellite around Jupiter. Since Italian astronomer and physicist Galileo Galilei's (1564-1642) discovery of its first four satellites in 1609, no one had ever seen another. It is the

largest of the four innermost satellites of Jupiter and eventually is named Amalthea after a variation of the Greek name of the foster mother of Zeus. It is also the last satellite to be discovered without the aid of photography.

(See also 1904)

1892 New and powerful mathematical models and techniques are first applied to celestial mechanics by French mathematician Jules Henri Poincaré (1854-1912). He publishes the first of his three-volume work *Les Méthodes Nouvelles de la Mécanique Céleste* this year and introduces the use of rigorous methods to the study of celestial mechanics.

1892 First clear evidence that a nova is an exploding star is given by American astronomer Edward Emerson Barnard (1857-1923). He studies a nova in the constellation Auriga and is the first to observe the puff of gaseous matter it gives off, realizing it is evidence of an explosion.

APRIL 1895 First observational proof that the rings of Saturn are not solid but consist of particles is offered by James Edward Keeler (1857-1900), American astronomer. He observes the spectrum of Saturn and its rings and shows that the rings do not rotate as a unit but that the inner rings have a shorter period. He demonstrates, using spectroscopic measurements of the radial velocities of its different parts, that it cannot be a solid ring by showing that the inner edge of the rings move slightly faster than the outer edge. He argues that a solid ring would have the reverse situation.

DECEMBER 1896 Piter Zeeman (1865-1943), Dutch physicist, discovers a magneto-optical effect that bears his name. He shows that a spectral line splits into several components in the presence of a magnetic field. This is later used in astronomy to measure magnetic fields on the sun and stars and in physics to study the fine structure of the atom.

1896 John Martin Schaeberle (1853-1924),

German astronomer, detects the very dim companion star of the star Procyon. This is the first indication that such dim or dwarf stars are not as rare as believed. In time, enough are found to form their own class.

AUGUST 13, 1898 First asteroid known to come within the orbit of Mars is discovered by German astronomer Gustav Witt. He discovers Asteroid 433, which later becomes known as Eros. Because Eros will come very close to the earth in 1930-1931, it later plays a very important role in determining solar parallax.

(See also 1931)

1899 First general atlas of variable stars is compiled by Johann Georg Hagen (1847-1939). Hagen publishes the *Atlas Stellarum Variabilium* and later becomes director of the Vatican Observatory, Specola Vaticana.

1899 First satellite to be discovered by the use of photography is Phoebe, the ninth of Saturn's satellites. Discovered by American astronomer William Henry Pickering (1858-1938), it is the outermost of Saturn's satellites. Pickering notes that its retrograde motion indicates that it may be a captured asteroid. Its diameter is only 99 mi (159 km).

1900 David Gill (1843-1914), Scottish astronomer, completes a photographic sweep of the southern stars for the first time. His *Cape Photographic Durchmusterung* catalogs over 450,000 stars. He advocates the use of astrographic (wide-angle) photography as the best method of preparing star catalogs since accurate, detailed charts can be produced by photography rather than the painstaking visual methods used before.

1901 First accurate, modern determination of the sun's distance from the earth is obtained by Scottish astronomer David Gill (1843-1914). Planetary transits prove incapable of providing astronomers with the desired level of precision to measure distances since both Mars and Ve-

nus have atmospheres that make their boundaries appear fuzzy. So instead of using Venus or Mars as the point from which he makes his computations, he uses a prominent planetoid. Although planetoids are farther from the earth, their star-like points of light provide the needed level of precision for Gill to make his landmark measurements.

1904 First indication that matter exists in interstellar space is offered by German astronomer Johannes Franz Hartmann (1865-1936). He studies the spectrum of the binary star Delta Orionis and observes anomalies in its spectrum lines that indicate they were not produced by the stellar atmosphere but rather in outer space. He then concludes that there is interstellar matter in the form of dust or gas. Astronomers eventually learn that interstellar matter exists throughout the universe, and although its density is extremely low, the volume of the space in a galaxy is so great that the total quantity of interstellar material is considerable.

1904 First of the outer satellites of Jupiter is discovered by Charles Dillon Perrine (1867-1951), American-Argentinian astronomer. While checking the new satellite's orbit, he discovers the seventh satellite the following year. Both satellites are so far from Jupiter that they need about 260 days to make one orbit around the mother planet.

1905 Notion of "absolute magnitude" for comparing the brightness of stars is first advanced by Ejnar Hertzsprung (1873-1967), Danish astronomer. With this technique, he can make judgments by comparing them against a visual magnitude that an astronomical body would have at a standard distance of 10 "parsecs." A parsec is the distance at which a star has a parallax of one second or 3.25 light-years. He then goes on to note the important relationship between color and luminosity of stars.

JUNE 30, 1908 Tunguska "event" is first recorded in the Tunguska region of Siberia. An explosive impact or detonation is heard over a distance of 621 mi (1,000 km) and millions of trees are flattened. The blast's consequences are similar to those of an H-bomb, as thousands of square miles are levelled. The first expedition to this uninhabited area finds no traces of a meteorite, and the cause of this incredibly powerful explosion is the focus of much speculation. Some think it was the icy nucleus of a small comet, while others speculate it may have been an impact with a small black hole or the crash of an extra-terrestrial craft. The most likely cause is the entry into the atmosphere of a piece of a comet which would have produced a large fireball and a blast wave. Since a comet is composed primarily of ice, the fragment would have melted and left no impact crater or debris. This event occurs during the earth's passage through the orbit of the Comet Encke.

1908 First detection of the magnetic field of an extraterrestrial body is achieved by American astronomer George Ellery Hale (1868-1938). He photographs sunspots using a spectroheliograph, which he invented in 1890 and which allows him to photograph the light of a single spectral line of the sun. While using photographic plates that are sensitive to wavelengths at the red end of the spectrum, he finds that hydrogen clouds above the sunspots appear to be moving like vortices in opposite directions on opposite sides of the solar equator. He then is able to show that the spectral lines from these clouds are modified in the same manner they would be if they were between two poles of a powerful magnet. He then realizes he has located strong magnetic fields inside sunspots. In physics, a similar phenomenon is called the Zeeman effect.

1912 First to recognize the red shifts in the spectra of galaxies is American astronomer Vesto Melvin Slipher (1875-1969). As the first to study the spectrum of the Andromeda Nebula and to measure the relative velocity of a nebula, he also determines that it is approaching rather than receding from the earth and calculates its

speed at 125 mi (201 km) per second. He later determines that nearly all other nebulae are receding from the earth. In making both discoveries, he studies the spectrum of the light from the nebulae and notes the Doppler effect. That is, if the light of the swirling nebula is shifted toward the blue end of the spectrum, the nebula is moving toward the observer. If it shifts toward the red end, it is moving away.

1912 Law of cepheids (variable stars that pulsate regularly) is discovered by American astronomer Henrietta Swann Leavitt (1868-1921). While studying the cepheids of the Minor Magellanic Cloud, she notes that the brighter stars have longer periods of variation and is able to determine a relation between brightness and period. This relation is used by American astronomer Harlow Shapley (1885-1972) and transformed into the relation of absolute brightness and period which becomes a powerful tool for determining large astronomical distances.

(See also 1913)

1913 First proof of the existence of star systems outside the Milky Way is given by Ejnar Hertzsprung (1873-1967), Danish astronomer, and Harlow Shapley (1885-1972), American astronomer. They use the caught law—discovered in 1912 by American astronomer Henrietta Swan Leavitt (1868-1921) and perfected by Shapley—to measure the distance of the Magellanic Clouds, which are two satellite galaxies of the Milky Way.

1914 First good attempt at working out the evolutionary life cycle of stars is made by Henry Norris Russell (1877-1957), American astronomer. He studies the relationship between the brightness of stars and their color and spectral class. He then plots a diagonal line for their comparative luminosity and says that this implies a definite life cycle that stars follow. Although his theory of a star's progressive sequence is overly simple, it is very significant for its time. He later analyzes the sun's spectrum and is the first to suggest (correctly) that the sun and stars are composed mainly of hydrogen.

1914 Seth Barnes Nicholson (1891-1963), American astronomer, locates the first of four satellites of Jupiter that he will discover. During his career he discovers two more in 1938 and another in 1941.

1915 Robert Thorburn Ayton Innes (1861-1933), Scottish astronomer, discovers a faint star near Alpha Centauri, which is a third companion to that known binary (double) star. It is also the closest star to earth, excluding our own sun, and is therefore named Proxima Centauri.

1916 First to offer an exact solution to Einstein's field equations of general relativity relating to the gravitational field of a point mass is German astronomer Karl Schwarzschild (1873-1916). He is also the first to calculate the gravitational phenomena in the environs of a star whose entire mass is concentrated in one point. This work introduces the "Schwarzschild limit," which is the maximum permissible density for a self-supporting body that, when exceeded, collapses to form a black hole. It also introduces the "Scwarzschild radius," which describes the critical area around a collapsed mass from which an emitted photon could not escape.

1918 First measurement of the size of our galaxy is made by American astronomer Harlow Shapley (1885-1972). Using the new 100-in (254 cm) reflecting telescope on Mount Wilson, he is able to determine the distances of the globular clusters. These are closely packed, spherically shaped groups of up to 100,000 stars. With this, he is able to make a three-dimensional model of them, which allows him to offer, for the first time, a picture of our galaxy that is close to its actual immense size. His model removes the sun from its supposed position as the center of the galaxy, but more importantly, it marks the real beginning of galactic astronomy.

1920 First measurement of the diameter of a star other than the sun is made by Albert Abraham Michelson (1852-1931), German-American physicist. Using a stellar interferometer of his own design, he determines that the star Betelgeuse in the Orion constellation has a diameter of 260 million mi (418 million km).

1921 Meghnad N. Saha (1893-1956), Indian astrophysicist, first demonstrates that the spectrum of a light source like a star depends as much on its temperature as on its chemical composition. Called the thermal ionization equation, its linking of the degree of ionization in a gas to temperature and electron pressure remains fundamental in all work on stellar atmospheres. He is one of the first to apply the new physics to the study of stellar spectra.

1922 First evolutionary model of the universe to contain an expansion element is published by Alexander Alexandrovich Friedmann (1888-1925), Russian mathematician. His mathematical analysis of the notion of an expanding universe proves to be of great significance in the mathematical derivation of cosmological models from German-Swiss-American physicist Albert Einstein's (1879-1955) general theory of relativity. He is also one of the first to postulate a "big bang" model for the evolution of the universe and is a founder of dynamic meteorology.

1922 William Weber Coblentz (1873-1962), American physicist, invents a thermocouple which enables astronomers to calculate the temperature of planets. Using a very sensitive thermocouple (a loop made of two different metals) in a vacuum placed at the focus of a telescope, Coblentz obtains the surface temperature of the planet Mercury.

1922 John Stanley Plaskett (1865-1941), Canadian astronomer, uses a 72-in (183 cm) reflecting telescope of his own design and discovers what comes to be known as Plaskett's twins or Plaskett's star. Previously thought to be a single star, this is revealed to be a spectroscopic binary star of unusually large size. Each component of this massive binary has an estimated mass 55 times that of our sun.

1924   First to consider the earth's age in terms of several billions of years instead of tens of millions is English astronomer and geophysicist Harold Jeffreys (1891-1989). This and other notions on the origin of the solar system are discussed in his *The Earth: Its Origin, History, and Physical Constitution*. He is also the first to hypothesize that the earth's core is liquid.

1924   First to locate stars in the Andromeda Nebula is Edwin Powell Hubble (1889-1953), American astronomer. Using his giant telescope, which is the largest of its time, he determines that some of these stars are Caught variables. His calculations of distance indicate that the Andromeda Nebula are outside our own galaxy, some 800,000 light-years away (he underestimates this to be eight times the distance of the farthest star of our own galaxy). He also provides proof that other galaxies are truly independent systems rather than part of our own Milky Way.

(See also 1952)

1924   First system for the classification of galaxies according to shape is developed by American astronomer Edwin Powell Hubble (1889-1953). Having discovered the first certain evidence that some of the nebulae are really separate galaxies of gas, dust, and stars beyond the Milky Way, he then shows a way to classify them. With this discovery, he begins what becomes his founding of the study of the universe beyond our own galaxy.

1926   First major work on stellar structure is published by Arthur Stanley Eddington (1882-1944), English astronomer and physicist. In his work, *The Internal Constitution of the Stars*, he makes a major contribution to theoretical astronomy with a mathematical description of processes within stars, combining the forces of gravitation with radiation pressure. He also explains that the expansive force of the sun's heat and radiation pressure counters the tremendous force of its gravity and keeps it from contracting into a tiny, compact mass.

1927   Modern big-bang theory of the origin of the universe is first formulated by Georges Edouard Lemaître (1894-1966), Belgian astronomer. He holds that in a backward extrapolation of time, the galaxies can be viewed as initially existing all crushed together in a kind of "superatom" or "cosmic egg." This "superatom" is supposed to have contained all the matter in the universe which then exploded. Following this "big bang," the galaxies began to recede from each other (as dots on a balloon that is being blown up move farther apart from each other.) What remains now, billions of years later, he says, is all tied to that original, super-explosion.

1927   Seth Barnes Nicholson (1891-1963), American astronomer, discovers that the surface temperature of the moon drops nearly 392°F (200°C) during an eclipse by the earth's shadow. He interprets this large, quick drop as indicating that any heat stored in its core rises very slowly to the surface. This leads to the idea that the moon's surface is covered with a layer of loose dust, since the vacuum between the dust particles would serve as an excellent heat insulator. He goes on to measure the maximum surface temperature of Mercury and to discover four more satellites of Jupiter.

1927   Henry Norris Russell (1877-1957), American astronomer, publishes a textbook in astronomy that for the first time shifts that subject's emphasis away from the solar system and toward the stars and astrophysics. The extensive coverage he gives in this popular textbook to astrophysics (the study of the physics and chemistry of heavenly bodies) and stellar astronomy changes the whole emphasis of the way the subject is taught and lays the groundwork for modern astrophysics.

1928   First to propose that matter is continuously being created throughout the universe is English mathematician and astronomer James Hopwood Jeans (1877-1946). This continuous creation theory is eventually embraced by those astronomers who subscribe to the "steady state" as opposed to the "Big Bang" theory of the universe.

1929   Edwin Powell Hubble (1889-1953), American astronomer, first formulates the law which states that galaxies recede from ours at speeds proportional to their distances. Hubble's law not only states the expansion of the universe, but it also implies that the expansion is uniform. Only the distance between galaxies change and not their positions relative to one another. Further, it does not imply that our galaxy is the center of the universe, but rather that no galaxy is its fixed center since each galaxy, including our own, is "running away" from every other galaxy. This marks the beginning of the theory of the expanding universe which proves of fundamental importance to modern cosmology.

FEBRUARY 18, 1930   Pluto, a new planet past Neptune, is discovered by American astronomer Clyde William Tombaugh. Using a photographic technique in which he compares two pictures of the same part of the sky taken on different days, he notices a slight shift in the position of one of the objects in the picture. None of the other objects moves upon comparison, since they are all stars. After this, he observes this object steadily for about a month and then announces on March 13, 1930, that he has discovered a new planet. He names it Pluto, after Jupiter and Neptune's brother.

1930   First indication that the earth's magnetic field influences cosmic radiation is offered by Arthur Compton (1892-1962), American physi-

cist. Compton conducts an experiment to determine whether cosmic rays are electromagnetic in nature or consist of charged particles. If electromagnetic, they would be unaffected by the earth's magnetic field and would strike every part of the earth equally. If made up of charged particles, they would curve in the magnetic field, with the poles receiving more radiation and the equatorial regions less. After a long series of measurements made around the world, Compton confirms that the "latitude effect" exists. He finds that cosmic radiation reaches a minimum at 0° geomagnetic latitude and rises to a maximum at latitude 50°, after which it remains constant.

1930   Coronograph is invented by French astrophysicist Bernard Ferdinand Lyot (1897-1952), who obtains direct photographs of the sun's prominences and of the entire corona. In principle, this appears to be a fairly simple solar telescope that has a disk to occult or artificially eclipse the sun's photosphere (its luminous surface). In practice, however, a very delicate and complex design is required to remove scattered photospheric light (since the photosphere is about a million times brighter than the corona). With this new device, astronomers no longer must wait for a total eclipse to study the sun's coronal spectral lines.

(See also 1931)

1931   First moving images of the complex activity of the sun are obtained by Bernard Ferdinand Lyot (1897-1952), French astrophysicist. He attaches a movie camera to his newly invented coronograph and obtains motion pictures of the sun's clouds, streamers, and arcs.

(See also 1930)

1931   Karl Guthe Jansky (1905-1950), American radio engineer, first detects radio waves coming from outer space and founds radio astronomy. During his research on static interference in radio communications for Bell Laboratories, he detects with his rotating antenna a new kind of weak static that he cannot identify. Eliminating the sun as its source (since the signal peaks about four minutes earlier each day), he decides it must lie beyond our solar system and theorizes that it comes from the Milky Way. Once astronomers learn to receive and interpret these microwaves (the shortest radio waves) that can penetrate through dust clouds, they realize they have a window to phenomena that ordinary telescopes could never see. Jansky does not pursue his discovery and leaves it for others to continue.

(See also 1937)

1931 Harold Spencer Jones (1890-1960), English astronomer, first begins a prolonged measurement of the parallax of the asteroid Eros (using 14 observatories in nine countries) to obtain the distance of the sun from the earth. In 1942, he finally announces his result of 93,005,000 mi (149,645,044 km). This generally correct figure is not improved until the late 1950s when radar is used.

1931 Bruno Benedetto Rossi, Italian astronomer, discovers the phenomenon of cascade showers in which a high-energy particle hits a nucleus in the upper atmosphere and generates a shower of secondary particles, which, in turn, create even more impacts. He also demonstrates the enormous energies of cosmic ray particles as he shows how they can penetrate more than a yard (3 m) of lead.

AUGUST 1932 First known form of antimatter, the positron, is discovered by Carl David Anderson (1905-1991), American physicist. While studying cosmic rays and gamma rays using a magnetic cloud chamber, he modifies it so as to be able to measure the energies of these rays in strong magnetic fields. He then discovers the tracks of an electron that appear to be curving in the wrong direction, which is what one would expect from electrons carrying a positive charge. He confirms this finding and suggests the name "positron" for this antimatter particle or anti-electron. Positrons prove to be stable in

a vacuum, but quickly react with the electrons of ordinary matter by annihilation to produce pure energy.

(See also 1936)

1932 First spectroscopic detection of a substance (carbon dioxide) in the atmosphere of Venus is made by American astronomers and colleagues Walter Sydney Adams (1876-1956) and Theodore Dunham, Jr. They make spectroscopic observations of the solar light reflected by the surface of Venus and obtain information about the composition of its atmosphere. They also find that Venus has an absence of oxygen and water vapor. Dunham continues this spectroscopic work and shows that the atmosphere of Venus is rich in carbon dioxide.

1937 Grote Reber, American radio engineer, builds the first radio telescope in the world. Inspired by the work of American radio engineer Karl Guthe Jansky (1905-1950), he single-handedly builds a spherical, concave, dish-shaped reflector with an electronic detector intended to capture and record these mysterious radio signals from space. Radio telescopes do not "see" or take photographs, but rather receive a signal that is amplified and recorded. Although initially unsuccessful, Reber upgrades his home-built equipment and tunes in to different wavelengths, and by the spring of 1939 is regularly picking up radio waves which he concludes are discrete sources like stars. In 1941 he begins making a complete survey of the sky and by the following year completes the first preliminary radio map of the sky, identifying radio sources that are not stars. His is an increasingly rare phenomenon in that one individual becomes almost single-handedly responsible for an entirely new field—in this case, for the complete beginning and early development of radio astronomy. It soon becomes clear to astronomers that radio astronomy is much less expensive than optical astronomy with its large, delicate telescopes, and that radio telescopes can be made more sensitive and often obtain patterns which contain important information about the size of the radio source.

(See also 1944)

1937 Interstellar hydrogen is discovered by Russian-American astronomer Otto Struve (1897-1963). After initially noting the prominent spectral lines indicating interstellar calcium, he is able to locate the spectral lines of hydrogen, in ionized form, which are extremely difficult to detect. The knowledge that interstellar hydrogen exists eventually leads astronomers to seek out and try to detect and trace the pattern of its radio "signature." This they eventually are able to do, resulting in the discovery of the spiral structure of our galaxy.

(See also 1951)

MAY 11, 1938 Adriaan van Maanen (1884-1946), Dutch-American astronomer, discovers a new type of variable star called a "flare" whose brightness varies at intervals of only a few minutes. This phenomenon indicates the existence of short-term processes in stars as well as the very slow process of stellar evolution. He goes on to discover more flare stars.

FEBRUARY 27, 1942 James Stanley Hey, British radio astronomer, and colleagues discover that the sun emits radio waves. This occurs during World War II, and Hey and his group initially believe that what they are detecting is the Germans jamming their radar sets. This fact is kept secret until it is pointed out that all the sets that had been "jammed" were in fact aimed in the direction of the sun. Although our sun qualifies as a radio star, it is not a very powerful one, especially during the quiet portions of its cycle.

1943 Difficult-to-see stars of the inner regions of the galaxy are studied and understood for the first time. Taking advantage of the wartime blackout in Los Angeles, German-American astronomer Walter Baade (1893-1960) uses the 100-in (254 cm) telescope at Mount Wilson

to peer at stars no one had studied before. He notes that there are reddish as well as blue-white stars, and he makes important distinctions between them. He notes that there are two populations of stars with different structure and history. One he classifies as Population I, made up of younger stars found in the spiral arms of galaxies. Population II, or older stars, are found in the nuclei of spiral galaxies. His classification proves to be a milestone in the history of stellar astronomy.

1944 First planetary satellite found to have an atmosphere is discovered by Gerard Peter Kuiper (1905-1973), Dutch-American astronomer. While studying Titan, the largest of Saturn's satellites, he discovers absorption bands from methane and a trace of ammonia in its spectrum. This is the first discovery of a satellite possessing an atmosphere. He goes on to establish a simple formula by which the stability of an atmosphere can be judged.

1944 Extragalactic radio astronomy is born as American radio engineer Grote Reber, using his own 33-ft (10 m), orientable, parabolic reflector, draws a radiomap of the galaxy showing three zones of very strong radio emissions. His wartime discovery passes almost completely unnoticed until confirmed in 1946 by English astronomers. By 1950, astronomers have embraced radio astronomy and the first catalog of radio sources is compiled.

(See also 1937)

1947 First galaxy to be identified as a radio source is Cygnus A by English astronomer Martin Ryle (1918-1984). He goes on to develop revolutionary radio telescope systems that accurately locate weak radio sources. It is under his leadership that the Cambridge (England) radio astronomy group compiles catalogs of radio sources.

1947 Bart Jan Bok (1906-1983), Dutch-American astronomer, discovers small, dark, circular dust clouds that prove to be phenomena associated with star formation. Visible only against a star background, these roundish, dark clouds are called Bok's globules. He suggests that they are precursors of stars or stars-in-the-making.

1948 First elaboration of the "Big Bang" theory of the origin of the universe is offered by George Gamow (1904-1968), Russian-American physicist. He revises and extends the ideas of Belgian astronomer Georges Edouard Lemaître (1894-1966), who stated in 1937 that the universe and all the elements were created from an extremely dense and hot fireball. He states that the elements were created by means of successive fusion reactions. This is based in part on the discovery of thermonuclear reactions by his colleague, German-American physicist Hans Albrecht Bethe. Gamow becomes an articulate spokesman and popularizer of this theory. He also predicts the residual background radiation or echo of the primal explosion that German-American physicist Arno Allan Penzias and American radio astronomer Robert Woodrow Wilson eventually discover in May 1964.

(See also May 1964)

1948 Innermost known asteroid is discovered by German-American astronomer Walter Baade (1893-1960). Named Icarus after the son of Daedelus in Greek mythology who flew too close to the sun, it comes within 19 million mi (30 million km) of the sun, nearer than any other known body in the solar system except for comets. Its orbit extends from beyond Mars' to within Mercury's, and it can come as close as 4 million mi (6 million km) to earth. In 1968, it is the first asteroid to be examined by radar, and it is found to have a diameter of about 0.5 mi (0.8 km) and a rotation of about 2.5 hours.

1948 Fifth and smallest satellite of Uranus is discovered by Dutch-American astronomer Gerard Peter Kuiper (1905-1973). Named Miranda, it has a 300-mi (483 km) diameter and an extremely varied landscape, with craters, grooved terrain, and fractures. Kuiper also dis-

covers a second satellite of Neptune the following year. It is named Nereid after the daughters of Neptune. More than any astronomer, Kuiper is responsible for restoring solar system astronomy to prominence in an era dominated by stellar and galactic research.

1949 First discrete radio sources to be identified with visible objects outside the solar system are discovered by J. G. Bolton of England and colleagues. The galactic radio source they discover is the remnant of a supernova.

1949 First use of the phrase "dirty snowball" to describe the composition of a comet is made by American astronomer Fred Whipple. He suggests that comets are essentially icy in nature, made up of a mixture of silicate dust and gravel mixed with frozen methane, ammonia, and water. When heated by the sun's closeness, cometary ice vaporizes explosively, and the dust it contains forms the haze of its "tail."

1950 Lyman Spitzer, Jr., American astronomer and physicist, invents the stellarator for containing very hot gases. His invention, which uses a twisting magnetic field that is wrapped around the hot plasma or gas, is a method for controlling fusion in experiments, and it grows out of his earlier research on fusion reactions in stars.

1951 First prediction of the existence of "solar wind" is made by German astrophysicist Ludwig Franz Benedikt Biermann after he studies cometary tails. The existence of this stream of plasma flowing from the sun is eventually verified in 1959 by Russian spacecraft. In that same year, American astronomer Eugene Newman Parker gives the name "solar wind" to the phenomenon he describes as the sun's constant emission in every direction of charged particles that drift outward through the solar system.

1951 First use of a high-speed electronic computer to help solve a problem in astronomy is made by Dutch-American astronomer Dirk Brouwer (1902-1966). This year he publishes the results of his study of the coordinates of the five outer planets for the years 1653 to 2060. Using the speed and power of new, all-electronic computers, he calculates the numerical integration of the planetary orbits from Jupiter to Pluto for over four centuries' time. He also determines a new value for solar parallax on the basis of perturbations in the orbit of the asteroid Eros (for the years 1926 to 1945).

1951   First optical identification of a discrete radio source, Cygnus A, is made by German-American astronomers Rudolph Leo B. Minkowski (1895-1976) and Walter Baade (1893-1960). They speculate that this radio source might be the result of colliding galaxies. They also locate wisps of gas in the constellation of Cassiopeia as the remnants of a long-exploded supernova.

(See also 1955)

1951   Alfred Charles Bernard Lovell, English astronomer, becomes the first professor of radio astronomy at Manchester University and begins work on the 250-ft (76 m), world's largest, fully steerable, "big dish" radio telescope at the Jodrell Bank Experimental Station. It is finished just in time to track the world's first artificial satellite, Russia's *Sputnik 1*, on October 4, 1957.

1951   First to demonstrate the actual spiral structure of our galaxy is American astronomer William Wilson Morgan. By detecting radio wave emissions with increasingly delicate instruments, he finds characteristic radio waves of ionized hydrogen coming from particularly hot, bright stars that are themselves characteristic of spiral arms in a galaxy. Finding several of these lines of ionized hydrogen coming from our galaxy offers strong evidence of its spiral structure. This same year Dutch astronomer Hendrik Van de Hulst elaborates on Morgan's work and also uses radio astronomy to confirm his spiral findings.

(See also 1937)

1952   Localized magnetic fields on the surface of the sun are first detected by American astronomers Harold Delos Babcock (1882-1968) and Horace Welcome Babcock (father and son). Using a solar magnetograph they invented the year before, they are able to record and measure solar magnetic fields despite their weakness and rapid variability. A few years later they establish the presence of magnetic fields in several stars, although these are easier to locate since they are very strong.

1952   Walter Baade (1893-1960), German-American astronomer, discovers that the distance of the Andromeda galaxy is over 2 million light-years away and not the 800,000 light-years that American astronomer Edwin Powell Hubble (1889-1953) had estimated in 1924. Baade uses the new 200-in (5 m) Palomar telescope in California and is able to observe many Population II stars that Hubble could not see. Baade's new information changes the picture of the universe in a number of important ways. It increases the entire universe size by 20 in volume and makes the age of the universe at least five or six billion years old. This also destroys the notion that our galaxy is a huge object among smaller galaxies. The greater size of the universe means that distant galaxies have to be much bigger and brighter than originally believed in order to be visible over such vast distances. Thus our Milky Way becomes only an average galaxy among millions of others.

1953   Synchrotron-emission theory is first proposed by Russian astrophysicist Iosif Samuilovich Shklovskii (1916-1985). He suggests that the high-energy particles emitted by the Crab Nebula also emit radio waves. This is called synchrotron radiation since it was first produced in giant accelerators known as synchrotrons. In these machines, particles (such as electrons) are accelerated to extremely large speeds, approaching the speed of light, with the help of strong magnets. The interaction of these fields with the electrons produces a radiation, and

Shklovskii suggests that similar processes may also occur in space.

JANUARY 1955 Strong radio-wave emissions from the planet Jupiter are first detected by American astrophysicists Kenneth Linn Franklin and Bernard Flood Burke. It is later learned that their source is the planet's immense magnetic field. Next to the sun, Jupiter proves to be the strongest source of radio waves in the solar system.

1955 Victor Amasaspovich Ambartzumian, Armenian astronomer, first suggests that the radio sources identified in 1951 by German-American astronomers Rudolph Leo B. Minkowski (1895-1976) and Walter Baade (1893-1960) are not the result of colliding galaxies but rather evidence of vast explosions within the core of galaxies. This analogy to a supernova on a galactic scale has become firmly established.

(See also 1963)

JULY 26, 1958 Radiation belts that encircle the earth are first understood by American physicist James Alfred Van Allen. He designs a special, lead-shielded cosmic ray counter that is launched aboard the United States satellite *Explorer 4* on this date. He correctly assumes that the counters launched earlier in the year, and which had unaccountably stopped working, had passed through radiation much too high for their instruments. His device not only proves him correct but discovers what comes to be called the "Van Allen Belt" or the magnetosphere. This is a region of very high radiation levels that widely encircle the earth around the equator and curve in dramatically near the poles. Scientists later learn that the belts are composed of charged particles that originated in the sun and became trapped in the earth's magnetic field.

1958 Nikolai Alexandrovich Kozyrev, Russian astronomer, observes the formation of a cloud or mist in the crater Alphonsus on the moon,

and takes a spectrum reading that indicates that it might be the result of something like volcanic activity. This is the first observation of the moon since the time of Galileo that even hints at the notion that it is anything but a cold, dead place. His apparent discovery of some type of volcanic-like activity on the moon is disputed, but it opens a new chapter in the study of the origin of the moon.

1960 First quasi-stellar objects, soon called quasars, are discovered. These are unusual objects since they are powerful radio sources but do not correspond with the locations of any known objects such as galaxies or nebulae. American astronomer Allan Rex Sandage and Australian astronomer Cyril Hazard soon pinpoint these sources but their spectra are unlike any seen before. They are eventually called quasars.

(See also 1963)

1962 John Archibald Wheeler, American physicist, publishes his *Geometrodynamics* in which he considers those aspects of general relativity that suggest the possibility of gravitational collapse. He theorizes conditions under which a collapse cannot be stopped, and first coins the term "black hole" to describe a collapsed mass whose gravitational field becomes so intense that not even light could escape it.

1962 Rotation of the planet Venus is first determined by American astronomers Roland L. Carpenter and Richard M. Goldstein. Using microwaves to penetrate the planet's thick cloud layer, they calculate the level and degree of wave distortion and are able to demonstrate that Venus has the amazingly slow period of rotation of about 250 days (latter refined to 243.09 days). They also discover that Venus rotates in retrograde fashion, from east to west rather than the earth's west to east.

1962 First radar observations of Mercury are made by the radio telescope built in a large natural basin in Arecibo, Puerto Rico. Most radio telescopes have a steerable dish which collects radio signals. At Arecibo, the enormous aluminum dish (1,000 ft [304 m] in diameter) is immobile on the earth while its receiving equipment (600 tons) hangs 50 stories in the air. It can be steered and pointed by remote control equipment. The Arecibo Observatory is designed so that one person can operate the entire facility. It is part of the National Astronomy and Ionosphere Center, a national center operated by Cornell University under contract with the National Science Foundation. Radar observations of Mercury are also made from Russia during this year.

1962   First x rays originating from deep space are discovered by Bruno Benedetto Rossi, Italian-American astronomer. He and his colleagues at the Massachusetts Institute of Technology mount a detector on a small rocket and discover an x-ray star in Scorpio—the first x-ray source to be detected outside the solar system.

1963   Allan Rex Sandage, American astronomer, discovers a suspicious object in galaxy M-82 that has a very strong radio source. His photographs show the galaxy to be undergoing an enormous explosion at its core—one that has been going on for 1.5 million years. This serves to confirm the theory of Armenian astronomer Victor Amasaspovich Ambartzumian.

(See also 1955)

1963   Quasars, or quasi-stellar objects, are first adequately explained by Dutch-American astronomer Maarten Schmidt. He explains that the unfamiliarity of their spectra is the result of an enormous red shift, and that the spectral lines are unfamiliar ones that ought to be in the ultraviolet section of the spectrum. This means that these radio emissions are not coming from a star at all but from an incredibly distant object like a galaxy—except that it is emitting more energy than an entire galaxy would. Quasars prove difficult to understand fully, and after 30 years of study, they appear to be better under-

stood in terms of black hole theory. They are now considered to be a massive black hole at the center of a galaxy.

MAY 1964   Background microwave radiation is discovered by German-American astronomer Arno Allan Penzias and American radio astronomer Robert Woodrow Wilson. After eliminating all possible sources, they eventually conclude that unexplained background radiation is coming from all directions with equal intensity. This is later explained by American physicist Robert Henry Dicke, who demonstrates that what they detected was in fact the remnants of the original big bang, the explosion that created the universe billions of years ago. As early as 1948, Russian-American astronomer George Gamow (1904-1968) suggested that it should be possible to detect the residual effects of this creative explosion. Most consider this to be evidence in favor of the Big Bang theory as opposed to its rival steady state model of creation.

FEBRUARY 3, 1966   First soft landing on the moon is achieved by Russian spacecraft *Luna 9*. It relays photographs to earth of the lunar surface taken as it is landing. The United States follows with its own series of soft landers which also relay surface photos back to earth.

(See also April 3, 1966)

MARCH 1, 1966   First man-made object to reach another planet is the Russian space probe *Venera 3*, which strikes the surface of Venus but does not transmit any data to earth.

APRIL 3, 1966   First spacecraft to be placed into orbit about the moon is Russian probe *Luna 10*. This is soon followed by a series of American lunar orbiters which map the entire surface of the moon in full detail.

AUGUST 6, 1967   First "pulsar" (pulsating star) is detected by Jocelyn Bell, English graduate student working with English astronomer Antony Hewish. She notes unexpectedly a rapidly fluctuating but unusually regular radio signal between the stars Vega and Altair. The incredibly short bursts last only one thirtieth of a second and follow each other with remarkable regularity—at intervals measured as 1.33730109 seconds. Hewish initially thinks they might be some form of deliberate communication from outer space, but the suggestion of Austrian-British-American astronomer Thomas Gold that they are the "signatures" of rapidly rotating neutron stars is later shown to be correct. Hewish goes on to establish some of the main properties of a pulsar.

(See also 1968)

1968 Thomas Gold, Austrian-British-American astronomer, first suggests that pulsars are rotating neutron stars. He says that these stars can be as massive as ordinary stars but would be composed of tightly packed neutrons, since their diameter would be about only 9 mi (14 km). They would also have enormously intense magnetic fields and would give off radiation following their curved paths. If this is so, he argues, pulsars would be losing substantial amounts of energy as they turned, meaning that the frequency of their pulsations should slowly be lengthening. Continued observation proves Gold correct, and pulsars become identified as rotating neutron stars.

JANUARY 1969 First optical identification of a pulsar is made by American astronomers Cocke, Taylor, and Disney. They locate the first example of a visual pulsar in the Crab Nebula, which is a well-known remnant of a supernova. It is found to be blinking on and off 30 times a second, in time to the microwave pulses.

JULY 20, 1969 First human beings set foot on another celestial body. United States spacecraft *Apollo 11* successfully lands its lunar module on the moon, and astronauts Neil Alden Armstrong and Edwin Eugene Aldrin, Jr. walk on its surface, taking samples, photographs, and setting up scientific experiments. After a total of 21 hours and 37 minutes on the lunar surface, they dock with the *Apollo 11* command

module piloted by astronaut Michael Collins, and return to earth on July 24.

1969 Formaldehyde is identified as the first of a constantly growing number of organic types of molecules being found in interstellar space by radio astronomy techniques. With the increasing ability to detect microwave radiation with great precision, astronomers find frequencies characteristic of water molecules and ammonia molecules as well as other, more complicated atom groupings in interstellar gas clouds. This marks the beginning of what comes to be called astrochemistry.

DECEMBER 15, 1970 First successful soft landing of a man-made object on another planet is made by the Russian Venus probe, *Venera 7*. After entering Venus orbit, the spacecraft drops an instrument package into the atmosphere which makes a soft landing. The instruments send data back to earth about the atmosphere and surface conditions on Venus for 23 minutes before the extreme pressure and temperature makes it inoperable.

NOVEMBER 13, 1971 First man-made object to be placed into orbit about another planet is the United States Mars probe *Mariner 9*. After successfully entering Mars orbit, the spacecraft photographically maps the entire planet. Photos reveal craters clustered in one hemisphere and volcanoes crowded into another. Its atmosphere proves to be almost entirely carbon dioxide and is about one-hundredth the density of earth's. Its temperature is too low for liquid water to exist at anytime, and its ice caps may contain frozen water and frozen carbon dioxide.

MARCH 2, 1972 First space probe intended to yield information about the outer solar system is launched. On its trip to deep space, the United States *Pioneer 10* takes close-up photos of Jupiter and continues into interstellar space as the first man-made object to leave the solar system.

1973 Complex molecules are discovered in Com-

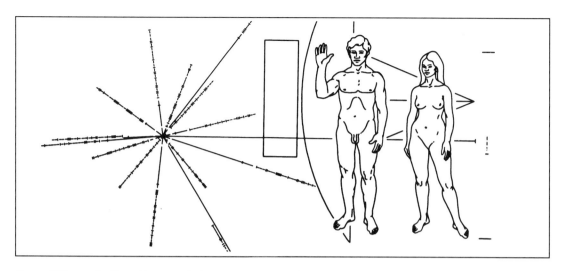

Pioneer F Plaque, the first object sent by humans into interstellar space.

et Kohoutek. First seen by L. Kohoutek of Hamburg, Germany when it was positioned near the orbit of Jupiter, this comet does not prove to be as spectacular as expected, but it is visible to the naked eye and proves scientifically interesting. Its orbit is essentially parabolic.

SEPTEMBER 10, 1974 Thirteenth satellite of Jupiter is discovered by American astronomer Charles T. Kowall. This tiny moon (about 5 mi [8km] in diameter) is named Leda after one of Jupiter's many loves in the Roman myths.

JULY 20, 1976 *Viking 1* first touches down on the surface of Mars. Two weeks later the other United States Mars probe, *Viking 2*, lands in a more northerly direction. During their descent, they find the Martian atmosphere to be chiefly composed of carbon dioxide, although they detect 2.7% nitrogen and 1.6% argon. Tests of the very rocky surface reveal no trace of organic material in the soil. Signs of old, dry riverbeds and tributaries lead astronomers to speculate that at some time there may have been liquid water on Mars. There are also no signs of life on a scale visible to the eye (camera).

MARCH 10, 1977 Rings of Uranus are discov-

ered by American astronomer James L. Elliot. Taking advantage of the planet moving in front of a ninth-magnitude star in the constellation of Libra, he observes Uranus while flying in an airplane (so as to be high enough to minimize the distorting and obscuring effects of our lower atmosphere). During a dimming-brightening pattern that occurs as Uranus passes the star, he notes that Uranus is surrounded by a series of thin concentric rings. While they are similar to those of Saturn, they are in no way as large and bright.

(See also 1978)

NOVEMBER 1, 1977 Charles T. Kowall, American astronomer, discovers an extraordinary object (Charon) while studying photographic plates of the area around the planet Jupiter. Searching for asteroids, he detects something that might be an asteroid but that is moving at only one-third the speed it should be near a planet as large as Jupiter. He finds that it is no ordinary object since the highly elliptical path it follows carries it out as far as Uranus at one end and as close to Saturn at the other, completing one orbit about the sun every 50.7 years. It is named after the most famous centaur in Greek mythology. It is still not known exactly what Charon is. It could be an exceptional asteroid (if so, it is the farthest one yet discovered), a planetismal, or even an escaped satellite of Saturn.

1977 Concept of an inflationary universe is first proposed. As put forth by American physicist Alan Goth, it suggests that in the instants after the big bang, the universe underwent a sudden and exceedingly rapid inflation. He states that although objects within the universe cannot move relative to one another faster than the speed of light, the universe itself, as a whole, could theoretically expand at any speed. This theory is used to explain some characteristics of the Big Bang theory that present problems, such as the uniformity of the microwave background.

JUNE 7, 1978 First natural satellite of an asteroid is discovered orbiting the asteroid Herculina. Also called Asteroid 532, it occults (temporarily obscures or hides) a star, permitting it to be better observed. Scientists obtain data indicating that the asteroid has a diameter of 135 mi (217 km) and has an orbiting companion whose diameter is 31 mi (50 km) and is about 620 mi (1,000 km) away.

JUNE 22, 1978 First satellite (Charon) of the planet Pluto is discovered by American astronomer James W. Christy. After examining photographs of Pluto, he notes what appears to be a distinct lump on one side that shifts position in later photos. Further analysis reveals it to be a separate body, located about 12,500 mi (20,112 km) from Pluto. The two are so close that they influence each other and revolve around a common center of gravity like two unequal parts of a dumbbell. Pluto is very small, only about 1,850 mi (2,977 km) in diameter, and Charon is about 750 mi (1,206 km) in diameter. Together they are the closest thing to a double planet known. Christy names the satellite after the ferryman who crosses the River Styx to Hades in Greek mythology.

DECEMBER 4, 1978 United States satellite *Pioneer Venus* first enters orbit around that planet. The unmanned spacecraft sends several probes into the dense atmosphere of Venus, finding it to be primarily composed of carbon dioxide. Radar beams are sent through its clouded surface, discovering that it seems to have a huge supercontinent covering five-sixths of the total surface. It has mountains, canyons, and what may be extinct volcanoes.

1978 First earth-based photo of the rings of Uranus are obtained by G. Neugebauer, American astronomer. He uses the 200-in (508 cm) Hale reflector at Palomar. Although the photo does not show the rings separately, it does confirm that there are nine narrow rings around the planet.

(See also March 10, 1977)

1979 First observation is made of a comet hitting the sun.

APRIL 1982 A. Wright and D. Jauncey, astronomers in New South Wales, Australia, discover the most remote object to date. They observe the quasar PKS 2000-330, estimated at 13,000 million light-years away. Because quasars are the most luminous objects in the universe, they can be seen to greater distances and hence farther back in time than anything else. They afford an excellent means of examining what might be called the youth of the universe.

OCTOBER 16, 1982 Halley's Comet is first recorded by Hale Observatory in Palomar, California. The famous comet is first sighted nearly four years before it completes its 76-year journey past earth.

(See also March 13, 1986)

1982 First "millisecond pulsar" is discovered. After pulsars were discovered in 1967, the fastest known was in the Crab Nebula and rotated some 30 times a second. This year, however, a pulsar is found that rotates more than 20 times as fast (at 642 times per second). Others are soon found that rotate in the range of a thousandth of a second and are called millisecond pulsars.

1983 First indication that extrasolar planets may exist is suggested by the International Astronomical Satellite (IRAS). The result of an international project (United States, the Netherlands, and United Kingdom), this satellite detects infrared radiation coming from the immediate neighborhood of the bright star Vega. This radiation is interpreted to be the result of a ring of particles surrounding the star, suggesting that such a belt might be in the process of condensing into planets. Some suggest it could also be an indication that planets already revolve around that star, as in our solar system.

JANUARY 24, 1986 First close look at Uranus is made as United States spacecraft *Voyager 2* makes a close flyby. Spacecraft instruments reveal that Uranus rotates in 17.24 hours and has a magnetic field. The existence of its rings is confirmed and its five satellites turn out to be larger than once thought. *Voyager 2* returns spectacular color images of these satellites and discovers 10 other small satellites all within the orbit of the satellite Miranda.

MARCH 13, 1986 Halley's Comet returns past earth and becomes the first comet to be studied from close range (by satellites). Five space probes are orbited, two Japanese, two Russian, and one European, to study Halley from space. On this date, the European spacecraft *Giotto* passes within 376 mi (605 km) of the comet's nucleus. Data obtained from these studies generally confirm the "dirty snowball" theory, stating that its nucleus is composed of rocky fragments held together with ices such as frozen methane, ammonia, carbon dioxide, and water. Unfortunately, its appearance is not very spectacular when viewed from earth, since it remains rela-

tively far away, even at its closest approach. It can only be seen well high in the sky from the southern hemisphere.

APRIL 24, 1990 First telescope to be placed in outer space, the Hubble Space Telescope (HST), is successfully launched. As the largest satellite (12 tons) ever launched, this boxcar-size, unmanned observatory is placed into orbit 381 mi (613 km) above the earth by the United States Space Shuttle *Discovery*. It is named for American astronomer Edwin Powell Hubble (1889-1953), who discovered some of the phenomena it is designed to explore. Its mission is to measure the size and therefore the age of the universe, to record distant objects that can now only be seen faintly, and to register the ultraviolet spectra of distant and ancient quasars. It would also probe for the possibility of a black hole at the center of the Milky Way, collect data about the birth of stars, and search for planets around stars similar to our sun. Although it is placed into orbit with a crucial flaw in its primary mirror, it is eventually repaired by American astronauts who revisit the telescope. Once repaired, HST proves to have resolution 10 times sharper than any earth-based telescope and is able to probe deeper into space than ever before possible.

OCTOBER 29, 1991 First good photographs of an asteroid in space are taken by the United States spacecraft *Galileo*. Launched into space during 1989 for an eventual rendezvous with Jupiter in 1995, the spacecraft comes within 1,000 mi (1609 km) of the asteroid Gaspra (between their orbits of Mars and Jupiter) and takes the first black-and-white photograph of an asteroid that has any kind of detail. It shows that Gaspra is irregular in shape, about 12 mi (19 km) long and 8 mi (13 km) wide, and is heavily pitted with craters. Its shape is described as resembling a dented football.

1991 Astronomers in England discover a new pulsar, PSR 1829-10, that suggests it could be an extrasolar planet.

1991 First map of the surface of Venus is completed by United States *Magellan* spacecraft. Using its penetrating synthetic aperture radar, the unmanned orbiting spacecraft is able to map about 84% of the cloud-covered surface, resolving details as small as 400 ft (121 m) across. These images show arid plateaus punctuated by mountains taller than Mt. Everest, and craters up to 30 mi (48 km) across. Volcanic domes 0.5 mi (0.8 km) high and 12 mi (19 km) wide dot the landscape, and volcanic ash covers the surface. Evidence also is found to indicate that its volcanoes are still active, and some mountains appear newly built as do many fault lines.

APRIL 1992 First evidence that potentially corroborates the theory of dark matter is obtained by American astrophysicist George Fitzgerald Smoot. As the head of the team analyzing data received from the November 1989 launch of the Cosmic Background Explorer (COBE) satellite, he notes fluctuations in background radiation indicating that after the big bang, there were variations in heat and density that caused a clumping together of matter (to form stars and other objects). This clumping data corroborates the theory of dark matter (invisible or transparent matter that does not interact with matter as we know it). Although it radiates little or no detectable energy, dark matter is thought to exert gravitational effects, and some believe it plays an important role in the formation of galaxies.

JUNE 1992 Fast neutrons produced in solar flares are revealed for the first time by the United States *Gamma Ray Observatory (GRO)* satellite. Placed into an orbit 280 mi (450 km) above the earth by the space shuttle *Atlantis*, its mission is to make the first survey of gamma-ray sources throughout the universe, and particularly to study such explosive energy sources as supernovas, quasars, neutron stars, pulsars, and black holes. Early data from the observatory suggest that the flashbulb-like bursts it detects (which

release more energy in a tenth of a second than the sun does in 10,000 years) result from collisions of neutron stars with comets, gas clouds, or other stars.

1992 David Jewitt and Jane X. Luu, American astronomers, discover what may be the first of a new class of objects at the edge of the solar system, beyond the orbits of Pluto and Neptune. Called a "planetismal," it is about 124 mi (200 km) in diameter and lies 42 astronomical units from the sun. During October 1993, they discover five more objects just outside Neptune's orbit. They may be comets moving toward the sun or they may be asteroids permanently located near Neptune.

AUGUST 1993 World's largest astronomical instrument first begins operations. The Very Long Baseline Array (VLBA) is spread over 5,000 mi (8,045 km) of United States territory, using an 82-ft (25 m) dish at each of 10 sites, which act together as a single radio telescope objects that are moving away from the earth (within an expanding universe) and whose light shifts out of the visible portion of the electromagnetic spectrum. VLBA is able to receive emissions from such objects.

JULY 16, 1994 Fragments from Comet Shoemaker-Levy 9 first crash into planet Jupiter. Beginning this day and continuing for slightly more than one week, 21 large fragments from this comet collide with Jupiter one after another. Close-up images taken by the United States Hubble Space Telescope show large impact zones punched into the upper atmosphere of Jupiter as the comets bombard and cause great localized damage to the surface of this giant planet.

SEPTEMBER 1, 1994 First "quasi-stellar" object or "mini-quasar" ever to be detected inside the

Milky Way is reported. This new-found object, which may be a black hole in a relatively nearby section of our galaxy, is shooting jets of material toward earth at 171,000 mi (275,139 km) per second—close to the speed of light. Scientists do not know exactly what causes the jets or "blobs" of electrons and other subatomic particles. They are believed to be coming from a double-star system in which one of the stars is either a black hole or an extremely dense neutron star (a collapsed, dead star that could eventually shrink into a black hole).

Biology

c. 2700 B.C. Tea as a beverage first appears in China, possibly because of the need to boil drinking water for health reasons. Adding tea leaves to boiled water would rid it of its flat taste.

(See also 800)

c. 1800 B.C. Process of fermentation is first understood and controlled by the Egyptians. They learn that if some fermenting bread is saved before it is baked and then added to dough that had not yet begun to ferment, the fresh dough would ferment in turn. Dough that ferments releases gases (carbon dioxide) that cause the bread to rise and grow spongy. The result is a soft "leavened bread" that is pleasant to eat. However, the flat, hard (unleavened) bread is just as nourishing.

c. 350 B.C. Aristotle (384-322 B.C.), Greek philosopher, is recognized as the founder of biology. A careful and meticulous observer of the living world, he studies over 500 animal species and make the first serious attempt to classify them according to a reasonable plan. He offers Greek words for genus and species and also considers the nature of reproduction and inheritance.

(See also 1472)

287 B.C. Theophrastus (c.371-287 B.C.), Greek botanist, dies after writing two botanical treatises considered the first systematic work on botany. Often called the "father of botany," he makes attempts to classify plants and describes over 500 species. As a pupil of Greek philosopher Aristotle (384-322 B.C.) he carries on his master's tradition of biology but concentrates primarily on the plant world. His writing is especially practical and offers many details on the gathering of plants for drugs. He brings a new, scientific attitude and outlook to the study of plants, and, because of his work, botany first appears as a distinct science. He writes in a logical manner, placing the facts in a coherent theoretical framework, and creates a true science of plants.

(See also 1483)

50 B.C. Crateuas, Greek physician, writes a work on pharmacology in which he is the first to make drawings of plants. These are the earliest known botanical drawings. His illustrations contain the name and medicinal properties of each plant but do not offer descriptions of them. His work pioneers plant illustrations and becomes very popular as it is reissued by others until the 3rd or 4th centuries A.D.

c. 77 First systematic pharmacopoeia is the

Materia Medica by Greek physician Pedanius Dioscorides. Discussing over 600 plants and nearly 1,000 drugs, his work survives the fall of Rome and the Dark Ages by being preserved by the Arabs. His work is mostly accurate and contains little of the fantastic claims of some of his peers. When it is rediscovered during the Renaissance and translated into Latin, it stimulates a great deal of botanical and herbal research.

(See also 1499)

c. 100 Aretaeus of Cappadocia (c.81-c.138), a Greek physician, first uses the term "diabetes." He chooses the Greek word for siphon, meaning to pass through, indicating the intense thirst created by the disease and the excessive urination that follows.

c. 180 The fact that nerves control muscles is discovered by Greek physician Galen (c.130-c.200) when he cuts the recurrent laryngeal nerve of a pig. He also shows the importance of the spinal cord, noting that when he cuts it at certain levels, differing degrees of paralysis results. Since dissection of humans is forbidden, he dissects animals and applies what he learns to the human body. He also is the first to observe that muscles work in contracting pairs. He becomes the greatest authority on physiology until the Renaissance, and his writings are taken literally as gospel, including his errors. He serves as physician to the gladiators and later as the personal physician to three different emperors. He becomes the last and most influential of the great medical practitioners.

c. 800 Cultivation of the tea plant in Japan first begins. Although drinking tea as a beverage began in China, an elaborate tea-serving ceremony arises in Japan that assumes social and religious significance.

1469 First scientific book to be printed after the invention of typographic printing is *Historia Naturalis*, by the Roman scholar Pliny (23-79). It becomes very influential and has scores of editions. Although primarily a natural history survey, it attempts to be a complete summary of ancient knowledge concerning the world and draws from 2,000 books written by nearly 500 writers. Pliny seldom attempts to separate fact from fiction, and because of this, many fantastic tales of the natural world are passed on to Renaissance believers. His steady emphasis throughout the work is a concern for practical use to mankind. If he could find no use to man for a plant or animal, he seldom deemed it important.

1472 First Latin translations of the biological works of the Greek philosopher and naturalist Aristotle (384-322 B.C.) appear in print. His *Historia Animalium*, *De Partibus Animalium*, and *De Generatione Animalium* sometimes appear under the overall title *De Animalibus*. Because of the availability of the writings of such an original biological thinker, he is a major influence and stimulus to Renaissance intellectuals. They greatly benefit from his unrivaled capacity to arrange his work and that of others, as well as his acute observations and judgments.

1478 First botanical painter of modern times is Italian artist Sandro Botticelli (1445-1510), who paints the picture titled *Primavera* (Spring) in Florence, Italy. Influenced by the naturalistic spirit of the Renaissance, he depicts Venus, with Cupid hanging over her, standing in a grove of orange and myrtle. Venus, who enters with Flora and Zephyr, is welcoming the approach of Spring. A garland of beautiful, yet accurately drawn, flowers are flowing from the mouth of Flora. The entire painting can be regarded as a botanical study, containing over thirty species of plants.

1483 First printed versions of the scientific botany of the Greek botanist Theophrastus (c.371-287 B.C.), titled *Historia Plantarum* and *De Causis Plantarum*, appear. From these works, Renaissance men learn that plants are worthy of serious study.

(See also 287 B.C.)

1499 First Greek text of the Greek physician Pedanius Dioscorides' *Materia Medica* is printed. It describes over 600 plants and their medicinal properties. As a surgeon who serves the Roman armies under the emperor Nero (37-68), his main interest in plants is their use as drugs.

(See also c.77)

c. 1505 Leonardo da Vinci (1452-1519), an Italian artist, makes the first known wax cast of the brain ventricles (of an ox). In his studies of the human skull and brain, Leonardo's drawings are more accurate than any made by previous anatomists, and he is the first to suggest that the cavities of the brain might be injected with wax, as he has done with this ox. This is the first suggestion of the injection of solid matter for the purpose of examining the form of bodily cavities. The method becomes widely accepted and practiced.

1530 Otto Brunfels (1464-1534), German humanist, physician, and botanist, writes his landmark herbal text, titled *Herbarum Vivae Eicones ad Naturae Imitationem*. Herbalism begins to change into scientific botany with this first of the modern herbals written by one of the "German fathers of botany." Brunfels is considered to be the first to produce a work on plants whose illustrations are based entirely on observation. Although the text is far from error-free, his herbal analysis of the plants of his native Germany is highly accurate. His pioneering work contains 135 superb wood engravings of living plants.

1533 First chair of botany in a university is instituted at Padua. Although the new chair is within the medical faculty, the separate title and specialization is equivalent to the appointment of the first professor of botany.

(See also 1545)

1542 Alvar Núñez Cabeza de Vaca (c.1490-c.1560), Spanish explorer, first describes a "black drink" made from a holly bush that is used as a purgative by Native Americans in southeastern America in his book, *La Relacion y Commentarios*. Núñez spends many years among the nomadic Indians as he and three companions are the only survivors of an original expedition of 400 men.

1542 First important modern glossary of botanical terms is established by German botanist Leonhard Fuchs (1501-1566). His work, *De Historia Stirpium*, presents plants in alphabetical order and gives, for each, a precise description and an account of its form and habitat. He also advises the best season to collect the plant along with what he calls its temperament and powers. His work contains woodcuts that are both beautiful and accurate and helps establish a tradition of plant illustration. His name is commemorated by a genus of plants called *Fuchsia*, which is a flowering plant as well as a bluish red color.

1545 First botanical garden in Europe affiliated with a university medical school is established at Padua, Italy. The establishment of this garden follows the appointment of a chair of botany at Padua. This first modern botanical garden is designed for the double purpose of teaching and the maintenance of a collection of plants for study. While the focus is primarily medicinal, the economic and scientific aspects of plants begin to be studied also. In the next twenty years many gardens are established throughout Italy.

(See also 1533)

1546 First scientific statement of contagion or of how infections are transmitted is proposed by Italian physician Girolamo Fracastoro (1478-1553). In his *De Contagione* he helps differentiate various types of fever and seeks to classify infectious diseases. He offers the idea that infection of all kinds, including fermentation, is the work of minute "seeds" that he calls "seminaria" or germs. This scientific germ theory of disease is proposed 300 years before it is empirically formulated by French chemist Louis

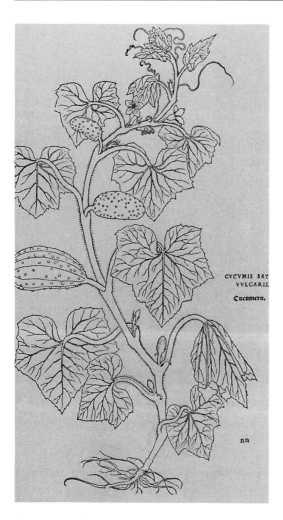

CVCVMIS SAT
VVLGARIS.
Cucumern.

nn

Line drawing of a cucumber plant from *De historia stirpium*, by Leonhard Fuchs.

Pasteur (1822-1895) and German bacteriologist Robert Koch (1843-1910). He states that each disease is caused by a different type of rapidly multiplying, minute body. These bodies are transmitted in three ways: by direct contact; by carriers such as clothing; and through the air. These are remarkably modern ideas.

1547 Portuguese sailors bring the first orange tree from China to Europe. Oranges are believed to be native to the tropical regions of Asia and are thought to have spread to India, to the east coast of Africa, and to the eastern Mediterranean.

1555 French naturalist Pierre Belon (1517-1564) is the first to notice the skeletal similarities of the various vertebrates. He publishes his *L'Histoire de la Nature des Oyseaux,* in which he first depicts the homologies between the skeleton of a bird and that of a man. In this landmark illustration, he compares them bone for bone. This is the very early beginnings of comparative anatomy. Belon's emphasis on anatomical similarities was a major contribution to the slow development of evolutionary theory that would occur over the next three centuries. He also is the first to illustrate a cetacean with its young still attached by the umbilical cord. Belon introduced the cedar tree into France. He was killed by robbers in the Bois de Boulogne in Paris.

1556 First tobacco seeds reach Spain from the New World. A few years later, a French diplomat, Jean Nicot (c. 1530-1600), sends tobacco seeds from Portugal to France. His name is later immortalized in the word for the active, poisonous ingredient in tobacco called "nicotine." In 1565, tobacco is introduced to England by English naval commander John Hawkins (1532-1595).

1561 Fallopian tubes are discovered by Italian anatomist Gabriello Fallopio (1523-1562), also called Fallopius. These are tiny tubes that lead the human egg or ovum from the ovary where it is formed, to the uterus where, if fertilized, it develops into an embryo. He also describes the inner ear and incompletely describes the cranial nerves.

1563 First description of a coconut is contained in *Coloquios dos Simples e Drogas e Cousas Medicinais da India* written by Garcia del Huerto (1490-1570) of Portugal. This book contains the first description of many other Indian plants and is considered the first book to be printed in India.

1569 Nicolas Monardes (1493-1588), Spanish botanist, first describes "Sassafras," which is given to him and named by a Frenchman from Florida. In his book, whose English title is *Joyfull Newes Out of the Newe Founde Worlde*, he offers descriptions of many other exotic plants and animals, including tobacco and the armadillo. He makes several medical claims for these new plants. The Sassafras or Ague tree has aromatic leaves, roots, and bark, and becomes a standard homemade medicine and tea. It is thought to have the property of dissolving bladder stones.

1576 First book on Spanish flora is written by French botanist Charles de L'Ecluse (1526-1609), also called Carolus Clusius. His *Rariorum Aliquot Stirpium per Hispanas Observatarum Historia* is also the first book to contain information on the tulip, which he introduces into Europe from Turkey. His collection and descriptions of new plant species contribute substantially to the advance of systematic botany.

1580 Italian physician Prospero Alpino (1553-1617) brings the first coffee into Italy. Having spent three years in the Orient studying the local customs, drugs, and diseases, he writes a botanical work and is credited with the introduction of coffee to the West.

1580 Tea is first brought into Italy by the Italian Maffei.

1583 First real scientific textbook on botany is *De Plantis* by Italian physician Andrea Cesalpino (1519-1603), also called Cesalpinus. It is considered by most to be the first serious textbook on botany because it defines the basic criteria of botanical taxonomy (the laws and principles of classification). Cesalpino offers an original system of arranging or classifying plants according to their flowers and fruits, thereby anticipating the Linnean system of classification. His work is a powerful contribution to the development of botanical theory.

1586 Thomas Harriot (1560-1621), English mathematician and astronomer, first takes potatoes from America to Sir Walter Raleigh's (1544-1618) Irish estate. The potato eventually becomes a staple in Ireland, and it is not until 1719 that it is reintroduced as a crop to North America (in New Hampshire).

1588 Thomas Harriot (1560-1621), English mathematician and astronomer, publishes *A Briefe and True Report of the New Found Land of Virginia* in London. This is the first book in vernacular English that is devoted to the flora and fauna of America. He is sent by Sir Walter Raleigh (1544-1618) as scientific advisor for the expedition of 1585-1586 to Roanoke Island off the coast of what is now North Carolina. Harriot is also considered the first Englishman to explore and describe the natural history of North America.

1592 Prospero Alpini (1553-1617), Italian physician, publishes his *De Plantis Aegypti*, which is the first treatise on the plants of Egypt. Having lived in Cairo for three years, Alpini makes an extensive study of its flora and introduces exotic plants to Europe upon his return. He is credited with giving the first botanical accounts of coffee, banana, and a genus of the ginger family that is later named *Alpinia*. He is also recognized as being the first to fertilize date plants artificially.

1607 First significant treatise on zoology in English is published by Edward Topsell (1572-c.1625), English naturalist. His book, *The Historie of Foure-Footed Beastes and Serpents*, is taken largely from the Swiss naturalist Konrad von Gesner (1516-1565), whose major work, *Historiae Animalium*, stands as the best purely zoological work of the Renaissance. It remains the standard reference work for two centuries. Gesner arranges the animals and then treats them alphabetically. He also includes over 1,000 woodcuts of beasts, both real and imagined. The inclusion of fanciful creatures indicates that scientific zoology is still in its infancy.

Gesner illustration of a rhinoceros.

Topsell's version of Gesner's classic work is thought of as the first and best English version.

1615 Santorio Santorio (1561-1636), Italian physician, publishes his *De Medicina Statica* in which he describes his physiological investigations on human metabolism. He is the first to subject the human body to quantitative measurement and creates several new devices to measure the factors of metabolism. As a proponent of the iatrophysical school of medicine, he attempts to explain the workings of the animal body on purely mechanical grounds. He experiments on himself, measuring his intake and output, and frequently eats, sleeps, and works on a large scale to study the fluctuations of his body. For this, he is considered the first to apply physics or mechanics to medicine.

1625 Santorio Santorio (1561-1636), Italian physician, describes his thermometer. This is the first clinical thermometer, as he obtains body temperature by placing it in the mouth.

1627 Lymph vessels or lacteals that take up the end products of fat digestion from the intestine are discovered by Italian physician Gaspare Aselli (c.1581-c.1626). He calls the new ves-

sels "venae albae et lacteae" (white and lacteal veins), and his discovery contributes to the growing knowledge of the circulation of body fluids. His discovery is published posthumously.

1630 First microscopical observations to be published are those of Italian naturalist and poet Francesco Stelluti (1577-1640). He publishes his realistic drawing of a honey bee in a book of poems. Having systematically observed a honey bee with a microscope in 1625, he made an engraving of it in great detail. Stelluti uses this engraving to dedicate a book of poems, titled *Persio*, to Cardinal Francesco Barberini (1597-1679). The cardinal's uncle Maffeo, Pope Urban VIII (1568-1644), was also a Barberini, and the family's crest shows three bees.

1646 Giovanni Battista Ferrari (1584-1655) of Italy publishes his *Hesperides sive de Malorum Aureorum Cultura et Usu* in Rome. This sumptuously illustrated book is the first work entirely devoted to citrus fruits.

1648 First quantitative experiments in plant physiology are performed by Johannes Baptista van Helmont (1577-1635), Flemish physician and alchemist. He plants a willow tree in a fixed amount of soil which he weighs. After five years, during which time he adds only water, the tree gains 164 lbs (74 kg) and the soil loses only 2 oz (57 g). Since Helmont believes that water is the basic element of the universe, he concludes that the water was converted by the tree into its own substance. Although he is not correct, his experiment is very important as it contributes to the new quantitative methods used during the Scientific Revolution.

1652 The first coffeehouse in London is established.

1653 Lymphatic vessels of the intestines are discovered by Swedish naturalist Olaf Rudbeck (1630-1702). The lymphatics resemble veins and capillaries but carry the clear, watery fluid portion of the blood (called lymph). Lymphatic

atto di caminare. | 7. Testa cō tutte le sue parti
supino | 8. Testa con la lingua ripie
he mostra il fianco | gata verso la gola
no. | 9. Lingua con le sue
ne dell'Ape | 4. linguette, o guaine
o tutto peloso. | che l'abbracciano

10. Aculeo, ouero Spin.
11. Gamba che mostra la
parte interiore.
12. Gamba dalla banda
esteriore.

Francesco Stelluti produced this detailed drawing of a bee in 1625 with the aid of the first microscope.

vessels gather in small knots (called lymph nodes or glands) in certain parts of the body and are useful in developing immunity to disease. Before Rudbeck, this vascular system had been unknown.

1657   First reference to a bronchocele (goiter) in English is made by Tomlinson.

1658   First to observe and describe red blood corpuscles is Dutch naturalist Jan Swammerdam (1637-1680). While studying the blood of a frog, he notices what come to be called erythrocytes, the cells that carry oxygen from the lungs to the tissues and give blood its characteristic color. The mature red blood cell in a mammal is small, flat, and round-shaped with a depressed center. Containing hemoglobin, its function is to carry oxygen from the lungs or gills to all body tissues and to carry carbon dioxide, a waste product of metabolism, back to the lungs, which expel it.

1659   Thomas Wharton (1610-1673) of England first uses the term "thyroid" in his *Adenographia*.

1659   First physiology textbook in English, *Natural History of Nutrition, Life, and Voluntary Motion*, is published by English physician Walter Charleton (1619-1707), a man of many interests who also publishes on physics, theology, natural history, and archaeology. Charleton's works are important for they mark a transition from the old, scholastic way of writing to a newer method of carefully recording observations and drawing conclusions known as the scientific method.

1660   First accurate description of the lung's structure is made by Italian physiologist Marcello Malpighi (1628-1694). As one of the earliest to use a microscope to study living tissue, Malpighi examines a frog's lungs and describes the thin air sacs surrounded by a network of tiny blood vessels. With this, he also discovers the capillaries, those microscopic vessels that connect veins and arteries. This also explains an essential part of the respiration process by which air diffuses from the lungs into these blood vessels to be carried throughout the body.

1662   Coffee is first introduced into France.

1664   René Descartes (1596-1650), French philosopher and mathematician, first puts forth his idea of reflex action. It is included in a French version of his work on animal physiology published several years after his death. He applies his mechanistic philosophy to the analysis of animal behavior and first uses the concept of reflex to mean any involuntary response the body makes when exposed to a stimulus. Descartes offers this notion of reflex action to distinguish between the automatic action of animals and the voluntary behavior of human beings.

(See also 1751)

1665 Robert Hooke (1635-1703), English physicist, improves the design of the compound microscope and uses his own device to observe insects and plants. He publishes his *Micrographia* this year. This highly popular work is also the first major collection of microscopical sketches.

1665 First documented drawing of the cell is made by English physicist Robert Hooke (1635-1703). While observing a sliver of cork under a microscope, Hooke notices it is composed of a pattern of tiny rectangular holes. He names these spaces "cells" because each looks like a small, empty room. Although he does not observe living cells, the name is retained. What Hooke saw were really the dead remnants of structures that, when alive, were filled with fluid. The word and the concept "cell" becomes essential and significant to biology. Hooke's discoveries and drawings are contained in his masterpiece, *Micrographia*, which has some of the most beautiful drawings of microscopic subjects ever made.

1667 John Ray (1627-1705), English botanist, founds systematic biology with the publication of his flora of the British Isles. This work is a catalog of the flowers of Cambridgeshire, *Catalogous Plantarum circa Cantabrigium Nascentium*, and it is the first book of local flora to be published. More importantly, it lays the groundwork for systematic classification and is therefore a precursor of Linnaeus's modern methods. Ray defined species and arranged

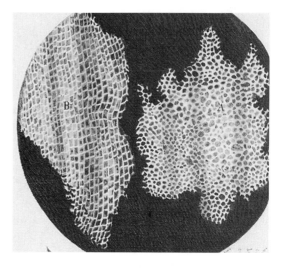

Illustration from Robert Hooke's *Micrographia*, 1665, comparing a slide of cork to honeycomb.

plants by their distinctive structural qualities, using characteristics taken from the root, stem, leaves, flowers, and fruit. During his life, Ray describes 18,600 different plant species.

(See also 1686)

1668 Dutch naturalist Jan Swammerdam (1637-1680) discovers that what had been considered the "bee-king" is really a queen bee. Although beekeeping is a very old art, few accurate details pertinent to the care and management of swarms of honeybees are known until Swammerdam. What had been regarded as the king bee who received obviously special treatment in the hive is shown to be a unique female who, once fertilized, is capable of laying 1,000 or more eggs a day. The queen usually lives about a year and rarely ever leaves the hive.

1669 First published reference to the preformation theory is made by Dutch naturalist Jan Swammerdam (1637-1680). This concept states that the embryo is preformed at conception and therefore only grows and develops during gestation. It also states that the egg contains a miniature individual that develops into the

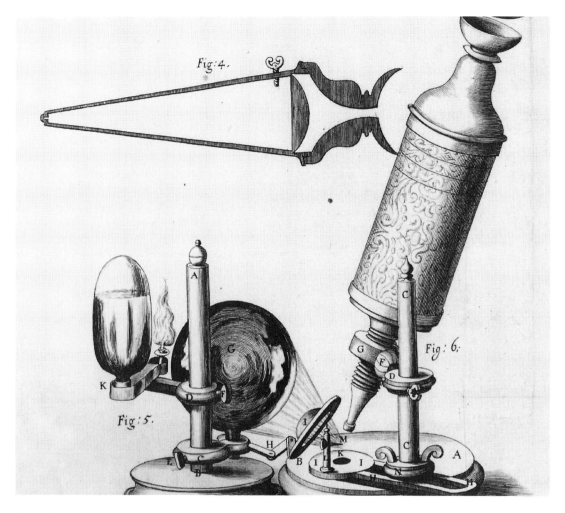

Robert Hooke's compound microscope.

adult stage in the proper environment. This theory is incorrect, since the egg is initially undifferentiated and develops in a series of major steps during which it changes considerably.

1669 Dutch naturalist Jan Swammerdam (1637-1680) begins the first major study of insect microanatomy and classification. During his short life, he collects and studies some 3,000 species of insects. He uses specialized microscopes and builds many different instruments that allow him to dissect insects under magnification. His work is never published during his

lifetime and it goes unrecognized until discovered by Dutch physician Hermann Boerhaave (1668-1738) a half century later.

(See also 1737)

1669 First tea is brought into England.

1669 First published work devoted entirely to the study of an invertebrate is written by Italian physiologist Marcello Malpighi (1628-1694). His study on the metamorphic development and internal structure of the silkworm marks the first, published examination of insect structure.

1669 Richard Lower (1631-1691), English physician, publishes his *Tractatus de Corde* in which he discovers that the brightness of arterial blood is due to its absorption of air. This is a major step toward understanding the nature of respiration. He also shows that the vagus nerve controls the heart.

1673 Marcello Malpighi (1628-1694), Italian physiologist, publishes his *De Ovo Incubato Observationes* and traces the first hours in the development of a hen's egg with his microscope. However, he concludes wrongly for preformation. In this first documented work on the embryology of the chick, Malpighi uses a microscope and discovers the aortic arches, neural folds, and somites (body segments) in the chick embryo. This pioneering study lays the foundation for microscopic embryology. Malpighi may also be regarded as the first histologist since he studied the minute tissue structures of plants and animals. Among his other discoveries are the papillae of the tongue and skin (receptors of the senses of taste and touch). He also clearly described the structure and function of the kidney and showed that bile is secreted in the liver and not the gall bladder.

(See also 1669)

1676 First use of the term "comparative anatomy" is made by Nehemiah Grew (1641-1712), English botanist and physician. In 1681, Grew publishes the first zoological book to contain the phrase "comparative anatomy" in the title. His work, *Comparative Anatomy of Stomach and Guts*, is also the first book to deal with one system of organs using the comparative method. He uses the human digestive system as the standard and compares all other animals to it.

1677 Dutch biologist and microscopist Anton von Leeuwenhoek (1632-1723) is the first to observe and describe spermatozoa. He details his discovery in a letter he publishes in London's *Philosophical Transactions* in 1679. In the same year, Johan Ham also sees them microscopically, but the semen he observes comes from a patient suffering from gonorrhea, and Ham concludes that spermatozoa are a consequence of that disease.

(See also 1683)

1681 Gerardus Blasius (c.1626-c.1692), Dutch anatomist who is also called Blaes, publishes his *Anatome Animalium Terrestrium Variorum*, which is the first general systematic treatise on comparative anatomy. After devoting many years to a comparative investigation of the structure of vertebrate animals, Blasius publishes this work in which he treats each animal in an orderly manner, organ by organ. It demonstrates the many anatomical similarities shared by the larger group of vertebrates, such as birds, rodents, and carnivores.

1683 First to discover one-celled animals now called protozoa is Anton von Leeuwenhoek (1632-1723), Dutch biologist and microscopist. He also discovers different types of infusoria (minute organisms found in decomposing matter and stagnant water) and observes the capillaries in living specimens of tadpoles and fish. In 1680, he discovers yeast cells, and in 1702 he discovers hydra (a genus of small, freshwater hydrozoan polyps). Using a single-lens microscope of his own making, the unschooled drapery shop owner and janitor outdoes the scientists of his time and is today considered by many as the father of microscopy.

1686 John Ray (1628-1705), English naturalist, publishes the first of his three-volume work *Historia Plantarum* (1686-1704). This work first introduces the idea of species, groups of individual plants with similar seeds, into botany and lays the groundwork for Linnaeus's modern form of systematic classification.

(See also 1667)

1689 Richard Morton (1637-1698), English physician, gives the first description of anorexia nervosa in his book *Phthisiologia; or, A Treatise of Consumptions*.

1689   First attempt at a simple and rational classification of plants into families is made by French botanist Pierre Magnol (1638-1715) in his *Prodromus Historiae Generalis Plantarum.*

AUGUST 25, 1694   Rudolph Jakob Camerarius (1665-1721), German botanist, first establishes by experiment the existence of sexual reproduction in flowers. He publishes his results in a letter to a colleague titled *De Sexu Plantarum* in which he establishes the male fertilizing role of pollen. His work is highly important in establishing a sexual theory of plant reproduction.

1706   First laboratory of marine zoology is founded in Marseilles, France.

1716   First written account of plant hybridization is given by the American natural philosopher and writer Cotton Mather (1663-1728). In a letter he describes a case involving red and blue kernels of *Zea mays* corn planted next to rows of regular yellow corn. The yellow rows closest to the *Zea mays* produced the most mixed color varieties.

1728   Jacopo Bartolomeo Beccaria (1682-1766), Italian physician, discovers gluten in wheat flour. This is considered the first protein substance of plant origin to be found.

1730   First botanical garden in America is founded by John Bartram (1699-1777), American botanist. He begins assembling native North American plants at his farm in Kingsessing, near Philadelphia in 1728. He collects and exports seeds and plants to Europe and establishes friendships with the great botanists of Europe. He conducts the first experiments in America on hybridizing flowering plants and is appointed botanist for the American colonies to King George III. Bartram eventually becomes known as the "father of American botany." He makes scientific trips through the Alleghanies, Catskills, Carolinas, and Florida.

Old watering-trough in Bartram's Garden.

His son, William, also makes major botanical contributions.

(See also 1791)

1730   Stephen Hales (1677-1761), English botanist and chemist, discovers that the reflex movements of a frog's legs depends upon the spinal cord. This is a major step toward an appreciation of the nervous system of vertebrates. Hales also experiments with hydrogen, carbon dioxide, carbon monoxide, and methane and is the first known person to collect different gases over water.

1734   René Antoine Ferchault de Reaumur (1683-1757), French physicist, publishes the first of his massive, six-volume study on insects (1734-1742). In this work containing 5,000 illustrations, he discusses the effect of heat on the development of insects and their larvae, and has much to say about leaf-boring and gall-forming insects. Throughout this work he emphasizes development, especially metamorphosis. He also emphasizes the practical aspects of beekeeping and designs the first egg incubator. Titled *Memoires pour Servir a L'Histoire des Insectes*, this work is of fundamental importance to insect biology and offers valuable contributions to the knowledge of the anatomical structure of insects.

1735   First to frame principles for defining the genera and species of organisms and to create a uniform system for naming them is Swedish botanist Carolus Linnaeus (1707-1778), also known as Carl von Linne. This year he publishes his *Systema Naturae* which offers a methodical and hierarchical classification of all living things and develops the binomial nomenclature that will be used in classifying plants and animals. In this system each type of living thing is given first a latinized generic name (for the genus or group to which it belongs), and then a specific (species) latinized name for itself. As the founder of modern taxonomy (classification), he produces a system that really works and offers a clear, concise, and natural method of describing exactly how one species differs from another. Eventually refined and improved, his work forms the basis of how science classifies organisms today.

1737   Entomology is founded with the posthumous publication of the work of Dutch naturalist Jan Swammerdam (1637-1680). Swammerdam's pioneering research on insects is discovered by Dutch physician Hermann Boerhaave (1668-1738) and published at his own expense as *Bybel der Natuure* 100 years after Swammerdam's birth. This major work contains detailed descriptions and illustrations of the life cycles of insects. It also demonstrates that they do not originate spontaneously but have reproductive systems as well as an internal anatomy.

(See also 1669)

1740   Regenerating powers of hydra are discovered by Swiss naturalist Abraham Trembley (1710-1784). He cuts this freshwater hydrozoan into several pieces and observes a complete individual regenerate from each part. His demonstration that the hydra is an animal proves the existence of reproduction by budding in the animal kingdom. He also is the first witness to

Published 100 years after his birth, Jan Swammerdam's *Bybel der Natuure* (Bible of Nature) contains remarkable details of insect anatomy, such as this gnat.

the multiplication of protozoans by division as well as cell division in algae.

1746   Jean Paul de Rome d'Ardene (1689-1769) of France publishes *Traité des Ranoncules,* which becomes the first work entirely about ranunculuses or plants of the crowfoot family. The buttercup is a member of this plant family.

1751   Robert Whytt (1714-1766), Scottish physiologist, publishes his book *The Vital and*

Other Involuntary Motions of Animals, in which he gives the first clear description of what is later called reflex action. Whytt demonstrates that such reflex movements as sneezing or the regulation of the eye's iris do not involve the brain. He uses a frog to show that in such cases, a sensory impulse fires off a motor impulse as soon as it reaches the spinal cord.

(See also 1664)

1757   Albrecht von Haller (1708-1777), Swiss physiologist, publishes the first volume of his eight-volume *Elementa Physiologiae Corporis Humani* (1757-1766) and distinguishes irritability or inherent muscular force from nerve or nervous force. In Haller's original notion of irritability, certain tissues have the power of responding to stimuli. In a series of 567 experiments, 190 of which he performs on himself, he establishes that pain is dependent on the nerves. This is the first documented time that a specific function is shown to be related to a specific tissue. His encyclopedic work marks the beginnings of modern physiology.

c. 1763   Joseph Gottlieb Kolreuter (1733-1806), German botanist, publishes his pioneering work on plant hybridization and gives the first account of his experiments on artificial fertilization of plants. He also investigates pollination and the production of hybrid plants and recognizes the relationship between insects and cross-pollination. He discovers the role that the wind plays in pollinating plants and states that in cases like the mistletoe where the wind cannot play a role, the birds and insects are the pollinating agents. He eventually proves that it is birds that effect dispersal for mistletoe plants.

1765   First drawing of cell division is made by Abraham Trembley (1710-1784), Swiss naturalist, who witnesses multiplication of protozoans by division and cell division in algae. In a letter to a colleague, Trembley illustrates the diatom *Synedra* dividing. Although he does not know that *Synedra* could be regarded as a single cell, his discovery of their mode of reproduction

(and consequently that of all single-celled creatures called protozoans) provides the foundation for understanding protozoans.

1768   Adam Kuhn (1741-1817), native of Germantown, Pennsylvania, and a student of the Swedish botanist Carolus Linnaeus (1707-1778), becomes the first professor of botany and materia medica in the American colonies at the College of Philadelphia.

1768   Captain James Cook (1728-1779), English navigator, makes the first of his three famous voyages into the Pacific. The first of the really scientific navigators, he sails under the auspices of the Royal Society.

1768   The first conservatory for plants in the Americas is constructed by American botanist Humphrey Marshall (1722-1801). He later builds at his home in Marshalltown, Pennsylvania, a hothouse and a botanical garden containing both foreign and domestic plants.

(See also 1785)

1771   Italian anatomist Luigi Galvani (1737-1798) discovers the electric nature of the nervous impulse, thereby establishing the science of electrophysiology. In a series of experiments in which he dissects frog legs, he finds that when he hangs up a frog's leg by the nerve from a brass hook with the toe touching a silver plate, the leg kicks convulsively. The leg would also kick during a thunderstorm. Galvani decides correctly that electricity is involved with the nerves, but he mistakenly attributes the muscle as the source of the electricity. He never learns that it is the two dissimilar metals that are producing the electric current.

1772   Joseph Priestley (1733-1804), English chemist, discovers that plants give off oxygen. As a proponent of the phlogiston theory, he believed that green plants restore used-up air to its original freshness by releasing oxygen into the air.

1773   John Walsh (1725-1795), English physicist, gives the first correct description of the torpedo fish and its electric organs. This cartilaginous fish has a flattened, disk-like body and a pair of electric organs near its head to shock its prey. Walsh gives his description in a letter to American statesman and scientist Benjamin Franklin (1706-1790) explaining his experiments with this unique fish. This is the first time its real, electrical nature is accepted as fact.

1774 Antoine Lavoisier (1743-1794), French chemist, discovers that oxygen is consumed by respiration. The discovery of oxygen leads him to investigate how plants and animals breathe. In his precise and quantitative experiments, he shuts up animals and weighs how much air they breathe. In this way he is able to demonstrate that oxygen is absorbed in the lungs and carbon dioxide is given off in exchange.

1776 Matthew Dobson (1732-1784), English physician, publishes his paper, "Experiments and Observations on the Urine in Diabetes," and first proves that the sweetness of diabetics' urine is caused by the presence of sugar.

1776 Marsh gas (methane) is discovered in the reeds beds of Lake Maggiore, Italy, by Italian physicist Alessandro Giuseppe Antonio Anastasio Volta (1745-1827). This colorless, odorless gas is found abundantly in nature as a product of the anaerobic bacterial decomposition of vegetable matter under water (and thus the name swamp gas or marsh gas). It burns readily in air and can be explosive.

1780 Fire or pear blight is first noted in New York by William Denning. This plant disease is caused by a bacterium which destroys pear and apple orchards in North America. Symptoms include a sudden, brown to black withering and dying of blossoms, leaves, and branches, giving an appearance of being scorched by fire.

1780 George Adams (1750-1795), English engineer, devises the first microtome. This mechanical instrument cuts thin slices of a sample for examination under a microscope and re-

places the imperfect method of using a hand-held razor.

1783 Chinch bug is first noted as a pest of wheat in the United States in North Carolina. Belonging to the insect family Lygaeidae, this pest is originally native to tropical America but eventually migrated to most of North America. These very tiny bugs suck a plant's sap.

1785 First American book devoted exclusively to botany is written by Humphrey Marshall (1722-1801), American botanist. His *Arbustum Americanum: The American Grove*, is published in Philadelphia. This work contains a list of trees and shrubs native to America.

(See also 1768)

1787 First special journal devoted to botanical science, the *Magazin für die Botanik*, is founded in Zurich, Switzerland. It becomes the *Annalen der Botanik* in 1791 and is soon transferred to Leipzig, Germany. It plays a considerable role in the development of botany, publishing original research papers and communications as well as abstracts and reviews of books. It is the forerunner of the many important botanical journals that follow in the next century.

1787 Robert Squibb writes the first separate book on horticulture in North America. His book, *The Gardener's Calendar for South Carolina and North Carolina*, is published in Charleston, South Carolina. Unlike botany, which is the science of plants, horticulture is a branch of agriculture dealing with garden crops, fruits, vegetables, and ornamentals.

1789 First measurements of human metabolic rate are made by the French chemists Armand Seguin (1765-1835) and Antoine Laurent Lavoisier (1743-1794). Lavoisier uses his young assistant, Seguin, as the experiment subject. Metabolism encompasses all the chemical changes that take place in the body's tissues when the cells are producing both energy and essential new organic materials. Lavoisier extends his

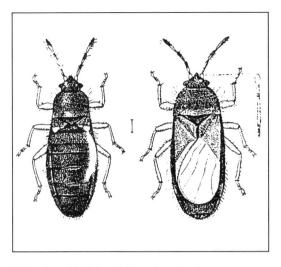

A wingless chinch bug (left) and a winged form.

experiments on combustion to biology since almost all the energy used by the body is eventually converted to heat. Although these experiments are crude, they are the first attempts at measuring the heat loss an individual undergoes.

1791 First significant American-born naturalist, William Bartram (1739-1823), publishes his *Travels through North & South Carolina, Georgia, East & West Florida*. As the son of the "father of American botany," John Bartram (1699-1777), he accompanies his father on a scientific journey through Florida in 1765-1766. His book is filled with generalized botanical, zoological, and anthropological descriptions and becomes especially popular in England, influencing English Romanticism. It also is the product of a true naturalist and a man who some call the first American ecologist.

1794 First paper on systematic zoology published in America is written by American naturalist William Dandridge Peck (1763-1822). It is titled "Descriptions of Four Remarkable Fishes Taken Near the Pisquataqua in New Hampshire." Peck is also considered the first teacher of entomology in the United States and is one

of the founders of the American Antiquarian Society.

(See also 1805)

1797 Georges Cuvier (1769-1832), French anatomist, founds comparative zoology. With the publication of his first book, *Tableau Elémentaire de l'Histoire Naturelle des Animaux*, Cuvier introduces his basic ideas and principles on how the functions and habits of an animal determine its form (and not the reverse). He also applies his views on anatomy to classification and extends and perfects Linnaeus' system by grouping related classes into even broader groups called phyla. He later divides the animal kingdom into four phyla (Vertebrata, Mollusca, Articulata, and Radiata), while always stressing the internal structure of animals rather than their superficial appearance.

(See also 1817)

1801 Cause of astigmatism is discovered by English physicist and physician Thomas Young (1773-1829). He studies physiological optics and learns that the lens of the eye changes shape in focusing on objects at different distances. This is called accommodation. He details the reason for astigmatism as arising from the irregularities of the curvature of the cornea.

1802 The word "biology," meaning the whole science of living things, is coined separately by French naturalist Jean Baptiste Pierre Antoine de Monet de Lamarck (1744-1829) and German naturalist Gottfried Reinhold Treviranus (1776-1837). Many, however, ascribe Treviranus priority since he published his "Biology, the Philosophy of Living Nature," in which he describes it as a new science which distinguishes between living and non-living material.

1803 André Michaux (1746-1802), French botanist, publishes the first book of American flora that is national in scope. His *Flora Boreali-Americana* is published posthumously in Paris and has many plates drawn by French artist

Illustration of *Isanthus coeruleus*, from the 1803 edition of *Flora Boreali-Americana*.

Pierre Joseph Redoute (1759-1840). After collecting plants in Persia for the French government, he is sent to North America where he collects and studies plants for twelve years. His mission includes sending back plants to be replanted in the depleted forest of Rambouillet, southwest of Paris. Most of the specimens he sends to France come from what are now the southeastern and midwestern states.

1804   Soybeans are first brought to the United States as ballast in ships.

1804   First bird-banding experiments in North America are conducted by the youthful French-American naturalist and artist John James Audubon (1785-1851). With these experiments, Audubon begins his serious study of American birds and initiates the study of bird migrations.

(See also 1827)

1805   William Dandridge Peck (1763-1822), American naturalist, is elected the first professor of natural history at Harvard University. A graduate of Harvard class of 1785, he had spent three years in Europe studying at various scientific centers and returned to America with many specimens and a large scientific library. Skilled at pen and ink drawings, he is reputed to be the most knowledgeable man in the new nation in the field of zoology.

(See also 1794)

1807   Harvard Botanic Garden is established in Cambridge, Massachusetts, by American naturalist William Dandridge Peck (1763-1822).

(See also 1794)

1808   First important American work on horticulture is *American Gardener's Calendar*, written by American nurseryman Bernard M'Mahon (c.1775-1816) and published in Philadelphia. Since 1802 he had been issuing one-page broadsides of nursery items for sale.

1809   Modern invertebrate zoology is founded by Jean Baptiste Pierre Antoine de Monet de Lamarck (1744-1829), French naturalist, who first introduces the term "invertebrates" in his *Philosophie Zoologique*. Until Lamarck, no one had given attention to this huge group of animals, and they were generally categorized by the single group simply called "worms." Lamarck eventually categorizes and classifies them, giving order to this incredibly diverse group. He also classifies organisms by function rather than by structure and is the first to employ the genealogical tree. This work also contains the first scientific theory of evolution, and although Lamarck is not correct in describing how it

Watercolor of a golden eagle, by artist J. J. Audubon.

occurs, he is the first to offer a scheme that rationalizes the evolutionary development of life. He also maintains that species are not fixed but do change and develop.

1811 Julien Jean-César Legallois (1770-1814), French physiologist, locates the first physiological center in the brain. The ancient Greeks knew that severing the spinal cord close to the brain resulted in a stoppage of breathing, and Legallois eventually locates the "respiration center" in the brain by experimenting with the brain stem and cutting it in several different

places. He demonstrates that it is the medulla oblongata that controls respiration. This indicates that the medulla is a major central nervous system area in the control of respiration.

1811 Iodine is discovered in water used to extract soluble material from the ashes of seaweed by French chemist and industrialist Bernard Courtois (1777-1838). As an essential microconstituent of the human body, it is necessary for maintaining normal metabolism. The ingestion of burnt seaweed was recommended as a treatment for goiter by the Spanish alchemist,

Arnold of Villanova (c.1235-1311) around 1300, and it is known to have been used by the Chinese as early as 1600 B.C.

MARCH 17, 1812 Academy of Natural Sciences is founded in Philadelphia. It eventually receives the nation's oldest plant collections and goes on to assemble one of the foremost botanical libraries in the United States.

1814 First American flora, or plant book, that is continental in scope and includes species from the Pacific Northwest is *Flora Americae Septentrionalis* by German-American naturalist Frederick Traugott Pursh (1774-1820). It is published in London. It is thought that Pursh was able to include botanical information about the American northwest in his book because he somehow obtained notes and specimens from the Lewis and Clark expedition.

1815 First bibliography of entomology is compiled by English naturalist William Elford Leach (1790-1836) and published in the *Edinburgh Encyclopaedia*.

1815 Michel Eugene Chevreul (1786-1889), French chemist, first proves that the sugar in diabetes is glucose.

1817 Georges Cuvier (1769-1832), French anatomist, publishes his major work *Le Règne Animal* which extends and perfects the classification system of Linnaeus by grouping related classes into broader groups called phyla. He is also the first to extend this system of classification to fossils.

(See also 1797)

1818 Marie François Xavier Bichat (1771-1802), French physician, publishes his major work, *Traité des Membranes en General*, in which he first propounds the notion of tissues. Although he does not use a microscope, he is able to show that each organ in the body is composed of different types of "tissues." He chooses this word to describe them because they are generally flat with delicate, thin layers. He is also the

first to draw both the anatomist's and the physiologist's attention to the fact that the body is a complex made up of simple structures. This work also establishes histology by distinguishing 21 kinds of tissue and relating disease to them.

1818 New York Horticultural Society is organized, the first of its kind in America. It is incorporated in 1822 with American physician David Hosack (1769-1835) as its first president. Hosack feels strongly that the society should be devoted to "practical improvements in the culture of plants," and argues that this demands the creation of a botanic garden. He also is instrumental in forming the first botanic garden in New York (on ground now occupied by Rockefeller Center).

1820 Jean François Coindet (1774-1834), Swiss physician, first uses iodine internally in the treatment of goiter and thyroid disorders.

1820 First United States *Pharmacopoeia* is published with the purpose "to select from among substances which possess medical power, those the utility of which is most fully established and best understood; and to form from them preparations and compositions, in which their powers may be exerted to the greatest advantage." This first edition contains almost all of the plants thought by the Native Americans and early settlers to have curative powers.

1821 First international congress of biology is organized by German biologist and physiologist Lorenz Oken (1779-1851), also called Ockenfuss. Oken advocates the annual meeting of scientists to share and publicize their views and begins a tradition that continues today in nearly all fields. Since this first meeting is in Germany, due to the political climate, it must meet in secret because any meetings are immediately considered to be political in character.

1821 Jean Louis Prévost (1790-1850), Swiss physician, jointly publishes a paper with French chemist Jean Baptiste André Dumas (1800-

1884) which demonstrates for the first time that spermatozoa originate in tissues of the male sex glands. With this, they show that it is the sperm and not the seminal fluid that is essential to reproduction. In 1824 they also give the first detailed account of the segmentation of a frog's egg although they are not able to account for this cleavage.

1823   First to recognize fingerprints as a means of identification is Czech physiologist Jan Evangelista Purkinje (1787-1869). Noting that fingerprints seem to follow nine general patterns and that the human palm has distinctive ridge patterns, he is the first to classify fingerprints according to type and to demonstrate their usefulness.

1824   First laboratory for the study of plant physiology is established by Czech physiologist Jan Evangelista Purkinje (1787-1869) at his home.

1824   First to demonstrate the general functions of the brain in vertebrates is French physiologist Marie Jean Pierre Flourens (1794-1867). In a series of experiments on pigeons, he concludes that the hemispheres are responsible for higher psychic and intellectual abilities, the cerebellum regulates all movements, and that the medulla controls vital functions, especially respiration. A pioneer of the idea of nervous coordination, he is also the first to recognize the role of the semicircular canals of the inner ear in maintaining body equilibrium and coordination.

1824   Karl Gustav Carus (1789-1869), German physiologist and psychologist, first sees polar bodies which are one of the minute cells which separate from the egg during its maturation. When a single egg cell is ready to mature, it undergoes cell division and its nucleus splits so that half of its chromosomes go to one cell and half to another. The larger of these two is called the secondary ovum, and it is the one that reaches maturation and moves into the fallopian

tubes to be fertilized. The other, smaller cell is the polar body which is not fertilized.

1825   Jan Evangelista Purkinje (1787-1869), Czech physiologist, first names and describes the "germinal vesicle" or nucleus in a hen's egg. Unable to obtain a microscope, he uses a hand lens and sees what countless earlier investigators had overlooked—the nucleus in the developing hen's egg. When eventually he obtains a microscope, he asserts that there are three fundamental structures in living organisms—fibers, fluids, and cells. In this, he anticipates the cell theory of Schleiden and Schwann.

(See also 1838)

1825   Caleb Hillier Parry (1755-1822), English physician, has his observation and first correct description of exophthalmic goiter published posthumously this year. It later becomes known as "Parry's disease."

1827   John James Audubon (1785-1851), French-American naturalist and artist, publishes the first volume of his *The Birds of America* (1827-38). Audubon's youthful goal is to paint all the known species of birds of North America, and his illustrated, complete work is considered among the most beautiful natural history studies ever done. Totaling four volumes and 435 hand-colored plates, his masterpiece is the first modern atlas of ornithology. Although Audubon usually painted from dead specimens, he is regarded as one of the first American conservationists. Today, many conservationists in the United States are members of a society named in his honor.

1827   Discovery of the mammalian ovum (egg) is first announced by German-Russian embryologist Karl Ernst von Baer (1792-1876) in his *De Ovi Mammalium et Hominis Genesi*. With his absolute identification and proof that mammals have eggs, he shows conclusively that mammalian development, including human development, is not fundamentally different from that of other animals. Baer thus establishes a basic biological concept and proves successful at a quest that had eluded the most brilliant and determined scientists before him.

1828   First to observe and describe what comes to be called "Brownian movement" is Scottish naturalist Robert Brown (1773-1858). In his studies of minuscule particles suspended in water he always witnesses the irregular movement of the particles. He is unable to account for this phenomenon, and it is not until a generation later that it is explained by the kinetic theory of gases. Brownian motion eventually becomes important to Einstein, for it is the first observable evidence of atomism—that water is composed of particles.

1828   Friedrich Wohler (1800-1882), German chemist, first synthesizes urea and lays the foundations (along with Berzelius, Liebig, and Bunsen) of organic chemistry and then later of biochemistry. His preparation of an organic compound from inorganic materials dispels the notion that organic compounds can only be produced by using living organisms.

1828   First American book dealing exclusively with flowers, Robin Green's *Treatise on the Cultivation of Flowers,* is published in Boston.

1828   First text on comparative embryology is the landmark work of German-Russian embryologist Karl Ernst von Baer (1792-1876), titled *Über die Entwickelungsgeschichte der Thiere.* In this work, which laid the foundation of comparative embryology, he offers his germ-layer theory, the law of corresponding stages in the development of embryos, and the discovery of the notochord. Baer's doctrine of germ layers states that the developing egg forms several layers of tissue out of which various specialized organs develop, with one given set of organs from one given germ layer. His law of stages says that in the early stages of development, vertebrate embryos are quite similar, even between animals that might be very different from each other in their mature stage. His

discovery of the notochord (the stiff rod running the length of the back that vertebrates have only early in their development but primitive creatures retain) indicates the relationship between vertebrate embryos and primitive prevertebrates. Baer's theory will provide evolutionists with significant ammunition.

1828   Luigi Rolando (1773-1831), Italian anatomist, achieves the first electrical stimulation of the brain. He does this in the context of experimenting with the brain's different centers to determine what region has control over what effect. All he is able to conclude, however, is that the cerebellum is a type of power station which he believes secretes electric fluid into the nervous system.

c. 1830   First herpetarium in Europe is founded by French zoologist André Marie Constant Dumeril (1774-1860) in France. He pioneers research on reptiles and amphibians and is considered the father of herpetology.

1830   *Proceedings of the Zoological Society of London* are first published.

1830   Boston Society of Natural History is founded with Thomas Nuttall (1786-1859), English-American naturalist, as its first president. The Society incorporates the C. J. Sprague and Thomas Taylor lichens and the Francis Boott collections. This and other similar organizations in other American cities were to play an important role in the development of natural history, mainly through their meetings.

1831   Scottish naturalist Robert Brown (1773-1858) is the first to name the cell's "nucleus" and to recognize it as a general and regular feature of all plant cells. Brown describes this "single circular areola" that is more opaque than the cell membrane by using the Latin word for "little nut" or "nucleus." With this breakthrough announcement, new discoveries begin to emerge quickly.

(See also 1838)

1833 French chemist Anselme Payen (1795-1871) discovers the first enzyme to be obtained in concentrated form. He separates a substance from barley which he finds has the property of hastening the conversion of starch into sugar which he calls diastase. This proves to be the first of many organic catalysts within living tissue (which eventually come to be called enzymes). Enzymes prove to be a large group of proteins that make possible the chemical reactions constituting metabolism.

1833 *Dispensary* is first published in United States. It concentrates on the natural products that are used as pharmaceuticals and contains information on drugs that are not yet official. As later editions appear, the natural drugs are increasingly replaced by chemically prepared compounds.

1833 Johannes Peter Müller (1801-1858), German physiologist, first offers his "law of specific nerve energies," which states that every sensory nerve gives rise to one form of sensation even if excited by abnormal stimuli. This means that the sensation produced by a stimulus depends on the receiving or end-organ and not on the nature of the stimulus. Thus the optic nerve when stimulated will produce a flash of light whether light is involved as the stimulus or not. Whether the stimulus is mechanical, electrical, or chemical, specific sensory nerves can interpret it only one way—eyes see light and ears hear noise.

1834 Cellulose is first obtained by French chemist, Anselme Payen (1795-1871). In his extensive analysis of wood and its components, he separates a substance from wood that has the same general properties as starch. He names it "cellulose" since he finds it in the cell walls. Cellulose proves to be the major component of the cell walls of all plants, and it strengthens the roots, leaves, and stems and makes them rigid. It is a polysaccharide similar to sugar and starch and proves to have enormous economic importance, being processed to produce paper and

chemically modified to yield other substances used in producing plastics and other synthetic products.

1835 Gabriel Gustav Valentin (1810-1883), German physiologist, publishes the first textbook on human development—*Manual of the Development of the Foetus*. He also discovers the tiny spot on the cell nucleus and names it "nucleolus" which is a diminutive form of the word nucleus. He further points out a significant analogy between plant cells and the skin cells of animals.

1836 First enzyme (pepsin) prepared from animal tissue is isolated by German physiologist Theodor Schwann (1810-1882). While investigating digestive processes, he shows that hydrochloric acid alone is not responsible for digestion, and by treating the stomach extracts with mercuric chloride, he isolates its active principle which he names "pepsin" from the Greek word "to digest." Its discovery marks a major turning point in the early history of biochemistry.

1837 Jan Evangelista Purkinje (1787-1869), Czech physiologist, first describes a large group of specific nerve cells with many branching extensions in the cerebellum, now known as "Purkinje cells."

1837 René Joachim Henri Dutrochet (1776-1847), French physiologist, publishes his research on plant physiology in which he is the first to conduct a systematic investigation of osmosis. He recognizes the importance of this phenomenon which is the spontaneous diffusion or passage that occurs through a semipermeable membrane separating two solutions and which tends to equalize their concentrations. He is also the first to recognize the importance of green pigment (chlorophyll) in the use of carbon dioxide by plants.

(See also 1851)

1837 Robert Remak (1815-1865), Polish-German physician, notes the true relationship of nerve cells to nerve fibers and first names the neurolemma (the myelin sheath around many nerve fibers). He also offers names that are still used for the three germinal layers in the developing embryo: "ectoderm" (outer skin which gives rise to skin and the nervous system), "mesoderm" (middle skin which produces muscles, the skeleton, and the excretory system), and "endoderm" (inner skin which differentiates to form the notochord, digestive system, and related glands.).

1838   First to formulate the cell theory for plants is German botanist Matthias Jakob Schleiden (1804-1881). In an article published this year, he argues that the different parts of a plant are composed of cells or derivatives of cells. He describes plants as a community of cells. However, Schleiden gets only halfway to modern cell theory since he incorrectly states that new cells bud out of the cell's nucleus which then disappears. Probably his main contribution is his communication of his cell theory for plants to his colleague, German physiologist Theodor Ambrose Hubert Schwann (1810-1882).

(See also 1839)

1838   First world expedition sponsored by the U.S. government sails. Lt. Charles Wilkes (1798-1877) leads a fleet of six naval vessels staffed with nine "scientific gentlemen." The fleet returns in June 1842, having criss-crossed the Pacific, mapped whaling grounds, and collected 9,674 species of plants and some animals.

1839   First known to use the term "protoplasm" is Czech physiologist Jan Evangelista Purkinje (1787-1869) who coins the word to describe living embryonic material. Taken from the Greek word for "first formed," it becomes accepted eventually as meaning the living material within the cell.

1839   First independent department of physiology is created by Czech physiologist Jan Evangelista Purkinje (1787-1869) at Breslau,

Prussia (now Wroclaw, Poland). Because of this, he is considered to be the founder of laboratory training in connection with university teaching.

(See also 1842)

1839   First full extension of the cell theory to animals is made by German physiologist Theodor Ambrose Hubert Schwann (1810-1882) in his book, *Mikroskopische Untersuchung*. When Schleiden tells Schwann of his cell theory of plants, Schwann immediately recognizes the similarity between Schleiden's plant-cell nuclei and the structures he had seen in animal nerve tissue. After some experimention, Schwann offers one of the greatest generalizations in the history of biology—that plants and animals are composed of units that are basically identical. Thus Schwann is the first to demonstrate the universality of the cell as a unit of life.

1840   First textbook of nervous diseases is written by Moritz Heinrich Romberg (1795-1873) in Germany. In 1853 it appears in English in two volumes as *A Manual of the Nervous Diseases of Man*.

(See also 1838)

1841   Carl August Trommer (1806-1879) of Germany first publishes his test for sugar in the urine.

1841   First systematic textbook in histology is written by Friedrich Gustav Jakob Henle (1809-1885), German pathologist and anatomist. His treatise, *Allgemeine Anatomie*, deals with the minute structure of animal and vegetable tissues. This book also includes the first modern statement of the germ theory of communicable diseases. Although he is unable to prove it, Henle states that "the material of contagions is not only an organic but a living one." He properly describes these germs as parasitic organisms.

1842   Charles Robert Darwin (1809-1882), English naturalist, writes the first abstract of what will become his theory of natural selection. Between this year and 1838 when he read the *Essay on Population* by English economist Thomas Robert Malthus (1766-1834), Darwin arrives at his principle of natural selection and first writes it down in a "Sketch" of his ideas.

1842   Theodor Ludwig Wilhelm Bischoff (1807-1882), German embryologist, publishes a textbook of comparative embryology titled *Entwicklungs-geschichte der Saugerthiere und des Menschen*. The systematic organization of this subject will make it especially valuable to the coming notion of evolution.

1842   First official physiological laboratory, the Physiological Institute, is founded by Czech physiologist Jan Evangelista Purkinje (1787-1868) at Breslau, Prussia (now Wroclaw, Poland). Such laboratories eventually become a standard branch of all medical schools.

1843   William Bowman (1816-1892), English anatomist, publishes the first of two volumes titled *The Physiological Anatomy and Physiology of Man*. This is the first work on physiology in which histology (the study of the minute structure of tissues) plays a major role.

1843   Claude Bernard (1813-1878), French physiologist, publishes his first physiological paper. Besides his many discoveries, he contributes a major concept called "milieu interior." This states that the complex functions of the body's many organs are intimately related and are all geared toward maintaining some kind of internal equilibrium. This is the first statement of the concept of homeostasis—that all living things maintain a constant internal environment through a controlled response to change. Although Bernard conceives of this idea as the struggle of an individual organism to survive, the concept is later extended to include any and all biological systems, including the entire biosphere.

1845   German zoologist Karl Theodor Ernst

von Siebold (1804-1885) is the first to recognize and describe protozoa as single-celled organisms. From now on, the new cell theory would have to state that organisms consist of one or more cells rather than simply "cells." He is also the first to study the one-celled cilia and to show that they maneuver about by a whipping motion from their hairlike projections.

1846   First book on pathology to link the study of diseased tissue with the physiology of normal tissue is written by German pathologist and anatomist Friedrich Gustav Jakob Henle. His two-volume work, *Handbuch der Rationellen Pathologie,* successfully describes diseased organs in relation to their normal physiological functions and signals the advent of modern pathology.

1848   Karl Theodor Ernst von Siebold (1804-1885), German zoologist, publishes his *Lehrbuch der Verwandtschafts-Verhaltnisse der Wirbellosen Thiere*, which becomes one of the first major texts on the comparative anatomy of invertebrates.

1848   Emil Heinrich Du Bois-Reymond (1818-1896), German physiologist, publishes his *Investigations into Animal Electricity* in which he first shows that the nerve impulse is accompanied by a change in the electrical condition of the nerve and has a measurable velocity. After inventing a muscle-measurer known as a myograph, he is able to establish firmly the electric nature of the nervous impulse. As a pioneer in the study of the electrical properties of animal tissues, Du Bois-Reymond is the founder of scientific electrophysiology.

1848   Antheridia (male reproductive organs) of ferns are discovered by Polish Count J. Leszczyc-Suminski. Since ancient times it had been thought that the spore dust of ferns were its "seeds," and with the invention of the microscope, spores were seen germinating a "proembryo" from which a new fern would grow. The amazing part of the Count's discovery is he was able to locate the sexual organs on a "proembryo" and not on the mature fern plant.

1849 Karl Friedrich von Gaertner (1772-1850) of Germany publishes the first comprehensive treatment of hybridization. His two works published this year are the most thorough and complete account of experimental investigation into the sexuality of plants written. He carefully details over 9,000 experiments and his work will be closely studied by Darwin and Mendel.

1851 From his studies on algae, German botanist Hugo von Mohl (1805-1872) is the first scientist to propose that new cells are created by cell division and publishes his findings in *Grundzuge der Anatomie und Physiologie der Vegetabilischen Zelle*. He is also the first to provide a clear explanation of the role of osmosis (the passage of a substance through a membrane from a region of higher concentration to a lower one) in the physiology of plants.

(See also 1837)

1852 Karl Friedrich Wilhelm Ludwig (1816-1895), German physiologist, begins studies of the heartbeat and muscular activity with his invention called the kymograph. With this rotating drum he is able to record graphically the variations in arterial blood pressure. More importantly, Ludwig's device gives a simple explanation for how blood actually moves throughout the body. Until now, it was explained by reference to some "vital force." Ludwig shows that the circulation of the blood can be explained in ordinary mechanical terms. In 1856, he becomes the first to keep animal organs alive after they are removed from the body. Using his knowledge of blood pressure and pumps, he sustains the organs by pumping blood through them (called perfusion).

1853 Gustave Adolphe Thuret (1817-1875), French botanist, gives the first account of fertilization by studying brown algae. Until now, no one had actually been able to observe the act of fertilization, but by using algae, Thuret is able to see it directly without having to disruptively intervene in any way. After collecting the large egg cells and swarming spermatozoids separately, Thuret brings them together experimentally and not only observes the act of fertilization, but is able to produce hybrids at will.

1854 Rudolph Carl Virchow (1821-1902), German pathologist, first names the neuroglia or supportive "glue cells" in the brain. As a professor of pathological anatomy, Virchow conducts microscopic study of tissues and discovers these cells located around the nerve cells in vertebrate nervous systems. This work is a prelude to his founding of cellular pathology.

(See also 1855)

1854 George Newport (1803-1854), English naturalist and anatomist, performs the first experiment on animal embryos. Studying frogs, he observes spermatozoa penetrate the gelatinous envelopes and become embedded in the internal wall. He notes that the point of sperm entry determines the planes of segmentation of the egg. This is a significant advance toward a knowledge of the sexual process in animals.

1855 Bartolomeo Panizza (1785-1867), Italian anatomist, first proves that parts of the cerebral cortex of the brain are essential for vision.

1855 Asa Fitch (1809-1879), American entomologist, issues the first of his famous fourteen reports (1855-1870) categorizing the insects of the state of New York.

1855 Modern cell theory is first stated by German pathologist Rudolph Virchow (1821-1902) as "Omnis cellula e cellula," meaning every cell comes from a cell. From this doctrine he proceeds to the generalization that all disease is ultimately disease of the cells. In his book *Die Cellularpathologie*, published in 1858, he demonstrates conclusively that the cell theory extends to diseased tissue and that cells of diseased tissue are descended from normal cells of ordinary tissue. This discovery modernizes biology and the entire medical field, as the cell is regarded correctly as the fundamental living unit in both healthy and diseased tissue. Although he states rightly that disease occurs because healthy living cells are altered or disturbed, he does not accept Pasteur's germ theory of disease. In fact, both are correct since disease can result from an invasion from outside (Pasteur) as well as from a breakdown of order within (Virchow).

1856 *Journal of the Linnaean Society* is first published.

1856 Nathanael Pringsheim (1823-1894), Silesian biologist, is the first to see the sperm enter the female egg (of a freshwater algae plant). Under proper conditions, he observes the wall of the female cell become thin at one point and finally open. He then sees the spermatozoids cluster around this opening and enter the female cell, which, immediately upon fertilization, forms a delicate membrane around itself to prevent any other sperms from penetrating. "Sex," says Pringsheim, "is a universal property of all organisms manifesting a wonderful analogy in the most highly organized animals, as well as in the simplest cellular plants."

1859 First zoological station is founded at Concarneau, France.

1860 Julius von Sachs (1832-1897), German botanist, first shows that plants can be grown from seed in water only if they are given the necessary chemical nutrients. He also works out the process of plant transpiration, in which water travels from its roots, up the stem, and eventually out from the leaves as vapor.

1860 Modern conception of the cell is established by Max Johann Sigismund Schultze (1825-1874), German anatomist, who first clearly points out the nature of protoplasm, showing that it is approximately the same for all life forms. Schultze defines the cell as a mass of protoplasm with a nucleus and demonstrates

that protoplasm with its nucleus is a fundamental substance found in both plants and animals.

1861 Rudolph Albert von Kolliker (1817-1905), Swiss anatomist and physiologist, publishes his *Entwicklungsgeschichte des Menschen und der Hoheren Thiere*, which is the first treatise to interpret the developing embryo in terms of cell theory. Considered one of the founders of modern embryology, he is very much ahead of his time because he considers the cell nucleus as the key to the transmission of hereditary characteristics.

1861 George Engelmann (1809-1884), German-American physician and botanist, publishes the first paper on plant pathology in the United States.

1862 Karl von Voit (1831-1908), German physiologist, and his colleague, German chemist Max Joseph von Pettenkofer (1818-1901), begin their metabolism experiments and build a chamber large enough to accommodate a person. Called a "respiration chamber," it enables them to study animal metabolism during states of activity, rest, and fasting. From this they are able to obtain the first accurate measurements of gross metabolism. For 11 years they make intensive experiments with the chamber and make the first accurate determination of human energy requirements. They also demonstrate the validity of the laws of conservation of energy in living animals and work toward the establishment of the concept that the basis of metabolism lies in the cells rather than in the blood. Their work helps establish the study of the physiology of metabolism and lays much of the foundation for modern nutritional science.

1862 Hematoxylin is first introduced. This colorless crystalline compound is extracted from the logwood plant and is used as a biological stain that turns a deep blue color upon oxidation.

1863 National Academy of Sciences is first incorporated in the United States with American botanist Asa Gray (1810-1888) as a founding member.

1864 Carl Fromman (1831-1892), German anatomist, introduces the first silver stain for neurons.

1864 Andrea Verga (1811-1895), Italian neurologist, publishes the first post-mortem report in a case of acromegaly (chronic hyperpituitarism marked by enlargement of the feet, face, and hands).

1864 Pasteurization, the process of slow heating that kills bacteria and other microorganisms, is invented by French chemist Louis Pasteur (1822-1895). In 1856, Pasteur is asked by a French industrialist to help save the wine and beer industry which suffers great financial losses when its products often go sour as they age. Pasteur finds that although the fermentation process does not require oxygen, it does involve living organisms and that the wrong ones cause the process to go bad. He discovers that gentle heating at about 120°F (49°C) effectively kills the unwanted microscopic organisms. Pasteurization is later used primarily to kill disease-causing microorganisms in milk as well as to prevent spoilage.

1865 First to stress the importance of using statistical methods in biology is English anthropologist Francis Galton (1822-1911) who begins his statistical investigations into the inheritance of human characteristics. Working in the general context of trying to improve the human species by selective parenthood, he measures mental ability and studies its transmission over time in families. He also is the first to study identical twins.

1865 Otto Deiters of Germany publishes the first realistic drawing of a neuron.

1865 Franz Schweigger-Seidel (1834-1871), German physiologist, is the first to prove that a spermatozoon possesses a nucleus and cytoplasm.

1865 First textbook of plant physiology, *Handbuch der Physiologischen Botanik,* is published by German botanist Wilhelm Friedrich Benedikt Hofmeister (1824-1877). His thorough work and original discoveries of plant structure make him a pioneer in comparative plant morphology (form and structure). Many consider him the father of modern botany.

(See also 1868)

1866 Gregor Johann Mendel (1822-1884), Austrian botanist and monk, discovers the laws of heredity and writes the first of a series of papers on heredity which formulate the laws of hybridization. His exact experiments, in which he carefully self-pollinates pea plants and studies the characteristics of each new generation, show that pairs of characteristics (like dwarf or tall plants) combine and sort themselves out according to fixed and understandable rules or mathematical formulae. He also realizes that both male and female equally contribute a determining factor and that pairs of factors in the offspring do not blend but remain distinct. Unfortunately, his work is disregarded until it is rediscovered by Dutch botanist Hugo Marie de Vries (1848-1935) in 1900. Unknown to both Charles Darwin and himself, Mendel had discovered the laws of inheritance that were later the proof Darwin's theory of evolution required.

(See also 1901)

1868 Wilhelm Friedrich Benedikt Hofmeister (1824-1877), German botanist, publishes his *Allegemeine Morphologie der Gewächse,* which becomes the first general work on plant morphogenesis (the study of the formation and differentiation of tissues and organs).

(See also 1865)

1869 Nucleic acid is first discovered by Swiss biochemist Johann Friedrich Miescher (1844-1895). He isolates a substance containing both nitrogen and phosphorous from the remnants of cells in pus, and when he separates it into a protein molecule and an acid molecule, it be-

Geneticist Gregor Mendel.

comes known as nucleic acid. This will later prove significant as he has isolated the nucleus in which deoxyribonucleic acid or DNA will later be found.

(See also 1944).

1869 Islets of Langerhans are discovered by German physician Paul Langerhans (1847-1888). Also called the islands of Langerhans, these patches of irregularly shaped endocrine tissue in the pancreas of most vertebrates manufacture, store, and secrete the hormone insulin. Insulin lowers the level of sugar (glucose) in the blood and also promotes the use of glucose as an energy source.

1869 Gypsy moth egg masses are first introduced into the United States from France by Massachusetts professor Leopold Trouvelot. He intends to breed the gypsy moth with the silkworm to overcome a wilt disease of the silkworm. When some moth eggs are accidentally blown away, they apparently survive, reproduce, and flourish. Ten years later, trees in the Medford region of Massachusetts are being defoliated by the caterpillar stage of the moth. Decades later, this pest has spread to 25 American states.

1870 John Hughlings Jackson (1835-1911), British neurologist, studies speech defects and muscle spasms caused by brain damage and epilepsy and first gives his name to "Jacksonian epilepsy," now generally called focal epilepsy. He also believes that the brain's main parts represent an evolutionary succession.

1870 First successful electrical stimulation of the brain is achieved by Gustav Theodor Fritsch (1838-1927), German anatomist and anthropologist, and Eduard Hitzig (1838-1907), German physiologist and neurologist. They also discover that electric shocks to one cerebral hemisphere of a dog's brain produce movement on the other side of its body. This is the first clear demonstration of the existence of cerebral localization. Until this, the brain was considered insensitive, since many had tried unsuccessfully to stimulate the rear of the brain (the easiest part to reach). Fritsch and Hitzig use weak currents and stimulate the front of the brain, discovering that a certain spot would make a lip curl and another spot would make an eye twitch.

1872 Ferdinand Julius Cohn (1850-1898), German botanist, publishes the first of four papers titled "Untersuchungen uber Bacterien" which mark the beginning of bacteriology as a distinct field. He systematically divides bacteria into genera and species and states that there are not, as believed, only a few species of very variable shapes, but that a very large number of different groups exist with very different forms. To make sense of them all, he proposes the first systematic approach based on morphology (form and structure).

1872 H. M. S. *Challenger* departs for what is the first oceanographic voyage (1872-1876) of scientific discovery. Researchers on board the ship voyage around the world's oceans to collect water samples, plant and animal specimens, and sea floor deposits.

1873 William Withey Gull (1816-1890), English physician, describes an eating disorder and names it anorexia nervosa.

1873 First to observe and describe systematically the behavior of chromosomes in the cell nucleus during normal cell division (mitosis) is German anatomist Walther Flemming (1843-1905). His work establishes the modern interpretation of nuclear division and introduces the science of cytogenetics, the study of physical changes in the cell's hereditary material.

(See also 1882)

1873 Franz Anton Schneider (1831-1890), German zoologist, first describes cell division in detail. His drawings include both the nucleus and the chromosomal strands.

1873 Camillo Golgi (1843-1926), Italian histologist, devises a way to stain tissue samples with inorganic dye (silver salts) for the first time. He applies this new method to nerve tissue and is able to see details not visible before. Until this, it is impossible to see and study the fine network of interconnections of nerve cells in the brain because they were regularly resistant to ordinary biological stains. Golgi introduces the use of silver salts or silver chromate, and finds that he can pick out nerve cells to the complete exclusion of all others. Their minute branches appear brownish black against a background of translucent yellow. Knowledge of the fine structure of the nervous system dates from his discovery.

1874 "Wernicke's area," that part of the brain associated with understanding words, is discovered by German neurologist Carl Wernicke (1848-1905). This region is located in the posterior third of the upper temporal convolution and is concerned with the comprehension of spoken or written language. This same year, he also subdivides aphasia (impairment of the ability to speak or to understand speech) into motor and sensory and describes them.

1874 First paper written on brain chemistry is

published by Johann Ludwig Wilhelm Thudichum (1829-1901), German-English chemist. In a series of highly original investigations, he extracts a series of new and complex substances from brain tissue: cephalin, sphingomyelin, phrenosin, kerasin, and glycoleucine. These are the cerebrosides which he discovers and names. However, his work is highly unpopular in his day, and he is ridiculed and his scientific reputation impugned. It is not until the 1940s that scientists find his analyses to be correct. Despite his difficulties, Thudichum laid the basis for the modern view of the brain as a chemical as well as an electrical device, and he is now considered the father of brain chemistry.

1875 First experiments on the electrical response of the brain are conducted by Richard Caton (1842-1926), English physiologist. Knowing that the brain can be stimulated electrically, he theorizes that it may have an electrical response. He therefore applies electrodes to the exposed surface of a rabbit's brain and observes on a galvanometer that when he shines a light in the animal's eye or pinches its ears, the current registers as varying. He continues and tries to demonstrate localization in the brain, but is not fully successful.

1875 Robert David Fitzgerald (1830-1892) of Australia publishes the first volume of his *Australian Orchids* (1875-1894) in Sydney. It is the first treatise on Australian orchids.

1875 First detailed account of how the nucleus divides is given by Belgian cytologist Edouard van Beneden (1846-1910). Based on his studies of rabbit cells, he shows how the nuclear material forms a disk across the middle of the cell prior to the separation of the split chromosomes. Later, he is able to learn much more about the chromosomes themselves.

(See also 1885)

1877 Wilhelm Pfeffer (1845-1920), German botanist, publishes his brilliant *Osmotische*

Untersuchungen in which he pioneers the study of semi-permeable membranes. These membranes have apertures or openings so tiny that only small molecules, and not large ones like proteins, can pass through. He is then able to measure osmotic pressure and correlate it to molecule size, meaning that by knowing osmotic pressure he also knows the molecular weight of a specific protein. In this manner he is the first to make reliable molecular measurements.

1879 Hermann Fol (1845-1892), Swiss physician and zoologist, is acknowledged as the first to observe a spermatozoon penetrate a starfish egg. He accomplishes for animals what Silesian biologist Nathanael Pringsheim (1823-1894) did for plants in 1856. Close observation reveals that only a single sperm enters the ovum, confirming the idea that only one sperm is needed for fertilization.

c. 1880 David Ferrier (1843-1928), Scottish neurologist, electrically stimulates the brains of a variety of animals and maps the brain's motor cortex and discovers its sensory strip. His work not only contributes to a knowledge of the localization of cerebral functions but also leads to the removal of brain tumors and other advances in brain surgery.

1880 Ludwig Rehn (1849-1930), German surgeon, carries out the first thyroidectomy (surgical removal) in exophthalmic goiter.

1880 Science of histogenesis, the study of the embryonic origins of different types of animal tissue, is founded by Swiss-German anatomist Wilhelm His (1831-1904). His *Anatomie Menschlicher Embryonen* is considered the first accurate and exhaustive study of the development of the human embryo.

(See also 1895)

1880 Causative factor of malaria is discovered by French physician and bacteriologist Charles Louis Alphonse Laveran (1845-1922). He finds that it is not a bacterium as thought, but a parasitic protozoan that infects human blood. This is the first case in which a protozoan, or a one-celled animal, is shown to be the cause of a disease.

1882 First clear delineation of the sequence of nuclear division is demonstrated by Walther Flemming (1843-1905), German anatomist, in his *Zellsubstanz, Kern und Zelltheilung*. In this work, he details his observations of the longitudinal division or splitting of chromosomes in animal (salamander) cells. As the first to observe this phenomenon, he realizes that the nucleus does not vanish at cell division. Flemming also names many features of the nucleus, calling the dark thread which appears at cell division, "chromatin." They are later renamed chromosomes.

(See also 1888)

1882 Friedrich August Johannes Loffler (1852-1915), German bacteriologist, and F. Schulze discover the bacterium causing glanders, a contagious and very destructive disease of animals, especially horses.

1882 Lyon first applies centrifugation to eggs. This technique, which uses rapid spinning to impose high centrifugal forces on whatever is being spun, causes separations of matter on the basis of differences in weight. It becomes an important tool in biochemical research.

1884 Phagocytosis in animals is discovered by Russian-French physiologist Elie Metchnikoff (1845-1916). In this phenomenon, amoeba-like cells engulf foreign bodies such as bacteria and act as the first line of defense against acute infection. He names them phagocytes from the Greek words meaning "devouring cells." In humans, these phagocytes are known as leukocytes or white corpuscles. His work becomes a fundamental tenet of the science of immunology.

1884 Tetanus bacterium is discovered and first described by German physician and bacteriologist Arthur Nicolaier (1862-1942). He does not obtain it in a pure culture, however. A serious infectious disease that causes severe contraction of the muscles, it is caused by a bacterium called *Clostridium tetani* that lives in soil and enters the bloodstream through breaks in the skin, especially deep puncture wounds. It can be fatal if not treated.

1884 Hans Christian Joachim Gram (1853-1938), Danish bacteriologist, invents the Gram staining method in which he stains bacteria using one of the methods of German bacteriologist, Paul Ehrlich (1854-1915), and then treats them with iodine solution followed by an alcohol wash which removes the stain from some but not from others. This becomes an important method of classifying bacteria and also works on antibiotics. Penicillin is Gram-positive (stain appears) and streptomycin is Gram-negative (no stain).

1885 Pierre Marie (1853-1940), French neurologist, first names the disease of giantism "acromegaly."

1885 Edouard van Beneden (1846-1910), Belgian cytologist, proves that chromosomes persist between cell divisions. He makes the first known chromosome count and discovers that each species has a fixed number of chromosomes. This means that the number of chromosomes in the various cells of the body is constant (humans have forty-six). He also discovers that in the formation of sex cells, the division of chromosomes during one of the cell divisions is not preceded by a doubling. Each egg and sperm cell has only half the usual count of chromosomes (resulting in a full count when they are united).

1888 Heinrich Wilhelm Gottfried von Waldeyer (1836-1921), German anatomist, first introduces the word "chromosomes." He combines the Greek words for "color" and "body" to describe the small, darkly-stained threads of chromatin material that appear in the process of cell division. He also is the first to argue that the nervous system is composed of separate cells and their delicate extensions. He calls this cell-plus-extension a "neuron."

1888 Theodor Heinrich Boveri (1862-1915), German zoologist, discovers and names "centrosome," the mitotic spindle that appears during cell division. He demonstrates that this structure is the division center for a dividing egg. He is also the first to note that chromosomes split as part of the mechanism of reproduction and that an individual chromosome does not affect each hereditary characteristic. Rather, a single chromosome is responsible for particular hereditary traits.

1889 Pasteur Institute first opens in Paris. Headed by French chemist and microbiologist, Louis Pasteur (1822-1895) after whom it is named, this research center is established for the purpose of undertaking fundamental research, prevention, and treatment of rabies.

1890 Conway Lloyd Morgan (1852-1936), English biologist and psychologist, publishes his major work on animal psychology titled *Animal Life and Intelligence* in which he is one of the first to use an experimental approach as a complement to observation.

(See also 1893)

1893 Liberty Hyde Bailey (1858-1954), American botanist, publishes the first detailed study of the growth of plants under artificial light. Bailey lives to be 96 and is ackowledged with making botany the basis of sound horticultural research, teaching, and practice. He is the first to elevate American horticulture from the level of a craft to a science.

1893 Comparative psychology is founded by English biologist and psychologist Conway Lloyd Morgan (1852-1936), who also makes the first modern enunciation of the principle of parsimony. Also called "Ockham's razor," after the 14th century English scholar, William of

Ockham (1280-1349), Ockham's rule was for simplicity in science, stating that "Entities must not needlessly be multiplied." Morgan offers a similar, modern version for the field of comparative psychology, saying that explanations for psychological activity should be sought in the most primitive faculties before interpreting them as the consequence of our most highly developed intellectual faculties.

1894 Japanese bacteriologist Shibasaburo Kitasato (1856-1931) and Swiss bacteriologist Alexandre Emile John Yersin (1863-1943) independently discover the plague bacillus. Yersin develops a serum against the disease.

1894 First journal of experimental embryology, *Archiv für Entwicklungsmechanick der Organismen*, is founded by German zoologist Wilhelm Roux (1850-1924).

1894 First proof that the law of conservation of energy can be applied to animal metabolism and nutrition is offered by German physiologist and hygienist Max Rubner (1854-1932). In his accurate caloric measurements of food, he finds that the energy produced by the consumption of food by the body is the exact same amount as if that quantity had been consumed by fire. He thus demonstrates that the laws of conservation apply to the living world as well as the inanimate.

1895 Heinrich Irenaeus Quincke (1842-1922), German physician, conducts the first lumbar puncture or spinal tap to study cerebrospinal fluid and for diagnosis and treatment.

1895 John William Harshberger (1869-1929), American botanist, first introduces the word "ethnobotany." This term means the systematic study of the botanical knowledge of a primitive people and their use of locally available plants in foods, medicines, clothing, or religious rituals. Increasingly, the drugs derived from plants used in folk medicine have been found to be beneficial in the treatment of many illnesses.

1895 First standardization of anatomical no-

menclature is achieved by Swiss-German anatomist Wilhelm His (1831-1904). He also is the founder of an institute of anatomy at the University of Leipzig and specializes as an embryologist.

(See also 1880)

1896 Anaerobic bacillus that causes botulism poisoning is discovered by Belgian physician and bacteriologist Emil Pierre Marie Van Ermengem (1851-1932). He finds that the *Clostridium botullinum* bacteria cannot survive in the presence of oxygen and normally live in the soil where they form heat-resistant spores that may contaminate food ready to be canned. These spores survive if the food is not cooked at a high enough temperature, and then secrete a toxin that is one of the most potent poisons known. Once ingested, the toxin damages the autonomic (involuntary) nervous system and can cause death.

1897 First hormone to be isolated is a form of epinephrine by John Jacob Abel (1857-1938), American physiologist and chemist. He extracts it from the adrenal gland and isolates it in a derivative form, calling it epinephrine. It is known to be a powerful heart stimulant that causes a noted rise in blood pressure. It is eventually isolated in pure form and given the trade name Adrenalin.

(See also 1901)

1898 John Newport Langley (1852-1925), British physiologist, first coins the term "autonomic nervous system." Also called the involuntary nervous system, it describes that part of the vertebrate nervous system that controls and regulates the internal organs without any conscious recognition or effort by an organism. Langley investigates the synapses in autonomic ganglia and eventually works out the details of the system.

1898 Mitochondria are first described and named by German physician Carl Benda (1857-1933). He chooses the name for these granular or rod-shaped bodies in a cell that function in the metabolism of fat, glycogen, and proteins from the Greek words for thread and granule.

1898   First to demonstrate the existence of a virus is Dutch botanist Martinus Willem Beijerinck (1851-1931). Since his tobacco-dealer father had gone bankrupt in part due to the devastating effects of the tobacco mosaic disease, he sets out to discover what type of bacteria or parasite causes the disease. In order to isolate the causative agent, he filters the sap of an infected plant to remove all known bacteria, but the fluid remains capable of infecting another plant and is seen to increase in contagiousness. He soon concludes that he is dealing with something that is too small to be seen under a microscope, is capable of passing through the sub-microscopic pores of the best filters, and can also multiply like a living thing. He considers it to be some sort of infectious fluid and he names it "virus," which comes from the Latin for slimy or poisonous liquid.

1901   First pure hormone to be isolated is produced by Jokichi Takamine (1854-1922), Japanese-American chemist, and T. B. Aldrich who isolate pure crystalline epinephrine from the adrenal gland. They patent the process and it is known by its trade name Adrenalin. It eventually is used as a powerful heart stimulant and a neurotransmitter.

(See also 1897)

1901   Santiago Ramon y Cajal (1852-1911), Spanish histologist, first proves the true nature of the connection between nerves and demonstrates that the nervous system consists of a maze of individual cells. He shows that the neurons do not touch but that the signal somehow jumps a gap (now called a "synapse"). He then states the principle that the nerve fibers he is studying are not a continuous network with every cell connected to every other cell, but that each cell is only connected with certain other specific cells and the rest communicate by "Protoplasmic kisses"—his early name for synapses.

He also develops a staining technique that allows him to make these discoveries and which is still in use.

1901   David Bruce (1855-1931), British bacteriologist, and Joseph Edward Dutton (1874-1905), British physician, discover trypanosome, a worm-like parasite, that causes African sleeping sickness. Two years later, this parasite is discovered to be transmitted to humans by the tsetse fly.

1901   Laws of inheritance are first published in the book, *Die Mutationstheorie*. Although the Dutch botanist Hugo Marie De Vries (1848-1935) arrives at these laws independently, he finds in checking the Austrian botanist Gregor Johann Mendel's (1822-1884) forgotten papers that he has in effect rediscovered Mendel's theories of heredity. He then announces Mendel's prior discovery and offers his own work only as confirmation.

(See also 1866)

1902   Robert Richet (1850-1935), French physiologist, and Paul Jules Portier (1866-1962), French biologist, discover anaphylaxis. This is a severe, immediate, and often fatal reaction to contact with an antigen from the first dose. Symptoms include difficulty in breathing because of swelling, an abrupt loss of blood pressure, and unconsciousness. Richet and Portier note that a dog previously injected with a poisonous substance would, on receiving a second small injection (of a quantity that would not harm a normal animal), suddenly become extremely ill and sometimes die. Richet names this phenomenon from the Greek words meaning "without protection."

1902   Hormones are first named and understood by Ernest Henry Starling (1866-1927) and William Maddock Bayliss (1860-1924), both English physiologists, when they isolate and discover the hormone "secretin" found in the duodenum. While studying how the pancreas secretes its digestive juices, they find that it is a chemical reflex and that the body has a chemical communication system as well as a nervous system. Starling suggests a name for all similar substances discharged into the blood by a particular organ, and he chooses "hormones" from the Greek word meaning to "rouse to activity."

1903 Almroth Edward Wright (1861-1947), English bacteriologist, discovers "opsonins" or antibodies in the blood of immunized animals. While studying the blood of these animals, Wright discovers it contains substances which greatly assist the white blood cells in destroying bacteria. Since they seem to prepare the bacteria for consumption by the white-cell phagocytes, he names them "opsonins" from the Greek word "to prepare food." Today we know them as antibodies.

1903 Archibald Edward Garrod (1857-1936), British physician, suggests that errors in genes lead to hereditary disorders. In his study of a rare metabolic disease, alkaptonuria, he lays the foundation for what will become the study of biochemical genetics. His 1909 book, *The Inborn Errors of Metabolism*, is the first study in biochemical genetics. In it, he demonstrates that a single gene mutation blocks a single metabolic step.

1904 Emil G. Racoviţa (1868-1947), Romanian biologist, publishes the first findings dealing with animal populations that dwell in caves. His work establishes modern biospeleology. He also is considered the founder of the first institute of speleology at Ciuj University in Romania.

MAY 1905 First proof that sex is determined by chromosomes is offered independently by American cytogeneticist Nettie Marie Stevens (1861-1912) and Edmund Beecher Wilson (1856-1939), American zoologist. Until their separate discoveries, no one really knew how the sex of a baby was determined. They find that there are differences between the chromosomes of a male and those of a female, and are the first to note the X chromosome and Y chromosome. They then discover that females always carry two X chromosomes (XX) and males one X and one Y (XY). This is a major breakthrough and marks the first time that any specific hereditary trait could be linked to a specific pair of chromosomes.

1905 John Sydney Edkins (1863-1940), English physiologist, first describes gastric secretin (gastrin). This is a digestive hormone secreted by the wall of the pyloric end of the stomach (where it joins the small intestine) that is released when food enters the stomach. This is the one of three digestive system hormones and it is different from Starling's secretin.

(see also 1902)

1905 Alfred Binet (1857-1911), French psychologist, devises the first of a series of tests (1905-1911) that make him the "father of intelligence testing." Binet's goal in devising these tests is to be able to measure an individual's innate ability to think and reason. He tests large numbers of French schoolchildren and grades his tests on a scale according to the chronological age of the children that pass them. This provides him with a large enough standard against which to measure an individual child. After Binet, further tests were developed and an IQ (intelligence quotient) was computed as the ratio of the subject's mental age to his or her chronological age, multiplied by 100.

1905 Bacterium responsible for syphilis is discovered by German zoologist Fritz Richard Schaudinn (1871-1906) and German dermatologist Erich Hoffmann. They discover the causal organism to be *Spirochaeta pallida*, later called *Treponema pallidum*, and contribute significantly to the control of this chronic, degenerative, venereal disease. Schaudinn also becomes the first to differentiate between *Entamoeba histolytica*, the cause of amebic dysentery, and its harmless counterpart, *Entamoeba coli*.

1905 Richard Willstatter (1872-1942), German organic chemist, discovers the structure of chlorophyll. Using chromatography, he finds that there are two major types of chlorophyll and that its function in plants is similar to that of hemoglobin (the red pigment in blood) in humans. He also finds that chlorophyll contains a magnesium atom, becoming the first to learn of the importance of magnesium as a plant nutrient.

1906 Howard Taylor Ricketts (1871-1910), American pathologist, first proves that a certain wood tick transmits Rocky Mountain spotted fever. He then is able to locate and identify the microorganism that causes the disease and finds that it is an unusual organism in that it has both bacteria- and virus-like qualities. It eventually is called Rickettsia. While studying typhus in Mexico City, Ricketts dies of that disease.

1907 Charles Horace Mayo (1865-1939), American surgeon, first uses the term "hyperthyroidism" to describe an overactive thyroid gland. This condition can cause weight loss, nervousness, and even protruding eyes. Mayo also pioneers modern procedures in goiter surgery.

1907 William Bateson (1861-1926), English biologist, first uses the word "genetics" to indicate the science of hereditary phenomena. After learning in 1900 of the discovery of Gregor Mendel's principles of heredity for plants, Bateson reinterprets those experiments in terms of animals and demonstrates that inherited traits are separate units, not blended, and are carried as individual packets of information. He then suggests that the study of the mechanism of inheritance be called genetics.

1907 Ross Granville Harrison (1870-1959), American zoologist, develops the first successful animal-tissue culture and pioneers organ transplant techniques. His cultivation of tadpole tissue is the first culture "in vitro" of animal cells, proving that cells can be induced to grow outside the body. His demonstration

that nerve fibers grow as projections that develop from the neuron body itself constitutes the foundation of modern nerve physiology and neurology.

1908 Margaret Lewis, American cytologist, first cultures mammalian cells in vitro. This means that for the first time, living cells from a mammal are kept alive and grown in a medium outside the body of the animal.

1909 Jean de Mayer, French physiologist, first suggests the name "insuline" for the hormone of the islet cells.

1909 N. Pierantoni begins the first studies of the mechanism of symbiosis between animals and microorganisms. His research introduces the notion that ecologically, all animals can be considered a part of the environment.

1909 Benjamin Minge Duggar (1872-1956), American botanist, publishes his *Fungous Diseases of Plants*, the first text to be published on this subject.

1909 Wilhelm Ludwig Johannsen (1857-1927), Danish botanist, credited with first offering the terms "gene" to describe the unit or carrier of heredity. He derives it from the Greek word meaning "to give birth to." He also suggests "genotype" to describe the genetic constitution of an organism, and "phenotype" for the actual appearance of the organism that is the product.

1909 Johann Paul Karplus (1866-1936) and Alois Kreidl (1864-1928), both Austrian physiologists, report on their first experimental studies on the hypothalamus. This important section of the brain controls hormone secretion by the pituitary gland, and regulates body temperature, certain metabolic processes, and portions of the involuntary nervous system.

1909 Charles Jules Henri Nicolle (1866-1936), French bacteriologist, proves that typhus fever is transmitted by the body louse. He is intrigued by the fact that while typhus is easily caught outside of a hospital, after patients are hospitalized they are found to be no longer contagious. He suspects body lice as carriers when he realizes that patients are stripped and scrubbed when admitted. His experiments prove him correct. He goes on to distinguish this typhus from murine typhus which is conveyed to humans by the rat flea.

1909   Phoebus Aaron Theodor Levene (1869-1940), Russian-American biochemist, having isolated the nucleotides (the basic building blocks of the nucleic acid molecule), first isolates the five-carbon sugar D-ribose from the ribonucleic acid (RNA) molecule. Twenty years later he discovers 2-deoxyribose (a sugar derived from D-ribose by removing an oxygen atom), which is part of the deoxyribonucleic acid (DNA) molecule. His work is the foundation for later breakthroughs in nucleic acid chemistry.

1910   Francis Peyton Rous (1879-1970), American pathologist, experiments with transmitting sarcomas in hens and discovers the first cancer-inducing viruses. He finds that he is able to transmit tumors to fowl of the same inbred stock by both grafting tumor cells and by injecting a submicroscopic agent taken from a tumor. The interval between his discovery and his receipt of the 1966 Nobel Prize for medicine or physiology is the longest on record.

1910   Harvey Cushing (1869-1939), American surgeon, and his team present the first experimental evidence of the link between the anterior pituitary and the reproductive organs. He also is the first to describe the disease known as Cushing's syndrome or Cushing's disease caused by a tumor in the pituitary and characterized by thin arms and legs but an obese face and trunk, atrophy of the skin, and accumulation of fluids in the body.

1910   First effective chemical treatment for syphilis is discovered by German bacteriologist Paul Ehrlich (1854-1915) and Japanese bacteriologist Sahachiro Hata (1872-1938). When Hata

finds that an arsenic-based compound appears to kill spirochetes, the microorganism that causes syphilis, they continue testing and find that the drug attacks the syphilis germs but does not harm healthy cells. Ehrlich names the new drug "Salvarsan," using the Latin words for "preserve" and "health," and it is later known as arsphenamine. Modern chemotherapy begins with Ehrlich and Salvarsan.

1911   William Morton Wheeler (1865-1937), American zoologist, publishes *The Ant Colony as an Organism* which examines for the first time the close correlation that exists between an organized colony of animals and a multicellular organism made up of cells. He also writes several books on the social insects and their behavior.

1912   First chair in Biochemistry in the United Kingdom goes to Frederick Gowland Hopkins (1861-1947) at Cambridge University.

1913   Plant quarantines are first imposed in the United States on a national basis. Employing the principles of exclusion and avoidance, these programs prove successful in containing certain plant diseases and avoiding others by keeping the pathogen or causal agent away from the healthy, growing plant. Quarantines or embargoes assure that areas free of disease remain that way.

1913   First chromosome map, showing five sex-linked genes, is produced by American geneticist Alfred Henry Sturtevant (1891-1970). He develops a technique for mapping the location of specific genes of the chromosomes in the fruit fly *Drosophila*. He also is the first to postulate that the genetic factors or genes are arranged linearly, like beads on a necklace.

1913   Vitamin A is discovered in butter and egg yolk by American chemist Elmer Verner McCollum (1879-1967). In his experiments with white rats, he finds that there is a factor essential to life found only in some fats, and he calls it fat-soluble A. His use of letters for vitamin names takes hold as more vitamins are discovered. In 1922, McCollum discovers Vitamin D in cod liver oil.

1914 Morris Simmonds (1855-1925), German physician, first describes pituitary cachexia or atrophy (named Simmonds' disease). Also called panhypopituitarism, this inadequate secretion of hormones by the pituitary gland can result in dwarfism, atrophy of the sex glands, and low metabolic rate and blood sugar.

1914 Mechanism of bioluminescence in marine animals is discovered by German zoologist Paul Ernst Christof Buchner. He demonstrates that it is the result of a symbiosis with bacteria, within which typical chemical reactions occur. This emission of light occurs sporadically in a wide range of organisms, from bacteria and fungi to insects and fish. The chemical reaction that takes place converts chemical energy to radiant energy totally and directly, with very little heat given off. For this reason it is also called cold light or cold luminescence. It appears to be associated with the protection and survival of a species.

1915 First evidence that a virus can infect and kill a bacterial cell is offered independently by Frederick William Twort (1877-1950), English bacteriologist, and Felix H. d'Herelle (1873-1949), Canadian-French physician. They independently discover the existence of bacteriophage—a virus that infests and kills bacteria—while attempting to grow bacteria in the laboratory. Both find their bacteria becoming transparent and dying, and upon isolating what was killing them, discover the agent to be a virus. D'Herelle names the virus bacteriophage meaning "bacteria-eater." Phages prove very important to the study of genetics and the control mechanisms of the cell.

c. 1915 Keith Lucas (1879-1916), British psychologist, first demonstrates the "all or none" response of stimulated neurons. This principle will eventually be recalled when the digital computer is invented later in the century, and its binary (0 or 1; on or off) system will contribute to the notion of the brain as a computer (and the computer as a brain). Physiologically, his work shows that the contraction response of ordinary muscle fiber is of an "all or none" variety, meaning that it responds to its fullest capacity regardless of the strength or weakness of the stimulus.

1916 First establishment of timing of ovulation in humans is accomplished.

1916 *Genetics,* the first purely genetics journal in the United States, is founded.

1917 Calvin Blackman Bridges (1889-1938), American geneticist, discovers the first chromosome deficiency. Working with the chromosomes of the fruit fly *Drosophila*, he conducts experiments and constructs gene maps that eventually prove the chromosome theory of heredity.

1918 Organizers at the cellular level are discovered by German zoologist Hans Spemann (1869-1941). He finds that an embryo develops according to the nature of the neighboring areas. By his development of new grafting techniques, he finds that a part of an embryo that has already begun to differentiate and develop as one particular thing can develop into something else if it is transplanted to another region. This shows that cells remain plastic until very late in development, and that it is the organizers in the embryo that bring about certain development nearby.

1921 Charles Foix (1882-1927), French neurologist, first locates the cause of paralysis agitans (Parkinson's disease) by locating the site of lesions in the midbrain's substantia nigra.

1921 Otto Loewi (1873-1961), German-American physiologist, discovers that acetylcholine functions as a neurotransmitter. It is the first such brain chemical to be so identified. In an ingenious experiment, he stimulates the vagus nerve of a frog's heart (being kept alive in a solution) and discovers that a chemical substance (acetylcholine) is liberated into the solution.

1921 Discovery and first isolation of the hormone insulin is achieved by Frederick Grant Banting (1891-1941), Canadian physician, Charles Best (1899-1978), Canadian physiologist, John James Richard Macleod (1876-1935), Scottish-American physiologist, and Canadian biochemist James Collip (1892-1965). They develop a method of extracting insulin from an animal pancreas and then injecting it into the blood of diabetics to lower their blood sugar.

1921 Boris Petrovich Uvarov (1888-1970), Russian-English scientist, studies migrating swarms of grasshoppers and is the first to realize that overcrowding of larvae in a restricted area can cause physiological and even morphological alterations that can modify behavior.

1923 First ultracentrifuge is introduced by Theodor H. E. Svedberg (1884-1971), Swedish chemist. This device spins so fast that it can remove particles from colloidal suspension because it produces an effect equal to hundreds of thousands of times normal gravity. This becomes a valuable research instrument for biologists since it is effective for particles that are larger than atoms or ordinary molecules but too small to be visible to the unaided eye.

1923 Adolph Melanchton Hanson, American surgeon, obtains the first really effective parathyroid gland extract from cattle.

1924 Richard Hesse (1868-1944) of Germany publishes his *Tiergeographie auf Oekologischer Grundlage,* which is considered the first real zoogeographic text that analyzes problems related to the environmental distribution of animals from a strictly ecological point of view.

1924 Efficiency of photosynthesis is first measured by German biochemist Otto Warburg

(1883-1970), who shows that it occurs with almost perfect thermodynamic efficiency. He also discovers the electron carrier, ferredoxin, in green plants, and how light energy is converted to chemical energy during photosynthesis.

1925 First human electroencephalogram (EEG) is recorded by Johannes Hans Berger (1873-1941), German neurologist. Pushing platinum wires into the scalp of his young son, he demonstrates that the electrical activity of the human brain can be recorded through the scalp. He later finds that metal plates strapped to the scalp serve just as well. He also finds that the waves produced by his device appear to change according to level of mental activity or stimuli and that the brain appears to be constantly active.

1925 Cytochrome is discovered and its function established by Russian-English biochemist David Keilin (1887-1963). He shows that what he names cytochrome is a respiratory enzyme that acts as a catalyst for the combination of oxygen with hydrogen within the cells. This explains what happens to oxygen after it is carried by the red blood corpuscles from the lungs to the cells.

1926 James Batcheller Sumner (1887-1955), American biochemist, isolates and crystallizes the first enzyme. He obtains an enzyme in crystalline form as he crystallizes urease (urea). Until this, enzymes were thought impossible to isolate. Sumner further proves that enzymes are proteins—an idea that was also thought impossible. This pioneering work marks the beginning of understanding enzymes which are catalytic agents that play a vital role in virtually all physiologic processes.

1926 Rudolf Magnus (1873-1927), German physiologist, discovers how the inner ear regulates balance. In his neurophysiological studies of posture and postural mechanisms, he demonstrates the function of the otoliths and semicircular canals of the inner ear in regulating the

Rudolf Magnus.

position and balance of the body. Vertebrates can sense rotation by the inertial lag of fluid in the semicircular canals of the ear, acting on sensory hairs. The three canals form loops lying in planes at right angles to each other, and by integrating signals from the canals, the central nervous system can detect rotation in planes other than those of the canals.

1927   Hermann Joseph Muller (1890-1967), American biologist, induces the first artificial mutations in the *Drosophila* fruit fly by using x rays. He reasons correctly that mutations concern only a single gene or a single point on a chromosome and not the entire chromosome. His work reveals that radiation has actually broken some of the chromosomes, and he produces scores of mutant flies. This discovery alerts him to the dangers of excessive x rays. Also from his studies a new discipline, radiation genetics, is founded.

1927   Lemuel Clyde McGee, American biochemist, first obtains an active extract of the male sex hormone from bull testicles. Although he has some success, he does not isolate the male sex hormones nor does he determine their structure.

1928   First experiment on "transformation"—a process by which heritable characteristics of one species are incorporated into another different species—is carried out by English bacteriologist Fred Griffith (1877-1941). He injects mice with a mixture of a living strain of pneumococci and the dead remains of a virulent strain of the bacteria. The mice then die from the infection caused by the live organism which somehow acquired the virulence of the dead strain. This seems to infer that genetic information is transmitted chemically.

c. 1928   Wilder Graves Penfield (1891-1976), Canadian neurosurgeon, first uses microelectrodes to map parts of the human brain's cerebral cortex and locates their different functions. He stimulates hundreds of unanesthetized patients in the course of brain operations and prepares a map of the brain that shows the relative amounts of brain activity available for different sensory and motor functions. His graphic figures correctly indicate that such sensation-rich areas like the lips, and motor-intensive areas like the hand occupy a disproportionately large area of the brain.

1928   Oliver Kamm and his team first isolate vasopressin and oxytocin. Both are hormones secreted by the pituitary gland. Vasopressin is an important anti-diuretic hormone and oxytocin plays a role during pregnancy.

1929   Edward Adelbert Doisy (1893-1986), American biochemist, first isolates estrone from the urine of pregnant women. Independent of Doisy, German chemist Adolf Friedrich Johann Butenandt (1903-1995) also isolates estrone this year. This is one of three hormones (estrone, estradiol, and estriol) that make up the estrogen group which primarily influences the female reproductive tract in its development, maturation, and function.

(See also 1935)

1929   Willard Myron Allen, American physician, and George Washington Corner, Ameri-

Canadian neurosurgeon Wilder G. Penfield.

Behavioral psychologist B. F. Skinner.

can anatomist, discover progesterone and demonstrate that it is necessary for the maintenance of pregnancy. With great difficulty, they extract a substance from the corpus luteum—the yellowish body in the ovaries. They test its role in pregnancy by injecting it into an animal whose corpus luteum has been removed. Without the corpus luteum, the pregnancy would likely abort or be prematurely delivered. They find that injection of this new substance they call "progesterone" ensures a normal delivery.

1929 P. Stricker and F. Grueter discover prolactin, a hormone secreted by the pituitary gland, that begins and maintains the secretion of milk in female mammals.

1929 First cure of hyperinsulinism by the removal of an islet cell tumor is achieved by Howland, Campbell, Maltby, and Robinson. This condition occurs when the islets of Langerhans overproduce insulin and can result in convulsions. Although rare, it is usually caused by a pancreatic tumor, as this group demonstrates.

1930 Burrhus Frederic Skinner (1904-1990), American psychologist, first describes operant

conditioning. This is a system of behavior or responses that operate on the environment to produce rewarding and reinforcing effects. With this method, he conducts highly original experiments in animal learning, notably with pigeons, and eventually comes to view human behavior in terms of physiological responses to the environment. Among his advances in behaviorist research and on conditioned responses are included his invention of teaching machines and programmed learning. Central to this approach is the concept of reinforcement or reward.

1931 Sewall Wright (1889-1988), American geneticist, presents the first useful picture of genetics and its role in evolution. As one of the founders of population genetics, he works extensively with populations of guinea pigs and finds that statistical problems of sampling are important in developing theories of evolution. He develops a mathematical theory of evolution as well as formulas for evaluating the statistical consequences of various mating systems and for measuring inbreeding. He also demonstrates that small, widely distributed populations frequently lose genes because of the limited opportunities to distribute the gene through mating. He argues that this phenomenon of gene loss can give rise to separate species no longer able to mate with one another.

1932 Hans Adolf Krebs (1900-1981), German-British biochemist, first describes and names the citric acid cycle (Krebs cycle). While studying carbohydrate metabolism and attempting to discover how lactic acid is broken down into carbon dioxide and water, he discovers a complicated series of chemical changes that is actually a chain or cycle of events. Called the Krebs cycle, this chemical sequence explains the metabolic pathways that carbohydrates, fats, and even proteins follow. The cycle also plays a fundamental role in virtually all cell metabolism and offers biochemistry a major leap forward in understanding.

1933 Mycological Society of America is first founded to promote the study of fungi, a group that includes mushrooms, molds, yeasts, and actinomycetes (bacterial parasites). Research in mycology becomes increasingly useful as it leads to the development of antibiotic drugs as well as industrial applications.

1934 First x-ray crystallography of a protein is made by John Desmond Bernal (1901-1971), Irish physicist. He pioneers the use of x rays to study the structure of crystals as a means of learning the atomic and molecular structure of a substance. Only with this knowledge could a substance be accurately synthesized. X-ray crystallography proves essential to the later synthesis of penicillin.

1935 Nature and structure of a virus is first defined by American biochemist Wendell Meredith Stanley (1904-1971). Working with the tobacco mosaic virus which is very stable and easily obtained, he crystallizes it in its pure form, showing that the virus is not a living thing. Since a crystal is a structure in which the molecules are arranged in regular rows, like bricks in a wall, it only can be formed out of

non-living units which are identical in shape and size. Stanley's argument that viruses are non-living is difficult for the scientific community to accept, since viruses can reproduce themselves—one of the criteria for life. It will later be shown that viruses consist mainly of protein and nucleic acid, and although they contain genetic material and are able to reproduce, they lack cell structures and the ability to metabolize food. Most important, they can only reproduce inside other living host cells.

1935 Edward Adelbert Doisy (1893-1986), American biochemist, first isolates the estrogen hormones directly from ovarian tissue.

(See also 1929)

1935 Testosterone is first isolated from testicular material by K. David and associates. They isolate a pure crystalline hormone and name it testosterone. As an organic compound belonging to the steroid family and occurring as the masculinizing hormone produced by the testis, it is isolated after it is realized that androsterone (isolated in 1931) is a by-product of testosterone.

1936 First non-organic agent to mimic an organizer is shown to be methylene blue by English biochemist Joseph Needham (1900-1995) and colleagues. In embryology, an organizer is a chemical substance secreted by cells in one part of an animal embryo which diffuses and affects the growth and differentiation of the cells in surrounding parts of the embryo. It essentially tells certain cells what they will become when they develop. Methylene blue is a bright greenish blue dye discovered in 1876, and its action on cells shows that the organizer function in embryology is a chemical one.

1936 Herbert McLean Evans (1882-1971), American anatomist and embryologist, and his group first isolate the interstitial cell stimulating hormone (ICSH). Also called luteinizing hormone, it is concerned with the regulation of the activity of the gonads or sex glands and is produced by the pituitary gland.

1937 H. S. Jacobs first describes the ability of alloxan (an oxidized product of uric acid) to produce hypoglycemia (in rabbits). This becomes an important new tool in the study of diabetes.

1938 First synthetic estrogen, stilbestrol, is discovered by English chemist Edward Charles Dodds and colleagues. This important group of hormones primarily influence the female reproductive tract in its development, maturation, and function.

1939 Johannes Friedrich Karl Holtfreter, German biologist, first shows selective reaggregation of cultured cells. This gathering together of cells outside the organism as a step toward the formation of something larger and specific is a sign that they are somehow receiving a message to do so. He also conducts studies on the development of amphibia and observes organization, structure, differentiation, and movement of cells.

1942 Klinefelter's syndrome is first described by physician Henry E. Klinefelter, Jr., endocrinologist Fuller Albright, and Edward C. Reifenstein. This relatively common (one per 500 live male births) condition is a disorder of the human sex chromosome and is characterized by testicular failure. Males with this congenital endocrine condition usually have an XXY genotype.

1944 "Blueprint" function of deoxyribose nucleic acid (DNA) is discovered by Canadian-American bacteriologist Oswald Theodore Avery (1877-1947), and Canadians Maclyn McCarty and Colin Munro Macleod (1909-1972). In their studies of the life-cycle and make-up of the bacterium that causes pneumonia, they prove conclusively that DNA is responsible for the development of a cellular feature. Their discovery—that the genetic in-

formation carried by the chromosomes is expressed in the structure of a molecule of an acid known since 1869 as deoxyribose nuclei acid (DNA)—is as revolutionary and startling to biology as was the discovery of the cell. This leads to further research on the structure and method of replication of DNA.

(See also 1953)

1944 Word "palynology" is first proposed by H. A. Hyde and D. A. Williams to replace the cumbersome and restrictive term "pollen analysis." Palynology, or the study of fossil pollen, focuses on acid-resistant microscopic plant and animal remains (such as pollen, algae, fungi, and spores). It becomes a valuable means for investigating past changes in the flora and the ecology of particular regions as well as the alteration in climate, drainage, and other environmental factors which gave rise to them.

1946 Genetic recombination is first demonstrated by Joshua Lederberg, American geneticist, and Edward Lawrie Tatum (1909-1975), American biochemist and geneticist, in *E. coli* bacteria. Although bacteria usually increase by simple cell division, Tatum and Lederberg bring together two different strains and show that they not only can reproduce sexually but that the offspring shows characteristics derived from both strains. This intermingling of the genetic material of bacteria (genetic recombination) greatly expands genetic research.

(See also 1953)

1949 Choh Hao Li, Chinese-American chemist and endocrinologist, and Herbert McLean Evans (1882-1971), American anatomist and embryologist, first isolate follicle stimulating hormone (FSH). Produced by the pituitary gland, it stimulates the development of the graafian follicle in the female and promotes development of the testes in the male.

1951 Fully saturated steroids such as cholester-

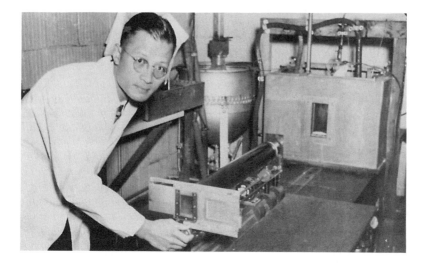

Dr. Choh Hao Li in his California laboratory.

ol, cortisone, and lanosterol are first synthesized by American chemist Robert Burns Woodward (1917-1979). He synthesizes cortisone from a coal-tar derivative since it has a similar structure (carbon rings and an attached methyl group). He goes through twenty separate steps and transforms one substance into another.

1951 Technique of electrode implantation is first developed and leads to the discovery of the brain's "pleasure centers." Although no specialized "sex center" has been located in the human brain, the parts most concerned with sexual response are the hypothalamus and the limbic system.

1952 Alan Lloyd Hodgkin and Andrew Fielding Huxley, both English physiologists, first work out the mechanism of nerve-impulse transmission, showing that a "sodium pump" system works to carry impulses. Using the fairly large nerve cells of a squid, they are able to elucidate the chemical and physical changes that occur along a neuron as it carries a nerve impulse. As an impulse reaches any particular point in the neuron, the membrane at that point becomes more permeable, allowing sodium ions to flow into and, shortly thereafter, potassium ions to flow out of, the neuron. As the impulse passes that particular point, sodium and potassium ions flow back out of and into the cell, restoring its original potential difference.

1953 Molecular structure of deoxyribonucleic acid (DNA), the gene substance that is the basis of heredity, is discovered by American biochemist James Dewey Watson and English biochemist Francis Harry Compton Crick. In trying to explain what blueprint or language gave the necessary instructions at the crucial time of cell division and growth, they focus on the relationship of the known chemical groups—adenine, cytosine, guanine, and thymine—that compose it. They then propose a double helix or twisted ladder model that links the chemical bases in definite pairs. This spiral staircase model is able to explain how the DNA molecule duplicates itself: since the two chains are complementary, if separated, each would serve as the template for the formation of the other. The Watson-Crick model is eventually confirmed, and today we know that DNA is the molecule that contains the essential set of directions that each cell needs to perform vital life functions.

1953 Henry Bernard Davis Kettlewell (1907-1979), English physician and entomologist, publishes his work on "industrial melanism" in moths, demonstrating the first case of evolution that can be witnessed as it occurs. He shows a classic example of evolution in the Pale Brindled Beauty moth which has a natural light coloring that matches the lichen trees on which it lives. But as its environment is changed by the factory smoke of the industrial revolution, it stands out conspicuously against the now-darkened and stained trees and is an easy target for predators. Kettlewell shows that since a mutant black form of the same species blends better with the changed trees, its coloring helps it to survive better and, by natural selection, to spread its genes more than its lighter counterpart.

1953 Transduction in *Salmonella* is discovered by American geneticists Joshua Lederberg and Norton David Zinder. In a mating experiment using different strains of *Salmonella*, they discover that a virus can convey or transfer genetic information from one organism to another. In unique experiments using a U-tube and a filter separating two bacteria strains, they find that a live virus (bacteriophage) would pass back and forth through the filter, multiply at the expense of the bacterium, and transfer genetic material from bacterium to bacterium. This startling and even unsettling phenomenon is named transduction.

1954 First known protein hormone ever synthesized is produced by American biochemist Vincent Du Vigneaud (1901-1978), who studies the hormones produced by the pituitary gland. After years of experimentation and study on the hormone oxytocin, he works out the exact order of its amino acid chain and is able to synthetically reproduce it. He then does the same for the hormone vasopressin. This leads the way to similar understanding of more complicated proteins.

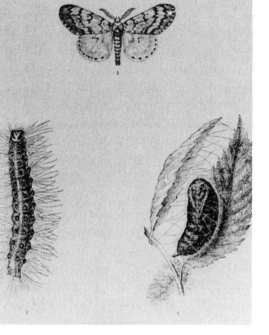

Gypsy moth drawing from U.S. Entomological Commission.

1954 Strychnine is first synthesized by Robert Burns Woodward (1917-1979), American chemist. This extremely complicated and poisonous alkaloid has a molecule built up of seven intricately related rings of atoms. Considered the father of modern organic synthesis, Woodward achieves an unprecedented number of similar syntheses and has a truly monumental scientific career.

1956 First of the tranquilizing drugs to be synthesized is reserpine by American chemist Robert Burns Woodward (1917-1979). Derived from the root of a shrub in India, this natural drug lowers blood pressure and has a sedative and calming effect. Woodward's synthetic version has the same calming effect without causing drowsiness.

1957 Maurice Solomon Raben, American physician, first develops a method for the extraction of human growth hormone (GH) from the pituitaries of cadavers. Also known as somatotropin, this hormone increases the rate of protein synthesis, promotes the release of fat from fat stores, and influences carbohydrate metabolism.

1957 First to synthesize deoxyribonucleic acid (DNA) molecules outside the cell is American biochemist Arthur Kornberg. Working with *E. coli* bacteria, he forms synthetic but not biologically active molecules of DNA by the action of an enzyme (now known as DNA polymerase) upon a mixture of nucleotides. This is also called "nonsense" DNA since the order in which the enzymes assemble the nucleotides seems meaningless.

1957 Avram Noam Chomsky, American psycholinguist, first claims in his book *Syntactic Structures* that the human brain is born programmed for learning language. He offers the revolutionary idea that all languages share a universal grammar that lies below the surface of all the different grammars and that language is a universal and innate facility in humans. He founds what is called transformational or generative grammar which is an innovative system of language analysis that revolutionizes the study of linguistics.

1958 Aaron Bunsen Lerner, American physician, and his group first isolate the pineal hormone melatonin, a methoxy derivative of serotonin. They call it melatonin because it is related to melanin and serotonin. Its function is not completely established.

(See also 1973)

1959 First disease characterized by chromosome aberration is discovered by French pediatrician Jérôme Lejeune and colleagues M. Gautier and R. A. Turpin. They discover that patients with Down's syndrome (also called mongoloids) have 47 chromosomes instead of the usual 46. The extra or 47th chromosome is a third copy of chromosome 21. Because of this, it is also called "trisomy 21." This condition results in retarded mental and physical growth and a shortened life span.

1960 Chlorophyll is first synthesized by American chemist Robert Burns Woodward (1917-1979). Chlorophyll catalyzes or promotes photosynthetic reactions in the presence of light, converting it into chemical energy.

1964 Chinese-American chemist and endocrinologist Choh Hao Li and colleagues discover a protein they later name beta-lipotrophin. A peptide of uncertain function, it is known to be manufactured in the pituitary gland and in the same cells as adrenocorticotropic hormone (ACTH).

1966 Choh Hao Li, Chinese-American chemist and endocrinologist, describes the structure of human growth hormone and first synthesizes it. He also shows that its structure is different from animal growth hormone and that it is far more complicated than other pituitary hormones. Made up of 256 amino acids this is the largest protein molecule synthesized in a laboratory to date.

1967 Charles Yanofsky, American biologist, shows for the first time that the codon (genetic code) sequence in a gene determines the sequence of amino acids in a protein. Working with *E. coli* bacteria, he is able to prove that there is a linear correspondence between the amino acid sequence of proteins and the sequence of nitrogen-containing bases that make up the structure of genetic material. He also is the first to demonstrate that certain mutant genes produce inactive proteins.

1968   First repressor genes are identified independently by Mark Steven Ptashne, American molecular biologist, and Walter Gilbert, American biophysicist and molecular biologist. A repressor gene inhibits the action of an operator gene or turns it "off." Gilbert uses a radioactive tracing technique and isolates a protein that turns off production of lactose-digesting enzymes in the *E. coli* bacteria.

1968   Roger Guillemin, French-American physiologist, Roger Cecil Burgus, American biochemist, Andrew Victor Schally, Polish-American endocrinologist, and Wylie Vale announce the isolation of the first known hypothalamic hormone, Thyrotropin-Releasing Hormone (TRH). They also sequence and synthesize this hormone that is the first in the series of hormones that stimulates the thyroid gland to produce thyroxine. TRH is later synthesized and used to measure pituitary gland function and to diagnose pituitary tumors.

1968   Restriction enzymes are discovered by Swiss microbiologist Werner Arber who shows that bacteria cells produce this special substance to defend themselves against viruses. This enzyme produced by bacteria counters the bacteriophage (virus) by splitting or cutting its deoxyribonucleic acid (DNA), thus leaving it without a complete set of genetic instructions and rendering the virus inactive. This understanding allows future genetic researchers and engineers to use restriction enzymes like DNA scissors and gives them the ability to copy, dissect, and even alter DNA and RNA.

1969   First synthesis of the enzyme ribonuclease or RNase from bovine pancreas is achieved by Stanford Moore (1913-1982) and William H. Stein (1911-1980), both American biochemists. They also discover how to chemically identify the active amino acids within the protein chain and determine the entire amino acid sequence. Some of the known roles of RNase are that it may play a part in forming an RNA (ribonucleic acid) molecule for a specific pur-

pose (such as messenger RNA) or it may break down RNA that has served its purpose so the components can be reused.

1970   First complete synthesis of a gene is announced by Har Gobind Khorana, Indian-American chemist. He demonstrates that he has made yeast and bacteria genes in the laboratory. In making the analine-transfer RNA, he does not use a natural gene for a template, but assembles the gene directly from its component chemicals.

1970   Howard Martin Temin, American virologist, and David Baltimore, American biochemist, independently discover reverse transcriptase in viruses. This is the enzyme that causes ribonucleic acid (RNA) to be transcribed into deoxyribonucleic acid (DNA). Before their discovery, it was assumed that the sequence of making a protein was a one-way system of DNA-RNA-protein. They show, however, that the reverse is possible, and that RNA can be transcribed into DNA, and that this phenomenon occurs in some tumor viruses.

1971   First synthesis of vitamin B-12 is achieved by American chemist Robert Burns Woodward (1917-1979). After collaborating with over 100 colleagues for eleven years on this problem and completing a sequence of more than 100 reactions, he synthesizes this complex co-enzyme. The scale of this concerted effort is unprecedented in the history of chemistry, and although it does not produce a practical source of B-12, the research adds greatly to the understanding of similar complicated compounds.

1973   First report is made claiming a circadian variation in blood melatonin levels (pineal hormone) in man. These variations affect mood and can cause the depression associated with seasonal affective disorder (SAD). Research indicates that light therapy may be useful in helping affected individuals through the shortened days of winter.

1973   Opiate or opioid receptor in the brain is

discovered by Candace Pert, American bio-chemist, and Solomon H. Snyder, American neuropharmacologist. The term opioid is used for any type of chemical that behaves in a manner similar to that of opium. They find that opiates attach themselves to receptor sites in the brain as their target cells. They then hy-pothesize that opiates slow the rate at which a cell can transmit a "pain" message, and that blocking the opioid receptor and slowing cell function may also be responsible for the sense of euphoria that accompanies opium use.

(See also 1975)

1974 Lars Y. Terenius, Swedish neurochemist, and Ageneta Wahlstrom discover the existence of naturally-produced small peptides that act upon the brain's opiate receptors. This discov-ery leads others to search for chemicals pro-duced by the body that also bind to opiate receptors.

(See also 1975)

1975 First known endogenous opioid peptide, enkephalin (popularly called "brain morphine"), is discovered by John Hughes of Scotland and his colleagues. This substance occurs naturally in the brain and suggests that the brain's chemi-cals can block the transmission of pain signals. These natural chemicals are an important part of the body's natural pain fighting mechanism.

1975 Monoclonal antibodies are invented by Argentinian biochemist César Milstein and German immunologist Georges Kohler. In re-sponse to the growing need of cellular scientists for large amounts of specific antibodies for their research, Milstein and Kohler discover how to extract large amounts of antibodies from clones (exact copies) of a cell. They do this by fusing two different natural cells to produce a hybrid that reproduces a specific antibody endlessly.

1976 New class of hormonal substances called endorphins are discovered by Roger Charles Louis Guillemin, French-American physiolo-

gist. This family of peptides are found to be manufactured within the central nervous sys-tem and have the natural ability to relieve pain. The name comes from the words "endogenously manufactured morphine-like substance."

1976 First commercial business to develop ge-netically-engineered products, Genentech, is formed in California.

1976 Martin F. Gellert, Czechoslovakian-Ameri-can biochemist, and colleagues discover the enzyme, gyrase, causes the DNA double helix to form a larger helix called a supercoil.

1977 Philip Allen Sharp and Richard John Rob-erts, both American molecular biologists, inde-pendently discover that DNA from higher or-ganisms has a substantial amount of seemingly meaningless information (called an "intron") that is discarded in making proteins. According to some geneticists, only about 5% of the base pairs in human DNA actually code for genes; the remaining noncoding regions are introns or "junk DNA."

1977 Floyd Elliott Bloom, American neuro-biologist, first pinpoints the location of endorphins in the pituitary gland. Endorphins are short chains of amino acids that seem to interact with the body's pain receptors. These opiate-like substances in the pituitary may someday be used as a form of natural pain control.

1978 Stephen Coplan Harrison, American bio-chemist, details the first high-resolution struc-ture of an intact virus while studying the toma-to bushy stunt virus.

1978 Tumors are first successfully produced in mice by the transfer of individual genes. Ameri-can molecular biologist Robert A. Weinberg transfers what are called "oncogenes" or tumor-causing genes into normal mice and causes tumors to grow in them. The oncogene is often identical to a normal gene except for a single amino acid along the chain which, after a certain cell division, is transformed into an oncogene.

1982 Gene from one mammal (rat growth hormone gene) functions for the first time in another mammal (a mouse). The mouse grows to twice its normal size, demonstrating that genetic engineering is achievable in living mammals.

1983 First artificial chromosome is created by Andrew W. Murray and Jack William Szostak, Canadian biochemist, who work with yeast.

1984 Allan Charles Wilson, American biochemist, and Russell Higuchi first clone genes from an extinct species as they take a gene from the preserved skin of a quagga, a type of zebra.

1985 First method of "fingerprinting" with DNA is developed by Alec Jeffreys of England who shows that there are certain core sequences of DNA that are unique to each individual. Using a technique called restriction fragment length polymorphism (RFLP), the DNA is divided into fragments which are separated by size. The separated fragments form a pattern of dozens of parallel bands that reflect the composition of the DNA. In principle, the pattern produced will always be the same for the same person. Although it is estimated that there are more than 10 billion possible patterns and that it appears virtually impossible that the DNA pattern of one person would match that of another, critics argue that patterns can stretch and shift, making it not entirely reliable.

1986 First gene known to inhibit growth is produced by an American team led by molecular biologist Robert A. Weinberg. The gene is able to suppress the cancer retinoblastoma.

1986 First license to market a living organism that was produced by genetic engineering is granted by the United States Department of Agriculture. It allows the Biologics Corporation to sell a virus that is used as a vaccine against a herpes disease in pigs.

APRIL 24, 1987 First test of free-living, genetically engineered bacteria is conducted as strawberry fields are sprayed with bacteria whose ice-forming gene has been eliminated.

1987 First criminal conviction are obtained on the basis of genetic information evidence. Following this landmark use of DNA or "genetic fingerprinting" in England, genetic evidence becomes increasingly admissible and accepted in a court of law.

1987 David C. Page and colleagues discover the gene responsible for male characteristics in mammals. It is a single gene on the Y chromosome that leads to the development of testes rather than ovaries.

1987 Genetically engineered plants are first developed.

APRIL 1988 First United States patent for a vertebrate is issued to the Harvard University Medical School for a strain of genetically-engineered mice it developed to discover the causes of cancer.

1991 First data ever gathered on insect oxygen consumption during flight are obtained by Charles P. Ellington and K. E. Mackin of England and Timothy M. Casey of the United States. Their findings indicate that hovering and forward-flying bumblebees consume about the same amount of oxygen. Ellington also uses high speed cinematography to study the mechanics of flight in bumblebees.

1991 First bacteria to be recovered from an extinct animal are located in the nearly complete remains of a mastodon found by a landscaping crew at Burning Tree Golf Course in Newark, Ohio. In the beast's partially-digested last meal are found live, 11,000-year-old bacteria from its intestinal tract that were kept in suspended animation by the bog's cold, dark, oxygen-free environment. Once revived by their exposure to light and air, they become the oldest living organism ever studied.

1992 Covering more than 9,000 species, *Distribution and Taxonomy of Birds of the World* changes bird arrangement from a system based on morphology (form and structure) to a new genetic system based on comparisons of the bird's DNA. This system is much more objective than the previous system which required some subjective assessment.

1992 First irradiated food (strawberries) becomes available to American consumers on their grocery-store shelves. Bombarding food with gamma rays is a process that not only slows ripening in fruits and sprouting in potatoes, but also disinfects herbs and teas and destroys parasites. Although approved by the Food and Drug Administration, irradiation is regarded by many critics as unsafe.

1993 First new living species of large mammal to be found in 50 years is discovered in the rain forests of Vietnam. Described as a member of the Bovidae family, it is identified by its recent remains. Although not yet seen alive by scientists, it is called a "forest goat" or "spindle horn" by local hunters and is estimated to weigh about 220 lbs (100 kg) as an adult. It has a rich brown coat with white and black markings and sharp, straight horns.

1993 First animal model on which to perform AIDS tests is a SCID-hu mouse. This mutant mouse lacks an immune system and cannot fight viruses, bacteria, or other foreign bodies. SCID stands for Severe Combined Immuno-Deficiency. By implanting the basics of a human immune system in these mice, researchers create the first animal that can be infected with the HIV virus that causes AIDS. The availability of SCID-hu mice may prove to be a practical vehicle for testing AIDS drugs. They may also contribute to some form of anti-AIDS gene therapy for humans.

SEPTEMBER 1994 First elimination of polio in the Western Hemisphere is announced by the Pan American Health Organization. This or-ganization states that the practice of mass immunization has wiped out polio as a health menace in North and South America. Polio is an enterovirus, or one that invades the body through the mouth, targets the neurons of the spinal cord and brain and destroys them, often resulting in muscle paralysis or death.

JULY 1995 First complete sequencing of an organism's genetic makeup is achieved by the Institute for Genomic Research in the United States. They sequence all 1.8 million base pairs—the rungs of the DNA double helix and the letters of the genetic alphabet—that make up the single circular chromosome of the bacterium *Haemophilus influenzae*. Foregoing the traditional method of mapping first and then sequencing, they sequence first and then attempt to piece it all together.

1996 New phylum is discovered by Danish zoologists Peter Funch and Reinhardt Kristensen. They discover a strange, new animal living on the mouthparts of Norwegian lobsters. They name it *Cycliophore*, Greek for "carrying a small wheel," which describes its funnel-like round mouth. This tiny symbiont eats crumbs from the lobster's meals and reproduces asexually when the lobster molts.

APRIL 24, 1996 First known complete genetic blueprint of an organism closely related to human cells is achieved by an international team of scientists coordinated by the National Institutes of Health in the United States. They identify all of the more than 6,000 genes that control reproduction, life, and death in yeast cells. For the first time, scientists have access to the full set of genetic instructions in a complex cell. The instructions for making and maintaining a yeast cell are encoded in 12,057,000 "letters" of code which, in turn, are packed into about 6,000 genes on 16 chromosomes. Human instructions are contained in about 3 billion such "letters" in about 80,000 genes on 46 chromosomes.

Chemistry

c. 6000 B.C. First metal known by man is probably gold, which appears in its metallic form in the sands of some rivers and whose luster and color would certainly attract attention. The earliest gold is probably obtained as small nuggets by washing alluvial deposits (soil deposited by running water). Gold is the most malleable and ductile of metals (1 oz [28 g] can be beaten out to 300 sq. ft [28 sq. m]) and gold ornaments dating back to this Neolithic Period have been found. As a rare metal found in a relatively pure state in nature, it is a pleasing and workable metal that does not tarnish or corrode. Copper is probably the next known metal since it, like gold, has an eye-catching color and is malleable. Many advances in early chemistry come from the alchemists' attempts to transform more common metals into gold.

c. 4000 B.C. First man-made glass takes the form of glass beads or ceramic glaze made in Egypt.

(See also c. 1370 B.C.)

c. 3400 B.C. Bronze, an alloy (mixture) of copper and tin, first appears in abundance in Sumeria. The Sumerians become expert in the working of gold, silver, copper, lead, and antimony. These oldest bronze specimens contain only one-third tin, while later ones contain

considerably more. By 2000 B.C., bronze is common enough to be used for weapons and armor.

c. 2000 B.C. Iron tools first appear in Egypt during the XII Dynasty. Although relatively abundant, it proves difficult to extract from the rock in which it is found, and it will be some time before it becomes a common metal. Since it is harder than bronze, iron proves desirable for tools and especially for weapons.

(See also c. 1500 B.C.)

c. 1700 B.C. Oldest record in the form of a pharmacopeia is a stone tablet prepared in ancient Babylonia. A pharmacopeia is prepared by some authority and attempts to give standardized and dependable information on drugs.

(See also 1498)

c. 1500 B.C. First to use iron routinely for tools are the Hittites, who build a great empire in Asia Minor. They come upon the secret of smelting iron which uses very high temperatures to separate the iron that is firmly bound in the ore.

c. 1370 B.C. First manufacture of glass on a large scale begins in Egypt. It is known that a glass factory at Tell el-Amarna existed in this period. The Egyptians can make nearly colorless glass, which is exported to all parts of the Roman Empire.

(See also c. 100 B.C.)

c. 900 B.C. First army to be equipped with good iron weapons in quantity is the Assyrian army. Located in what is now northern Iraq, their clearly superior armament helps them build a mighty empire.

c. 600 B.C. First chemical theorist is the Greek philosopher Thales (624-546 B.C.) who poses the question, "Of what is the universe made?" He concludes that there exists one basic material, and that this fundamental "element" is water. This is the first clear expression of the idea of an element (or fundamental "stuff" out of which everything is made or to which everything can be reduced). Later, other Greek philosophers will dispute Thales and suggest that the basic element is air (Anaximenes [c.570-c.500 B.C.]) or fire (Heraclitus [c.540-c.475 B.C.]).

c. 450 B.C. Empedocles (c.492-c.432 B.C.), Greek philosopher, first offers his concept of the composition of matter, being made of four elements—earth, air, fire, and water. This notion is adapted by Aristotle who views the elements as combinations of two pairs of opposed properties: hot and cold, dry and moist. He also adds the idea that each element has its own innate set of properties. This notion of four elements becomes the basis of chemical theory for over 2,000 years.

c. 450 B.C. First to state the notion of atomism is Greek philosopher Leucippus. He argues that upon continuous division of a substance, eventually a point would be reached beyond which further division was not possible. His disciple, the Greek philosopher Democritus (c.470-c.380 B.C.), ultimately names these small particles "atomos," meaning indivisible.

c. 400 B.C. Chemical name for "elements"

("stoicheia") is first used by Greek philosopher Plato (c.427-c.347 B.C.). He states that things are produced from a formless primary matter that takes on forms. The minute particle of each element has a special shape, he says (fire is a tetrahedron, air an octahedron, water an icosahedron, and earth a cube). His dialogue, *Timaeus*, includes a discussion of the compositions of organic and inorganic bodies and can be considered a rudimentary treatise on chemistry.

c. 200 B.C. First important Greek writer on alchemy is the natural scientist Bolos of Mendes also called Bolos-Democritos or pseudo-Democritos. In his works *Physica et Mystica* and *De Arte Sacra Magna*, he devotes himself to the one of the great challenges of "khemeia," or chemistry, which was the changing of lead or iron into gold.

c. 100 B.C. Glass-blowing first appears. Glass objects are made using a hollow tube with the molten glass being blown into decorative molds. This method later evolves into free blowing. It is believed to be an invention of Syrian craftsmen who blow glass vessels for everyday and luxury use and export them to all parts of the Roman Empire. Syrian gaffers (blowers) eventually do away with their molds once they perfected the technique that is still in use today. The molten glass with a consistency of molasses is gathered on the end of a hollow pipe, inflated to a bubble, and formed into a vessel by blowing, swinging, or rolling on a smooth stone or iron surface. While still soft, it can be further manipulated with hand tools.

(See also c. 350)

c. 220 First crude mirrors appear. They are usually slightly convex disks made of buffed or polished copper, brass, and bronze.

c. 350 Free glass-blowing first appears. This important innovation does away with the older method of blowing molten glass into decorative molds. It allows the gaffer or blower to shape the object being made simply by blowing the molten glass at the end of a hollow tube and sometimes by swinging and rolling it.

c. 600 First appearance of bells made of cast bronze occurs. Until now, bells had been made with sheet iron and were forged and riveted. They serve a basic purpose of signalling and are used to mark significant points of ritual by calling to worship, tolling hours, and announcing events, as well as rejoicing, mourning, and warning.

c. 650 Oldest medieval source of written technical information, *Compositiones ad Tingenda Musica*, is produced. Its text contains scores of "recipes" for coloring as well as instructions on how to melt and work metals.

c. 670 First documented use of "Greek fire" is made by the Byzantines to repel the Arabs attacking Constantinople. They are able to destroy many Arabs ships by using a chemical mixture that bursts into flame when it contacts water. The ingredients for this chemical weapon are kept secret for centuries.

(See also c. 1100)

c. 775 Considered the first of the important Arab alchemists is Abu Musa Jabir Ibn Hayyan (c. 721- c. 815), known in the West as Geber. He discovers the principal salts of arsenic, sulphur, and mercury and is known as the "father of Arab chemistry" because of his contributions to distillation, sublimation, and his discovery of such facts as heating a metal adds to its weight.

821 The title of a medieval recipe book, *Mappae Clavicula*, first appears in the library catalog of the Reichenau monastery on Lake Constance. It is a randomly assembled text of formulas for pigments and painting, metallurgy, and glassmaking. It also contains information on how to make incendiary devices.

(See also c. 1100)

c. 1050 New sugar-refining process that replaces the ancient method of pressing cane is invented by the Arabs. They extract the cane juice after treating it with lime and ashes. After filtering, sugar is obtained by evaporation and crystallization.

c. 1100 First detailed description of "Greek fire," an incendiary used during battles, is written by Marcus Graecis in his *Liber Ignium ad Comburendos Hostes*. Although its exact composition is still not known today, it appears to have been a petroleum-based mixture. Other writers also mention such ingredients as saltpeter, pitch, naptha, sulfur, and charcoal.

c. 1100 Updated version of *Mappae Clavicula* contains the first description and recipe for a liquid, distilled from wine, that will catch fire. It is later identified as alcohol.

1144 First translations of Arabic alchemical manuscripts begin in Spain by English scholar Robert of Chester, also known as Robertus Cataneus or Robertus Retinensis. His Latin translations introduce European scholars to alchemy. By 1200, it was possible for European scholars to absorb the alchemical findings of the past and to try to go beyond them.

c. 1150 Theophilus the Monk (who may have been Roger of Helmarshausen) writes his original text called *Diversarum Artium Schedula*, which offers much detail about several different technical processes. It provides information on painting, glassmaking, metalwork and contains the first direct reference to paper or "Byzantine parchment."

c. 1200 Venetians develop a method for silvering glass or adding a metallic backing to glass to make a it more reflective and useful as a practical mirror. The high quality and decorative, bevelled edges of these mirrors make them very expensive.

c. 1245 Considered the first important European alchemist is German scholar Albertus Magnus (1193-1280), also known as Albert of Bollstadt. He teaches at the University of Padua and

This Arab physician, known in the west as Geber, is considered the father of Arab chemistry.

brings the Latin translations of the Arabs' work northward to Paris where he begins lecturing this year. He conducts many alchemical experiments and is the first to isolate metallic arsenic.

1247 Roger Bacon (c.1220-1292), English scholar, is one of the first in the West to mention gunpowder and give recipes on how to make it (in a letter written this year). He says it is made up of seven parts of saltpeter, five of wood charcoal, and five of sulphur. This supposedly contains too little saltpeter to work well. Bacon also divides alchemy or chemistry

into theoretical and practical, or what he calls speculative and operative, and states that it is a science somewhere between physics and biology. He also advocates using chemicals in the practice of medicine.

c. 1275 Arnold of Villanova (c.1235-1311), Spanish alchemist, first prepares pure alcohol. He also discovers carbon monoxide when he describes how wood burning under conditions of poor ventilation can give off poisonous fumes.

c. 1310 An unknown Spaniard, known only as

Engine for throwing Greek fire.

the "False Geber" because he publishes under that Arab's name, writes four books on alchemy and is the first to describe sulfuric acid. The discovery of sulfuric acid is considered the greatest chemical achievement of the Middle Ages since it makes possible all manner of chemical changes that vinegar, the strongest acid known to the ancients, cannot achieve. Even today it is among the most important industrial chemicals.

1498   First true pharmacopeia in Europe is the *Nuovo Receptaris*, published in Florence, Italy. This book provides standards of quality and strength for drugs that are used by physicians. (See also 1618)

c. 1530   First European to describe zinc is the Swiss physician and alchemist Paracelsus (1493-1541), whose real name is Theophrastus Bombastus von Hohenheim. He calls it a "bastard metal." Paracelsus is best known for founding the school of iatrochemistry, which applies chemistry to medicine.

1540   Giovanni Ventura Rosetti writes his *Plictho dell'arte de Tentori*, which is the first book on dyeing fibers and fabrics.

1540   First published work on metallurgy is *De la Pirotechnia* written by Italian chemist and metallurgist Vannuccio Biringuccio (1480-c.1539). This original work is the first of its kind in that it concentrates solely on metallurgy. It also offers the first description of the amalgamation process used to extract silver from its ores and describes casting, glassmaking, smelting, ores and minerals, assaying, and chemical processes.

1540   First written description of how to prepare ether is recorded by German physician Valerius Cordus (1515-1544). He tells how alcohol is distilled with mineral acids to make ether. It soon becomes an official part of every pharmacologist's repertoire, and many physicians gain fame by adding it to their medicinal concoctions.

c. 1548   Bernard Palissy (c.1510-1589), French potter and writer, discovers the secret of producing Italian *maiolica* and begins his production of what is called rustic ware. The tin-glazing technique known as majolica made its way to Italy from the Muslim east via Spain. Italian maiolica came to represent the best of its kind, since painting was done on dry but unfired tin-glaze which absorbed colors like a sponge and made mistakes uncorrectable. Considered one of the most eminent chemists of his time, his experiments with pottery enamels develops a distinctive French style.

1557   Julius Caesar Scaliger (1484-1558), Italian-French physician and scholar, writes his *Exercitationes Exotericae de Subtilitate* in which he makes the first specific reference to platinum as a noble metal with refractory qualities that is found in Central America. The Spanish call this new metal "platina" or "little silver" because of its resemblance to silver.

1574   Lazarus Ercker (c.1530-1594) of Germany publishes his *Beschreibung aller Furnemisten Mineralischen Ertzt und Bergkwercks Arten*, which is the first manual of analytical, metal-

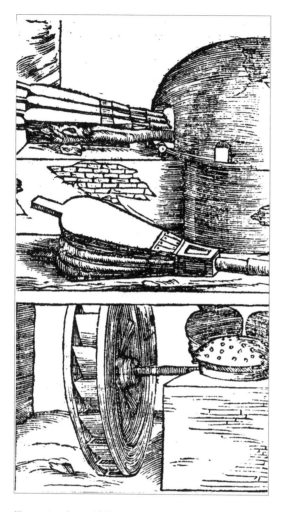

Illustration from 1540 text *De la Pirotechnia* by Vanuccio Biringucci, showing techniques for separating metals from ores.

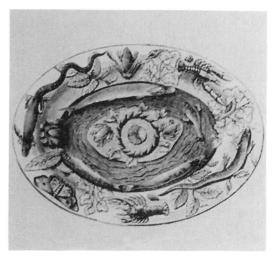

Palissy dish displaying plants and animals of France.

lurgical chemistry. His text is especially valuable to the practicing assayer. This highly practical book also describes the making of brass and the fabrication of crucibles in wooden molds in a screw press. It also gives many recipes for cement powders.

c. 1590 Zacharias Janssen (1580-c.1638), Dutch optician, helps his father Hans invent the first known compound microscope. This multiple lens design increases magnification consider-

ably. Two or more lenses are housed in a long tube, and although none individually is very powerful, each lens magnifies the image produced by the other.

1597 First textbook on chemistry is *Alchemia* by the German alchemist Andreas Libavius (c.1540-1616). A follower of Paracelsus, (See also c. 1530) he believes in the importance of the medical applications of chemistry. He also is the first to describe the preparation of hydrochloric acid, tin tetracholoride, and ammonium sulfate and is credited with planning the first true chemical laboratory. He also describes the preparation of aqua regia or "royal water," a mixture of nitric acid and hydrochloric acid that is named for its ability to dissolve gold.

1609 Johann Hartmann (1568-1631), Bavarian mathematician and iatrochemist, is nominated professor of "chymiatria" at Marburg and is thus considered to be the first professor of chemistry in Europe.

1610 First chemistry textbook in French is *Tyrocinium Chymicum* written by French chemist Jean Beguin (c.1550-c.1620) and published in Paris. This popular text goes through 50 edi-

Frontispiece sketch of Palissy, from *Palissy the Potter; or The Huguenot, Artist, and Martyr*.

Practicing metallurgy in an assay laboratory.

tions by 1669. He is also the first to mention obtaining acetone by distilling lead acetate.

1612 First book on glassmaking is Antonio Neri's *L'Arte Vetraria* published in Florence.

1618 First national pharmacopeia, *Pharmacopeia Londinesis*, is published in England. This is a book published by the government that provides standards of identity, quality, and strength of drugs prescribed.

1619 Daniel Sennert (1572-1637), German physician, publishes his *De Chymicorum*, which is the first application of Greek atomic theory to chemistry. Sennert speaks of atoms and even "second-level atoms" or molecules, and attempts to explain chemical changes in terms of these minute particles. This revival of the atomism of the Greek philosopher Democritus (c. 470-c. 380 B.C.) marks the beginning of a real scientific theory of atomism.

1646 Johann Rudolf Glauber (1604-1670), German chemist, publishes the first of his five-volume *Furni Novi Philosophici* (1646-1649), which includes his discovery of "Glauber's salt." This work gives his recipes for mineral acids

and salts including "sal mirabile"—the sodium sulfate residue that forms by the action of sulfuric acid on ordinary salt—that becomes known as "Glauber's salt." He touts his discovery as a cure-all and sells it as a laxative.

1648 First to use quantitative methods in connection with a biological problem is the Flemish physician and alchemist Johann Baptista van Helmont (1580-1644) whose writings, *Ortus Medicinae*, are published posthumously this year. Called the "father of biochemistry," he is also the first to recognize that more than one air-like substance exists, and he names this vapor or non-solid "chaos," whose phonetic sound in Flemish is "gas." He classifies several types of gases, and describes carbon dioxide as "gas sylvestre" or "gas from wood."

1658 First university chemistry laboratory is built by Dutch physician Franciscus Sylvius De La Boe (1614-1672) in Leyden, Germany, where he teaches from 1658 to 1672. He is also one of the first to appreciate the idea of chemical affinity (attractions).

1669 Johann Joachim Becher (1635-1682), German chemist, publishes his *Physica Subterranea*

Johann Joachim Becher.

in which he is the first to attempt the formulation of a general theory of chemistry. He divides all solids into three kinds of earth, one of which he calls "terra pinguis" or "fatty earth." This concept of "terra pinguis" as the substance in air that burns, forms the basis of the later phlogiston theory.

1674 First discovery of an element that was not known in any earlier form is made by German chemist Hennig Brand (c.1630-c.1692), who discovers phosphorous. By distilling concentrated urine, Brand obtains a white, waxy substance that glows in the dark. He calls it phosphorous from the Greek for "light-bearer." This mysterious and attractive glow is the result of the slow combination of phosphorous with the air. It undergoes a slow combustion even at room temperature.

1674 John Mayow (1641-1679), English physiologist, publishes his *Tractatus Quinque Medico-Physici* in which he compares respiration (breathing) to combustion. He is the first to state that the volume of air is reduced in respiration, and he proves it with his ingenious experiments. He is also ahead of his time when he correctly asserts that the blood carries air from the lungs

to all parts of the body and that oxygen is a distinct atmospheric entity.

1702 Boric acid is discovered by German physician and chemist Wilhelm Homberg (1652-1715). He prepares it by the action of iron sulfate on borax, and notes that it forms leaf-like crystals that are soft and soapy to the touch. Soluble in water, alcohol, glycerin, and boric acid can be used as a mild antiseptic. Used early on as a food preservative, the acid proves poisonous. It is later used in solutions for electroplating nickel, for tanning leather, and as an ingredient in catalysts for numerous organic chemical reactions.

(See also 1778)

1709 English physician and chemist John Freind (1675-1728) publishes *Praelectiones Chymicae* in London, which is one of the earliest attempts to use Newtonian principles to explain chemical phenomena. He argues that there exists an attractive force between particles that acts over a very small distance and decreases with distance faster than the inverse square rule. He also states that this attraction may be greater on one side of a particle than another.

1723 Georg Ernst Stahl (1660-1734), German chemist, publishes his major work, *Fundamenta Chymiae Dogmaticae et Experimentalis,* in which he develops his phlogiston theory. Although incorrect, it is the first comprehensive explanation of the phenomenon of combustion, and its popularity dominates for the next century. To Stahl, air is only the carrier of phlogiston, which he says is what is burning during combustion. Combustible materials are rich in phlogiston, and what is left after combustion is without phlogiston and therefore unable to burn. The name phlogiston is taken from the Greek word meaning "to set on fire."

1733 Swedish chemist Georg Brandt (1694-1768) publishes the first accurate and complete study of arsenic and its compounds. Not until his work is the elemental nature of arsenic understood.

1735 First person to discover a metal entirely unknown to the ancients is Swedish chemist Georg Brandt (1694-1768) who discovers cobalt. Swedish miners had retrieved this bluish mineral that resembled copper ore, but could not extract copper from it and believed it bewitched by earth spirits they called "kobolds." When Brandt demonstrates that this is in fact a new mineral, and isolates it in 1743, he names it "cobolt" after the earth spirits.

1736 Henri Louis Duhamel du Monceau (1700-1782), French chemist and botanist, first uses the term "base" to indicate those substances that combine with acids to form salts. He is also the first to make a distinction between soda and potash (potassium).

1737 First chair of chemistry is established at the faculty of medicine in Bologna, Italy, and is assigned to the German chemist Johann B. Becher.

1738 Daniel Bernoulli (1700-1782), Swiss mathematician, publishes his *Hydrodynamica* containing his kinetic theory of gases. This treatise becomes a work of major importance in both physics and chemistry. This lays the foundation for a theory of the movements of the molecules in a gas and is the first attempt at an explanation of the behavior of gases which, he assumes, are composed of a vast number of tiny particles. He uses mathematical methods and employs the known probability techniques of his time. This pioneering work of formulating principles that treat gases as groups of particles will later become the basis for atomic theories.

1739 Johann Andreas Cramer (1710-1777), German metallurgist, publishes his *Elementa Artis Docimasticae*, which is the first textbook on assaying and chemical analysis. This important technique or process determines the proportion of metal, particularly precious metal, that is in ores and other metallurgical products.

This work contains the first description of the use of a blowpipe. A blowpipe is a conical brass tube curved at the small end and terminating in a round orifice. It is used to analyze ores by placing a substance in the flame of the blowpipe and observing the changes it undergoes.

1740 First large chemical industry is founded by Joshua Ward (1685-1761) at Richmond, England, to produce sulfuric acid. The acid is obtained by heating sulphur with saltpeter in iron capsules and condensing the vapor and byproducts into large balls. Chemistry will prove to be the science most directly tied to industry.

1747 Sugar is first extracted from beets by German chemist Andreas Sigismund Marggraf (1709-1782). Although beets were known to the Egyptians and later civilizations as a source of sweets, it is not recognized as a source of sugar until Marggraf obtains sugar crystals from the beet. This leads to the development of the sugar industry.

(See also 1801)

1748 Osmosis or osmotic pressure is discovered by French physicist Jean Antoine Nollet (1700-1770). Osmosis is a process by which a solvent (the liquid that dissolves another substance) in solution passes through a barrier or permeable membrane. He covers a glass tube containing sugar water with a piece of paper and then places the tube, paper end down, into the water. He finds that the level of liquid in the tube rises since the pure water passed through the paper faster than the sugar water could. The pressure of the water passing through the membrane is called osmotic pressure. He discovers that a liquid does not soak through a membrane in both directions equally, but that there is usually a greater flow in one direction than the other. At this point, however, he is unable to explain why this is happening.

(See also 1826)

1749 Formic acid is discovered by German chemist Andreas Sigismund Marggraf (1709-1782).

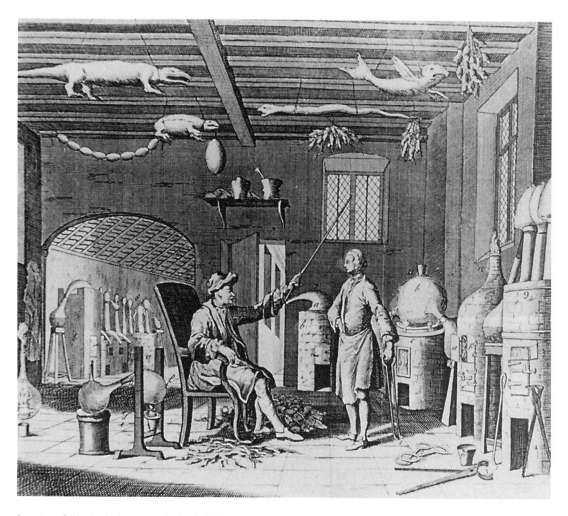

Interior of chemical laboratory, England, 1748.

Also called methanoic acid, it is the simplest of the carboxylic acids and comes to be used in processing textiles and leather. He prepares this colorless, fuming liquid with a pungent odor by distilling red ants and then redistilling the product. It also occurs in the stings of bees and in some plants. The name is derived from "formica," the Latin for ant.

1750 First complete account of platinum and its properties appears in *Philosophical Transactions*. English physician William Brownrigg (1711-1800) gives a thorough study of platinum,

noting that it is heavier and even more chemically inert than gold. This metal was long-known by the natives of South America, and knowledge of it had made its way to Europe by mid-16th century.

(See also 1557)

1751 Nickel is discovered by Swedish chemist Axel Fredrik Cronstedt (1722-1765). While studying a type of false copper that Swedish miners called "Kupfernickel," or "Old Nick's (meaning the devil's) copper," he initially obtains green crystals, which, when heated with charcoal, yield a white metal. Resembling both iron and cobalt, he finds that nickel can be attracted weakly to a magnet, and it is the first time any metal but iron is found subject to magnetic attraction since the times of the Greeks. He calls this new metal nickel after a shortened version of the miner's old name.

1756 First to show that carbon dioxide can be formed by the decomposition of a mineral as well as by combustion and fermentation is Scottish chemist Joseph Black (1728-1799). He publishes his paper "Experiments upon Magnesia Alba, Quicklime, and Some Other Alcaline Substances" in which he describes his discovery of carbon dioxide. He shows that carbon dioxide is a normal component of the atmosphere and that it is what humans exhale. His experiments are a model of logical, quantitative work as he carefully weighs and measures, and they are regarded as the first work on quantitative chemistry.

1762 Joseph Black (1729-1799), Scottish chemist, first introduces the term as well as the notion of "latent heat." He is the first to realize that there is a distinction between the quantity of heat and the intensity of heat, and that it is only the latter that is measured as temperature. He shows that when ice is heated, it slowly melts but its temperature remains the same. The ice therefore absorbs a quantity of "latent heat" in melting, thus increasing the amount of heat it contains but not increasing its intensity. The concept of latent heat will prove important to the conservation of energy laws and the invention of the steam engine.

1765 Swedish chemist Karl Wilhelm Scheele (1742-1786) discovers prussic acid. Known also as hydrogen cyanide, this highly volatile, colorless liquid is extremely poisonous and is eventually used in industrial chemical processing. Scheele had the habit, as did many of his contemporaries, of testing many of the chemicals he discovers and uses on himself. His early

Karl Wilhelm Scheele.

death is thought to have been the result of slow, cumulative poisoning. During his career, Scheele makes over 20 major chemical discoveries, many of them acids, making him one of the most prolific chemists of all time.

1766 Pierre Joseph Macquer (1718-1784), French chemist, publishes his *Dictionnaire de Chymie*, which is the first dictionary of chemistry.

1770 Johan Gottlieb Gahn (1745-1818) and Karl Wilhelm Scheele (1742-1786), both Swedish chemists, independently discover phosphorous (phosphoric acid) in bones, noting that it is an essential component. All vertebrates possess phosphorous in their systems and about 85% of it is in their bones. It is present in the fluids within cells of living tissues as a phosphate ion, and the body's bones contribute to the regulation of the concentration of this ion.

1771 English chemist Peter Woulfe (c.1727-1803) obtains picric acid by treating indigo with nitric acid. Named after the Greek word for "bitter," picric acid is produced by the action of concentrated nitric acid on various organic substances such as silk, leather, or wool. It crystallizes in yellow plates and has an intensely bitter taste. Although it is used initially as a dye, it comes to be used in explosives. As this substance is used to dye wool yellow, it is considered to be the first artificial organic dye.

1772 Louis Bernard Guyton de Morveau (1737-1816), French lawyer and chemist, demonstrates for the first time that metals gain weight on calcination (the chemical change that occurs when metals are heated to just below their fusion or melting point).

1772 Nitrogen is discovered by Scottish chemist Daniel Rutherford (1749-1819). After burning a candle in a closed container until it goes out, he then absorbs all the carbon dioxide produced by combining it with certain chemicals. He finds that there is still a great deal of gas left that will support neither combustion or animal respiration. He does not know what this gas is. Nitrogen is rediscovered and named shortly after Rutherford by Priestley, Scheele, and Cavendish.

(See also 1790)

1774 Oxygen is discovered by English chemist Joseph Priestley (1733-1804). He finds that mercury, when heated in air, forms a brick-red compound or calx we now call mercuric oxide. He then heats this in a tube by using sunlight and a lens, and finds that the compound breaks up, liberating mercury. Besides this liquid mercury, a gas is also given off in which combustibles burn more brightly than in ordinary air. Breathing it himself, he feels "light and easy." Attempting to explain this new gas in terms of the prevailing phlogiston theory, he calls it "dephlogisticated air." Oxygen had actually been isolated sometime between 1770 and 1773 by Swedish chemist Karl Wilhelm Scheele (1742-1786), who called it "fire air." Priestley, however, publishes his results first.

(See also 1775)

1774 Eudiometer is invented by Italian naturalist and physiologist Felice Fontana (1730-1805). This is a finely graded and calibrated tube for the volumetric measurement and analysis of gases. With the discovery of oxygen this year by English chemist Joseph Priestley (1733-1804), Fontana finds he is able to measure what Priestley calls "the freshness of air" or its oxygen content.

1775 Antoine Laurent Lavoisier (1743-1794), French chemist, publishes his "Memoire" paper which contains his first major disavowal of the phlogiston theory as well as a revision of his combustion theory. After a meeting in Paris during October 1774 with English chemist Joseph Priestley (1733-1804), who discusses his experiments of "dephlogisticated air," Lavoisier realizes that Priestley's gas is really oxygen after all.

1777 German chemist Carl Wenzel (1740-1793) discovers the reaction rates of various chemicals and first states that they are roughly proportional to the substance concentration. Experimenting on the composition of salts and on the rate of solution of metals in acids, he states that the rate of solution of a metal is proportional to the concentration of the acid.

1778 Tobern Olof Bergman (1735-1784), Swedish mineralogist, offers the first comprehensive analysis of mineral water. After introducing many new reagents and analytical methods for chemical analysis, he discovers hydrogen sulphide in mineral springs.

1778 Hubert Franz Hofer of Germany discovers that boric acid exists naturally in a hot spring in Tuscany (Italy). The following year, Italian anatomist Paolo Mascagni (1752-1815), discovers solid boric acid (sassolit) in Montecerboli and Castelnuovo in his native land.

(See also 1702)

SEPTEMBER 5, 1779 Antoine Laurent Lavoisier (1743-1794), French chemist, first proposes the name "oxygen" for that part of the air that is breathable and which is responsible for combustion. With the downfall of the theory of phlogiston, a new name is required, and Lavoisier notes that when elements like carbon and sulphur are burned in this new gas, they produce compounds with acidic qualities. For a short time he uses the term "principe acidifant" or "acidifying principle," and then switches to "principe oyxgine." He devises "oxygine" from the Greek words "oxys" for "acid," and "geinomai" for "I beget." He then changes the spelling to "oxygene," which becomes "oxygen" in English in 1790.

c. 1780 Claude Louis Berthollet (1748-1822), French chemist, establishes the composition of ammonia and conducts the first thorough study of chlorine, introducing it as a bleaching agent. Although he incorrectly concludes that chlorine is a compound of oxygen rather than an element, he correctly determines the composition of ammonia and is the first to show its composition of nitrogen and hydrogen. Until chlorine, the only way to bleach cloth is to leave it in the sun for weeks.

1780 Lactic acid is discovered by Swedish chemist Karl Wilhelm Scheele (1742-1786). Found in soil and in fermented milk products, it eventually is used in food processing and for tanning leather and dyeing wool. It is also found in the blood and muscles of animals, since lactic acid is formed in a muscle when it performs work. When a muscle is not allowed to rest and continues to work, a build-up of this acid causes pain and eventually its shutdown.

1781 Molybdenum is discovered by Swedish chemist Peter Jacob Hjelm (1746-1813). He is able to reduce molybdic acid to a metal state by using carbon as a reducing agent. Molybdenum is a tough, malleable, silver-white metal used in alloys.

1782 French chemist Antoine Laurent Lavoisier (1743-1794) studies metals that have calcinated or rusted and observes that their total weight is not changed but merely shifted around. This establishes an early version of the first law of conservation of matter. Lavoisier confronts what

Antoine Lavoisier's apparatus for measuring weight changes.

appears to be the loss of mass during some chemical reactions by conducting a classic experiment. Using a sealed glass container which he weighs before and after the reaction, he shows that although there appears to be less material inside, it was simply transformed into a gas and therefore weighs the same as before the reaction.

1782 Tellurium is discovered by Austrian mineralogist Franz Joseph Müller (1740-1825), also known as Baron von Reichenstein. Working with gold ore, he extracts a substance that seems to resemble antimony. Müller sends a sample to German chemist Martin Heinrich Klaproth (1743-1817), who confirms the existence of a new element and names it after the Latin "tellus" for our own planet, Earth. Although there are adequate supplies of this element, no single use has been developed that creates a large demand for it.

1783 Nicolas Leblanc (1742-1806), French chemist, invents a process for the manufacture of soda (sodium hydroxide and sodium carbonate) from salt (sodium chloride). He soon opens a factory to manufacture soda that is much-

Antoine Lavoisier.

needed in France for a variety of chemical purposes and for the soap, glass, and paper-pulp industries. This is the first chemical discovery that has an immediate commercial use. It is also the first major example of a technology being born out of an advance in theoretical science. Leblanc's process is not perfect, however, requiring hand labor and giving off hydro-chloric acid as a by-product of producing only 28% pure soda.

1783 Juan Jose D'Elhuyar (1754-1796) and his younger brother Don Fausto D'Elhuyar (1755-1833), both Spanish mineralogists, analyze a mineral called wolframite that is obtained from a tin mine and discover the new metal wolfram. The Swedish chemist Karl Wilhelm Scheele (1742-1786) had investigated it in 1781 but believed it to be a compound. He names it "tungsten" for "heavy stone" in Swedish (because of its high density), and the names catches on.

1783 Glycerine is discovered by Swedish chemist Karl Wilhelm Scheele (1742-1786). Also called glycerol, it is obtained by the hydrolysis of animal fats or vegetable oils. This thick, clear, sweet-tasting liquid is eventually used in

making resins and gums for paints and as a softener in other products.

1784 Citric acid is discovered by Swedish chemist Karl Wilhelm Scheele (1742-1786). This colorless, crystalline organic compound belongs to the family of carboxylic acids and is present in nearly all plants and in many animal tissues and fluids. It is eventually manufactured by the fermentation of cane sugar or molasses in the presence of a fungus, and is used as a flavoring agent in soft drinks. It also improves the stability of foods and can be used as a metal-cleaning product. The name is derived from the Latin "citrus," which includes such fruits as lemons and oranges.

1786 Claude Louis Berthollet (1748-1822), French chemist, and French mathematicians Gaspard Monge (1746-1818) and Alexandre Theophile Vandermonde (1735-1796) publish a paper, "Memoire sur le Fer" which first establishes that the difference between iron and steel is due to carbon. They demonstrate that it is the addition of carbon to iron that toughens and hardens it.

1788 Blagden's law is first enunciated when English physician and secretary of the Royal Society Charles Blagden (1748-1820) discovers that the lowering of the freezing point of a solution is proportional to the concentration of the solute.

1789 First French journal for chemistry, *Annales de Chimie,* is founded by French chemist Antoine Laurent Lavoisier (1743-1794) and his colleagues. This journal, which continues today, publishes some of the most important papers in the history of chemistry.

1789 Antoine Laurent Lavoisier (1743-1794), French chemist, publishes *Traité Elémentaire de Chimie,* which defines the new science and becomes the first modern chemical textbook. In it, he unifies the subject and states his law of the indestructibility of matter (conservation of mass). He also popularizes chemistry and puts an end to the old phlogiston theories. Lavoisier further revives the old notion of element as something that cannot be broken down, and lists 33 of them (only two are incorrect).

1789 Uranium is discovered by German chemist Martin Heinrich Klaproth (1743-1817), who obtains a yellow compound from the heavy black ore called pitchblende. This substance is actually the oxide of the metal (which has to be further reduced to get the actual metal). He names the new metal after the newly discovered planet Uranus. Later in the same year he suggests that the orange-red gem named zircon is not a silicate of alumina but a new element he names zirconia.

(See also 1842)

1790 Adair Crawford (1748-1795), Irish physician and chemist, and Scottish surgeon and chemist William Cumberland Cruikshank (1745-1800) first suggest that the mineral found in lead mines at Strontian in Argyllshire, Scotland contains a new element they call strontianite. In 1808, it is first isolated by the English chemist Humphry Davy (1778-1829) who calls it strontium. It is a soft metal like lead which has a silvery lustre when cut, but reacts to the air and takes on a yellowish color.

1790 Nitrogen is first named by French chemist Jean Antoine Claude Chaptal (1756-1832). Responding to the experiment by English chemist and physicist Henry Cavendish (1731-1810), who converts air into "nitre" or potassium nitrate by mixing it with an excess of oxygen and wood ash (potassium carbonate), he names the new gas "nitrogen" for nitrogen-producer.

1791 Titanium is discovered by English mineralogist William Gregor (1761-1817). He describes his analysis of a mineral obtained from a valley in Cornwall, England, reporting that the mineral contains a "reddish brown calx" that he cannot identify. He thinks it may be a new substance but does not name it.

(See also 1795)

1794 A new mineral named ytterbite is discovered by Finnish chemist Johan Gadolin (1760-1852). He finds it in a quarry at the little town of Ytterby near Stockholm, Sweden. In the following century, this complex element will be shown to contain over a dozen different elements which come to be called the rare earth elements.

1795 Martin Heinrich Klaproth (1743-1817), German chemist, isolates the new metal titanium and names it after the Titans of Greek mythology. He also gives full credit to English mineralogist William Gregor (1761-1817), who first discovered it in 1791. Titanium proves to be a common metal that is highly resistant to acids, and although it is the strongest of the metals, it is remarkably light. It proves highly suitable for supersonic aircraft whose outer skin must resist high temperatures yet be light and strong.

1796 Smithson Tennant (1761-1815), English chemist, conducts a series of quantitative combustion experiments and establishes for the first time that a diamond is chemically identical to carbon. He completely burns a diamond and, by measuring the carbon dioxide produced in this process, proves that a diamond does not only contain carbon but that it consists entirely of carbon.

1798 Chromium is discovered by French chemist Louis Nicolas Vauquelin (1763-1829). While examining a mineral from Siberia, he obtains a yellow solution (potassium chromate) which yields a red precipitate with a mercury salt and a yellow one with a lead salt. He concludes that he has found a new element and is able to isolate it a year later. Chromium gets its name from its multicolored compounds. It proves to be a hard, steel-gray metal that takes a high polish and is used in alloys to increase strength and corrosion resistance.

MAY 1800 First demonstration that an electric current can bring about a chemical reaction is made by English chemist William Nicholson (1753-1815) and English anatomist Anthony Carlisle (1768-1840). They discover that water can be decomposed (hydrogen and oxygen separated) by an electric current. Using the new voltaic cell, they place wires attached to it in water and find that, when current runs through the wires, bubbles of gas (hydrogen and oxygen) are given off. They thus have used electricity to produce a chemical reaction and have "electrolyzed" the water (breaking up its molecules into their individual elements). Their discovery of electrolysis opens up the new field of electro-chemistry.

1800 Johann Wilhelm Ritter (1776-1810), German physicist, repeats the Nicholson/Carlisle water electrolysis experiment but arranges to collect the two gases (hydrogen and oxygen) separately for the first time. In further experiments this year, he discovers electroplating when he passes an electric current through a solution of copper sulfate and produces metallic copper. The electric current separates the molecules of copper and allows it to be plated to other metals.

1801 German chemist and physicist Franz Karl Achard (1753-1821) develops compatriot chemist Andreas Sigismund Marggraf's sugar-from-beets discovery and opens the first European beet sugar refinery in Silesia. He improves on Marggraf's methods, which are expensive and have a low yield, and successfully transforms a crop used traditionally as animal feed into a new industry. Later, he opens a school devoted to teaching and promoting the production of beet sugar.

(See also 1747)

1801 Niobium is discovered by English chemist Charles Hatchett (1765-1847). While studying a black mineral that had been sent to England from the Massachusetts area of the American colonies, Hatchett declares that it is a new mineral and names it columbium after the poetic synonym for the new United States.

His claim is challenged by English chemist and physicist William Hyde Wollaston (1766-1828), who later states it is the same element as tantalum. This view prevails until 1846 when it is finally proven that Hatchett is correct. An international commission gives it the name niobium after Niobe, the daughter of Tantalus (the son of Zeus).

1801 Vanadium is discovered by Spanish-Mexican mineralogist Andreas Manuel Del Rio (1764-1849). His studies of what is called "brown lead" help him uncover a new element, which he names erythronium (also the name of the spring-blooming plant called the dog's tooth violet). However, when colleagues convince him that his new element is really chromium, he drops his claim of discovery. Later, when Swedish chemist Nils Gabriel Sefström (1787-1845) discovers vanadium in 1831, it proves to be identical with Del Rio's erythronium. It is usually Sefström who gets credit for this discovery.

1801 Robert Hare (1781-1858), American chemist, invents the oxy-hydrogen blowpipe. It is the ancestor of the modern oxyacetylene torch used to weld or cut steel. Hare uses his invention to melt sizeable quantities of platinum.

1802 Joseph Louis Gay-Lussac (1778-1850), French chemist, shows for the first time that all gases expand equally or by equal amounts with the same increase in temperature. While most scientists of this time know that heat makes gases expand in volume, it is not until Gay-Lussac's series of careful experiments that they are able to predict that expansion with precision. This principle becomes known as both "Gay-Lussac's law" and "Charles' law," after French physicist Jacques Alexandre Charles (1746-1823) who actually discovers it before Gay-Lussac but never publishes his experiments. Gay-Lussac later discovers that gases combine with each other in simple proportions, an elaboration that becomes known as Gay-Lussac's law of combining volumes.

1802 First to use letters as chemical symbols is Scottish chemist Thomas Thomson (1773-1852) who publishes his *System of Chemistry* and introduces a system of symbols for individual minerals using the first letters of their names. In 1813, Swedish chemist Jöns Jakob Berzelius (1779-1848) suggests that these chemical symbols be based on the first letter of the Latin name of the elements. When two or more elements possess the same initial, a second letter from the body of the name could be added. This system is eventually adopted by mid-19th century.

1802 William Hyde Wollaston (1766-1828), English chemist and physicist, examines a candle flame through a prism and is able to distinguish dark lines that cross the spectrum. He dismisses them as natural boundaries between the colors, and does not know that he is one of the first to observe the different wave lengths of light.

1802 Tantalum is discovered by Swedish chemist Anders Gustaf Ekeberg (1767-1813). He is able to find the same new element in minerals taken from Kimito, Finland, and Ytterby, Sweden. He names it after the mythical son of Zeus Tantalus, who was condemned for revealing secrets of the gods, to stand in water up to his chin with fruit hanging above him. Whenever he tried to drink the water, it would recede, and the fruit would evade his grasp as well. The English verb "tantalize" comes from his frustration. Ekeberg apparently names his new element after Tantalus because of the difficulty he experienced trying to discover the metal. It is also an allusion to the metal's inability to absorb any acid although immersed and saturated in it.

c. 1803 William Hyde Wollaston (1766-1828), English chemist and physicist, discovers a method of making platinum malleable, thus allowing it to be hammered and molded into certain shapes for making laboratory apparatus. He keeps this process a secret and amasses great wealth, making him financially independent.

He arranges to have his methods released and publicized after his death, and this information forms the basis of modern powder metallurgy. This is the fabrication of metal forms out of a powder rather than molten metal, and it is often more economical than traditional methods. It also is used when melting is impractical because of the very high melting point of a certain metal or because an alloy is desired from two materials that cannot be fused (like copper and graphite).

1803 Jöns Jakob Berzelius (1779-1848), Swedish chemist, and Swedish mineralogist Wilhelm Hisinger (1766-1852) work together and discover cerium. It is also discovered independently this year by German chemist Martin Heinrich Klaproth (1743-1817). A fairly abundant rare earth metal, it is named after the asteroid Ceres, which was discovered in 1801. It later comes to have several industrial applications.

1803 Claude Louis Berthollet (1748-1822), French chemist, discovers that the manner and rate of chemical reactions depended on more than affinities or the attraction of one substance for another. He shows that reactions are affected by relative quantities and temperature, and that many reactions are reversible. This is a foreshadowing of the very important chemical law of mass action and can be regarded as the first attempt to consider the physics of chemistry.

1803 Jöns Jakob Berzelius (1779-1848), Swedish chemist, first suggests that acids and bases have opposite electrical charges. He discovers this while experimenting with a new voltaic pile and links this discovery to the concept of electrical polarity, which he then extends to the elements. He states that oxygen is the most electronegative element. He arranges the elements in a series with oxygen at one end and potassium, the most electropositive element, at the other end.

1804 New elements, iridium and osmium, are discovered by English chemist Smithson Tennant

(1761-1815) in the black powder that remains when platinum ore is chemically treated. Although many chemists suspect that this material contains other elements, it is not until Tennant dissolves platinum in dilute aqua regia that he finds two new elements with distinct properties in the leftover insoluble, black residue. He calls one iridium, from the Greek "iris" for rainbow, since it has a "striking variety of colors." The other he names osmium from the Greek word "osme" for smell, since it has a peculiarly pungent odor.

1805   Modern atomic theory first begins with English chemist John Dalton (1766-1844), who publishes the first table of atomic weights and invents a new system of chemical symbols. His chemical research leads him to believe that all the elements are composed of extremely tiny, indivisible atoms that are indestructible. Further, he says that the atoms of any one element are all exactly alike, but atoms of different elements are different from each other. However, one substance could be transformed into another by rearranging one particular combination and forming a new one. The key to his theory is his argument that atoms differ from each other only in mass which is something that can be known because it can be measured. He then offers a method whereby the weights of atoms can be determined and thus becomes the first to offer a quantitative atomic theory. It becomes quickly accepted by the majority of chemists.

1806   First amino acid to be discovered is asparagine which Louis Nicolas Vauquelin (1763-1829), French chemist, isolates from asparagus. This is the first of more than 100 naturally occurring organic acids. The twenty or so most important amino acids are found in proteins (as their building blocks).

1806   Humphry Davy (1778-1829), English chemist, performs experiments on electrolysis (the interactions of electric currents with chemical compounds) that lead to his first statement of the electro-chemical theory. He concludes that the production of electricity in simple electrolytic cells results from chemical action, and that chemical combination occurs between substances of opposite charge. He then reasons that electrolysis offers the most likely means of decomposing all substances to their elements. Davy also discovers potassium by passing an electric current through molten ash, and one week later isolates sodium from soda.

1807   Jean Antoine Claude Chaptal (1756-1832), French chemist, publishes his *Chimie Appliquée aux Arts* which is the first book devoted specifically to industrial chemistry. In 1781, he established a plant at Montpellier for the first commercial production if sulfuric acid in France. The sulfuric acid industry grows as the demand for organic chemicals increases, and it becomes one of the most essential products for modern industry.

1808   First to identify the sugar in grapes as glucose is Joseph Louis Proust (1754-1826), French chemist, who distinguishes among the different sugars in plants. Earlier he had proposed an ambitious generalization that eventually became known as Proust's law.

1808   English chemist Humphry Davy (1778-1829) first isolates barium, strontium, calcium, and magnesium.

1809   Joseph Louis Gay-Lussac (1778-1850), French chemist, collaborates with German naturalist Friedrich Wilhelm Heinrich Alexander von Humboldt (1769-1859) and states the law of combining volumes or gaseous volume. They discover that in forming compounds, gases combine in proportions by volume that can be expressed in small whole numbers. An example is water in which two parts of hydrogen unite with one part of oxygen. Gay-Lussac's law leads directly to what is called "Avogadro's hypothesis."

(See also 1811)

1811 Joseph Louis Gay-Lussac (1778-1850) and Louis Jacques Thenard (1777-1857), both French chemists, determine the elementary composition of sugar for the first time. This is also the first efficient method for analyzing organic matter. They take sugar or any other organic substance, mix it with potassium chlorate, form this solid into a tiny loaf, and drop it through a large stopcock into a heated tube set vertically. They then analyze the gas that develops.

1811 Avagadro's hypothesis is first proposed. Stimulated by the discovery by French chemist Joseph Louis Gay-Lussac (1778-1850) that all gases expand to the same extent with the rise in temperature, Italian physicist Amedeo Avogadro (1776-1856) first offers his theory of particles or molecules, which is confirmed much later by modern chemistry. He states that equal volumes of all gases contain the same number of particles if they are under the same pressure and temperature.

1811 Iodine is discovered by French chemist Bernard Courtois. Working in the family business of manufacturing saltpeter from seaweed ashes treated with acid, he accidentally adds too much acid and produces a violet-colored vapor cloud. When this condenses, it forms dark, shiny crystals. Feeling that this might be a new element but unsure of his abilities, he passes it on to English chemist Humphry Davy (1778-1829) and French chemist Joseph Louis Gay-Lussac (1778-1850), who prove him to be correct. Gay-Lussac names it after the Greek word for purple, "iodes," but Davy's anglicized version "iodine" proves more acceptable since it is conveniently analogous to chlorine and fluorine.

1811 Nitrogen trichloride is discovered by French chemist Pierre Louis Dulong (1785-1838). Experimenting with compounds of chlorine and nitrogen, he discovers that nitrogen trichloride is violently explosive and loses an eye and two fingers. Despite this, he continues to work on the compound.

1813 Humphry Davy (1778-1829), English chemist, publishes his *Elements of Agricultural Chemistry*, a pioneering work that becomes the first textbook dealing with the applications of chemistry to agriculture. He also introduces a chemical approach to mineralogy and the tanning industry.

1815 Joseph Louis Gay-Lussac (1778-1850), French chemist, first establishes the quantitative composition of the organic radical called cyanogen and describes its properties, preparation, and compounds in a paper entitled, "Recherche sur L'Acide Prussique." His research on cyanides offers conclusive proof that prussic acid, or hydrogen cyanide, contains no oxygen. This shows once and for all that acids can be acids without the presence of oxygen.

1815 First safety lamp for coal miners is invented by English chemist Humphry Davy (1778-1829). Each year hundreds of miners are killed by explosions in mines when their candle-lamps ignite "fire-damp" gas. Davy spends three months studying mining conditions, and solves the problem by understanding the properties and nature of methane gas and oxygen. After confirming that it is mainly methane gas that causes the explosions and that it would ignite only at high temperatures, he designs a safety lamp whose flame is surrounded by wire gauze. This dissipates the heat and prevents ignition of this explosive gas. This invention is the first major step toward safety taken by the coal mining industry.

1816 Limelight or first spotlight in the history of the theater is invented by Thomas Drummond (1797-1840), an English engineer. He discovers that heating lime or a block of calcium in an alcohol flame burned in oxygen-rich air produces a soft but stunningly bright artificial light that can be directed and focussed. Drummond uses a parabolic mirror to reflect this light and turn it into a spotlight. Impractical for street lighting because it must be constantly tended by an individual operator, it is first used to

DeChangy's 19th century incandescent lamp for miners, inspired by Humphrey Davy's earlier safety lamp.

for stone, "lithos," since unlike the previously known alkalai metals (sodium and potassium), it is discovered in the mineral kingdom. Sodium and potassium were found in vegetable ashes.

1817   Cadmium is discovered by German chemist Friedrich Strohmayer (1776-1835). While in an apothecary's shop, he notices that a bottle labeled zinc oxide actually contains zinc carbonate. Pursuing this, he finds that zinc carbonate yields an orange-yellow color when heated, as if it contained iron as an impurity. Unable to find any iron, he traces the yellow substance to an oxide of a new element that is chemically similar to zinc. He names it for the Kadmean ore in which it is usually found with zinc.

1818   Selenium is discovered by Swedish chemist Jöns Jakob Berzelius (1779-1848) and Swedish mineralogist Johan Gottlieb Gahn (1745-1818), who investigate a red deposit that forms on the floors of a sulfuric acid factory. They note that the sulphur used there comes from copper pyrite found in a particular mine in Sweden. After many stages of precipitating, separating, and evaporating this red sediment, they distill a residue that shows by its unusual behavior that it is a new element. Berzelius names it after the Greek word for Moon, "selene" (since tellurium which it resembles was named after the Earth).

1819   Law of isomorphism is first stated by German chemist Eilhardt Mitscherlich (1794-1863). He discovers that compounds known to have similar compositions tend to crystallize together, as if the molecules of one became linked with the similarly shaped molecules of another. From this knowledge, he states the law which says that if two compounds crystallize together and the structure of only one of them is known, then it is safe to assume that the structure of the second is quite similar. This eventually serves as a guide to the correct atomic weights of elements.

illuminate a theater stage in 1837. By the 1860s, it is widely used, and the expression "to be in the limelight" refers originally to front and center stage which was the most brilliantly illuminated spot.

1817   Lithium is discovered by Swedish chemist Johan August Arfvedson (1792-1841). While examining a newly discovered mineral called perlite, he finds that this alkali metal is not a silicate of sodium as expected, but an entirely new element. It is named after the Greek word

1820 Coumarin is discovered by German chemist Heinrich August von Vohel (1778-1867). This crystalline and colorless organic compound smells like new-mown hay and is found in the tonka bean native to Guyana. It comes to be used as a vanilla substitute for many decades before its health hazards are recognized. It is also used in perfumery, being the first natural perfume to be synthesized from coal tar. Its name comes from the native name in Guyana for the tonka tree.

1821 Solanine is discovered by Rene Desfosses of France. He obtains this glucosoidal alkaloid from the berries of *Solanum nigrum*, a species of the genus *Solanum*. This species is related to many other species of the common nightshade or potato family of flowering plants (such as eggplant, pepper, tomato, and tobacco). This class of plants contains well-known alkaloids that come to have considerable significance because of their powerful physiological effects. Besides solanine, other well-known alkaloids are morphine, strychnine, quinine, and nicotine.

1821 Caffeine is discovered by German chemist Friedlieb Ferdinand Runge (1795-1867). He extracts it from the coffee bean and names it after the French word for coffee, "cafe." A nitrogenous organic compound of the alkaloid group, it also occurs naturally in tea, guarana, mate, kola nuts, and cacao. It is a stimulant of the central nervous system and also acts as a diuretic.

1822 Adsorptive properties of activated charcoal are first put to use by French chemist Anselme Payen (1795-1871). While attempting to remove coloring impurities from the sugar made from sugar beets, he notices the filtering properties of charcoal and begins its regular use. Charcoal has since found numerous applications for its filtration and adsorption properties, and is later used during World War I in gas masks to replace standard carbon filters which absorb dangerous organic gases.

1823 First to produce laboratory temperatures below 0°F (-18°C) is English physicist and chemist Michael Faraday (1791-1867), who devises methods for liquefying gases (such as carbon dioxide, hydrogen sulfide, hydrogen bromide, and chlorine) under pressure. Because of this landmark work on gas liquefaction, he also can be considered a pioneer in the modern branch of physics called cryogenics or the study of extreme cold.

1823 Chemical nature of fats is first investigated by French chemist Michel Eugene Chevreul (1786-1889), who publishes his classic work dealing with oils, fats, and vegetable colors. He shows that fat is a compound of glycerol with an organic acid. This is one of the first works addressing the issue of the fundamental structure of this large class of compounds and it has a revolutionary effect on the soap and candle industries. Chevreul becomes a pioneer in both the analysis of organic substances and in gerontology (the scientific study of the phenomena of aging). He publishes his final scientific paper at the age of 102 and dies a year later.

1823 Silicon is first isolated in a reasonably pure form by Swedish chemist Jöns Jakob Berzelius (1779-1848), who describes it as a new element. Making up 27% of Earth's crust, silicon is the second most abundant element. Since it never occurs naturally by itself, however, many scientists doubt that it exists until Berzelius proves it does. Although it has little use on its own, its most practical form is as a compound, and silica and silicates come to have all manner of industrial uses. The name is derived from the Latin "silica" for pebble or flint.

1823 Waterproof fabric is first patented by Scottish chemist and inventor Charles Macintosh (1766-1843). While trying to find uses for the waste products of the Glasgow gasworks—one of which is a volatile liquid called naptha—he discovers that it dissolves rubber. This gives him the idea to try to waterproof fabric, and he paints one side of wool cloth with this new

An 1846 wood engraving of Michael Faraday lecturing.

rubber solution and puts another layer of wool over it. He finds that the rubber interior makes a sandwich of waterproof cloth. He later produces rainproof cloth which does not prove practical until the invention of vulcanized rubber. His company eventually produces the "Mackintosh" or the world's first raincoat.

1825 Benzene is first isolated by English physicist and chemist Michael Faraday (1791-1867). He discovers this hydrocarbon in a liquid which condenses in the cylinders of an illuminating gas made from fish oil. A clear, colorless highly flammable liquid, it can produce severe or fatal irritation of the mucous membranes if inhaled. Many regard this discovery as Faraday's greatest single contribution since it is benzene that plays a key role in the development of a means of representing molecular structure achieved by German chemist Friedrich August Kekulé von Stradonitz (1829-1896). This, in turn, proves to be the starting point for understanding organic chemistry.

(See also 1861)

Michael Faraday.

1826 First quantitative experiments on osmosis are conducted by French physiologist René Joachim Henri Dutrochet (1776-1847), who both discovers and names osmosis (the passage of solvent through a semipermeable membrane). Using experiments, he determines that the pressures involved during the diffusion of solutions are proportional to the solution concentrations. He constructs an osmometer (a device to measure osmotic pressure) and develops a technique to detect heat production in plants. He is also the first to observe the motion of particles suspended in a liquid, later called Brownian motion, and one of the first to recognize the importance of individual cells in the functioning of an organism.

1826 Otto Unverdorben (1806-1873), German chemist, first prepares aniline by distilling the indigo plant. He finds that the alkaline oil in the indigo distillate forms beautiful crystals when combined with sulfuric acid, and he names it "kristallin." It is later named "aniline" by the German chemist Carl Julius Fritzsche (1808-1871) because indigo is called "anil" in Spanish. This organic product proves to be useful in

making dyes, drugs, explosives, plastics, and photographic and rubber chemicals.

(See also 1856)

1826 Bromine is discovered by French chemist Antoine Jerome Balard (1802-1876). While studying the brine from a salt marsh in his native Montpellier, he obtains a dark red liquid with a foul smell after he chlorinates water from which the sodium chloride has been removed. Upon analysis, he realizes it is a new element and demonstrates its properties, showing that they are analogous to iodine and chlorine. He names it muride after the Latin "muria" for brine, but most chemists oppose this name since it is easily confused with muriatic acid. Its name is eventually changed to bromine after the Greek "bromos" for stench or evil smell.

1827 First classification of the components of food into carbohydrates and proteins is made by English chemist and physiologist William Prout (1785-1850). He uses the words saccharinous, oleaginous, and albuminous for the three respective groups.

1827 Friedrich Wohler (1800-1882), German chemist, first isolates metallic aluminum and describes its properties. He is able to extract the aluminum by heating a mixture of potassium and aluminum chloride in a platinum crucible. He uses a similar technique to discover beryllium in 1828. Beryllium is a steel-grey metal that does not occur free in nature and is very brittle at room temperature.

1829 Thorium is discovered by Swedish chemist Jöns Jakob Berzelius (1779-1848) while studying a mineral obtained from Norway. Named after Thor, the Scandinavian god of war, this silvery white metal turns gray or black when exposed to air. It is eventually discovered to be radioactive and has the highest boiling point of any oxide (6,134° F or 3,390°C).

(See also April 1898)

1830 Paraffin is discovered by German naturalist Karl von Reichenbach (1788-1869). While experimenting with wood tar, he obtains a white, waxy substance that is a mixture of hydrocarbons, and which he names paraffin from the Latin "parum" for little or not very and "affinis" for affinity. Paraffin is generally an unreactive substance. It comes to be called paraffin wax and is distinguished from paraffin oil. This is a distilled product of lignite or petroleum and has a high boiling point. It is also called kerosene.

(See also 1852)

1831   Pierre Jean Robiquet (1780-1840), French apothecary, and Jean Jacques Colin (1784-1865), French chemist, discover the red dye alizarin, which they isolate from the root of the common madder plant. The root of the madder plant had been ground and used for dyeing in ancient India, Persia, and Egypt. Isolation of the dye in its purified form eventually leads to its synthetic production.

(See also June 26, 1869)

1831   Thomas Graham (1805-1869), Scottish physical chemist, first states the law that bears his name. He declares that a gas diffuses at a rate inversely proportional to its molecular weight. Based on this understanding, mixtures of gases could now be separated on the basis of their rates of diffusion, which indicates their specific gravity. Because of Graham's law, he is widely held to be one of the founders of physical chemistry.

1832   Jean Baptiste André Dumas (1800-1884), French chemist, and his assistant Auguste Laurent (1807-1853) discover anthracene which they isolate from coal tar. It is eventually used as the starting compound for the manufacture of dyestuffs.

1832   Codeine is discovered by French apothecary Pierre Jean Robiquet (1780-1840). The isolation of morphine from opium after the turn of the century showed the way to obtain new drugs from the older, crude drugs. Codeine is an alkaloid (a naturally occurring constituent) also present in opium, and Robiquet's discovery shows it to be a methyl-derivative of morphine. It is used initially as a sedative and for reducing irritation of the respiratory passages, and goes into production this same year. Today it is often used in combination with aspirin or acetaminophen to relieve pain and as a cough suppressant. It is addictive but not so much as morphine.

1832   Carotene (carotin) is discovered in carrots by German chemist Heinrich Wilhelm Ferdinand Wackenroder (1798-1854). This highly unsaturated hydrocarbon is an organic compound usually found as pigment in plants, giving them a yellow, orange, or red color. It is converted in the liver of animals into vitamin A, of which it can be regarded as a precursor. Carotenoids are increasingly regarded as essential to a diet that will minimize the risk of cancer.

1833   Creosote is discovered by German naturalist Karl von Reichenbach (1788-1869). He obtains this oily liquid by distilling wood tar. Used as an insecticide, germicide, and disinfectant, in such applications as sheep dips and the preservation of timber, the name comes from the Greek "kreas" for flesh and "sozo" meaning to save or preserve.

1834   Kyanol is first isolated from coal tar by German chemist Friedlieb Ferdinand Runge (1795-1867). He finds different kinds of alkaline, organic substances in coal tar by digesting the tar with water solutions of acid. When this solution is later neutralized, an oil separates out which, when distilled, yields kyanol, pyrrol, and leukol. Kyanol becomes the primary industrial source of aromatic compounds and remains so until after World War I.

1834   Nitrobenzene is first synthesized by German chemist Eilhardt Mitscherlich (1794-1863). He treats benzene with fuming nitric acid to produce this compound which is later used in perfumes.

1836 John Frederic Daniell (1790-1845), English chemist, invents the electrochemical cell. Called the Daniell cell, this new battery is made of copper and zinc and is the first reliable source of electric current. By introducing a barrier between the zinc and copper, he is able to stop the formation of hydrogen which had been impairing battery function. The current of his battery does not decline rapidly.

1836 Acetylene is discovered by Irish chemist Edmund William Davy (1785-1857). He obtains this gaseous hydrocarbon by the action of water on the black residue (potassium carbide) that results from the preparation of potassium from cream of carbon and carbon. It is an extremely unstable gas and is highly poisonous when inhaled. Stored and transported in cylinders for safety, it is later pressurized for use in welding torches. Davy is the cousin of the famous English chemist Humphry Davy (1778-1829).

1839 First fuel cell using hydrogen and oxygen is constructed by English physicist William Robert Grove (1811-1896). He uses gaseous hydrogen and oxygen rather than their volatile liquid counterparts as well as platinum plates and an electrolyte solution. This process converts chemical energy into electrical energy and produces a weak electric current. Fuel cells remain a curiosity until the space age when they become the best way to provide electricity to spacecraft.

1839 Lanthanum is discovered by Swedish chemist Carl Gustav Mosander (1797-1858). Working with the fairly new element cerium, he finds that by heating it and treating the product with nitric acid, he obtains an entirely new substance. As a rare earth element difficult to find, it is named from the Greek "lanthano," to lurk or lie hidden. A silvery-white metal soft enough to cut with a knife, it decomposes in water and corrodes in oxygen. Its compounds impart special optical qualities to glass, and come to be

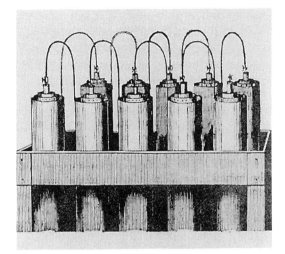

Sketch of J. Frederic Daniell's constant battery from his book, *An Introduction to the Study of Chemical Philosophy.*

used in the manufacture of certain specialized lenses.

1840 Jöns Jakob Berzelius (1779-1848), Swedish chemist, first introduces the name "allotropy" to describe the existence of different varieties of an element. He converts charcoal into graphite and declares that the same element may have different forms. Allotropy thus comes to be used to describe any element that might exist in two or more forms with different chemical properties.

1840 Ozone is discovered by German-Swiss chemist Christian Friedrich Schonbein (1799-1868). As early as 1785, it was known that oxygen gives off a peculiar smell when sparked with electricity, but most attributed it to the electrical equipment. When Schonbein studies this strong, pungent odor, he finds that he can reproduce it by electrolyzing water or by allowing phosphorous to oxidize. He believes that the resulting smell comes from a gaseous substance that is a new element, and upon studying its properties, is convinced he is correct. He finds that it bleaches litmus paper and is present in rainwater after lightning. He names this new gas ozone from the Greek "ozo" to smell. He does not recognize for some time that it is a form of oxygen. The ozone molecule consists of three oxygen atoms, whereas a normal oxygen molecule has two atoms. Unlike oxygen, it is toxic and very explosive. It is later used as a commercial bleaching agent and as an effective means of destroying microorganisms from drinking water.

1841 Chemical Society is founded in London with Scottish physical chemist Thomas Graham as its first president. This is the first such society to be composed of professional chemists and to have a continuous existence.

1842 Justus von Liebig (1803-1873), German chemist, publishes his *Die Organische Chemie in ihrer Anwendung auf Physiologoie und Pathologie* which is the first formal treatise on organic chemistry as applied to physiology and to pathology. His work shows that chemistry can be productively linked or applied to biology and gives impetus to the growth of biochemistry.

1842 First commercial artificial fertilizer is developed by English agricultural scientist John Bennet Lawes (1814-1900), who experiments with artificial fertilizer and develops superphosphate. He patents a method for its manufacture (first from animal bones and then from minerals) and sets up a factory.

(See also 1843)

1842 Eugène Melchior Peligot (1811-1890), French chemist, first isolates the metal uranium. He heats uranium chloride with potassium metal and obtains a black powder whose properties indicate that it is almost pure uranium. The turn-of-the-century discovery that uranium is radioactive opens a whole new era for chemistry as well as physics.

(See also 1789)

1843 First agricultural laboratory is established by English agricultural chemist John Bennet Lawes (1814-1900). He continues his chemi-

CHEMICAL SOCIETY OF LONDON.

House of the Society of Arts, John Street, Adelphi.

23rd February, 1841.

A MEETING was convened to take into consideration the formation of a Chemical Society, at which meeting a Provisional Committee was appointed for carrying that object into effect.

The Provisional Committee having issued a printed circular inviting a number of gentlemen engaged in the practice and pursuit of Chemistry to become original members, the following gentlemen communicated their written assent :—

Aikin, Arthur.	Graham, Prof. Thos.	Pepys, W. H.
Andrews, Dr. Thos.	Graham, John.	Phillips, Richard.
Barron, Rev. J. A.	Griffin, J. J.	Playfair, Dr. Lyon.
Blake, Jas.	Griffiths, Thos.	Porrett, Robert.
Blythe, Wm.	Grove, W. R.	Potts, Dr. L. H.
Brande, Prof. W. T.	Heisch, C.	Rees, Dr. G. O.
Brayley, E. W., Jun.	Hennell, H.	Reid, Dr. D. Boswell.
Brooke, H. J.	Henry, T. H.	Richardson, Thos.
Button, Chas.	Herapath, Wm.	Scanlan, Maurice.
Clark, Dr. Thos.	Hope, Dr. T. C.	Sims, Ollive.
Cock, W. J.	Hughes, F. R.	Smith, Denham.
Cooper, J. T.	Johnson, Percival A.	Solly, E., Jun.
Cooper, J. T., Jun.	Johnston, Prof. Jas.	Stenhouse, Dr. J.
Crosse, Andrew.	Leeson, Dr. W. B.	Taylor, Richard.
Crum, Walter.	Longstaff, Dr. G. D.	Tennant, John.
Cumming, Prof. J.	Lowe, Geo.	Teschemacher, E. F.
Daniell, Prof. J. F.	M'Gregor, Dr. Rob.	Thomson, Dr. Thos.
Daubeny, Dr. C.	Macintosh, Chas.	Thomson, Dr. R. D.
Davy, Dr. E.	Mercer, John.	Turner, Dr. Wilton.
De la Rue, W.	Miller, Prof. W. H.	Warington, Rob.
Everitt, Thos.	Moody, Col. Thos.	West, Wm.
Ferguson, Wm.	Mushet, David.	Wheeler, Jas. Lowe.
Fownes, G.	Paris, Dr. J. A.	Wilson, John.
Frampton, Dr. A.	Pattinson, H. L.	Wilson, Dr. G.
Gassiot, J. P.	Pearsall, Thos. L.	Yorke, Col. P.
Gill, Thos.	Penny, Prof. F.	

Chem. Proc.—No. 1.   B

The first members of the Chemical Society of London, 1841.

cal experiments with artificial fertilizers and extends them to animal feed. His "nitrogen balance" experiments help establish the concept of essential dietary components.

(See also 1842)

1843 Erbium is discovered by Swedish chemist Carl Gustav Mosander (1797-1858). Working with the rare earth element called ytterbite, he isolates three other rare earth elements, erbium, terbium, and didymium. As a soft, malleable solid with a bright, silvery luster, erbium has a pink color as a compound that makes it desirable as a tinting agent. It has also found limited application in the nuclear power industry.

1844 Ruthenium is discovered by Estonian chemist Karl Karlovich Klaus (1796-1864), sometimes called Claus. He is able to extract a new metallic element from platinum residues. It proves to be a hard and lustrous metal whose name is taken from the ancient name for Russia, "Ruthenia." When alloyed with platinum and palladium, it forms very hard, resistant contacts for electrical equipment that resist wear.

1845 Acetic acid is first synthesized from inorganic materials by German chemist Adolph Wilhelm Hermann Kolbe (1818-1884). According to the prevailing ideas of the time, this should be impossible. However, Kolbe is convinced that organic compounds can be synthesized from inorganic materials, and he converts carbon disulfide, through various steps, into acetic acid. This is a major chemical breakthrough. Kolbe is also the first to apply galvanic current or electrolysis to organic compounds and produces "double acids," which, in turn, leads to an inexpensive method of producing aspirin.

1845 Guncotton is discovered by German-Swiss chemist Christian Friedrich Schonbein (1799-1868) while experimenting with nitric and sulfuric acids. After spilling them on the floor, he uses a cotton apron to wipe them up and hangs it above a hot stove to dry. Once dry, the apron bursts into flame. Schonbein experiments further and discovers that the nitric acid has bonded with the cellulose in the cotton to form nitrocellulose. He eventually develops a secret process that results in a fine powder that dries into slabs. Although highly unstable and very dangerous, it proves useful on the battlefield as a replacement for gunpowder. Since guncotton is nearly smokeless, it is much preferable to the large clouds of obscuring, black smoke given off by gunpowder.

1846 Louis Pasteur (1822-1895), French chemist, discovers molecular asymmetry and demonstrates the existence of isomers becoming among the earliest to deal with the three-dimensional structure of molecules. Pasteur is an unknown this early in his career, and his announcement that not only are certain crystals asymmetrical but this asymmetry exists all the way down to the molecular level is a major departure from what had been believed. His discovery adds significantly to the science of polarimetry, in which measurements of the manner in which the plane of polarized light is twisted can be used to help determine the structure of an organic substance.

1846 William Robert Grove (1811-1896), English physicist, offers the first experimental evidence for thermal dissociation. In chemistry, dissociation is the breaking up of the molecules of a compound into simpler constituents that are usually capable of recombining under other conditions. He shows that water (steam) in contact with a strongly heated wire, will absorb energy and break up (dissociate) into hydrogen and oxygen. Awareness of this important phenomenon contributes to the advancement of the science of physical chemistry.

1846 Nitroglycerine is first produced by Italian chemist Ascanio Sobrero (1812-1888) who slowly adds glycerine (commonly used in skin lotion) to a mixture of nitric and sulfuric acids. He is so impressed by the explosive potential of a single drop of this colorless, oily liquid in a heated test tube and so fearful of its use in war, that he makes no attempt to exploit it. He waits a year before publishing his results, and even then does it almost in secret. It is another 20 years before Swedish inventor Alfred Bernhard Nobel (1833-1896) learns the proper formula and puts it to use.

(See also 1867)

1847 James Young (1811-1883), Scottish chemist, discovers what he describes as a "petroleum spring" at a coal mine in Derbyshire. Within a few years, he is producing what is eventually called kerosene by distilling it from coal and shale oil, although he is not the first to do so.

(See also 1852)

1850 Ludwig F. Wilhelmy (1812-1864) of Germany constructs one of the first mathematical equations for describing the rate of progress of a chemical reaction when he offers an algebraic formula to describe the laws according to which the action of acids on sugar (hydrolysis of sugar) takes place.

1850 Thomas Graham (1805-1869), Scottish physical chemist, studies the diffusion of a substance through a membrane (osmosis) and first distinguishes between crystalloids and colloids. Noticing that some chemicals diffuse more slowly through parchment than might be expected, he is able to distinguish two classes of substances. Those that pass through easily can also be crystallized, so he calls them crystalloids. Since glue is typical of the second group, he calls them colloids. He becomes the founder of colloidal chemistry and contributes most of the nomenclature eventually used by that discipline.

1852 Kerosene is first obtained by Canadian physician Abraham Gesner (1797-1864), who distills it from thick, crude oil or petroleum. He names the pale, liquid fuel kerosene after the Greek word "keros" for wax. Slightly heavier than gasoline, it belongs to the family of hydrocarbons called alkanes or paraffins, and is sometimes referred to as paraffin oil. It also becomes known as coal oil, lamp oil, and illuminating oil. With the discovery in 1859 of huge oil deposits in the United States, kerosene becomes the chief product of American refineries and replaces whale oil for indoor lighting.

(See also 1830)

1852 Alexander William Williamson (1824-1904), English chemist, publishes his study which shows for the first time that catalytic

action clearly involves and is explained by the formation of an intermediate compound.

1853 First chemical-pharmaceutical company, the Carlo Erba Company, is established in Milan, Italy. The E. R. Squibb Company follows in Brooklyn, New York, in 1858.

1854 Lyon Playfair (1818-1898), English chemist, first proposes the use of incendiary gases and poisonous gases in the Crimean War. His suggestions are not followed. During World War I, however, the Germans first release chlorine gas in January 1915 against the Russians, but it has little effect. On April 22 of that year, it proves the difference in a skirmish against the British and French. When the Germans use the more deadly phosgene gas, the Allies respond in kind, and the Germans escalate with their use of mustard gas in 1917. By 1918, both sides use this new gas on a large scale. During World War II, both sides produce but do not use poison gas.

1855 Henri Etienne Sainte-Claire Deville (1818-1881), French chemist, first produces aluminum in a pure state. Although he produces the metal in moderate quantities using his original method of heating aluminum chloride with metallic sodium, it remains very expensive to produce. At this point in time, however, aluminum is used more for show than for practical reasons, such as for the cap at the top of the Washington Monument in Washington, D.C., or for the rattle of Napoleon III's infant son.

1855 First Bunsen burner is invented by German chemist Robert Wilhelm Bunsen (1811-1899), who uses a burner that is perforated at the bottom so that air is drawn in by the gas flow. This device burns a mixture of gas and air to produce a hot, scarcely luminous flame. It burns steadily with little light and no smoke and becomes known to most beginning students in a chemistry laboratory.

1856 First synthetic dye is produced by English

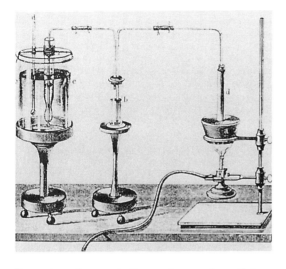

Illustration of an early Bunsen burner, from Robert Bunsen's *Gasometrische Methoden*, 1858.

chemist William Henry Perkin (1838-1907). While trying to synthesize quinine, he accidentally discovers how to produce "mauve" from the impure aniline in coal tar. After treating aniline with potassium dichromate, he notices a purplish glint in the result. Adding alcohol proves to turn the entire result a beautiful purple. Suspecting he has created a dye, he leaves school and uses family money to start a factory. Within six months he creates "aniline purple" (that French dyers name "mauve"), and creates an entire dyestuffs industry. His discovery also stimulates the development of synthetic organic chemistry.

1858 First to develop symbols to represent the valency or combining power of the carbon atom are German chemist Friedrich August Kekulé von Stradonitz (1829-1896) and Scottish chemist Archibald Scott Cooper (1831-1892). They discover independently that the carbon atom is always tetravalent or has a valence of four (meaning that it always combines with four other atoms). Kekule's work further defines and refines the concept of valency and shows a way of arranging the various atoms of a particular element and those of other elements with

which it might combine, as represented in a chemical formula. This work lays the foundation for modern structural theory in organic chemistry.

(See also 1861)

1859 "Kolbe reaction" is discovered by German chemist Adolph Wilhelm Hermann Kolbe (1818-1884) when he succeeds in using phenol and carbon dioxide to produce salicylic acid. This new method leads to the large scale production of acetylsalicylic acid (aspirin) and to cheaper production costs for this new wonder drug.

1859 James Clerk Maxwell (1831-1879), Scottish mathematician and physicist, studies the rings of Saturn and produces the first extensive mathematical development of the kinetic theory of gases. He shares this discovery of the distribution of molecular speeds in a gas with Austrian physicist Ludwig Edward Boltzmann (1844-1906), who accomplishes the same independently. Both analyze the behavior of what come to be called perfect gases on the assumption that they are an assemblage of randomly moving particles (kinetic meaning motion). Using this theory, it becomes possible to predict the behavior of certain gases. It has come to be known as the Maxwell-Boltzmann theory of gases.

MAY 10, 1860 First element discovered using the newly developed spectroscope is cesium. Robert Wilhelm Bunsen (1811-1899), German chemist, and German physicist Gustav Robert Kirchhoff (1824-1887) name their new element cesium after its "sky blue" color in the spectrum. They first detect cesium in the salts of mineral waters from Durkheim, Germany, and they derive the name from the word "caesius" used by the Romans to describe the blue sky. The discovery of cesium marks the beginning of a new era in increasing the number of known elements. In the previous three decades, only five new elements had been discovered.

SEPTEMBER 1860 First International Congress of Chemistry is held in Karlsruhe, Germany and attracts 140 delegates from around the world. Italian chemist Stanislao Cannizzaro (1826-1910) publishes the forgotten ideas of Italian physicist Amedeo Avogadro (1776-1856)—about the distinction between molecules and atoms—in an attempt to bring some order and agreement on determining atomic weights. Until now, chemists did not agree on empirical formulas, and there was considerable confusion about the use of equivalent weight, atomic weight, and molecular weight. For oxygen, the equivalent weight is 8, the atomic weight is 16, and the molecular weight is 32. Cannizzaro realizes that Avogadro's hypothesis can be used to distinguish among them and clarify matters. After his speech, he eventually wins over the chemical world to his opinion.

1860 Jean Servais Stas (1813-1891), Belgian chemist, begins work that leads to an accurate method of determining atomic weights. By 1865, he produces the first modern table of atomic weights using oxygen as a standard. He gives oxygen a value of 16 and is able to demonstrate that Prout's hypothesis, that all atomic weights are exact multiples of the density of hydrogen, is not correct.

(See also 1929)

1861 Friedrich August Kekulé von Stradonitz (1829-1896), German chemist, publishes the first volume of *Lehrbuch der Organischen Chemie* in which he is the first to define organic chemistry as the study of carbon compounds. Three years earlier, he demonstrated that carbon is always tetravalent (any one carbon atom will always form four bonds in a compound) and that it can link with itself to form long chains. This lays the groundwork for the modern structural theory in organic chemistry. In 1865, Kekule discovers the puzzling structure of the benzene molecule when he dreams of a snake biting its tail while whirling. From this vision, his concept of the six-carbon benzene ring is

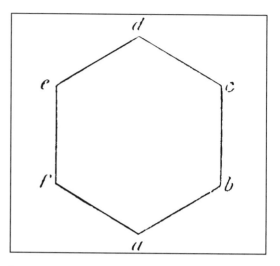

August Kekulé's benzol diagram.

born and with it, the known facts of organic chemistry fall into place.

(See also 1858)

1861 Robert Wilhelm Bunsen (1811-1895), German chemist, and German physicist Gustav Robert Kirchhoff (1824-1887) discover the metal rubidium using their new spectroscope. They detect this new element in lepidolite (a composite of lithium and aluminum) by noticing two prominent red lines in the spectrum. They name it rubidium after the Latin, "rubidis" for dark red.

1861 The term "chemical structure" is first coined by Russian chemist Aleksander Mikhailovich Butlerov (1828-1886). His basic concept of chemical structure means that the chemical nature of a molecule is determined not only by the number and type of atoms but also by their arrangement. Thus, although ethyl alcohol and dimethyl ether have the same empirical formula, their widely different properties are attributed to the fact that their structural formulas are different. For ethyl alcohol, one of the six hydrogen atoms is attached to an oxygen atom, while all six hydrogen atoms are attached to carbon atoms in dimethyl ether.

1861 Thallium is discovered by English physicist William Crookes (1832-1919) while examining some selenium mud. Using the newly invented spectroscope, he finds a green line in the spectrum that does not correspond with any known element. He consequently concludes that this is a new element and names it after the Greek word "thallos" for green shoot. A bluish gray metal soft enough to be cut by a knife, it is poisonous in its pure form and is used as a rodenticide and ant poison. Other thallium compounds prove useful for electronic equipment and as an infrared detector. The following year, it is isolated by French chemist Claude August Lamy (1820-1878).

1861 Practical and inexpensive way of producing sodium bicarbonate is discovered and patented by Belgian chemist Ernest Solvay (1838-1922). He solves the practical problems of large-scale conversion by devising carbonating towers which allow large amounts of ammonia, salt solution, and carbon dioxide to be mixed. His process even allows for the recovery of expensive ammonia which can then be reused. Sodium bicarbonate or soda ash is in great demand by the glass and soap industries, and Solvay's process is so efficient and economical, that by 1913 he is producing virtually the entire world supply. He uses his great wealth to endow schools and provide others with the education he himself never obtained.

1862 Charles Friedel (1832-1899), French chemist, prepares isopropyl, the first secondary alcohol and one which is later used as a solvent and as rubbing alcohol. Friedel is the first to conduct any thorough research on alcohols, and his synthesis of isopropyl alcohol is easily oxidized into acetone, of which it is a major industrial source.

FEBRUARY 7, 1863 John Alexander Reina Newlands (1837-1898), English chemist, announces his arrangement of the elements in the order of their atomic weights. Arranging his elements into vertical columns of seven, he sees that similar elements tend to fall into the same horizontal rows. For example, potassium would fall next to sodium (to which it is very similar), selenium falls into the same row as the similar sulfur, and calcium would be next to the similar magnesium. After discovering that their properties seem to repeat themselves in each group of seven elements, he calls this the Law of Octaves (after the musical scale). Although his arrangements are sometimes forced, he is the first to do this and is a precursor of Mendeleev.

(See also March 6, 1869)

1863 Law of mass action is first formulated by Norwegian chemist and mathematician Cato Maximilian Guldberg (1836-1902) and his brother-in-law Norwegian chemist Peter Waage (1833-1900) in a paper they publish this year. They state that the direction taken by a reaction is dependent not merely on the mass of the various components of the reaction, but upon the concentration or the mass present in a given volume. In other words, the higher the concentration of substances, the greater the likelihood of collisions among them. Published in Norwegian, it is not translated and acknowledged until much later.

1863 Ferdinand Reich (1799-1882), German mineralogist, and his assistant Hieronymus Theodor Richter (1824-1898) examine zinc ore spectroscopically and discover the new, indigo-colored element indium. Reich obtains a yellow precipitate from zinc ore and suspects it may contain a new metal. He is, however, handicapped in using the new spectroscope since he is color blind and would be unable to identify any new color. Richter examines it spectroscopically and finds two dark blue lines on the spectrum that tell Reich he has discovered a new element. The name indigo is derived from the Latin "indicum," which is after the Greek "indikon" meaning Indian dye. In the next century, it is used in the making of transistors.

1863 Barbituric acid is discovered by German

chemist Adolf Johann Friedrich Wilhelm von Baeyer (1835-1917). In his extensive investigations of the derivatives of uric acid, he finds what turns out to be the parent of all the barbiturate drugs (sedatives and hypnotics). It is believed that Baeyer names this compound after a friend named Barbara.

1864 Aleksander Mikhailovich Butlerov (1828-1886), Russian chemist, prepares butyl alcohol, the first tertiary alcohol. It is later used as a solvent and in synthetic rubber.

1865 First plastic material, celluloid, is produced by English metallurgist and chemist Alexander Parkes (1813-1890). After working since the 1850s with nitrocellulose, alcohol, camphor, and castor oil, he finally obtains a material that can be molded under pressure while still warm. Having worked earlier with rubber solutions and cellulose nitrate, he discovers that pyroxylin or partially nitrated cellulose, if dissolved in alcohol and ether in which camphor also has been dissolved, will harden upon evaporation and soften slightly when heated. Parkes is unsuccessful at marketing his product, however, and it is left to American inventor John Wesley Hyatt (1837-1920) to make it a success.

(See also 1869)

1867 First table of the relative strength of acids is compiled by Danish chemist Hans Peter Jörgen Julius Thomsen (1826-1909).

1867 Dynamite is discovered by Swedish inventor Alfred Bernhard Nobel (1833-1896) while searching for a safer, more controllable version of nitroglycerin. Convinced that a safe explosive is greatly needed by the mining, engineering, and transport industries, he learns that he can turn the highly volatile and dangerous-to-transport liquid nitroglycerine into a manageable solid by soaking it up into a porous material. He finally finds that "kieselguhr" or earth containing silica will soak up nitroglycerine without altering its chemical makeup or prop-

erties. The result is a dough-like substance that can be made into hard cakes or sticks and exploded only by using a detonating cap (which he also invents). While dynamite does prove highly useful for all manner of engineering and mining projects around the world, its major users are the military forces of the world. Since Nobel actually believed that his invention would shorten wars and eventually make them outlawed because of its horrible effects, he provides over $9 million for the establishment after his death of annual prizes (Nobel Prizes) to reward excellence in certain scientific areas.

MARCH 6, 1869 First periodic table of the elements is published by Russian chemist Dmitri Ivanovich Mendeleev (1834-1907). In his search for some organizing principle on which to base an arrangement of the 63 known chemical elements, he makes a card recording each element's atomic weight, valence, and other chemical and physical properties and constantly arranges them to find a pattern. Eventually he finds that arranging them in ascending order, according to their weights, allows their properties to repeat in a predictable, orderly manner. When laid out in sequence, left to right, he sees that the properties of the tenth element (sodium) are similar to those of the second element (lithium), those of the eleventh element (magnesium) are similar to the third (beryllium), and so on. When this system appears to break down, Mendeleev has the insight to leave spaces for undiscovered elements and to rearrange others (assuming correctly that their atomic weights were wrong). During this time, the German chemist Julius Lothar Meyer (1830-1895) offers a similar table of elements, but not only is he a year late, but he does not predict the existence of undiscovered elements. Mendeleev gives chemists an order for the elements that will eventually lead them to understand the internal structure of atoms.

JUNE 26, 1869 William Henry Perkin (1838-1907), English chemist, obtains a patent for the synthesis of the red dye alizarin. He gets the

```
                              Ti = 50     Zr = 90      ? = 180.
                              V = 51      Nb = 94      Ta = 182.
                              Cr = 52     Mo = 96      W = 186.
                              Mn = 55     Rh = 104,4   Pt = 197,4
                              Fe = 56     Ru = 104,4   Ir = 198
                           Ni = Co = 59   Pl = 106,6   Os = 199.
  Ik..= 1                        Cu = 63,4  Ag = 108   Hg = 200.
          Be = 9,4   Mg = 24    Zn = 65,2   Cd = 112
          B = 11     Al = 27,4   ? = 68     Ur = 116   Au = 197?
          C = 12     Si = 28     ? = 70     Sn = 118
          N = 14     P = 31     As = 75     Sb = 122   Bi = 210
          O = 16     S = 32     Se = 79,4   Te = 128?
          F = 19     Cl = 35,5  Br = 80     I = 127
  Li = 7  Na = 23    K = 39     Rb = 85,4   Cs = 133   Tl = 204
                     Ca = 40    Sr = 87,6   Ba = 137   Pb = 207.
                     ? = 45     Ce = 92
                     ?Er = 56   La = 94
                     ?Yt = 60   Di = 95
                     ?In = 75,6 Th = 118?
```

Illustration of Dimitri Mendeleev's first periodic table, published in 1869.

patent one day after the German chemist Karl Graebe (1841-1927). It is Graebe who discovers its molecular structure, but Perkin who first finds a practical way to produce it. Graebe's synthesis process is workable but far too costly. Perkin develops a cost-effective process that allows him to produce 220 tons (200 metric tons) of alizarin annually by 1871.

1869 First synthetic plastic is produced by American inventor John Wesley Hyatt (1837-1920), who improves upon Alexander Parkes' invention of celluloid in 1865. Searching for an ivory substitute, he duplicates Parkes' steps but also uses pressure and produces a new plastic material he names "celluloid." He markets it successfully, first making billiard balls and later baby rattles, combs, and photographic film. Its major drawback, however, is its ability to ignite in a flash. Research in plastics is greatly stimulated by his breakthrough.

1869 Chemical concept of "critical temperature" is first proposed by Irish physical chemist Thomas Andrews (1813-1885). While working with carbon dioxide (which can be liquefied at room temperature by pressure alone) and

observing its behavior when pressurized, he suggests that every gas has a precise temperature, which he calls its critical temperature, above which it cannot be liquefied even under greater pressure. His concept proves to be a crucial discovery and leads the way to a breakthrough in the liquefaction of the so-called permanent gases (whose temperature must be lowered before pressure is exerted).

1873 Dichloro-diphenyl-trichloroethane (DDT) is first synthesized by German chemist Othmar Zeidler. He produces this colorless, odorless, insoluble compound by the reaction of chloral with chlorobenzene in the presence of sulfuric acid. He is not aware of its insecticidal properties.
(See also 1939)

1874 Jacobus Henricus Van't Hoff (1852-1911), Dutch physical chemist, and French chemist Joseph Achille Le Bel (1847-1930) independently discover the theory of the relationship of optical activity to molecular structure. Ever since Pasteur, chemists were puzzled by the fact that some chemicals can exist in two forms that are identical except for the way they affect polarized light. Both scientists explain this phenomenon of optical isomerism by showing that a carbon atom can exist in only two forms that are mirror images of each other and that this exists in the molecules themselves. This asymmetric, three-dimensional molecular structure of carbon suggests to each scientist a theory of stereochemistry (or the three-dimensional structure of molecules).
(See also 1899)

c. 1875 Friedrich Otto Fritz Giesel (1852-1927), German chemist, is the first in Germany to produce radium salts commercially. A pioneer in radiology and radiochemistry, he dies from lung cancer caused by radiation.

1875 Gallium is discovered by French chemist Paul Emil Lecoq de Boisbaudran (1838-1912). As a pioneer in the new science of spectroscopy, he analyzes the spectral lines of one element

Worker manufacturing celluloid, circa 1912.

after another, searching for an element whose properties had been predicted to exist by the famous Russian chemist Dmitri Ivanovich Mendeleev (1834-1907) in 1869. While analyzing a substance he separates from a zinc blend obtained from the Pyrenees area, he observes two violet lines never before seen. After proving by chemical means that it is a new element, he names it apparently after his native France, which is "gallia" in Latin (although it could also be derived from the Latin "gallus" for cock and thus his own name, Lecoq).

c. **1876** Josiah Willard Gibbs (1839-1903), American physicist, applies the laws of thermodynamics to chemical reactions and first develops the modern concepts of "free energy" and "chemical potential" as the driving force or fuel behind chemical reactions. He states that when a chemical reaction occurs, the free energy of the system changes. When it decreases, the entropy always increases and the reaction is spontaneous. He also calls the rate at which free energy changes as the concentration of a particular substance changes, the chemical potential of that substance. So he is able to show that as heat flows spontaneously from a point of high temperature to one of low, a chemical reaction moves spontaneously from a point of high chemical potential to one of low. This, combined with his other work, lays the mathematical foundation of chemical thermodynamics.

1876 Phase rule is discovered by American physicist Josiah Willard Gibbs (1839-1903). By applying thermodynamic principles to equilibria between different phases (liquid, solid, and gas), he works out a simple equation, called the "phase rule," that relates and allows prediction of the variables (like temperature and pressure) to different phases (like solid, liquid, or gas). This work is very thorough and establishes chemical thermodynamics. Although Gibbs' work is generally ignored in Europe, the phase rule is later put into practical application by Dutch physical chemist Hendrik Willem Bakhuis Roozeboom (1854-1907).

1877 James Mason Crafts (1839-1917), American chemist, and French chemist Charles Friedel (1832-1899) discover that aluminum chloride is a versatile catalyst for reactions tying together a chain of carbon atoms to a ring of carbon atoms. While studying the effect of metallic aluminum on certain chlorine-containing organic compounds, they notice that a reaction occurs only after a period of inactivity, after which hydrogen chloride is formed. They find that during this inactive period, aluminum chloride is formed which is the agent responsible for the reaction. Known as the Friedel-Crafts reaction, this insight into catalysis becomes a major tool of future chemical synthesizers.

1878 Adolf Johann Friedrich Wilhelm von Baeyer (1835-1917), German chemist, produces the first synthesized indigo blue dye. Searching for the parent compound from which natural indigo is derived, he continues to break it down until eventually his removal of both oxygen atoms leaves a substance he calls "indole." With this, he is able to create indigo from

phenylacetic acid and succeeds in producing an entirely synthetic dye. His distillation process has great impact on the dye industry and becomes widely used to produce dyes from other materials.

1878 Rayon is invented by French chemist Louis Marie Hillaire Bernigaud de Chardonnet (1839-1924). He begins by producing fibers which are made by forcing solutions of nitrocellulose through tiny holes and allowing the solvent to evaporate. These fibers are produced by treating cotton with nitric and sulfuric acids and then dissolving the mixture in alcohol and ether. The fibers are formed as this solution is passed through glass tubes and allowed to dry. Although the end product is highly flammable unless properly treated, it causes a sensation as it resembles silk. Rayon is the first artificial fiber to come into common use.

1878 Jean Charles Galissard de Marignac (1817-1894), Swiss chemist, analyzes the rare earth erbium and finds that it consists of two parts. He calls one erbia and the other ytterbia.

1879 Lars Fredik Nilson (1840-1899), Swedish chemist, shows that ytterbia is a mixture of two oxides. He calls one ytterbia and the other scandia. The latter is demonstrated to be a new element and its name eventually becomes scandium in honor of his native Scandinavia. Later in this year, scandium is recognized as being an element whose properties were predicted by Russian chemist Dmitri Ivanovich Mendeleev (1834-1907).

1879 Saccharin is first prepared by German chemist Constanin Fahlberg (1850-1910) and American chemist Ira Remsen (1846-1927). Also called thibenzoyl sulfimide, it has an obviously sweet taste which they discover while investigating the derivatives of toulene, an ingredient found in coal tar. They trace toulene back to this new compound that is 300 to 500 times sweeter than sugar. Saccharin comes on the market as a sugar substitute as early as the turn of the century, and although there are some doubts about its safe consumption, it is not until the early 1970s that tests indicate a link between saccharin and bladder cancer. Canada prohibits the use of saccharin in prepared foods in 1977.

1879   Samarium is discovered by French chemist Paul Emil Lecoq de Bosibaudran (1838-1912). While analyzing didymium, he finds that it produces unknown spectral lines and concludes that it is a new element. He names it after the host mineral, samarskite. The mineral had been named after the Russian engineer named von Samarsky who originally found it in Russia.

(See also 1901)

c. 1880   Carl Oswald Viktor Engler (1842-1925), German chemist, begins his studies on petroleum (commonly known as crude oil). He is the first to state that it is formed from the fat of prehistoric animals and is thus organic in origin, and he can be considered the founder of petroleum science. He also builds a device for determining the combustion danger of petroleum (called an Engler viscosimeter).

1882   Thiophene is discovered by German organic chemist Viktor Meyer (1848-1897). While demonstrating a color test for benzene to his class, he finds that he is unable to reproduce a deep blue color as he had always done. Determined to find out why his test failed, he eventually discovers that the benzene used in his failed test had been prepared from benzoic acid and was very pure. All of his past tests had used ordinary coal tar benzene. The latter, he finds, gives the proper blue reaction because it contains a new substance he names thiophene. The name is formed from the Greek "thio" for sulphur and "phene" which forms the basis of derivatives of benzene. A colorless, insoluble liquid, it is sometimes used in antihistamines.

1883   Johann Gustav Christoffer Kjeldahl (1849-

1900), Danish chemist, first devises a method for determining the amount of nitrogen in an organic compound. This consists of transforming all the nitrogen in a weighed substance into ammonium sulfate (by digestion with sulfuric acid), and then freeing the ammonia by alkalizing the solution and measuring its amount by distilling it into a measured volume of standard acid. This simple, fast method becomes known as "Kjeldahl's method" and is used mostly for estimating the nitrogen content of foodstuffs and fertilizers.

1884   Theory of electrolytic dissociation is offered by Swedish chemist Svante August Arrhenius (1859-1927), who first proposes the concept of ions being atoms bearing electric charges. He offers this theory as part of his dissertation on electrolytes (substances that conduct electricity when in solution) in which he argues that molecules of some compounds break up (dissociate) into charged particles when put in a liquid. His ideas are eventually proved correct and he becomes one of the founders of modern physical chemistry.

1885   First to produce a large quantity of liquid oxygen is Scottish chemist and physicist James Dewar (1842-1923). His device produces liquid oxygen in a sufficient quantity for him to study its properties. He is able to show that liquid oxygen is attracted to a magnet, as is liquid ozone which is a variety of oxygen.

(See also 1898)

1885   Incandescent gas mantle is invented by Austrian chemist Karl Auer, Baron von Welsbach (1858-1929). Assuming that gas flames would be brighter if they heated up something which itself would glow brightly, he experiments with many substances. He finally finds that if he soaks a cylindrical fabric with thorium nitrate (to which a tiny bit of cerium nitrate is added), and burns off the fabric to leave an incandescent shell of metal oxide, the result is a brilliant white glow. Called the Welsbach mantle, it becomes very popular and actually gives gas lighting an edge with the new and feeble electric light systems now being introduced. Auer later applies his knowledge to improving the electric light bulb, and invents a filament made of osmium that paves the way for the tungsten filament. Auer is the first to find practical uses for the new rare earth metals.

1886   Fluorine is first isolated by French chemist Ferdinand Frédéric Henri Moissan (1852-1907), who succeeds in preparing a pure sample of it with the help of an electric furnace of his own devising. Although fluorine was deduced to exist by the English chemist Humphry Davy (1778-1829) as early as 1813, isolating it proved very difficult. At least one investigator is seriously injured and another dies from inhaling toxic gas. Moissan succeeds in isolating fluorine, a pale yellow gas, by decomposing hydrogen fluoride electrolytically with his electric furnace. He also chills the electrolytic solution to reduce fluorine's reactivity. Fluorine can combine, sometimes violently, with nearly any element, and water will even burn in an atmosphere of fluorine gas. Scientists eventually learn how to handle it safely, and fluorine-containing compounds called CFCs (chlorofluorocarbons) are used in refrigerators and air conditioners.

1886   Germanium is discovered by German chemist Clemens Alexander Winkler (1838-1904). While analyzing a new silver ore called argyrodite, he finds he can account for only 93% of its composition. Working steadily on this problem for four months, he succeeds in separating a new base from it and finally isolates a new metallic element. He names it germanium after his native Germany. Germanium is soon recognized as the "eka-silicon" predicted to exist by Russian chemist Dmitri Ivanovich Mendeleev (1834-1907), and its discovery serves to confirm the validity of his periodic table.

1886   New and inexpensive way of making aluminum is invented independently by American chemist Charles Martin Hall (1863-1914) and French metallurgist Paul Louis Toussaint Héroult (1863-1914). Both devise an electrochemical process for extracting aluminum from its ore. The new method makes aluminum cheaper almost overnight and forms the basis of a huge aluminum industry. Aluminum combines strength and lightness, and becomes second only to steel as a structural material. Hall makes his discovery in February of this year, and Héroult achieves his in April. Oddly, both are born and die in the same years.

1887   Wilhelm Ostwald (1853-1932), Russian-German physical chemist, founds the first academic journal devoted exclusively to physical chemistry. *Zeitschrift für Physikalische Chemie* becomes an influential journal, and it is by Ostwald's efforts that physical chemistry becomes an organized, independent branch of chemistry.

1889   Smokeless, explosive cordite is first developed by English chemist Frederick Augustus Abel (1827-1902), and Scottish chemist and physicist James Dewar (1842-1923). Their new mixture borrows from previous discoveries, as they mix nitroglycerine and nitrocellulose to which some petroleum jelly has been added. The resulting gelatinous mass can be squirted out into cords (from which it gets its name) which are then carefully dried and measured out in precise amounts. The use of cordite in wartime in place of gunpowder removes the stench and thick, blinding smoke that always clouded a battlefield, and is thus adopted by all military forces.

1890   First crystallization of a protein, albumin, is obtained by German chemist Franz Hofmeister (1850-1922). At this point in time, the biological function of proteins has not been established, and they are classified primarily according to their solubility in a number of solvents. Albumin is one of these classes of proteins that are soluble in water and in water half-saturated with ammonium sulfate.

1891 Carborundum is discovered by American inventor Edward Goodrich Acheson (1856-1931). While trying to convert carbon into a diamond, he finds that carbon heated with clay yields an extremely hard substance. Believing that this new substance is a compound of carbon and alundum, he calls it carborundum. It is eventually seen to be silicon carbide, a compound of silicon and carbon, and for decades it remains as the hardest known substance, excepting the diamond. It eventually proves to be very useful as an abrasive for scouring and polishing.

1892 James Dewar (1842-1923), Scottish chemist and physicist, works on the liquefaction of gases and improves the Violle vacuum insulator by constructing a double-walled flask with a vacuum between the walls. A vacuum prevents the transfer of energy that occurs through conduction or convection, and with this he overcomes the problem of how to keep gases cold long enough to study them while liquefied. He then coats all sides with silver so heat will be reflected and not absorbed. His Dewar flask becomes the first Thermos bottle.

1892 Artificial silk is first made by English chemists Charles Frederick Cross (1855-1935) and Edward John Bevan (1856-1921). Their method involves changing cellulose to a viscous, yellow solution they call viscose, and ejecting it into a sulfuric acid solution. Their discovery, viscose, eventually forms the basis for the thin plastic, transparent sheets that become known as "cellophane."

(See also 1911)

1894 John William Strutt Rayleigh (1842-1919), English physicist, and Scottish chemist William Ramsay (1852-1916) succeed in isolating a new gas in the atmosphere that is denser than nitrogen and combines with no other element. They name it "argon," which is Greek for idle or inactive (since it is non-reactive). It is the first of a series of rare gases with unusual properties whose existence had not been pre-

dicted. The two collaborate after Rayleigh notices that nitrogen extracted from the air is always slightly heavier than nitrogen obtained from chemical compounds. They obtain argon by passing atmospheric nitrogen over hot magnesium until no more nitrogen is absorbed and a leftover gas remains. They use a spectroscope on it to verify that it is a new element.

1895 Helium is discovered by Scottish chemist William Ramsay (1852-1916), in a mineral named cleveite. It was speculated that helium existed only on the Sun, but Ramsay proves that it also exists on Earth. While studying a mysterious bubble of gas left over when nitrogen from the air is combined with oxygen, Ramsay speculates that this gas could explain why nitrogen obtained from the air is heavier than that obtained from chemical compounds. This mystery gas turns out to actually be a family of five gases, one of which is helium. Ramsay proves this when he analyzes its spectral lines and finds them identical to those observed on the Sun. Helium is an odorless, tasteless, and colorless gas that is also insoluble and incombustible. It is discovered independently this same year by Swedish chemist and geologist Per Theodore Cleve (1840-1905).

(See also 1908)

1895 Paul Walden (1863-1957), Russian-German chemist, discovers he can alter the effect a substance produces without changing the substance itself, producing what comes to be known as the "Walden inversion." He finds that he can take malic acid which rotates polarized light in a clockwise manner, change it into something else, and when changed back again into malic acid, it rotates polarized light counterclockwise. Somewhere in the process of reactions that occurred, the compound had been "inverted." The Walden inversion leads to important understanding of all stages of reaction mechanisms. This optical phenomenon comes to be used to advance the modern theory of chemical reaction processes.

1897 Nickel catalysis is first developed by French chemist Paul Sabatier (1854-1941), who discovers it via a failed experiment. Although he is unable to produce a volatile nickel ethylene compound as intended, he saves what gases do form and finds that ethane is present. This means that nickel has acted as a catalyst and brought about the addition of hydrogen to ethylene to form ethane. The industrial and commercial implications of this are major since before this, hydrogen addition is a very expensive process using platinum or palladium. With plentiful and cheap nickel instead, hydrogenations can be used on a large, industrial-sized scale, making possible the formation of edible fats like margarine and shortening from inedible plant oils.

1897 New and cheaper way of producing sodium is first introduced by American chemist Hamilton Young Castner (1859-1899). Until the Castner process, sodium was very expensive to produce, and its scarcity in turn, held up the production of aluminum (since it was used, along with potassium, as a reducer). Castner invents an electrolytic method of isolating sodium and chlorine from brine. His electrochemical process passes an electrical current through molten sodium hydroxide.

APRIL 1898 Radioactivity of the element thorium is discovered by Polish-French chemist Marie Sklodowska Curie (1867-1934). She makes this discovery just after the discovery of radioactivity itself. Gerhard Carl Schmidt (1865-1949) of Germany arrives at the same conclusion this year independently of Curie.

JULY 1898 New radioactive element called polonium is first isolated by Polish-French chemist Marie Sklodowska Curie (1867-1934) and her husband, French chemist Pierre Curie (1859-1906). This is the first success in their deliberate search for new radioactive elements. It is found after refining and purifying a large quantity of pitchblende, and its discovery leads to their finding evidence of a second radioactive

element, radium, in pitchblende that is even more potent than polonium. The Curies name their new element in honor of Marie Curie's native Poland.

DECEMBER 1898 Marie Sklodowska Curie (1867-1934) and her husband Pierre Curie (1859-1906) of France discover the radioactive substance radium. The actual amount of radium that they obtain is so small, however, that it can only be detected by its spectral characteristics. Wanting to obtain radium in visible, weighable quantities in order to study it further, they exhaust their life savings and spend the next four years refining 8 tons (7.3 metric tons) of the uranium-bearing ore called pitchblende in order to obtain a full gram of radium. Their discoveries in radioactivity inaugurate the investigation of this new field. Marie also coins the word radioactive.

1898 Krypton is discovered by Scottish chemist William Ramsay (1852-1916), who obtains this inert gas only after months of boiling down liquid air. In its natural state, krypton is colorless, odorless, and tasteless and is found in the atmosphere in very small quantities. Ramsay names this gas after the neuter form of the Greek adjective "kryptos," meaning hidden. Krypton proves useful for lighting, and is used to fill electric arc lamps that can pierce through fog at airports. Ramsay also obtains the inert gases xenon and neon with the same process.

1898 First to liquefy hydrogen is Scottish chemist and physicist James Dewar (1842-1923). He builds a large-scale machine with which he is able to carry this out. The following year he lowers the temperature of hydrogen to -432.4°F (-258°C) and is the first to solidify it.

1899 William Jackson Pope (1870-1939), English chemist, discovers the first optically active (able to polarize light) compound that contains no carbon atoms. He demonstrates that this optical isomerism is applicable to other atoms like sulfur, selenium and tin. This proves that the Van't Hoff theory applies to atoms other than carbon.

(See also 1874)

1899   German physicists Julius Elster (1854-1920) and Hans Friedrich Geitel (1855-1923) demonstrate together that external effects do not influence the intensity of radiation. They are also the first to characterize radiation as being caused by changes that take place within the atom. Until their work, it is not known whether radiation is an effect caused only by uranium or whether it can be found elsewhere in nature. They show that radiation can be found at varying levels nearly everywhere in the universe. Individually, each is also productive in physics, as Elster produces the first practical photoelectric cell and Geitel builds the first cathode ray tube.

1899   Radioactive element actinium is discovered by French chemist Andre Louis Debierne (1874-1949). He continues the work of the Curies and finds that actinium is precipitated out with other rare earth elements when pitchblende, an ore of uranium, is treated with ammonium hydroxide. He names the new element after the Greek words "aktis" and "aktinos" for ray and beam. It is later independently discovered by German chemist Friedrich Otto Fritz Giesel (1852-1927).

(See also 1902)

DECEMBER 28, 1900   First direct-arc electric furnace for producing steel is put into operation at La Paz, France. This heating chamber with electricity as the heat source achieves very high temperatures that melt and alloy metals. The electricity has no electrochemical effect on the metal but simply heats it.

1900   Friedrich Ernst Dorn (1848-1916), German physicist, analyzes the gas given off by (radioactive) radium and discovers the inert gas he names radon. This is the first clear demonstration that the process of giving off radiation

transmutes one element into another during the radioactive decay process. This gas also proves to be the final member of the family of inert gases that was begun with the discovery of argon in 1894. Nearly one hundred years later, the threat of exposure to radioactivity from radon is brought to the public's attention in the United States.

1900   Moses Gomberg (1866-1947), Russian-American chemist, discovers triphenylmethyl, the first example of a free carbon radical. Carbon has four valences, but triphenylmethyl uses only three valences of the central carbon atom, leaving one free or unfilled. Free radicals eventually prove very important to chemical reactions since the manner in which they are formed and destroyed helps determine the nature of the reaction.

1900   Paul Karl Ludwig Drude (1863-1906), German physicist, proposes the first model for the structure of metals. His physics of metals is based on electronic theory and the model he offers explains the constant relationship between electrical conductivity and the heat conductivity in all metals. His theory also supposes that a dense gas of free electrons interacting with one another exist in a metal. This theory becomes important to later research on semiconductors.

1900   Vladimir Nikolaevich Ipatieff (1867-1952), Russian-American chemist, discovers the role catalysts can play when he finds that organic reactions taking place at high temperatures can be influenced in their course by varying the nature of the substance with which they are in contact. Until Ipatieff, it was believed that such influence was impossible since organic molecules randomly broke into pieces at high temperatures. His discovery, however, shows that they can be directed, and he works out the effect of different catalysts and the details of various reactions. He later uses his pioneering work on high-pressure catalytic reactions and knowledge of hydrocarbons to produce high octane gasoline from low-quality fuel for the United States during World War II.

1901 Europium is discovered by French chemist Eugène Anatole Demarcay (1852-1904). Examining a compound of the element samarium (which itself comes from the complex mineral ytterite), he finds that it also contains a previously unrecognized new element that he names europium after the continent of Europe. It is the most reactive of the rare earth elements and can even catch fire spontaneously. It also proves to be one of the most efficient of all the elements in the capture of neutrons, and this property makes it useful in the control systems of nuclear reactors.

(See also 1879)

1901 "Grignard reagents" are discovered by French chemist Francois August Victor Grignard (1871-1935) as he learns of the highly useful catalytic role that organic manganese halides can play in preparing organic compounds. He finds that their very high chemical reactivity makes them one of the most valuable classes of synthetic reagents (any substance which, from its capacity for certain reactions, is used in detecting, examining, or measuring other substances). Further work produces an entire series of such reagents, and Grignard gives the synthetic chemist a powerful new tool.

1902 Friedrich Otto Fritz Giesel (1852-1927), German chemist, first isolates actinium, element 89 on the periodic table. This discovery of actinium is independent of that of French chemist André Louis Debierne (1874-1949) who first discovered it in 1899. This atomic number makes it the first member of the second family of rare, earth-like elements known as actinides. All the elements in this family, including actinium, are radioactive.

(See also 1961)

1903 Ernest Rutherford (1871-1937), British physicist, and English chemist Frederick Soddy (1877-1956) explain radioactivity by their theory

of atomic disintegration. They discover that uranium breaks down and forms a series of new intermediate substances as it gives off radiation. Soddy will continue one line of this research and eventually put forth the notion of isotopes. Rutherford will show that each intermediate element breaks down at a particular rate so that half of any quantity is gone in a fixed period. Rutherford will call this fixed period "half life."

1904 Joseph John Thomson (1856-1940), English physicist, formulates an atomic model and is one of the first to suggest a theory of structure of the atom (as a sphere of positive electricity in which negatively charged electrons are embedded). This model becomes known as the "plum pudding atom" because of its similarity to this typically English dessert which has raisins embedded in a pound cake. Thomson's model will soon be replaced by the far more useful model offered by English physicist Ernest Rutherford (1871-1937), a student of Thomson's.

1904 First silicone is prepared by English chemist Frederic Stanley Kipping (1863-1949). Although he is never really successful in synthesizing a double-bonded silicone compound, his 40 years of pioneering research and the papers he produces on silicones lays the foundation for the post World War II boom in the use of silicones as greases, water repellents, and synthetic rubbers.

1906 Chromatography is invented by Russian botanist Mikhail Semenovich Tsvett (1872-1920), also called Tswett. Working with plant pigments whose similar organic compounds make them difficult to separate and study singly, Tsvett devises a convenient means of separation. He trickles a solution or mixture of dissolved pigments down a glass tube packed with calcium carbonate powder. As the solution washes downward, each pigment sticks to the powder with a different degree of strength, creating a series of colored bands, with each band of color representing a different substance. Since color plays an important role in this process, he chooses the Greek words for "writing in color," chromatography, for the name. Soon chromatography is improved by using better absorbing powders and proves useful on almost all kinds of mixtures, including colorless ones. The chromatograph eventually becomes an essential piece of biochemical laboratory equipment and chromatography becomes a universally used method for separating and identifying organic substances.

1906 Third law of thermodynamics is first stated by German physical chemist Hermann Walther Nernst (1864-1941). In his studies on the nature of chemical reactions, he knows that they behave very differently near absolute zero. In attempting to get as close to absolute zero as possible, his research indicates that at a temperature above absolute zero -459.4°F (0K or -273°C), all matter tends toward random motion and all energy tends to dissipate. He holds that this is the third law of thermodynamics and that it proves that absolute zero can never be attained. Nernst's law is found to be a powerful tool for predicting chemical equilibrium and thus for determining the feasibility of many chemical reactions. He is regarded as one of the founders of modern physical chemistry.

1906 Alfred Wilm (1869-1937) of Germany discovers that aluminum-copper alloys can be hardened by aging. Once he finds that the properties of aluminum can be greatly influenced by other metals, he produces a hardenable alloy (duraluminum) that contains 4% copper, 0.5% magnesium, 0.6% manganese, about 1% silicon, and 0.3% iron. The hardening process produces fine crystals of this compound, and this breakthrough contributes to the increased recognition of the industrial applications of aluminum.

1907 Lutetium, the last of the stable rare earth elements is discovered by French chemist Georges Urbain (1872-1938). He shows that the compound called ytterbia, thought to be an oxide of the element ytterbium, is actually a

mixture of two parts, one of which is an element he calls lutetium after the ancient Latin name "Lutetia" for his native Paris. There is little commercial demand for lutetium except for some applications as a catalyst in the petroleum industry.

1907 Histamine is first synthesized by German chemist Adolf Otto Reinhold Windaus (1876-1959). This important compound is found in animal tissues and has major physiological effects. This powerful stimulant does not attract much research attention until it is discovered that histamine exists in the body and produces such effects as dilation of blood vessels.

1907 Leo Hendrik Baekeland (1863-1944), Belgian-American chemist, produces synthetic resins on an industrial scale and creates Bakelite. This breakthrough is the first of the "thermosetting plastics," or plastics that once set, will not soften under heat. He discovers what comes to be called Bakelite while searching for a synthetic substitute for shellac (a natural and expensive product). This work results only in a resinous substance that forms by a condensation reaction of formaldehyde with phenol, which proves impervious to solvents. He initially tries to find a solvent that will work, but then realizes that the substance itself is potentially valuable. He continues to experiment and produces a liquid that once it solidifies into the shape of its mold, becomes both water and solvent resistant. He names it Bakelite and patents it this year. It is the first totally synthetic plastic.

1908 Helium is first liquefied by Dutch physicist Heike Kamerlingh-Onnes (1853-1926). An expert in experimental cryogenic techniques, he builds an elaborate device that will cool helium intensively by means of evaporating liquid hydrogen. Now able to study helium and use it more intensively, he eventually discovers that certain metals, such as lead and mercury, undergo a total loss of electrical resistance at very low temperatures. This leads to his discovery of the phenomenon of superconductivity.

(See also 1911)

1908 Ammonia is first synthesized by German chemist Fritz Haber (1868-1934). Using iron as a catalyst, he combines atmospheric nitrogen and hydrogen under pressure and forms ammonia. Using this method (the Haber process), which then easily converts ammonia into nitrates, he is able to fix nitrogen and potentially create plentiful supplies of it for fertilizer or explosives.

1909 Alfred Stock (1876-1946), German chemist, first synthesizes boron hydrides (compounds of boron and hydrogen). He purifies and characterizes several of these compounds, but is never able to obtain enough for practical use. Later when methods are found that produce greater yields of boron hydrides, they prove useful to space exploration as high-energy additives to rocket fuel. Boron is very hard and proves to have many industrial uses, and its compounds are today used in detergents, toothpastes, and similar products.

1909 Søren Peer Lauritz Sørensen (1868-1939), Danish chemist, first introduces the expression pH to denote the negative logarithm of the concentration of the hydrogen ion present. It becomes a standard measure of alkalinity and acidity. Since the hydrogen ion is always present in any system that contains water, it means that it is of almost universal interest to chemists who know that many reactions vary greatly in speed and in nature according to the concentration of hydrogen ions present. Sørensen's pH also simplifies many mathematical and graphical representations and makes many chemical relationships easier to understand.

1911 Modern concept of the atom is first stated by English physicist Ernest Rutherford (1871-1937). He experiments with alpha particles, firing them at a sheet of gold only one fifty-thousandth of an inch thick and finds that most

of them pass through with little or no deflection. However, when he notices that about 1 in 8,000 alpha particles is deflected sometimes more than 90 degrees, he interprets this to mean that the positive charge in gold atoms in the sheet is concentrated in a very small volume. He then calls this small concentration of charge the atom's nucleus. From this evolves his theory of the nuclear atom which states that atoms are made of a positive nucleus surrounded by electrons. This modern concept of the atom comes to replace the notion of the featureless, indivisible spheres that dominated atomistic thinking for 23 centuries since the Greek philosopher Democritus (c.470-c.380 B.C.).

1911 Fritz Pregl (1869-1930), Austrian chemist, first introduces organic microanalysis. He invents analytic methods and devices that make it possible to determine the empirical formula of an organic compound from just a few milligrams of the substance. He constructs tiny and very precise pieces of measuring equipment made of blown glass and creates a real breakthrough in microchemistry.

1911 First continuous sheets of cellophane are produced. It is developed by Jacques E. Brandenberger who initially produces a solution called viscose rayon, a synthetic textile from cellulose acetate, which he then forces through thin slots to form sheets. After perfecting this process, he patents it this year and begins production of his thin, plastic film. Initially used for eyepieces on gas masks, it later proves to be an ideal wrapping material for food, and forms the basis of an entirely new industry for the packaging of goods in an attractive and hygienic fashion. It is also the first material that allows the contents of a package to be seen.

(See also 1892)

1911 Phenomenon of superconductivity is discovered by Dutch physicist Heike Kamerlingh-Onnes (1853-1926). After developing tech-

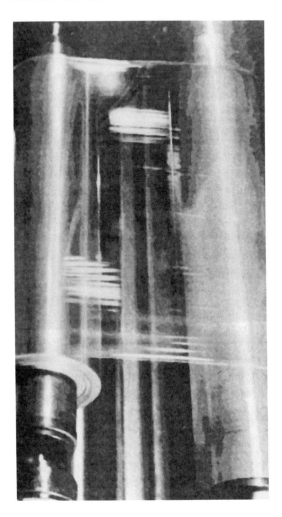

Cellophane in production.

1911 Chaim Weizmann (1874-1952), Russian-English-Israeli chemist, discovers how to use a particular strain of bacterium present in fermenting grain to synthesize acetone. The compound acetone is essential to the production of the explosive cordite, and because of his discovery, Great Britain will have adequate supplies of cordite for its munitions during World War I. This discovery is the forerunner of using microorganisms to accomplish a wide variety of synthetic purposes.

1912 Gasoline is first produced from coal and heavy oil by German chemist Friedrich Karl Rudolf Bergius (1884-1949), who treats it with hydrogen. Called the hydrogenation process, this conversion process is especially practical because it produces gasoline without an excess of undesirable by-products. It takes 12 years, however, for Bergius to translate laboratory success into a practical, industrial operation. During World War II, Germany uses the Bergius process to keep itself supplied with gasoline.

1913 First experimental demonstration that the major properties of an element are determined by the atomic number and not the atomic weight is made by English physicist Henry Gwyn Jeffreys Moseley (1887-1915). He studies the x-ray frequency characteristics of the various elements and discovers a technique to determine the atomic number of the nucleus. He finds that the positive charge of the nucleus indicates the number of electrons present in every neutral atom. This firmly establishes the relationship between atomic number and the charge of the atomic nucleus. Knowing that the atomic number for hydrogen, the smallest atom, is one and it is 92 for uranium, the most complex atom known, chemists now could be sure how many new elements remained to be discovered and where they would fall on the periodic table. Known as Moseley's law, this fundamental discovery concerning atomic numbers is a milestone in advancing knowledge of the atom. Unfortunately, Moseley is killed at

niques to liquefy helium using very low temperatures, he finds that certain substances tend to lose all resistance to the flow of electrical current when subjected to extremely cold temperatures. Continuing to experiment, he finds that each metal becomes superconducting at a given temperature which he calls its "transition temperature." He also finds that exposing a metal to a large enough magnetic field eliminates its superconductivity. Despite the obvious practical implications of this phenomenon, research in superconductivity has made only small strides in the years since its discovery.

27 when he is sent to Turkey during World War I and dies at Gallipoli.

1913 William Lawrence Bragg (1890-1971), Australian-English physicist, and his father William Henry Bragg (1862-1942) make the first determinations of the structures of simple crystals and demonstrate the tetrahedral distribution of carbon atoms in diamonds. The recent use of x rays to probe the molecular structure of crystals leads the Braggs to develop a mathematical system whereby the interference patterns of crystals can be examined. Their perfection of x-ray crystallography leads to the later examination of the molecular structure of thousands of crystalline substances, and is instrumental in the synthesis of both penicillin and insulin.

1913 Jean Baptiste Perrin (1870-1942), French physicist, publishes *Les Atomes* in which he offers systematically-obtained evidence of the size and number of atoms and molecules in a given volume of gum resin. For the first time, science has real evidence through observation of the existence of these tiny entities. Perrin's work finally proves that atoms are real.

1913 First use in the United States of chlorination to purify water. Pure chlorine is an extremely poisonous, yellowish green gas with a sharp, penetrating smell. Used in proper amounts, it kills bacterial germs in water. The first time it was used to purify water was during a typhoid epidemic in London in 1905. It is now routinely added to drinking water in the United States and many other countries around the world.

1913 Rust-resistant steel (stainless steel) is first introduced by English metallurgist Harry Brearly. He accidentally discovers that the alloy of nickel-chromium steel has anti-corrosive qualities when he finds a still-shining sample of it among a pile of discarded and rusted experimental scraps of metal. He soon uses a ratio of up to 8% nickel and 18% chromium to produce

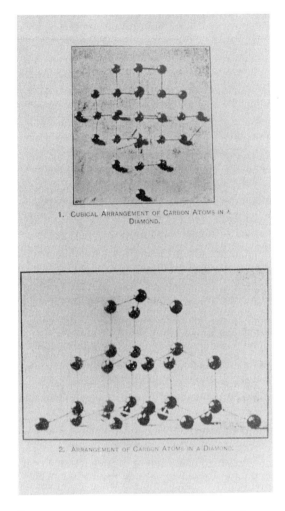

1. CUBICAL ARRANGEMENT OF CARBON ATOMS IN A DIAMOND.

2. ARRANGEMENT OF CARBON ATOMS IN A DIAMOND.

Two arrangements of carbon atoms in a diamond.

stainless steel that becomes essential in the food, cutlery, pharmaceutical, and chemical industries. He obtains an American patent on it in 1915.

1916 Kotaro Honda (1870-1954), Japanese metallurgist, discovers that the addition of cobalt to tungsten steel produces an alloy capable of forming an extremely powerful magnet. Called alnico, this new magnet is not only more strongly magnetic, but it resists corrosion, is immune to vibration and temperature changes, and costs less to build than ordinary steel magnets. This

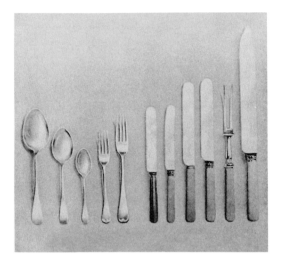

Examples of stainless steel table cutlery.

discovery, in turn, opens the way to still stronger magnetic alloys.

1917   Otto Hahn (1879-1968), German physical chemist, and Austrian-Swedish physicist Lise Meitner (1878-1968) discover the new element protactinium (atomic number 91). It is independently discovered by Polish chemist Kasimir Fajans (1887-1975) as well as by the English team of Frederick Soddy (1877-1956), John Cranston, and Alexander Fleck. Occurring in all uranium ore, it is recognized as the parent of actinium which it forms by the emission of an alpha particle. It later becomes specially important as the progenitor of the fissile uranium isotope uranium-233 in the production of nuclear fuel from thorium.

1919   First human-made transmutation or nuclear reaction is produced by English physicist Ernest Rutherford (1871-1937). He is the first person ever to change one element into another when he shows that nitrogen, under alpha particle bombardment, ejects a hydrogen nuclei. What he does is combine a helium nucleus (an alpha particle) and a nitrogen nucleus to form a hydrogen nucleus (a proton) and an oxygen nucleus, converting one type of atom

into another by subatomic bombardment. In a very tiny way, this is also the first human-made nuclear reaction.

1921   Anti-knock properties of tetraethyl lead for gasoline are discovered by American chemist Thomas Midgely, Jr. (1889-1944). In the early 1900s, higher compression automobile engines improve performance but also often cause a metallic knocking sound in the engine which means that it is not burning gasoline efficiently. Severe knocking can also damage an engine. After much trial and error, Midgely develops a mixture of tetraethyl lead and ethylene dibromide and dichloride that prevents an engine's knock and improves its octane rating. It is soon adopted throughout the world. Much later it is discovered that this breakthrough has severe health-related side effects since poisonous lead compounds are put into the air when gasoline is burned. Leaded gasoline is eventually banned in the United States and tetraethyl lead is replaced by new additives.

1922   Polarography is first introduced by Czech physical chemist Jaroslav Heyrovský (1890-1967). This is an extremely delicate analytical system for determining the concentration of ions in a solution of unknown composition. He builds a device that contains a mercury electrode arranged so that a small drop of mercury falls repeatedly through a solution into a mercury pool below. An electric current then flows through the solution, and as the potential is increased, the current reaches a plateau, the height of which depends on the concentration of certain ions in the solution. Measuring this plateau allows one to know the concentration of ions in the solution and thus be able to analyze its composition. Heyrovský spends many more years perfecting this technique which he names polarimetry.

JANUARY 1923   Georg von Hevesy (1885-1966), Hungarian-Danish-Swedish chemist, and Dutch physicist Dirk Coster discover the new element hafnium (atomic number 72). They

Thomas Midgely, Jr.

mixture in which tiny particles of one or several substances are scattered in another substance, most often water. Since Svedburg wishes to identify the exact size of the colloids he is studying, he improves the regular centrifuge and builds a device that is capable of rotating fast enough to create a force hundreds of thousands of times that of gravity. With these high rates of spin, the centrifugal effect can force different protein molecules to separate out at their own rates. His machine has far-reaching effects and enables biologists and chemists to examine and study viruses. This is a major advance in colloid chemistry.

JUNE 1925 Walter Karl Friedrich Noddack (1893-1960) and Ida Eva Tacke (1896-1979), both German chemists, discover the new element rhenium (atomic number 75). They soon extract one gram of rhenium from 1,455 lb (660 kg) of the mineral molybdenite. They are unaware that this discovery is the eighty-first and last element that possesses stable isotopes. Rhenium proves useful for medical instruments, photoflash lamps, and as a catalyst in the petroleum industry. Noddack and Tacke marry in 1926 and continue to work together.

1926 First crystallization of an enzyme is achieved by American biochemist James Batcheller Sumner (1887-1955), who proves that enzymes are proteins. In extracting the enzyme urease from the jack bean, he realizes that the crystals he isolates are both enzymes and proteins. This goes against the prevailing scientific opinion on enzymes, and it takes some years and support from others to demonstrate unequivocally that Sumner is correct. Sumner's urease is therefore the first enzyme shown incontrovertibly to be a protein.

prove what for years many chemists had suspected, that ordinary zirconium contained another element. It is very difficult to separate out and isolate, but they finally succeed and name it after the Latin name for Copenhagen, the city in Denmark where they discovered it. Hafnium is used in nuclear reactors and in light bulbs filaments and electrodes.

1923 Ultracentrifuge is invented by Swedish chemist Theodor Svedberg (1884-1971). Ordinary centrifuges are not effective in separating very small colloidal particles. A colloid is a

1926 Understanding of the true nature of polymers is first achieved by German chemist Hermann Staudinger (1881-1965), who shows that small molecules form long, chain-like structures (polymers) by chemical interaction and not simply by mere physical aggregation.

This proves that the various early plastics being produced at this time are not random conglomerates of small molecules but rather are similar polymers with their units arranged in a straight line. His discovery suggests that such linear molecules can be synthesized by various processes and still maintain their individuality despite chemical modification. Staudinger's work removes many of the uncertainties about polymers and leads to their eventual proliferation after World War II.

1928   Otto Paul Hermann Diels (1876-1954) and Kurt Alder (1902-1958), both German chemists, discover a technique of atomic combination that can be put to use in the formation of many synthetic compounds. It comes to be known as the Diels-Alder reaction. The reaction involves a method of joining two compounds so as to form a ring of atoms. Also called diene synthesis, it involves the use of dienes (organic compounds with two carbon-to-carbon double bonds) to effect syntheses of many organic substances under conditions that throw light on the molecular structure of the products obtained. It proves especially important in the production of synthetic rubber and plastics.

1929   First commercially successful synthetic rubber, neoprene, is discovered by Belgian-American chemist Julius Arthur Nieuwland (1878-1936). While researching the polymerization of acetylene, he notices the strong similarities between the structure of polymerized acetylene and natural rubber. He shares his research with the DuPont Chemical team headed by American chemist Wallace Hume Carothers (1896-1937), and they develop and patent a rubber-like solid called chloroprene. DuPont calls it neoprene and begins marketing it in 1930. Neoprene proves superior since natural rubbers which tend to be soft at high temperatures and brittle at low ones. It also makes a nation much less dependent on external supplies of natural rubber.

1929   William Francis Giauque (1895-1982), American chemist, discovers that oxygen is a mixture of three isotopes. He finds that of every 10,000 oxygen atoms, 9,976 have a mass or atomic weight of 16 (which was the accepted atomic weight of oxygen). Of the remaining 24 atoms, however, he also finds that 4 have a mass of 17 and 20 have a mass of 18. He calls these oxygen-17 and oxygen-18. Since oxygen atoms served for a century as the standard against which atomic weights could be determined, the fact that the standard itself was not constant suggested that all atomic weights would have to be slightly adjusted. This discovery leads to a debate between chemists and physicists concerning an atomic weight standard that is not resolved until 1961. The commonest isotope of carbon, carbon-12, is eventually accepted by all as the new standard since it would entail the smallest of adjustments.

1930   Thomas Midgely, Jr. (1889-1944), American chemist, first prepares Freon (difluorodichloromethane) as a refrigerant gas. Also called a chlorofluorocarbon (CFC), it comes to be used in all refrigerators, freezers, and air conditioning systems. Although neither poisonous nor inflammable (like all other refrigerants of this time), it is later believed to play a major role in the depletion of the earth's ozone layer. In the 1980s, many countries begin phasing out the use of this gas.

1931   First electron microscope is attributed to German electrical engineer Ernst August Friedrich Ruska (1906-1988). More powerful than a regular light microscope, it obtains a magnification of 12,000 times. The electron microscope revolutionizes observation by giving clear, detailed images that are magnified much more than had been possible.

1931   First synthetic fiber (nylon) is made by a group headed by American chemist Wallace Hume Carothers (1896-1937). Searching for a synthetic replacement for silk, the team moves from polymer fibers to polyester which, when drawn out into strands has a silky appearance. Since polyester melts at temperatures too low for use in the textile industry, Carothers' team switches to a polyamide called nylon which has a higher melting point. When stretching it to form fibers, they find its strength increases by more than 90%. In 1938, nylon stockings become the first commercially successful product made of synthetic fiber. After World War II, it proves highly versatile, entering the marketplace in guitar strings and tires, apparel and carpets.

1931 Deuterium, one of the heavy isotopes of hydrogen, is discovered by American chemist Harold Clayton Urey (1893-1981). He reasons that if "heavy hydrogen" does exist, it would evaporate from a liquid state more slowly than would normal hydrogen, and this difference might make it easier to separate. He then devises an experiment that allows liquid hydrogen to evaporate very slowly until only one millimeter is left out of four liters. Examining the spectral pattern produced by this liquid reveals spectral lines that are slightly different from those of hydrogen, indicating that it is heavy hydrogen. He names this heavy isotope after the Greek word "deuteros" for second. Deuterium comes to be used in particle accelerator research and also proves to be an ideal tracer in biological studies.

1933 Gilbert Newton Lewis (1875-1946), American chemist, is the first to prepare a sample of water in which all the hydrogen atoms consisted of deuterium (the heavy hydrogen isotope). Many scientists had suspected that hydrogen had a heavy isotope (one with an atomic weight of 2 rather than the 1 of the usual hydrogen atom). Although Lewis is not the first to discover it, he is the first to prepare water composed of deuterium. Called "heavy water," this will later play an important role in the production of the atomic bomb. It serves to slow down neutrons and make them more effective in setting up a chain reaction.

1933 Polyethylene is discovered accidentally by English chemists, J. Swallow and M. Perrin while researching the effects of high pressure on chemical reactants. After nearly three years of trying unsuccessfully to produce a waxy solid from ethylene and benzaldehyde, they inadvertently allow oxygen into the process via an unknown leak, and suddenly produce 0.28 oz (8 g) of polyethylene. Further investigation reveals that oxygen is essential to the polymerization process. Although this early polyethylene can find no immediate practical uses, it comes to replace the natural product, gutta-percha, as an insulator of undersea communications cables during World War II. Postwar research eventually leads to its refinement and improvement, and it becomes an everyday product in our lives. (See also 1953)

1934 First method of separating isotopes (different nuclear forms of the same element) of artificial radioactive elements is developed by Hungarian-American physicist Leo Szilard (1898-1964) and English physicist T. A. Chalmers. They discover that in producing atoms with high energy levels, nuclear transformations are followed by chemical effects (the breaking of bonds). This gives birth to the chemistry of excited states or hot-atom chemistry.

1934 Australian physicist Marcus Lawrence Elwin Oliphant, Austrian chemist Paul Harteck, and English physicist Ernest Rutherford (1871-1937) discover tritium, one of the heavy isotopes of hydrogen. They bombard deuterium itself (the hydrogen isotope number 2) with high-energy deuterons (the nuclei of deuterium atoms), and find that an unknown atom of a complicated form of hydrogen, hydrogen-3, is formed. They call this new hydrogen isotope "tritium" (from the Greek "tritos" for third), and also find that it is weakly radioactive.

1934 Dicumarol is first isolated by American biochemist J. P. Lent, who finds it in sweet clover hay that has spoiled. An organic compound closely related to the sweetener coumarin, it is used in medicine to reduce the tendency of the blood to clot (an anticoagulant).

1934 American chemist and inventor Arnold Orville Beckman first develops a pH meter. He initially devises an "acidmeter" or pH meter to help fruit growers accurately analyze the degree of acidity in their fruit juices. The measurement of pH is a degree of acidity or alkalinity of a solution, and almost every industrially important product has a characteristic pH that must be measured and controlled. Beckman's pH meter uses electricity to accurately measure a solution's acidity or alkalinity.

1937 Cyclamate is discovered by American chemist Michael Sveda. While working with the salts of cyclohexysulfamic acid, he notices a sweet taste on his fingers and finds that this odorless, white crystalline powder has about 30 times the sweetening power of sucrose. He works for the next decade to perfect this artificial sweetener and it is eventually marketed successfully.

(See also June 1950)

1937 First element to be prepared artificially that does not exist in nature is achieved by Italian-American physicist Emilio Segre (1905-1989) and Carlo Perrier who search and find technetium, the lightest element yet discovered. They demonstrate that a sample of molybdenum which has been bombarded with deuterons and neutrons produces the "missing" element 43. They name it technetium from the Greek "technetos" for artificial (since it is the first element to be made artificially).

APRIL 6, 1938 Teflon is accidentally discovered in the residue of refrigerant gases by American chemist Roy J. Plunkett at the labs of the DuPont Company. While studying fluorocarbon gases after the war in search of new refrigerants, he discovers that tetraflouroethylene gas can polymerize into a slippery, white, waxy substance. It proves insoluble in everything, unwettable by anything, a good electrical insu-

Arnold O. Beckman.

lator, heat-resistant, and most importantly, fat will not stick to it. Its first commercial use is for non-stick frying pans. Later, it becomes useful in many aspects of the space program, in surgery for heart valves, and in joint replacements.

1938 Vitamin B-6 (pyridoxine) is first isolated from skim milk by Austrian-German chemist Richard Kuhn (1900-1967). A member of the family of "carotenoids" (fat-soluble, yellow coloring agents widely distributed in nature), this vitamin functions in the formation and breakdown of amino acids, and therefore of protein, in living tissues. It is also involved in fat and carbohydrate metabolism processes. Although there is no disease related specifically to a deficiency of B6, it is believed that a severe lack leads quickly to hardening of the arteries.

1939 Francium, a very rare natural radioactive isotope, is first isolated by French chemist Marguerite Perey (1909-1975). She positively identifies this as element number 87, one of the breakdown products of uranium. She names it "francium" after the country in which the element is discovered.

1939 Dichloro-diphenyl-trichloroethane (DDT)

is first used as an insecticide by Swiss chemist Paul Hermann Müller (1899-1965), who discovers its insect-killing properties. During World War II, DDT proves effective against lice, fleas, and mosquitos (the carriers of typhus, plague, and malaria and yellow fever). After the war, it is used against insects that attack crops. However, the high stability and durability of DDT (it proves effective for 12 years, cannot be washed away by water, and resists breakdown by light or air) leads to its accumulation first in insects, and then in the fishes and birds that eat them, eventually infiltrating the entire food chain. In light of the mounting evidence of the adverse impact on the environment and human health, the use of DDT is banned in the United States in 1972.

Martin D. Kamen.

1940 First of the transuranic elements (those with a higher atomic number than uranium), neptunium, is prepared by American physicist Edwin Mattison McMillan (1907-1991) and American physical chemist Philip Hauge Abelson. In analyzing the fission products of uranium, they find element 93 and prove that it is an entirely new element. They name this element, which lies beyond uranium, after the planet Neptune whose orbit around the sun lies beyond that of Uranus.

1940 Radioactive isotope carbon-14 is discovered by Canadian-American biochemist Martin David Kamen. While searching for radioactive isotopes of the light elements, he eliminates oxygen and nitrogen but is able to isolate carbon-14. Further tests indicate that it has a half-life of 5,700 years, a property that makes it one of the most useful of all isotopes in biochemical research. It is later put to archaeological use for dating ancient artifacts and sites.

(See also 1947)

1941 Spectrophotometer is first developed by American chemist and inventor Arnold Orville Beckman. This device measures light in the visible spectrum, wavelength by wavelength, at the electron level and can be used for many

kinds of chemical analysis. His quartz photoelectric spectrometer has a precision accuracy to within 99.9% which is compared to the 25% accuracy of what existed previously. Since it also gives results in minutes rather than weeks, it revolutionizes the chemical analysis of biological assays. Beckman's company goes on to become one the largest suppliers of scientific and medical instrumentation in the United States.

NOVEMBER 1942 Frank Harold Spedding, American physicist, develops the necessary methods to produce pure uranium in very large quantities for the atomic bomb effort for the United States. As an expert in reducing individual rare-earth elements to a metallic state at low cost, Spedding produces two tons of uranium. This is later used for the first "atomic pile" and makes possible the first self-sustaining nuclear chain reaction experiment at the University of Chicago on December 2, 1942.

1943 Dow Corning of the United States begins production of the first silicones. This diversified class of chemical compounds belongs to the family of polymers. Although they are partially organic in composition, their molecules differ from those of most polymers in that the backbone of their structure does not contain carbon but is a polysiloxane chain in which atoms of silicon and oxygen alternate. Silicone fluids and rubbers are exceptionally stable and impervious to many things, and are used during World War II in electronic equipment and aircraft engines. Dow Corning takes the lead in the production of silicone rubber which has a useful temperature range of -178°F (-117°C) to 600°F (315°C). Among today's applications, silicone is used for sealants and lubricants, electrical products, and space technology. It is no longer used for breast implants, however, since a link with cancer has been established.

1944   Quinine is first synthesized by American chemists Robert Burns Woodward (1917-1979) and William von Eggers Doering. They use the technique of total synthesis which means they begin with simple compounds that can be easily synthesized from their constituent elements, and from these they synthesize quinine. At no point in this synthesis is it necessary for them to make use of some intermediate that is related to some living or once-living organism. The United States immediately begins synthesizing tons of quinine for American soldiers fighting in North Africa and the South Pacific who are being plagued by malaria.

1944   Paper chromatography is first developed by English biochemists Archer John Porter Martin and Richard Laurence Millington Synge. Their method of using porous filter paper to separate and identify the nearly identical but different types of amino acids proves an instant success. Where the old, slow method of chromatography involves dripping a large quantity of solution through a column of absorbent powder, this quick and economical technique allows the solution to creep up a column of absorbent filter paper. Also called partition chromatography, it is very fast and works with only drops of a solution. It proves to be an excellent technique for separating the complex

mixtures that are formed when a protein molecule is broken into fragments.

(See also 1952)

1947   Radioactive dating is invented by American chemist Willard Frank Libby (1908-1980). Knowing that all living things carry traces of carbon-14 (since plants absorb it during photosynthesis and animals absorb it by eating plants), he reasons correctly that as long as a creature is alive, it will constantly replenish its supply of that isotope. When dead, however, the supply will begin to undergo radioactive decay. He eventually constructs devices using Geiger counters that make carbon-14 easy to detect and can make accurate measurements of how much it has decayed. His technique of radiocarbon dating soon becomes an accepted tool to judge the age of ancient objects, sites, and artifacts. Its effective range improves to 70,000 years and it has an uncertainty factor of plus or minus 10%.

1948   Germanium crystals are used by the Bell Telephone Company in the United States to build the first transistors.

1949   First to use an electronic computer in direct application to a biochemical problem is English biochemist Dorothy Mary Crowfoot Hodgkin (1910-1994). In attempting to learn the atomic and molecular structure of penicillin (so it could be synthesized by others), she uses a punch-card style computer to work out x-ray data on this very complex molecule. Even with the computer, it takes her nearly five years to map the atomic structure of penicillin crystals.

JUNE 1950   Artificial sweetener cyclamate is first introduced to the American market by Abbot Laboratories under the name Sucaryl. Thirty to fifty times sweeter than real sugar, it does not have saccharin's aftertaste and becomes essential to the low-calorie soft drink industry. Like saccharin, however, it is later linked to bladder cancer and is banned in 1970 by the United States.

(See also 1965)

1950 DuPont Company puts the first acrylic fiber on the market, calling it Orlon. Used for sweaters and pile fabrics, the trademark name Orlon consists mainly of acrylonitrile, made by a combination of acetylene with hydrogen cyanide.

1950 First total synthesis of carotenoids is achieved by Swiss chemist Paul Karrer (1889-1971) and H. H. Inhoffen. This group of non-nitrogenous yellow, orange, or red biological pigments are most conspicuous in such food items as carrots, tomatoes, and sweet potatoes. All animals cells also contain carotenoids. In synthesizing carotenoids, Karrer proves that vitamin A is related to them in structure. Today, increasing claims for the health benefits of carotenoids are being made, especially as cancer deterrents.

1952 Gas chromatography is first perfected by English biochemist Archer John Porter Martin. Also called vapor-phase chromatography, it allows the separation of chemical vapors by differential absorption on a porous solid. This means that chemists can separate a mixture of gases. Martin and his colleague L. T. James first use gas chromatography to microanalyze fatty acids. Gas chromatography is widely used in nearly every branch of the chemical industry, especially in the production of petrochemicals from oil and natural gas.

1953 Giulio Natta (1903-1979), Italian chemist, develops the first "isotactic" polymers. While studying the role of catalysts in the techniques of polymerization, he finds that he can use a resin, to which ions of metals are attached, as a catalyst in the production of polyethylene. This results in a new, tougher product with a higher melting point than the older version. His technique allows for their production at will.

1953 First hydrometallurgical plant to use high pressure and temperature for the processing of nickel, copper, and cobalt is built in Canada. Hydrometallurgy is the extraction of metal from ore by the use of aqueous solutions. The steps involved are leaching, or dissolution of the metal by water or acid separation of the waste and purification of the leach solution and the precipitation of metal from the leach solution by chemical or electrolytic means.

1953 Karl Ziegler (1898-1973), German chemist, discovers a method of producing high density polyethylene. While attempting to polymerize ethylene using catalysts at a much lower pressure than usual, he obtains unexpected results. After finding that trace amounts of nickel had accidentally entered the experiment, he continues and succeeds in producing a new, tougher product that has a much higher melting point by using a resin catalyst. This allows polyethylene to be made at low temperatures and normal pressure, making the industrial process simpler and cheaper. High density polyethylene is recyclable and is used in dishes, squeezable bottles, and other soft plastics

(See also 1933)

DECEMBER 16, 1954 First synthetic diamonds are produced in the General Electric Laboratories in the United States. Working with a high-pressure device invented by American physicist Percy Williams Bridgman (1882-1961), a lab team is able to convert graphite powder into tiny diamond crystals. Bridgman's diamond anvil cell produces a temperature of more than 4,800°F (2,700°C) and a pressure of more than 1.5 million pounds per square inch (100,000 atmospheres). Most of the synthetic diamonds produced today are used not as gemstones but for industrial purposes.

(See also 1957)

1954 First synthesis of a polypeptide hormone is achieved by American biochemist Vincent Du Vigneaud (1901-1978). Working with oxytocin, a hormone produced by the pituitary gland, he breaks it down, discovering that it is a small protein molecule made up of eight amino acids. He then works out the exact order in which these acids appear in the chain and is consequently able to produce synthetic oxytocin that has all the properties of its natural counterpart. This is the first protein hormone ever synthesized and it opens the door to the synthesis of many larger, more complex hormones for medical use.

1954   Strychnine is first synthesized by American chemist Robert Burns Woodward (1917-1979). This extremely complicated and highly poisonous alkaloid is found to have a molecule composed of seven intricately-related rings of atoms. In this same year, Woodward also synthesizes lysergic acid which is structurally related to LSD.

1957   First industrially produced synthetic diamonds are put on the market by the General Electric Company.

(See also December 16, 1954)

1958   Transuranium element nobelium is discovered by American physicist Glenn Theodore Seaborg. After being involved as a student with the discovery of neptunium—the first of the transuranic elements—Seaborg goes on to discover plutonium, and is involved with the discovery of six more elements (americum, curium, berkelium, californium, mendelevium, and now nobelium).

1960   Chlorophyll is first synthesized by American chemist Robert Burns Woodward (1917-1979). He achieves this extremely difficult task and succeeds in synthesizing the complex chlorophyll molecule. As the green coloring matter visible in leaves and plants, it converts sunlight into chemical energy.

1961   First sample of element 103, lawrencium, is produced by American physicist Albert Ghiorso. This is the fifteenth and last of the actinides, the second family of rare earth-like elements.

(See also 1902)

1962   First known case of an inert gas atom forming a bond with any other atom or group of atoms is demonstrated by English chemist Neil Bartlett. He successfully attempts to get the heaviest of all the noble (inert) gases, xenon, to react with a very active chemical. He immerses platinum fluoride in xenon gas and the two substances combine to form xenon fluoroplatinate. After this, such gases are no longer referred to as "inert" but rather as "noble" gases.

1964   Researchers at the Joint Nuclear Research Institute in Dubna, Russia, report they have produced an isotope of element 104 by bombarding plutonium-242 with neon-22 nuclei. They propose the name kurchatovium in honor of Soviet nuclear physicist Igor Kurchatov (1903-1960). Element 104 is the first transactinide element or the first element to begin a new, rare-earth-type row in the Periodic Table. In 1969, scientists at the University of California at Berkeley also report the production of element 104 and suggest the name rutherfordium. This name controversy still continues as the International Union of Pure and Applied Chemistry (IUPAC) has recommended the naming of this and other transactinide elements with the Latin terms for their numbers. 104 would be called unnilquadium.

1965   Artificial sweetener aspartame is discovered by American chemist James M. Schlatter who combines two amino acids and notices its extremely sweet taste. This new chemical proves to be about 200 times sweeter than sugar and is marketed under the name aspartame. Approved initially in 1974 by the United States Food and Drug Administration for use in dry foods like cereal and chewing gum, it is withdrawn the following year but reapproved again in 1981. In 1983 it is cleared for use in carbonated beverages. As an artificial sweetener with no aftertaste, it becomes the most widely used sugar substitute.

1967 Discovery of element number 105 is first reported in the United States and is named hahnium. This artificially produced transuranium element becomes a matter of dispute with Russian scientists who claim to have done the same.

1969 Gerald Maurice Edelman, American biochemist, first works out the structure of a gamma globulin, a type of protein that exists in the blood and out of which various antibodies are formed.

1970 Megavitamin therapy is first proposed by American chemist Linus Pauling (1901-1994). Recommending massive doses of vitamin C (ascorbic acid) for health, he argues that higher doses of vitamins are required for full health. His ideas are still popular although not always accepted by most biologists.

1972 Vitamin B-12 is first synthesized by American chemist Robert Burns Woodward (1917-1979). After ten years of work, he successfully synthesizes the most complicated nonpolymeric molecule of all. Although his process does not prove to be a practical source of this important vitamin, the research adds greatly to an understanding of complicated compounds.

1974 Ozone layer damage from chlorofluorocarbons (CFCs) is first pointed out by F. Sherwood Rowland and Mario Molina of the United States. The phenomenon called ozone depletion is caused by a buildup of CFCs in the atmosphere. CFCs are chlorine-based compounds used primarily as aerosol propellants, refrigerants, and coolants. After being released into the air, they are broken down into smaller elements by intense ultraviolet radiation, and chlorine ions attack and destroy ozone molecules. Since the ozone layer is essential for the protection of life on Earth because it filters out all forms of incoming ultraviolet radiation and acts as a protective screen, any depletion in that layer is considered serious.

1985 First "buckminsterfullerene" (carbon 60) is discovered by American chemist Robert F. Curl, English chemist Harry W. Kroto, and American physicist Richard E. Smalley. After preparing an experiment intended to recreate conditions that exist in the outer reaches of the universe (by vaporizing graphite with a laser inside a chamber filled with helium gas), they find a previously unknown form of carbon. This molecule of carbon, carbon 60, is a cage-like molecule made up of 60 carbon atoms elegantly arranged in the shape of a soccer ball. It also resembles a geodesic sphere, and is named buckminsterfullerene after the creator of the geodesic dome, American architect and inventor Richard Buckminster Fuller (1895-1983). It is now believed that fullerenes may be among the most common of molecules. Now science knows how to create them and modify them to create new chemicals. Proposed uses are for specialized automotive lubricants, light materials for airplane wings, and rechargeable batteries.

1991 John V. Budding, an American chemist, leads a group that compresses iron and hydrogen in a diamond cell, making the first observations of the chemical combination of iron and hydrogen under ultrahigh pressures to form iron hydride.

1991 Researchers obtain the first evidence of medium-range structural order in glass. This is a breakthrough since glass is one of the most disordered of all solid materials. Using a sensitive neutron-scattering technique, they study the atomic structure of calcium silicate glass and discover calcium ions distributed in a manner similar to crystals. This discovery may lead to ways for optimizing the mechanical, optical, and other properties of glass by modifying this newly-discovered structure.

1991 First molecular magnet that retains magnetic properties at room temperature or above is developed by Arthur J. Epstein and Joel S. Miller of the United States. Molecular magnets as opposed to those based on atomic constituents, have long been sought after, since they have many advantages. If molecular magnets could be made out of organic polymers, they would have many of the desirable properties of plastics. This development may point the way to more stable room temperature organic magnets with practical applications.

1992 First solid compound of the noble gas, helium, is produced by researchers in the United States. A solid is formed when helium and nitrogen are mixed together and subjected to about 77,000 times normal (atmospheric) pressure. Researchers think that it may be the first of a new family of chemical compounds that are formed at high pressure and held together by weak interactions known as van der Waals forces. Studies of such compounds may be important in understanding the structure and properties of matter in the interiors of the gas-giant planets of our solar system.

1993 First experimental observation of a state of ultralow friction is reported by American chemist Jacob Israelachvili and his co-workers. They observe this phenomenon during studies of the so-called stick-slip motion of specially treated mica surfaces. Stick-slip motion is an inter-rupted motion that occurs in such phenomena as friction, fluid flow, and sound generation, and is the major cause of friction damage to moving surfaces. After coating mica surfaces with single-molecule hydrocarbon layers, a superkinetic state of ultralow friction is attained that may have applications in aerospace components, computer disk heads, and many other devices.

1993 First synthesis of the largest all-hydrogen molecule ever made is achieved by American chemists Jeffrey S. Moore and Zhifu Xu.

MAY 3, 1996 Biochemical signals are first identified to be signals for a developing organism to determine if webbing in its feet will occur. American researchers Hongyan Zou and Lee Niswander experimentally block the molecular signal of bone morphogenetic proteins (BMP) and find that chickens develop duck feet. Research suggests that BMPs are important elsewhere in the developing body and that the signaling system uncovered in chicks is also active in people. BMPs may also be involved in programmed cell death, indicating that they may do different things at different stages of embryonic development.

Communications

c. 40,000 B.C. Oldest documented "art" found in Africa are paint slabs in Namibia attributed to this time period.

c. 20,000 B.C. Earliest recorded animal engravings are red and black cave paintings found in northeastern Spain and believed to have been made around this time period. This is a tangible sign of early humans' pictorial communication attempts. Although considered works of art today, it is not clear whether these works tell a story of an incident that actually occurred, or whether they were part of a magic ritual believed to aid in a coming hunt.

c. 13,000 B.C. First known artifact with a map is a piece of bone that shows a region in the Ukraine where it was found.

c. 8000 B.C. Forms of fired clay first are used by Neolithic people to keep track of agricultural products. They are unmarked and have geometric shapes. Around 4500 B.C., similar clay tokens with marks on them are fired and used for agricultural record-keeping.

c. 4000 B.C. Earliest known printing by impression was in use in Babylon. Engraved stamps or seals were impressed or pushed into moist clay which was then dried. This was "inkless" as well as "paperless" printing.

c. 4000 B.C. Earliest known form of writing is cuneiform developed by the Sumerians in the Near East around this time. Cuneiform derives its name from the Latin "cuneus" for wedge, and refers to the wedges that are used to inscribe this writing onto soft clay tablets. Cuneiform symbols originated as pictograms, which were simple pictures of the actual thing represented. This marks the first step in the evolution of pictures into words.

c. 3100 B.C. The earliest recorded examples of Egyptian hieroglyphic writing are dated to this time. This form of stylized picture-writing is a stage beyond the pictograms of cuneiform writing, and although it sometimes represents the object it depicts, it usually denotes particular sounds or groups of sounds.

c. 3000 B.C. First known crude paper is made in Egypt from the papyrus plant, a water reed growing along the Nile River. It is made by stripping thin coats from the stalk of the plant, spreading them out in a sheet, and placing another sheet of strips crosswise to form a thicker sheet that is pressed and dried. The English word paper comes from the word papyrus. The papyrus is then pasted into strips and rolled. The Greeks call the papyrus reed "bublos" or "biblos," from which derived "biblion" meaning book. English words for Bible and bibliography are attributed to "biblion" as well.

c. 2300 B.C. Oldest known surviving map of a city is a map carved in stone depicting the Mesopotamian city of Lagash.

c. 2000 B.C. Earliest reference to a postal system is made by the Egyptians who use this term to describe the system of communication that links their large empire. Postal systems do not arise in China until the Chou dynasty around 1000 B.C.

c. 1700 B.C. First recorded alphabet is developed by the peoples of Syria and Palestine. Borrowed largely from the Egyptians' stylized hieroglyphics, it is later modified by the Phoenicians and transmitted to the Greeks in the tenth century B.C. Two hundred years later, the Greeks replace all the Phoenician symbols with their own, add several new symbols, and change the direction of their writing so words and sentences are read from left to right. The Greek alphabet is adopted by the Etruscans who teach it to the Romans, who modify it even further.

c. 1580 B.C. Oldest known surviving papyrus of any length, the Egyptian *Book of the Dead* is dated to this time. This collection of mortuary texts is composed of magic spells and formulas and is placed in tombs to protect the dead. Egyptian scribes copy these texts on rolls of papyrus, often colorfully illustrate them, and then sell them to individuals for burial use.

c. 1400 B.C. Olmec people on the east coast of Mexico first develop a numeration system and calendar as well as the beginnings of a written language. These are eventually adopted by the Mayans and the Zapotec people of Monte Alban, Mexico who develop it into a useful hieroglyphic script.

c. 1300 B.C. Earliest evidence of writing in China is the incised divination bones discovered in Hunan Province in 1899 and dated to

this time. These bones contain short sentences addressed to spirits.

c. 550 B.C. First known map of the inhabited Earth is drawn by Anaximander (c.610–c.546 B.C.), Greek philosopher. Recognizing that Earth's surface must be curved (which accounts for the change in the position of the stars as one travels), he therefore puts the map on a cylinder to show that Earth's surface is curved.

540 B.C. First public library is founded in Athens by the Greek dictator Peisistratus (600–527 B.C.).

c. 500 B.C. Standardized silver coins are introduced for the first known time at Aegina and Corinth as an economy based on coin money begins to develop. Many of these early coins are certified with a punchmark indicating their weight or fineness, and are also made of electrum which is a natural mixture of gold and silver. For the next four centuries, the Athenian "drachma" has a nearly constant silver content (between 67 and 65 grains).

c. 400 B.C. Mo-tzu (c.470–c.391 B.C.), Chinese philosopher also called Mo ti, observes that the reflected light rays of an illuminated object passing through a pinhole into a darkened enclosure result in an inverted but otherwise exact image of the object. This may be the earliest account of what will come to be called the "camera obscura."

(See also 1521)

c. 300 B.C. Demetrius Phalereus, Greek philosopher and governor of Athens, first recorded the practice of collecting copies of all known books to stock the pharaoh's library. At the height of Alexandrian culture, the two royal libraries contained between 500,000 and 750,000 books on papyrus scrolls.

250 B.C. Meng T'ien, general of the Ch'in dynasty, invents the camel's hair brush. It is used as a pen and eventually revolutionizes the writing of Chinese characters.

(See also c.225 B.C.)

c. 225 B.C. Writing with a hair brush first begins in China. Before the brush, writing had been done on slips of bamboo or wood with a bamboo pen, using ink of lacquer made from tree sap. The brush makes writing on silk rolls possible.

(See also 250 B.C.)

c. 200 B.C. A new writing material called parchment is made from the skins of animals and perfected in the Greek state of Pergamum. Later in Rome, the use of a notebook or "codex," which consists of parchment leaves sewn together, becomes popular.

c. 100 B.C. Romans first develop the "cursus publicus," which is the most highly developed postal system of the ancient world. It uses relay stages at convenient intervals along its roads and is capable of speedy and reliable communications throughout the sprawling empire.

59 B.C. Gaius Julius Caesar (100-44 B.C.), Roman dictator, orders the publication of the "Acta Senatus" which contain the debates of the senate. He is also said to have invented the first newspaper as he had announcements and news items inscribed on wax tablets and posted on Rome's main buildings for the public to read. These bulletins are variously called "Acta Urbis" (Acts of the City), "Acta Diurnis," (Daily Acts), or "Acta Populi Romani" (Acts of the Romans). A group of tablets hinged together is called "liber," meaning the bark of a tree, and it is from this word that we derive the word for "library."

c. 50 B.C. First known organized system of shorthand writing or stenography is invented by Marcus Tullius Tiron, secretary to the Roman orator Cicero (106-43 B.C.). His method employs brief strokes to represent the charac-

ters of the alphabet. It is later taught in Roman schools and is used to record speeches.

105 Modern papermaking from fiber is first traced to Ts'ai Lun (c.50-118), an official in the Imperial court of China. He describes an inexpensive way of making sheet paper using mulberry and other bast fibers along with fish nets, old rags, and hemp waste. Whether he is the actual inventor or merely reporting it to the emperor is not known, but it is the first description of papermaking from disintegrated fiber.

c. 110 Method of using a grid to locate points on a map is first developed in China.

c. 150 Rather than being rolled into scrolls, parchment is folded into pages to make books for the first time. Parchment is the processed skins of certain animals, usually sheep, goats, and calves, on which writing is inscribed. A more thorough method of cleaning, stretching, and scraping the skins is invented in Pergamum in Asia Minor about this time, making possible the use of both sides of a manuscript leaf. This leads, in turn, to the bound book or codex which replaces the rolled manuscript.

175 First noted form of lithographic printing comes into use in China. The classics of Confucius are cut into stone, insuring their permanency and accuracy, and permitting students and scholars to make exact copies by a simple process of rubbing paper on the inked stone. This method of duplicating texts continues for a thousand years.

c. 190 Nazca lines and figures—giant drawings of animals and figures on the desert ground in Peru that are only wholly visible from the air— are first spotted.

450 First true ink is made from lampblack in China. Prepared by mixing lampblack (formed by burning certain types of oil) with a solution of gum or gluten, it is then rubbed in water on a smooth stone and is used for writing or to make a clean, neat impression. It is not suitable for printing under real pressure.

c. 600 The Chinese print entire pages with woodblocks.

610 Papermaking in Japan begins. It is believed to have originated from Korea, which is part of China at this time.

c. 625 First known written reference to a quill pen is made by Spanish scholar Isidore of Seville (c.560-636). The quill pen is believed to have been used earlier than this.

c. 700 The earliest printed text known, a Buddhist charm scroll, is made around this time. Woodblocks are used to produce the images.

c. 750 Geber (c.721-c.815), Arab alchemist, first notes the darkening effect of light on silver nitrate. This will eventually prove to be the basis for photography.

(See also 1727)

c. 800 The Japanese first develop the phonetic script called "kana" for writing. This simplified script fosters literacy in Japan.

800 Modern papermaking in Egypt begins to replace rolls of Egyptian papyrus and parchment as writing material. The art is transferred from the East to the Middle East around 751 when the city of Samarkand is captured by the Arabs. Among the captives are some Chinese who teach their abductors how to make paper.

MAY 11, 868 *Vajrasekhara-sutra* or the "Diamond Sutra" in Sanskrit is completed. This Tantric Buddhist text is considered the oldest complete printed book extant. It is actually a 16 ft (4.8 m) long scroll and was discovered in 1907.

883 First clear reference to block printing and printed books in Chinese literature is given in an account written by Liu Pin.

c. 950 Earliest use of paper in Spain occurs.

969 First clear reference to playing cards is made during the Liao dynasty in China. Playing cards are believed to have originated in China as a development of the game of dice, and playing cards, called "sheet dice," are one of the earliest forms of block printing in China. Playing cards are introduced into Europe around 1375, and by 1397 card playing is so popular in Paris that an edict is passed forbidding workmen from playing cards on work days.

c. 970 Use of printed paper money in China becomes commonplace. Called "convenience money," paper money is the first form of Chinese printing encountered by visitors from the West.

1034 Pi Sheng, Chinese blacksmith and alchemist, invents movable (printing) type made of firebaked clay. He bakes ideographs out of clay, sorts and stores them, and arranges them however desired in a frame for printing.

(See also 1221)

1107 The Chinese invent multi-color printing which makes their paper currency harder to counterfeit.

1109 Earliest existing European manuscript on paper is a deed of Roger II of Sicily (1095-1154) written in Arabic and Greek.

1116 First stitched books are made by the Chinese. They print on only one side of the paper, which is then is folded and sewn with linen and cotton thread.

c. 1122 Theophilus Presbyter of Germany writes his *Diversarum Artium Schedula*, in which he first mentions "Byzantine parchment" (paper) and "plummets" (early pencils). These early pencils are thin rods of lead with tin added. Plummets are pointed at one end for writing and flattened at the other for drawing straight lines on paper.

1189 First recorded paper mill in the West is set up at Herault, on the French side of the Pyrenees mountains. It is built by a crusader who was captured and escaped from his slavery in such a mill in Damascus.

1221   First movable printing type made of wood appears in China.

1228   Earliest known use of paper in Germany is noted.

1259   First known use of the term "stationarii" in Italy. It is used to describe special scribes who keep and rent a supply of authorized copies of books for the University of Bologna. Book dealers at the university are also called "Stationarii," and it is from this that the English words for stationer and stationery are derived. As universities grow in size and the practice of renting books changes to selling them, the stationarii become "librarii" or real book sellers.

1282   First known authentic watermarks on paper are made in Italy. A watermark is a barely noticeable distinguishing mark impressed on paper as a trademark. The early marks or devices are usually emblems of all sorts, and the earliest use of the fleur-de-lis as a watermark occurs about this time.

1309   Earliest known use of paper in England occurs.

1373   Charles V (1338-1380), king of France, first opens what becomes the National Library in Paris.

1390   First paper mill in Germany is built by Ulman Stromer at Nuremberg. This mill is depicted in the early book, *Liber Chronicarum,* published in 1493. It is also the site of the first recorded labor strike in the paper industry.

1418   Xylography, the art of engraving on wood and printing from these engravings, first appears in Europe. Begun in China, this process is used by manuscript copyists to reproduce the outline of ornamental initial capital letters. Cutting letters from wood proves a delicate

operation, however, especially when used for smaller letters in the text; this process is a technological dead-end for printing.

1423   Earliest dated European woodblock print is made. It uses the ancient Chinese technique of carving in relief on wood from which inked impressions are made on paper. The woodblock shows a print of St. Christopher bearing the infant Christ. Another woodblock print, "Madonna with Four Virgin Saints in a Garden," is often dated 1418, but some claim its date to be 1450.

1450   Johann Gutenberg (c.1398-c.1468), German inventor, first invents printing with movable type. To the Chinese inventions of paper and printing with blocks, he adds two innovations—a method of casting metal type and an ink that adheres to the cast metal—and lays the technical foundations for what will become a revolutionary printing industry. It is thought that Gutenberg may have spent as many as 20 years perfecting a practical printing process. Developing the proper techniques for forming tiny metal letters that are easily interchangeable is especially challenging and difficult, but eventually it allows him to reuse the cast letters over and over. In this way, a limited number of such movable type can be used to print an unlimited number of different books or copies of those books. Mechanical printing eventually results in affordable books, and it is estimated that by 1500, up to 9 million printed copies of 30,000 individual works are in circulation. Books spread literacy, broaden the base of scholarship, and disseminate information and knowledge. It is not surprising that the printing revolution is soon followed by the Scientific Revolution.

(See also c.1454)

1450   Earliest known use of an "ex libris" is made. This is a bookplate or label printed on paper and pasted in the front of a book to indicate ownership.

c. 1454   First book printed from movable type is

Johann Gutenberg.

published by German inventor Johann Gutenberg (c.1398-c.1468). He produces a Bible with double columns and 42 lines to a page. His "42-line Bible" establishes the era of mechanical printing in Europe. This beautiful work is also considered, by some, to be the greatest of all printed books, setting standards in all its technical aspects, from book design to the quality of paper and ink, that have seldom been matched. Of the more than 200 copies originally printed, only 48 are known to exist.

1454   First dated document printed from movable type is a 30-line Indulgence granted by Pope Nicolaus V (1397-1455) to those who donated money for the struggle against the Turks. Although indulgences are usually done by hand in manuscript form, these are printed with spaces left blank for the donors' names.

1456   First known medical work to be printed is a calendar referred to as the "Bloodletting Calendar," printed for the year 1457. Such calendars told the best days to bleed a patient or to take medicine.

1457   First known printed book to give both the name of the printers and the date of printing is

the *Psalterium Latinum*. This can also be described as the first printed colophon (meaning from the Greek, "finishing stroke"). Early printers continue the manuscript tradition of giving the name of the responsible person, the date, and the place of printing. This work is also distinguished as being the first book printed in two colors.

OCTOBER 4, 1458   First contemporary reference acknowledging Johann Gutenberg (c.1398-c.1468) of Germany as the inventor of printing is recorded in the French royal mint describing the king's order to send someone to Mainz, Germany to learn the art of printing from Gutenberg.

1460   First large, non-religious book to be printed is the *Catholicon*, an encyclopedic work written by Johannes Balbus (also known as Giovanni Balbi). This large book has 748 pages of double columns and is basically an elaborate Latin grammar and an etymological dictionary. It is believed to have been printed by German inventor Johann Gutenberg (c.1398-c.1468).

1461   First dated book with woodcut illustrations is *Der Edelstein*, a book of fables by Ulrich Bonner of Switzerland. It is also the earliest dated book in the German language. It is printed by Albrecht Pfister.

1463   First recorded use of a separate title page at the beginning of a printed work is a Papal Bull, *Bul zu Deutsch*, issued by Pope Pius II (1405-1464). This example is not followed until several years later.

1465   First printed book in Italy is Cicero's *De Oratore*, printed by German printers Conrad Sweynheym and Arnold Pannartz at Subiaco, a village near Rome.

1466   First edition of the printed Bible in a modern language is Johann Mentelin's version printed in German at Strasbourg.

1472   First illustrated book on a technical subject is *De Re Militari*, a book on warfare by Roberto Valturio. It contains nearly 100 woodcut illustrations of contemporary engines of war for land and sea.

1473   Earliest known printed book to contain music and musical notes is the *Collectorium super Magnificat* by Jean Gerson. It is printed at Esslingen, Germany by Conrad Fyner.

(See also 1480)

1474   First known book printed in the English language is the *Recuyell of the Histories of Troy*, printed in Belgium by William Caxton (c.1422-1491) of England. This and most of his following books are printed in English rather than Latin.

1475   First dated book printed in Hebrew is published at Reggio di Calabria, Italy by Abraham ben Garton. It is a *Commentary on the Pentateuch* by Rabbi Salomon Rashi.

DECEMBER 13, 1476   First known piece of printing done in England is a Letter of Indulgence printed by William Caxton (c.1422-1491) of England.

1476   First dated book printed in the French language is *La Légende Doree*. It is printed at Lyons, France by Guillaume Le Roy and Barthelemy Buyer.

1476   First separately displayed title page in a printed book appears in the *Kalendario* of German astronomer Regiomontanus (1436-1476). It takes several more years before the practice of putting the title accompanied by the date, place, and printer's name on a separate page becomes common.

NOVEMBER 18, 1477   First dated book printed in England is *The Dictes or Sayengis of the Philosophhres*, printed by William Caxton (c.1422-1491) of England. It is translated from the French, which, in turn, was translated from Latin.

1477   First copper-plate engravings to be printed as an integral part of a book appear in *Monte Santo de Dio* by Antonio Bettini (1396-1487), bishop of Foligno, Italy. The printer is Niccolo de Lorenzo. Copperplate engravings do not come into general use until the close of the 16th century.

1478   First notice of errata in a printed book appears in an edition of Juvenal, *Enarrationes Saturarum Juvenalis*, printed by Gabriele di Pietro of Venice, Italy. Di Pietro lists the mistakes contained in the book on two, two-column pages and apologizes for the carelessness of a workman.

1480   First printed book dealing entirely with music is the *Theoricum Opus Musicae* by musical theorist Franchinus Gafurius (1451-1522). It is printed in Naples, Italy by Francesco di Dino. Although devoted entirely to music, it contains no printed musical examples, but rather has spaces left for the notes to be inserted by hand.

1481   First illustrated book produced in England is *The Mirrour of the World*, printed by William Caxton (c.1422-1491). This book covers an enormous range of natural subjects and contains 27 crude woodcuts.

1482   First printed book to be illustrated with geometric figures is the *Elements* by Greek mathematician Euclid. Published by Erhard Ratdolt, German printer and typographer, it is also the first printed edition of Euclid's classic work.

1484   Anna Rugerin of Germany, who takes over her late husband's business, is the first recorded female printing press owner and operator.

MARCH 22, 1485   Book censorship begins in Mainz, Germany with the first decrees prohibiting certain "dangerous" books. This is initiated by Archbishop Berthold von Henneberg (1484-1504) who asks the town council of

Frankfurt to carefully examine the printed books to be exhibited at the Lenten fair and to collaborate with ecclesiastical authorities in suppressing dangerous publications. The Catholic Church will eventually institutionalize this practice and issue its *Index* of prohibited books.

(See also 1559)

1485   Erhard Ratdolt, German printer and typographer, publishes the first printed book to use more than two different colored inks on the same page. The book is *De Sphaera* by Johannes de Sacrobosco (c.1200-1256), an English mathematician and astronomer also referred to as John of Holywood.

1490   Running heads appear for the first time in the *Philosophia Pauperum*, by German scholar Albertus Magnus (1193-1280). This now-common device repeats a book's title or a chapter's heading at the top of successive pages of a book.

1492   First globe map of Earth is produced by Martin Behaim (c.1436-1507), Spanish navigator and geographer. Although it does not contain the yet-undiscovered American continent and Pacific Ocean, his globe is important as a correct method of depicting Earth and its features. It is also important for the insight it provides into the common geographical assumptions on the eve of the discovery of North America.

MARCH 1495   Aldus Manutius (1449-1515), Italian scholar and printer, issues the *Erotomata* of Constantine Lascaris, his first dated, printed book. He later produces the first printed editions of many of the Greek and Latin classics and also originates the inexpensive "pocket-size" edition. His Aldine Press is one of the most famous in the history of printing, and his great idea is to use the printing press to perpetuate the works of the classical writers of Greece and Rome. His books are of both a

Printing a book using an early printing press.

convenient size and affordable price, putting them within the reach of all scholars and popularizing classical literature.

1495   First printed book to contain musical notation or lines with printed notes is the *Polychronicon*, printed by Wynkyn de Worde of England. The passage in which the printed musical notes occurs describes the consonances of Greek philosopher Pythagoras (c.582- c.497 B.C.).

c. 1496   First printed book to deal with dancing

is *L'Art et Instruction de Bien Danser* by Michel de Toulouze.

1500   First use of italic type is in the *Epistola Devotissime da Sancta Catharina da Siena*, printed by Italian scholar and printer Aldus Manutius (1449-1515). Designed by Italian type designer and cutter Francesco Griffo, this type is modeled on an informal, handwritten style, and its sloping, light, and compact letter form inspires all the major roman type faces that are designed during the next three centuries.

1507   Name "America" is first recorded on a map by Martin Waldseemüller (c.1470-c.1518), German cartographer, who publishes a map showing a new continent he names "America." He concludes that what Columbus found in 1492 was in fact an entirely unknown continent as the Italian navigator Amerigo Vespucci (1454-1512) had claimed. He names it America after Vespucci.

1516   First printing press on the African continent is established at Fez, Morocco by Jewish refugees who had worked in Lisbon, Spain for the Rabbi Eliezer Toledano.
(See also c.1520)

c. 1520   First book printed in Africa is a handbook in Hebrew printed at Fez, Morocco by Jewish exiles from Spain.

1521   First published treatment of the camera obscura is made by Cesare Cesariano, student of Italian artist Leonardo da Vinci (1452-1519), who publishes his master's observations of this phenomenon. During the Renaissance, experimenters use a dark room or chamber ("camera obscura" in Italian) to experiment with this phenomenon in which a small hole in the wall of a dark room produces an inverted image (of the view outside the room) on the wall opposite the hole. The camera obscura becomes the direct ancestor of the modern photographic camera.

1534   First complete edition is published from

Biblical scholar and reformer Martin Luther's (1483-1546) German Bible translation. This immediately becomes the most widely read book in Germany and influences decisively the development of the German language and its literature.

1538   The first practical dictionary, *Dictionarium Latino-Gallicum* is compiled and printed at Paris by French publisher Robert Estienne (c.1503-1559).

1539   First book printed in the New World is a Spanish missionary booklet of 12 pages printed in Mexico City by Italian-Spanish printer Juan Pablos (also called Giovanni Paoli).

1545   First known pictorial representation of a type-caster at work is in the *Corte Instruccye ende Onderwijs* by Cornelius van der Heyden.

1545   First published illustration of a camera obscura is made by Regnier Gemma Frisius (1508-1555), a Dutch mathematician also called Reinerus, in his *De Radio Astronomico et Geometrico*. The picture commemorates the observation of the solar eclipse of January 1544.
(See also 1521)

1558   Giambattista della Porta (c.1535-1615), Italian natural philosopher, publishes his *Magia Naturalis* in which he gives an account of the camera obscura that first brings it to the attention of the general public. He is also the first to suggest that it be used in drawing and painting, and he gives instructions on how images can be traced and how the image can be righted using mirrors.

1559   First list of forbidden books, *Index Librorum Prohibitorum*, is published by Papal authorities. It becomes a regularly published book listing titles and authors that are regarded as "dangerous to faith and morals" and that can be read by Roman Catholics only with special permission. The last full edition of the *Index* is published in 1948, and in 1966, it is declared that no more will be published.

Using the camera obscura.

c. 1563 First book printed in Russia is an anonymous work titled *Apostol* published in Moscow.

1565 First known image of a pencil is shown in a book on fossil collecting by German naturalist, physician, and bibliographer Konrad von Gesner (1516-1565). He shows a hollow tube into which a tapered piece of graphite has been inserted.

(See also 1662)

1568 Gerardus Mercator (1512-1594), Flemish geographer, first uses his "cylindrical projection" method which solves the map-making problem of how to depict a spherical surface on a flat piece of paper. Also called "Mercator projection," this method involves making projections from the center of the Earth onto a cylinder surrounding and touching it at the equator. All meridians (vertical lines of longitude) are straight and are at right angles to the parallels (horizontal lines of latitude). Although this seriously distorts the features near the poles, the Mercator projection made it easier to relate geographic locations to each other. In

1595, his book *Atlas, sive Cosmographicae* is published posthumously. The cover depicts the Greek Titan, Atlas, holding the world on his shoulders, and ever since, all future books of maps are called "atlases."

c. 1580 First paper mill in Mexico is built in Culhuacan.

1583 First book printed by Europeans in China is a *Catechism* printed in Chinese.

1584 First printing press in Peru is established at Lima by Italian publisher Antonio Ricardo. His first product is *Programatica sobre los Diez del Año*, which contains instructions from the King of Spain on the recent calendar reforms.

1585 Juan Gonzalez de Mendoza (1545-1618), bishop, publishes his *Historia de las Casas Mas Notables, Rito y Costumbres del Gran Reyno de la China*, which contains the first known appearance of Chinese characters in a book printed in the West.

1585 First published account of a camera obscura fitted with a lens is written by Italian physicist and mathematician Giovanni Battista Benedetti (1530-1590), in his *Diversarum Speculationum Mathematicum et Physicum*.

1590 First book printed in Japan with European characters and in the Japanese language is titled *Flos Sactorum o Vida de los Santos*.

1599 First known book auction catalog is issued in Holland and describes the library of Dutch statesman and writer Phillippe van Marnix (1538-1598).

c. 1605 First true newspaper (combining miscellaneous topical information with regular, periodic publication) is the *Nieuwe Tijdinghen* published in Antwerp, Belgium by Abraham Verhoeven.

(See also 1620)

1609 One of the first recognized weekly news-

papers to be regularly published starts in Strasbourg, France.

1613 François d'Aguilon (1567-1617), Belgian mathematician and optician also called Aguillon or Aguilonius), publishes his *Opticorum* in which he first coins the name "stereoscopic." Stereographic principles had been known since the time of the Greek astronomer Hipparchus (c.190-c.120 B.C.), but it had no name until Aguilon. Based on the phenomenon of binocular vision in which each eye sees slightly different images, it eventually forms the basis of stereoscopic photography and three-dimensional films.

1620 Belgian newssheet, *Nieuwe Tijdinghen*, is the first newssheet publication to contain illustrations.

SEPTEMBER 24, 1621 First regularly published newspaper in England is the single-sheet *Corante, or News From Italy, Germany, Hungarie, Spaine and France*.

(See March 11, 1702)

1624 First printed book in an African native tongue is a catechism in Kixicongo dialect printed in Lisbon, Spain.

1631 First major French newspaper is *La Gazette de France*, published by French physician and administrator Theophraste Renaudot (c.1586-1653) . It is a weekly sheet that relates government-sanctioned news.

1634 Phonetic symbols are used for the first time, in English, in the book *The History of Bees* by Charles Butler of England.

1636 The first officially recorded bookbinder in the American colonies is John Sanders of Boston, Massachusetts. He opens a shop the following year.

1638 First printing press in North America is

attributed to English-American printer Stephen Day (c.1594-1668). He establishes his printing house in Cambridge, Massachusetts.

(See also January 1639)

JANUARY 1639 First piece of printing in North America is issued by English-American printer Stephen Day (c.1594-1668). It is titled *The Freeman's Oath*, a broadside of which no copy exists. His third piece, *The Whole Booke of Psalmes* (edited by Richard Mather and commonly known as the *Bay Psalm Book*, 1640), is considered to be the first real book published in North America. Of the 1,700 copies printed only 11 have survived.

1646 "Magic lantern" is first described by Athanasius Kircher (1601-1680), German scholar, in his *Ars Magna Lucis et Umbrae*. This candle-lit device is the earliest written record of what can be described as a primitive slide projector. With this device, light falling through a small opening in a box can be directed onto a screen to create a precise image of something outside the box or room.

1651 First book in France to espouse a system of shorthand is the *Méthode pour Ecrire aussi Vite Qu'On Parle* by Jacques Cossard.

1653 First public library in the American colonies is founded in Boston, Massachusetts.

1662 First commercial pencil-making venture is that of a pencil factory in Nuremberg, Germany, run by Frederick Staedtler. The pencils are made by fitting a thin strip of graphite into a grooved piece of wood and then covering the groove and sealing it in place with a glued wooden strip.

(See also 1761)

1668 Robert Hooke (1635-1703), English physicist, develops a device which reflects images from a mirror through a convex lens. The images pass through a large hole in a wall onto a white screen in a lighted room. This device comes to be called the "camera lucida," or light room, as opposed to the "camera obscura," or dark chamber. The former allows an artist to work in light.

(See also 1807)

1677 First illustrated book from an American press is the *Narrative of the Indian Wars* by William Hubbard (1621-1704).

1683 First comprehensive treatise on typefounding and printing is contained in Joseph Moxon (1627-1700) of England's *Mechanick Exercises*. In this work, Moxon clearly and succinctly describes the tools and techniques used in the traditional practice of printing.

1690 First paper mill in the American colonies is built in Germantown, Pennsylvania, by William Rittenhouse. The mill makes paper out of linen rags and produces about 250 lbs (113 kg) a day.

1690 First newspaper in the American colonies, *Publick Occurrences Both Foreign and Domestick*, appears in Boston, Massachusetts. This radical, three-page paper is printed without a license and is immediately suppressed by the authorities.

(See also 1704)

MARCH 11, 1702 First daily newspaper in England, *The Daily Courant* appears in London. It is begun by Elizabeth Mallet who mails copies to colonial subscribers across the Atlantic. This single-sheet daily lasts until 1735.

APRIL 24, 1704 First regular newspaper in the American colonies is the *Boston Newsletter* founded by John Campbell. It becomes the *Boston Gazette* in 1719. Since Campbell is also a postmaster, his postriders deliver his papers without any additional expense. It is eventually put out of business in 1776 by American rebels.

1704 First alphabetical encyclopedia in English

is credited to John Harris (c.1666-1719) of England. Titled *Lexicon Technicum, or An Universal English Dictionary of the Arts and Sciences*, it contains all the elements of a modern encyclopedia. It has engraved plates, clear practical text, bibliographies to important subjects, and emphasizes the arts and scientific and technical subjects over the biographical.

1709   First modern copyright law, the Copyright Act, is passed in England.

1710   Jakob Christoph Le Blon (1667-1741), German-French painter and engraver, invents three-color printing. He is the first to make use of several metal plates, each of an individual color, for making prints with continuous gradations of color. In a few years, Le Blon is producing four-color printing.

1710   First scientific treatise on printing types is *De Germaniae Miraculo* by Paul Pater (1656-1724) of Germany.

JANUARY 1714   First recorded patent for a typewriter is issued in England to English engineer Henry Mill. He describes his device as "An artificial machine or method for the impressing or transcribing of letters singly or progressively." No drawing or model exists, and it is not known if his machine really worked.

1718   First color printing in the American colonies is accomplished by Andrew Bradford of Philadelphia.

1719   René Antoine Ferchault de Reaumur (1683-1757), French physicist, first suggests the use of wood as a papermaking fiber. Despite this early suggestion, making paper from wood pulp and other vegetable pulps does not occur for another century.

1720   William Caslon (1692-1766), English typefounder, first begins his design of the typeface that bears his name. His work helps to modernize the book, making it a separate creation rather than a printed imitation of the old, hand-produced book.

1721   First music textbook printed in America is *An Introduction to the Singing of Psalm-Tunes* by John Tufts (1689-1750). No original editions are extant, although it goes through many later editions.

1725   William Ged (1690-1749), Scottish inventor, invents stereotypy. This process of making stereotype plaster plates for printing enables printers to make molds of printed surfaces so that metal plates for future editions can be cast from the original mold.

1727   Johann Heinrich Schulze (1687-1744), German physician, accidentally discovers that silver nitrate (silver salts) darkens when exposed to sunlight. He realizes that this change is the result of exposure to light and not to heat. This discovery becomes crucial to photography. (See also 1777)

1737   Pierre Simon Fournier (1712-1768), French engraver and typefounder, publishes his *Table des Proportions Qu'il Faut Observer entre Les Caracteres*. Dealing with punch cutting and typefounding, this work first introduces the point system for measuring type size.

1740   First circulating library is established in London, England by a bookseller named Wright.

1744   First printed book in England intended specifically for children (apart from schoolbooks) is *A Little Pretty Pocket Book* by English merchant and publisher John Newbery (1713-1767). To commemorate the "first genuine children's publisher," the American journal *Publishers Weekly* establishes the Newbery Medal in 1922 for the best children's book of the year.

1749   Jacques Francois Rosart (1714-1777), Belgian typefounder, invents the first means of printing music from movable type. He devises characters that represent the basic elements of music (such as complete notes, separate note heads, stems, tails, and so on) that are cut in such a way that they can be put together in any combination as easily as letter type.

c. 1750 William Tiffin of England first introduces a new form of shorthand called phonetic stenography which employs symbols for sounds. This innovation makes stenography much faster than the old method of using symbols for letters.

1750 John Baskerville (1706-1775), English printer and inventor, first develops a fine quality paper which he calls vellum. His new, high-quality paper has a special finish and is made from wood pulp and rags. It is not to be confused with early medieval vellum which is made from the delicate skins of calf, lamb, or kid. He also improves printers' ink and invents a typeface still in use.

(See also 1759)

1750 First printing press manufactured in North America is built by Christopher Sauer, Jr. at Germantown, Pennsylvania.

MARCH 23, 1752 First work printed in Canada is the *Halifax Gazette* by John Bushell. This sheet appears weekly until 1766.

1759 John Baskerville (1706-1775), English printer and inventor, publishes *Paradise Regained*, the first book to be completely printed on wove paper. Wove paper is made with a revolving roller that is covered with wires, eliminating any fine lines from running across the grain. The use of this paper combined with his own truly black ink results in an especially bold quality of printing.

1761 First major pencil-making factory is established in Nuremberg, Germany, by Kasper Faber.

1764 Pierre Simon Fournier (1712-1768), French engraver and typefounder, publishes his *Manuel Typographique,* which is devoted largely

to type specimens and is considered to be the first book on punch-cutting and typefounding.

1768 First edition of the *Encyclopaedia Britannica*, the oldest and largest continuously published English-language general encyclopedia, is published in Edinburgh, Scotland.

1769 First person to cut and cast printing type in the American colonies is Connecticut silversmith and lapidary Abel Buell (1742-1822). Buell later participates in pulling down the lead statue of King George III in New York and melting it down to be used for type metal.

JUNE 15, 1770 Patent is granted to John Tetlow of England for the first machine for the ruling of music paper and paper for bookkeeping. Previously, all such ruling was done by hand.

1777 First daily newspaper in France is the *Journal de Paris.*

1777 Karl Wilhelm Scheele (1742-1786), Swedish chemist, first publishes the results of his experiments on the light sensitivity of silver chloride. He discovers the rapid action of the blue and violet end of the spectrum of visible light in the blackening of silver chloride. These experiments later inform others a half century later who will advance the development of photography.

1780 Steel pens for writing art are first introduced. They will come to replace quill pens.

1784 First mail coach in England departs Bristol for its 17-hour trip to London. In 1830, trains begin to transport mail in England.

1784 Valentin Hauy (1745-1822), French physician and educator, founds an institution for the blind and first introduces a system for their education.

(See also 1793)

JANUARY 1, 1785 First edition of the *Daily Universal Register* is issued in London by Eng-

Typesetting in a mid-eighteenth century printing shop.

lish journalist John Walter (1739-1812). Three years later, he renames his paper *The Times*.

1793   Valentin Hauy (1745-1822), French physician and educator, first uses a method of utilizing raised letters and embossed paper for teaching the blind to read. It has only limited success.

(See also 1830)

1793   Claude Chappe (1763-1805) of France invents a visual telegraph system that he builds between Paris and Lille. This optical-relay system called a "semaphore" consists of tall vertical posts whose arms can be moved to a variety of positions that represent numbers and letters of the alphabet. His system uses a signalling code and a telescope to communicate between telegraph stations. Chappe coins the word "telegraph" from the Greek "tele" (far) and "graphien" (to write). By 1844, the Chappe visual telegraph network connects 29 cities in France through its 500 stations.

1795   The Typographical Society of New York is formed and is the first organization of working printers. The new union is able to secure an increase of wages to $1 a day.

1795 Nicolas Jacques Conte (1755-1805) of France discovers a way to reduce the graphite content of pencils (graphite being an English monopoly). He combines small amounts of powdered graphite with clay which is then molded into thin rods and fired. The result is a pencil whose hardness or firmness can be calibrated, and it essentially the same as we use today.

1796 Aloys Senefelder (1771-1834), Czech writer and inventor, invents lithography. His new process of "writing on stone" first uses a greased stone slab and ink but soon progresses to a metal plate and acid. With the early method, the design to be reproduced is drawn directly onto a stone slab or plate. The slab is then dampened completely, but the water is repelled from the grease-covered areas. When a coating of ink is applied to the plate, it washes away from the wet areas but adheres to the grease marks. The plate is then pressed to paper for an exact copy. He later finds that a solution of gum arabic and nitric acid is even more effective than water in completely repelling ink from the non-greased areas. He also replaces the stone with metal plates. This new method produces an exceptionally faithful reproduction and is quickly embraced by printers as a quick and effective method of mass-producing commercial images. Mass-market art eventually becomes possible because of lithography.

AUGUST 1799 Rosetta Stone is discovered by a Frenchman named Bouchard or Boussard near the town of Rosetta (Rashid), northeast of Alexandria, Egypt. This ancient Egyptian stone provides the key to deciphering Egyptian hieroglyphics.

(See also 1822)

1800 First printing press made of iron is invented by English inventor Charles Stanhope (1753-1816). Iron will eventually replace wood for presses. Stanhope's early version is hand-operated. The greater stability and strength of its iron construction makes possible the printing of a whole sheet with one pull of the bar. This greater pressure is due to a combination of a lever and screw mechanism.

1800 The Original Society of Papermakers is founded in England. It is the first organized union of this particular trade.

1802 Thomas Wedgewood (1771-1805), English physicist and inventor, makes some of the first attempts at "fixing" or making permanent the images created by exposing silver salts to sunlight. He is not successful and the silhouettes he produces continue to darken unless kept from the light.

1803 A machine for the continuous manufacture of paper is used for the first time in England. It is built for the English brothers Henry and Sealy Foudrinier, following a process developed by Nicolas-Louis Robert in 1798. Their machine can also cut the paper to specific sizes.

(See also 1816)

1803 First mechanical (steam-powered) printing press is produced by Friedrich Koenig (c.1774-1833), German engineer and printer. This early version does not work well, and it is not until he moves to London and abandons the traditional flat platen for a cylinder-shaped platen in 1811 that he is successful. His new press can print over 1,000 pages an hour. Steam-powered printing eventually swells both readership and the variety of printed material, making possible magazines, newspapers, and the novel as a literary form.

(See also November 29, 1814)

1805 William Wing of Connecticut patents the first type-casting machine invented in the United States.

1807 John Thomas Smith (1766-1833) of England publishes his *Antiquities of Westminster*, which is the first book in England with lithographic illustrations.

1807   William Hyde Wollaston (1766-1828), English chemist and physicist, develops an optical device as an aid to artists that he calls the "camera lucida." Consisting of a four-sided prism mounted on a small stand above a sheet of paper, it facilitates the accurate sketching of an object. Technically it is not any form of a camera, since it does not project an image into a chamber.

(See also 1668)

1812   Samuel Thomas von Soemmering (1755-1830), German naturalist and anatomist, invents a multi-wire electric telegraph system. His system uses electrolysis for detecting electrically transmitted signals. He uses 25 wires (one for each letter) that terminate in a tank, with the wire that produces bubbles (after receiving current) indicating a letter.

1812   William Monroe of the United States is the first to make pencils in commercial quantities in America.

(See also 1861)

NOVEMBER 29, 1814   Friedrich Koenig (c.1774-1833), German engineer and printer, witnesses his steam-driven printer go into actual production for the first time at the London *Times*. His cylinder press can print up to 1,200 impressions an hour which is four times the rate of the old, manual Stanhope presses.

1816   First steam-powered paper mill in the United States begins operations in Pittsburgh, Pennsylvania.

1816   Foudrinier continuous paper-making machine is first used in the United States.

(See also 1803)

1816   Joseph Nicephore Niepce (1765-1833), French inventor, uses a camera obscura to capture an image of Paris on paper that is treated with silver chloride. This is considered the first photochemical (negative) print and is the first true attempt at photography. His exposure time is at least one hour, but he cannot adequately fix the image. He continues to experiment to find a substance that will keep the image from fading away.

(See also 1822)

1817   The first paper-making machine to be built in the United States is erected by Thomas Gilpin near Philadelphia, Pennsylvania.

1818   Cornelius S. Van Winkle (1785-1843) of the United States, publishes *The Printers' Guide*, which becomes the first recognized American printers' manual.

1822   First known permanent photographic image is achieved by Joseph Nicephore Niepce (1765-1833), French inventor, by "fixing" transfers of engravings on pewter plates coated with bitumin. Over the next five years he produces what he calls heliographs or sun-drawings using light-sensitive material, and lays the groundwork for others who will take his invention further. In 1827 he meets French artist and inventor Louis Jacques Mandé Daguerre (1789-1851), and they agree to pursue together the problem of obtaining a permanent image using a camera obscura and light sensitive chemicals. They become partners in 1829.

1822   Jean François Champollion (1790-1832), French historian and linguist, deciphers Egyptian hieroglyphics using the Rosetta Stone. He is the first to realize that some of the signs are alphabetic, some syllabic, and some determinative (standing for a whole idea or object previously expressed).

1824   Peter Mark Roget (1779-1869), English physician and author of the famed *Thesaurus*, first discusses the optical phenomenon arising from the rotation of wheels and gives an early scientific account of the "persistence of vision" with regard to moving objects. This will eventually come into play in motion picture photography and projection.

1825 First regular newspaper in Russia, *Severnaya Pchela* is founded at St. Petersburg by Gretsch and Bulgarin.

1826 Newspaper *Le Figaro* is founded at Paris, France.

1827 Charles Wheatstone (1802-1875), English physicist, first coins the term "microphone" for an acoustic device he develops to amplify weak sounds.

1829 William Austin Burt (1792-1858), American inventor, patents his "Typographer," which is recognized as the first usable typewriter. The letters on his table-size printer are set around a circular carriage which is rotated by hand. It is extremely slow and does not attract any financial backing.

1830 Louis Braille (1809-1852), French inventor, first devises an entirely new system of 12 raised dots which allows blind people to both read and write by touch. His system is later simplified to six dots by Charles Barber, and the Braille system is adopted after 1850. It comes into widespread use after the International Congress in Paris in 1878. Braille is totally blind by the age of five as the result of an accident, and attends the National Institute of Blind Youth founded in 1784 by French physician and educator Valentin Hauy (1745-1822). While there, he works on a dot-code system, based on the 26 letters of the alphabet, which can be identified with the touch of a single finger.

1831 Joseph Henry (1797-1878), American physicist, improves the electromagnet and sends a current through a wire, opening and closing a circuit and ringing (and therefore inventing) the first electric bell. More importantly, Henry's experiment demonstrates the principle behind the electromagnetic telegraph. Henry is

later visited by American artist and inventor Samuel Finley Breese Morse (1791-1872), to whom Henry freely explains the science behind the phenomenon. Morse goes on to invent the electric telegraph.

(See also May 24, 1844)

1832 William Savage (1770-1843) of England publishes his *Preparation of Printing Ink*, which is the first practical treatise on the subject.

1832 Charles Wheatstone (1802-1875), English physicist, invents the stereoscope. His device demonstrates that two pictures can be visually combined to create the illusion of depth and three dimensions. He uses two different cameras simultaneously to simulate two eyes and, by spacing their lenses the right distance apart, the resulting images appear to have depth.

1833 Joseph Antoine Ferdinand Plateau (1801-1883), Belgian physicist, invents what he calls the Phenakistocscope. When this apparatus— which is made up of a cardboard disk around which sequential drawings are attached—is rotated, the drawings appear to move.

(See also 1834)

1833 Erich von Stampfer of Austria invents a device whose lights flash at controlled intervals and appear to freeze motion. Called a stroboscope, it demonstrates a phenomenon that underlies the principle of motion pictures.

1834 William George Horner of England invents the "wheel-of-life," or zoetrope, which is an improvement of Belgian physicist Joseph Antoine Ferdinand Plateau's (1801-1883) device. His design allows several people, instead of a single viewer, to enjoy the moving pictures.

1835 Louis Jacques Mandé Daguerre (1789-1851), French artist and inventor, invents a system of developing images on metal plates coated with silver oxide. Building on the work of his dead partner, French inventor Joseph Nicephore Niepce (1765-1833), Daguerre suc-

Helen Keller reading a braille book.

ceeds in reducing the long-exposure time through his discovery of a chemical process for developing (making visible) the latent (invisible) image formed upon brief exposure. He still cannot fix the image permanently.

(See also 1837)

1835 Joseph Henry (1797-1878), American physicist, first establishes the basic scientific principles at work behind the telegraph but does not put them into actual practice.

1835 William Henry Fox Talbot (1800-1877),

English inventor, produces his first "photogenic" drawings that he obtains by placing leaves, lace, or translucent engravings against sensitized paper and exposing both to light. He also produces a one-inch square negative image of his home that he makes by inserting sensitized paper in a very small camera with a short focal length. He fixes the image with either potassium iodide or salt. He uses this method until he learns of Herschel's hyposulfate method.

(See also January 29, 1839)

JUNE 1837 First patent for a telegraphic signal is

granted to Charles Wheatstone (1802-1875), English physicist, and William Fothergill Cooke (1806-1879), English electrical engineer. Their patent is for a five-needle telegraph system in which each letter is pointed at. This is the first English telegraph and links Liverpool with Manchester in 1839.

1837 Louis Jacques Mande Daguerre (1789-1851), French artist and inventor, finds a way to arrest the action of light on his images by using a bath of sodium chloride (common table salt in hot water) and makes what becomes the first Daguerreotype. Still in existence, it is an image of a still-life in his studio. He uses this method until he learns of English astronomer John Frederick William Herschel's (1792-1871) hyposulfate discovery.

(See also January 29, 1839)

1837 Isaac Pitman (1813-1897), English inventor, publishes his *Stenographic Sound Hand* in which he invents the first system of shorthand based on a scientific analysis of sounds that comprise speech. His system discards former orthographic principles (spelling) and is based solely on phonetics. The relative speed and accessibility of his system makes it the preeminent shorthand system in the English-speaking world.

(See also 1888)

1837 Samuel Finley Breese Morse (1791-1872), American artist and inventor, first applies for a patent on the single-wire telegraph that uses an electromagnet for transmitting signals. He also describes a new alphabet type that comprises the marking of dots and dashes instead of a zigzag line as previously used.

1837 T. M. Lucas of England first creates an alphabet of phonetic symbols for blind people and publishes a transcription of the New Testament. This system is later improved upon by the Englishman James H. Frere, who invents the return line. This system runs the text right

to left and then left to right, so that the finger does not lose its place.

MARCH 17, 1838 The first effective mechanical type-casting machine is patented by David Bruce, Jr., of the United States. This machine both melts the metal and pours it into the appropriate type molds.

JANUARY 29, 1839 John Frederick William Herschel (1792-1871), English astronomer, discovers that sodium thiosulfate (called sodium hyposulfate) is a solvent for silver salts, which makes it possible to stop light-sensitive silver salts from darkening any further. He also is the first to make use of the terms "photographic negative" and "photographic positive."

MARCH 13, 1840 Patent is granted to James H. Young and Adrian Delcambre, both of Lille, France, for what is considered as the first practical composing or typesetting machine. An operator sits at a keyboard and depresses the appropriate key releasing a given character which slides down a runway to an assembly place. A second operator justifies or spaces out the words to a given distance, and a third keeps the magazines filled with type. Named the Pianotype, this machine is capable of setting about 6,000 letters and spaces an hour. It later becomes the first composing machine to be used in a printing office.

1840 Prepaid postage stamp is first introduced in England. Using the "penny black" stamp, the item gets to its destination already paid for instead of the recipient having to pay as before. The United States begins its own system this year based on the "penny black."

1840 Alexander Bain (1810-1877), Scottish inventor, applies for the first patent for "transmitting copies over a distance, by means of electricity." His crude device, which some regard as the first fax (facsimile) machine, uses matching pendulum transmitters and receivers that send and respond to electrical impulses. It does not work on any practical basis, however, and is essentially unpredictable.

1840 John Benjamin Dancer (1812-1887), English optician and microscopist, makes the first daguerreotype microphotographs. He reduces a 20-inch (51 cm) document to 1/8 of an inch using a camera with a microscope lens. He also displays a daguerreotype of a flea magnified to 7 x 5 inches.

1840 New method of making paper from wood pulp is first introduced in Germany by the German weaver Friedrich Gottlob Keller, who invents a wood-grinding machine. This device defibers blocks of wood by pressure against a revolving wet grindstone and mixes the wood fiber with 40% rag fiber. In 1846 Keller's process is acquired by German papermaker Heinrich Volter and this marks the beginning of commercial wood pulp production for paper-making.

1840 John William Draper (1811-1882), American chemist, takes a daguerreotype portrait of his sister Dorothy. This is the oldest surviving photograph of a person.

FEBRUARY 8, 1841 First photographic negative is produced by English inventor William Henry Fox Talbot (1800-1877). This year he patents his Calotype process, which he perfected in 1839. Also called the Talbotype, this method improves on the Daguerreotype and produces for the first time a photographic negative from which any number of positive prints can be made on paper. By putting his negative (with its typical black-and-white reversal) in contact with a blank sheet of light-sensitive paper and exposing it, he obtains a positive print.

1841 Robert Hunt (1807-1887) of England publishes *A Popular Treatise on the Art of Photography* and emphasizes the technical aspects of this new field. This is the first manual in English on the subject.

1841 Horace Greeley (1811-1872), American editor and publisher, founds the *New York Tribune*, which eventually becomes the *Herald Tribune* after a merger in 1924.

MAY 1842 First use of photography for news and documentary is made by Friedrich Stelzner and Hermann Biow who take a set of daguerreotypes of the ruins of Hamburg after the great four-day fire in Germany.

1842 First underwater telegraph cable is laid in New York Harbor by Samuel Finley Breese Morse (1791-1872), American artist and inventor. Constructed of copper and insulated in part with India rubber, it works for only a short time and fails due to a lack of proper insulation materials.

1842 *London Illustrated News* is founded. It uses etchings that are made from Daguerreotypes.

1843 Mathew Brady (1823-1896), American photographer, is one of the first students of American artist and inventor Samuel Finley Breese Morse (1791-1872), who teaches Daguerre's photographic process at $50 per student. The following year, Brady opens the first "Daguerrean Miniature Gallery" on lower Broadway in New York. He goes on to become one of the most famous 19th-century American photographers and is best known for his photographs of the American Civil War.

1843 Alexander Bain (1810-1877), English inventor, first conceives the basic principles of transmitting pictures by electricity and invents the electro-chemical telegraph. His transmitter and receiver are based on a pendulum system which scans a metal version of the image to be transmitted. The pendulum transmits an electrical impulse each time it crosses the metal, and a matching pendulum at the receiving end transmits the impulses to treated paper. Although it is never employed as a practical system, many call Bain the "father of facsimile."

MAY 24, 1844 Samuel Finley Breese Morse (1791-1872), American artist and inventor, successfully transmits the first message over a telegraph circuit between Baltimore and Washington. The series of dots and dashes conveys the words "What hath God wrought." It takes Morse several years to obtain the necessary funding from the United States Congress. When he realizes that an underground Baltimore-Washington cable will not work, he strings his wires above ground on poles. The message is successfully transmitted through the rhythmic interruptions of an electrical current that consists of long and short intervals or dots and dashes. Morse devises the code that is used, allocating different short signals to the most frequently used letters and the longest electric signals to those least used. After Morse's success, telegraph lines proliferate throughout the country, and the telegraph remains the standard way of communicating within and between cities until the widespread availability of the telephone.

1844 First book to be illustrated with photographs is *The Pencil of Nature* by William Henry Fox Talbot (1800-1877), English inventor.

1845 First newspaper to use a mixture of rag and wood pulp is the *Kreisblatt* of Frankenberg in Saxony, Germany.

1846 Aimé Laussedat (1817-1907), French inventor, develops photogrammetry, a method of photographing buildings and structures from different angles in order to measure them. He also is the first to use photography in map production. Although handicapped by narrow lenses and poor equipment, he is the first to perceive the potential use of photographs for measurements in surveying and map making.

1847 First successful rotary printing press is patented by American inventor and manufacturer Richard March Hoe (1812-1886). Seeking to increase speed, he discards the old flatbed press and places the type on a revolving cylinder. Around this central cylinder revolve four to ten impression cylinders. This revolutionary system is first used by the *Philadelphia Public Ledger* this same year, and it produces 8,000 sheets per hour printed on one side. In 1853, his improved Type revolving press can produce 20,000 sheets an hour. By 1871, he devises a rotary web perfecting press that is fed by continuous rolls of paper or webs, and can print on both sides in one motion. Hoe presses are used around the world.

1847 Claude Felix Abel Niepce de Saint Victor (1805-1870) of France first uses light sensitive materials on glass for photographs. He coats a glass plate with albumen containing iodide of potassium which, after drying, is coated with aceto-silver nitrate, washed in distilled water, and exposed. He is the cousin of French inventor Joseph Nicephore Niepce (1765-1833).

1847 William Siemens (1823-1883), German-English inventor, develops the first machines for coating telegraph cables with newly discovered gutta-percha from Malaysia. This yellowish or brownish leathery material from the latex of certain trees becomes plastic upon heating and is very resistant to water. The insulating properties of this new material proves capable of protecting underwater wires.

1848 Six New York newspapers first form an association or news agency to share telegraph costs. It later becomes known as the Associated Press.

1849 Considered as the first practical three-dimensional viewing apparatus is an improved stereoscope invented by David Brewster (1781-1868), Scottish physicist. It features a system of lenses and prisms that would overlap the two photographs without straining the viewer's eyes. Like the kaleidoscope which Brewster also invented, this becomes enormously popular and is marketed as "refined amusement" suitable for every home.

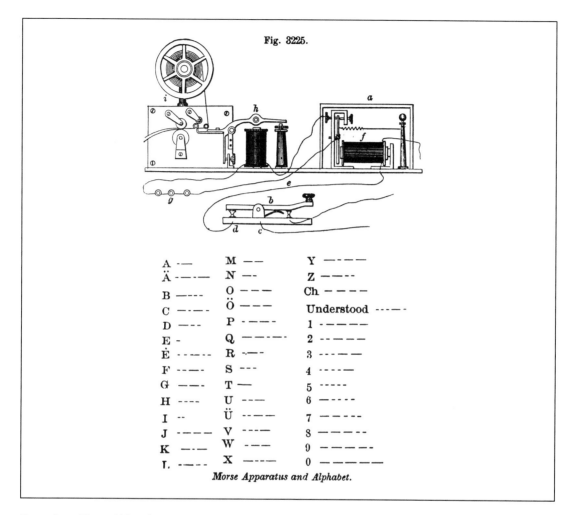

Fig. 3225.

A — -	M — —	Y — - - —
Ä - — - —	N — -	Z — — - -
B — - - -	O — — —	Ch — — — —
C — - — -	Ö — — -	Understood - - - - — -
D — - -	P - — — -	1 - - — — —
E -	Q — — - —	2 - - — — —
Ė - - - - -	R - — -	3 - - - — —
F - - — -	S - - -	4 - - - - —
G — — -	T —	5 - - - - -
H - - - -	U - - —	6 — - - - -
I - -	Ü - - — —	7 — — - - -
J - — — —	V - - - —	8 — — — - -
K — - —	W - — —	9 — — — — -
Ĺ - — - -	X — - - —	0 — — — — —

Morse Apparatus and Alphabet.

Illustration of Samuel Morse's coded alphabet and the parts of the telegraph.

MAY 27, 1850 Albumen on paper for photographic prints is first introduced by Louis Desiré Blanquart-Evrard of France. On this date, he communicates his process to the French Academy. This new paper is in great demand since it retains the sharpness and detail of a glass negative.

1850 Frederick Scott Archer (1813-1857), English sculptor, offers the first practical directions for using collodion (a derivative of guncotton that becomes liquid in alcohol) as a binder for light-sensitive silver salts. Use of this "wet plate" process considerably shortens a photograph's exposure time. Archer's wet collodion process is the fastest photographic process yet devised and supplants all existing methods—daguerreotype, calotype, and albumen. It remains popular until replaced by the more convenient gelatine dry plate nearly 30 years later. Despite his great invention, Archer dies in nearly complete poverty.

1850 John Watkins Brett (1805-1863), English engineer, and his brother Jacob lay the first submarine telegraph cable between England

and France. It is not strong enough and breaks, but they try again next year.

(See also 1851)

JUNE 1851 William Henry Fox Talbot (1800-1877), English inventor, takes the first successful photograph by electric light. The point of his experiment, however, is not so much to learn about the use of artificial light as it is to determine the possibility of recording fast moving objects by the flash of a Leyden battery.

1851 Great Exhibition of the Industry of All Nations is held at the Crystal Palace in London. This first official international exhibition also marks the first important public display of photography. About 700 photographs from six nations are displayed.

1851 First photographic society, the Société Héliographique, is founded in France. Among its 40 members are prominent artists and scientists.

1851 First illustrated news-magazine in Russia, the *Russkii Khudozhestvennyi Listok,* is founded by Vassili Timms.

1851 Continental Code or International Code is compiled at the first telegraph conference held in Berlin, Germany. It uses the best features of all known codes and takes eleven letters from the American system of Morse Code.

1851 First successful submarine telegraph cable between England and France is laid by John Watkins Brett (1805-1863), English engineer, and his brother Jacob. After the rupture of their first cable in 1850, they use a cable made of four copper wires, each .06 in (1.65 mm) in diameter. The cable's wires are twisted into a single strand that is covered with tar-cloth and reinforced with ten galvanized iron wires, each .28 in (7 mm) in diameter. This successful connection of Dover and Calais leads to thinking about a transatlantic cable.

1851 Paul Julius Reuter (1816-1899), German publisher, first opens a telegraph office near the London stock exchange. He soon expands his business from commercial telegrams to a wire service for all types of news and information.

1851 The New York *Daily Times* is first established. In 1857 it becomes *The New York Times.*

NOVEMBER 1853 First war photographer is the Hungarian artist, Karoly Pap Szathmari (1812-1888), who photographs camp scenes after the outbreak of the Crimean War and later photographs Turks fighting in the Danube Valley.

1853 The Royal Photographic Society of England is formed by English photographer Roger Fenton (1819-1869). It is known as the oldest existing photographic society.

1854 Cyrus West Field (1819-1892), American financier, forms the New York, Newfoundland and London Telegraph Company and first proposes to lay a transatlantic telegraph cable.

(See also 1857)

FEBRUARY 1856 Hamilton Lanphere Smith (1819-1903), professor of natural science at Kenyon College, patents the first tintype in the United States. Also called the ferrotype, it is invented by French teacher and amateur photographer Adolphe Alexandre Martin (1824-1896), who first describes his process to the Académie des Sciences. In 1856, it is patented in England as well. This one-of-a-kind image on a metal plate is both lightweight and cheap and becomes the preferred version for travellers or people who want to carry a photograph with them. Since they do not offer great pictures and look as cheap as they are, they are mainly popular in America, with Europeans generally disdaining them.

1856 First notable general treatise on the various applications of photography to the printing press is *Photographie auf Stahl, Kupfer und Stein* by Georg Kessler of Germany.

1856 *Illustrated London News* is the first periodical to include regular color plates.

(See also 1895)

1856 First thread sewing machine for bookbinding is invented by David McConnell Smyth of the United States. It will eventually replace hand-sewing as human labor costs rise and make machines more economical.

1857 First attempt to lay a transatlantic telegraph cable fails. The American Navy ship *Niagara* and the Royal Navy ship *Agamemnon* fail in their attempt to lay cable between Ireland and Newfoundland. The cable breaks at a depth of 6,000 ft (1,800 m).

(See also August 16, 1858)

AUGUST 16, 1858 First partially successful transatlantic cable is opened with an exchange of greetings between English Queen Victoria (1819-1901) and United States President James Buchanan (1791-1868). Several weeks later, a telegraph operator applies too much voltage and ruins the cable connection. It will be some time before success is achieved.

(See also July 27, 1866)

1858 Hyman L. Lipman, American inventor, first patents a method of gluing a rubber eraser on top of a wooden pencil, giving the pencil its definitive form.

1858 First aerial photographs are taken from a balloon by French writer, caricaturist, and photographer Nadar, whose real name is Gaspard Felix Tournachon (1820-1910). Aware that aerial photography would be invaluable to map making and military reconnaissance, he undertakes the exceedingly difficult task of attaching his camera to the side of the balloon and developing his wet collodion plates in the basket of the balloon. He achieves spectacular, never-before-seen, bird's-eye view of Paris.

MAY 11, 1860 First use of the term "snapshot" is found in this issue of *Photographic News*.

1860 Charles C. Harrison of the United States first introduces the wide-angle Globe lens for cameras.

1860 First aerial view photographs in the United States are taken by American photographer James Wallace Black (1825-1896) and Samuel A. King. They photograph the city of Boston from a balloon at 1,200 ft (366 m).

1860 First camera to be considered the forerunner of the modern small camera is built by Adolphe Bertsch. Called a "chambre noire automatique," it is a 4 in (10 cm) square metal box without a shutter that makes exposures by removing a lens cap. It has a fixed focus and renders sharp all objects beyond a distance of 40 ft (12 m).

1861 First single lens reflex camera is patented by Thomas Sutton (1819-1875) of England. It is based on the use of a mirror to redirect the light rays to a horizontal ground-glass focusing surface.

1861 James Clerk Maxwell (1831-1879), Scottish mathematician and physicist, and Thomas Sutton (1819-1875) of England, demonstrate the first color reproduction using a photographic process. They develop the additive three-color process which becomes a standard method of color photography.

1861 John Eberhard Faber (1822-1879), German-American manufacturer, builds the first large-scale pencil factory in the U. S. in New York City.

1861 Popular press begins in England with the publication of the *Daily Telegraph*.

1862 Ammonia is first used for developing photographs.

1863 Popular press begins in France with the publication of *Le Petit Journal*.

1863 William Bullock (1813-1867), American

Illustration of an early newspaper printing press.

inventor, is credited with successfully building the first rotary self-feeding and self-perfecting press which prints on a continuous roll of paper on both sides. It can also cut newsprint before or after printing and revolutionizes the printing of newspapers.

1865 Johann Philipp Reis (1834-1874), German physics teacher, constructs a rudimentary apparatus for transmitting musical sounds over a distance. He uses acoustic waves to vibrate a membrane which moves a lever which interrupts an electric current. Although this is rec-

ognized as the first electric telephone, it is used as little more than a scientific toy.

1865 Pastor Hansen of Denmark first produces a "writing ball" typewriter that foreshadows the modern "golf ball" typesetter of 1961. He designs a porcelain ball of block letters with plungers on top. Resembling a pin cushion, the ball is placed above a half-cylinder which holds the paper.

1865 Forerunner of the International Telegraph Union (ITU) is established as 20 countries

Paper rolls for a self-feeding printing press.

agree to cooperate on their telegraphic communications.

JULY 27, 1866 First completely successful transatlantic telegraph cable is laid. Due primarily to the skill and persistence of American financier Cyrus West Field (1819-1892) and his reorganized Anglo-American Company, this ambitious goal is finally achieved. West uses the huge English steamship *Great Eastern* to lay the cable (insulated with gutta-percha), and Europe and the American continent are united electronically for the first time. West himself sends the first message indicating success.

1866 Mahlon Loomis (1826-1886), American dentist and inventor, is the first to use an aerial for wireless telegraphy. He sends telegraph messages using radio waves between two mountains in West Virginia using aerials kept in the air by kites.

(See also July 30, 1872)

1867 First commercially successful typewriter is produced by American editor and inventor Christopher Latham Sholes (1819-1890) and his colleagues Carlos Glidden and Samuel W.

Soule. Together, they design a type-bar machine with a carriage that automatically moves one space to the left when a letter is printed, and whose keys work on the piano principle, all striking the platen at exactly the same point. They also solve the colliding-bar problem caused by very fast typists and devise today's QWERTY keyboard which spaces out the most-used letters. The Remington company buys Sholes's patent and markets its Model 1 in 1876. One of these is bought by American author Mark Twain (1835-1910), who uses it to produce the first typed manuscript for a publisher.

MAY 1872 Eadweard Muybridge (1830-1904), English photographer, uses a series of cameras to take sequential pictures of a running horse and demonstrates for the first time that a horse does have all four feet off the ground at certain times.

JULY 30, 1872 Mahlon Loomis (1826-1886), American dentist and inventor, obtains the first United States patent for a wireless system that uses an aerial. He describes correctly how setting up "disturbances in the atmosphere would cause electric waves to travel through the at-

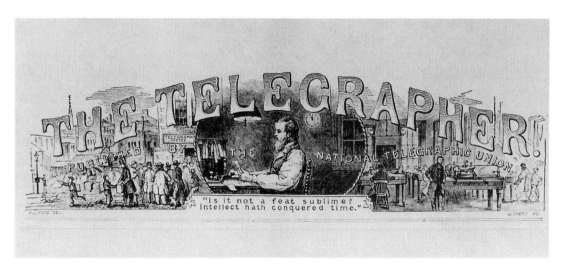

Masthead of *The Telegrapher*, a publication of the National Telegraphic Union, forerunner of the International Telegraphic Union.

mosphere and ground." He does not receive adequate financial backing to carry on his work.

1872   Thomas Alva Edison (1847-1931), American inventor, first obtains a patent for an electric typewriter that proves to be the prototype of future teletype machines used by press agencies. In 1874, he perfects his quadraplex telegraph which is able to transmit two messages over one telegraph line or four messages in each direction over two wires.

1873   First orthochromatic plates are produced by Hermann Wilhelm Vogel (1834-1898), German photochemist. By a process of adding dyes to silver bromide emulsions—called optical sensitization—he makes photographic plates that are sensitive to all but red and oversensitive to blue light. Thus Vogel transforms what was a color-blind emulsion into one which is more accurately sensitive for most colors (except red and blue). His discovery paves the way for correct reproduction of colors in paintings, landscapes, and portraiture, doing away with retouching.

1873   William Moon (1818-1894) of England

publishes his *Light for the Blind*. Using the first practical type for blind people, this book contains Moon's simplified outlines of the ordinary alphabet which many find easier to learn than Braille.

1874   First automatic camera for chronophotography is designed by French astronomer Pierre Jules César Janssen (1824-1907). Called a "revolver," this design involves a revolving camera attached to a telescope with which he is able to successively photograph the planet Venus. He then makes the first display of successive images on celluloid tape, a precursor of the moving picture. His work is an early contribution to the photographic analysis of movement.

1874   Karl Ferdinand Braun (1850-1918), German physicist, discovers that certain crystals transmit electricity more easily in one direction than another. Such crystals are eventually used as rectifiers for wireless systems. They come into their own again during the 20th century in the form of solid-state systems.

JUNE 2, 1875   Alexander Graham Bell (1847-

Union telegraph operator tapping rebel telegraph lines during the Civil War.

1922), Scottish-American inventor, transmits sounds over electric cables. While working with his assistant, Thomas Watson, on the harmonic telegraph, Bell finds he can hear Watson who is in another room. He investigates and finds that it is because an intermittent transmitter accidentally was allowed to transmit continuous instead of intermittent current. This discovery leads to continuous work and a fairly quick design by Bell for an electric telephone.

(See also February 14, 1876)

1875 William Crookes (1832-1919), English physicist, first devises an improved vacuum tube in which the air is almost completely removed. This is a great improvement over the old Geissler tubes and allows radiation to be more effectively studied. The rays he discovers are later named "cathode rays" by German physicist Eugen Goldstein (1850-1930), and the cathode ray tube becomes the basis of television in the 20th century.

FEBRUARY 14, 1876 Alexander Graham Bell (1847-1922), Scottish-American inventor, files his first patent for an apparatus designed for "transmitting vocal or other sounds

Advertisement for the first successful transatlantic telegraph cable.

telegraphically...causing electrical undulations." This patent is filed only two hours before American electrician and inventor Elisha Gray (1835-1901) files a patent for a similar device. On March 10 of this year, Bell succeeds in transmitting intelligible words ("Come here, Watson, I want you."), and receives a patent the next day. After years of litigation with his rival Gray, the court grants priority to Bell, primarily because of this two-hour gap.

MAY 10, 1876 Alexander Graham Bell (1847-

1922), Scottish-American inventor, first demonstrates the telephone. Later in the year, he conducts the first two-way telephone conversation over an outdoor line (2 mi [3.2 km] between Boston and Cambridgeport).

1876   American Library Association is first established in Philadelphia, Pennsylvania, by American librarian Melvil Dewey (1851-1931), the founder of the decimal system of library classification.

1876   Thomas Alva Edison (1847-1931), American inventor, builds his laboratory at Menlo Park, New Jersey. Called the "invention factory," this facility can be considered the first industrial research laboratory and begins the concept of the modern research team.

APRIL 27, 1877   Thomas Alva Edison (1847-1931), American inventor, files a patent for a telephone system that uses a carbon microphone. Edison is the first to design a transducer using granules of carbonized hard coal (the basis for modern microphones). A transducer is a device that is actuated by power from one system and supplies power, usually in another form, to a second system.

AUGUST 12, 1877   First phonograph is born as Thomas Alva Edison (1847-1931), American inventor, sends a free-hand drawing of his new invention to his Menlo Park "invention factory" workshop and orders it made. While working himself on improving the recently invented telephone, Edison attaches a sharp point to the phone receiver and finds that it vibrates strongly. He then surmises that a similar point can be used to indent the impression of a sound onto a moving sheet of tin foil, and suspects that this sound can be reproduced by retracing the point's path with another point attached to a diaphragm. The key to his device is a rotating cylinder on which sound grooves are traced and then replayed. He estimates that a metal cylinder four inches long and four inches in diameter can be made for $18.00. By December of this year the first new machine has been made and Edison applies for a patent.
(See also February 17, 1878).

1877   First photographic studio to use electric lights opens in London. It is owned by Henry Van der Wyde.

1877   Modern telephone mouthpiece is invented by Emile Berliner (1851-1929), German-American inventor. Upon discovering that the resistance of a loose contact varies with pressure and will act as a telephone receiver, he applies this principle and constructs an improved voice transmitter with a variable-pressure contact that comes to be known as a microphone.

1877   David Edward Hughes (1831-1900), English electrician, first designs an extremely sensitive inertia transmitter that earns him the honor of being considered the inventor of the microphone. His carbon microphone for a telephone uses a membrane that is in contact with carbon granules that are enclosed between two electrical contacts. Vibration of the membrane causes the electrical resistance between the two contacts to vary accordingly. His invention is a forerunner of the modern carbon microphone. American inventor Thomas Alva Edison (1847-1931), however, is the first to use this system on a telephone.

JANUARY 28, 1878   First commercial telephone exchange opens in New Haven, Connecticut.

FEBRUARY 17, 1878   Thomas Alva Edison (1847-1931), American inventor, receives a patent for the first cylindrical phonograph. He records sound using a cylinder covered with tin foil and a free-floating needle. As the cylinder turns, the needle is moved by sound waves and makes a track on the foil. Following that track with the needle reproduces the sound waves. This is Edison's favorite of all his inventions, and once he proves it is possible to record and reproduce sound mechanically, others begin to develop and refine his idea. By the early 20th century, the phonograph becomes a common accessory in most American homes.

1878   First halftone process for printing photographs is developed by Frederic Eugene Ives (1856-1937), American printer and inventor. He is able to translate the gradations of shading in photographs by breaking down photos into dots of various sizes to convey shades or "halftones." By 1885, he has developed the standard halftone technique in which the illustration to be reproduced is placed behind a glass screen gridded with fine black lines and photographed. The resulting image, composed of thousands of dots of various sizes, is then photoengraved. His process revolutionizes printed illustrations.

(See also 1881)

1879   Karel Klic (1841-1926) of Czechoslovakia invents the photo-mechanical intaglio process known as photogravure. Using a polished copper plate to which fine resin dust has been made to adhere with heat, he takes a piece of carbon tissue printed under a dispositive and transfers it to the grained copper plate. The soluble portions of the carbon image are then removed by washing the plate with warm water. The plate is then etched with perchloride of iron in successive baths of varying strength and, after cleaning off the gelatine, the plate is etched in different depths in proportion to the tones of the picture (the shadows are the deepest and thus hold the most ink). He makes his first successful prints with this method in 1895.

AUGUST 30, 1881   Clement Ader (1841-1923), French aeronautical pioneer and inventor, receives a German patent for the first stereophonic system. He uses two groups of microphones on either side of a stage. These are connected to headsets worn by the audience.

(See also 1933)

1881   First movie camera is invented by Etienne Jules Marey (1830-1904), French physiologist and inventor. In attempting to investigate the mechanics of animal locomotion, he develops a photographic gun with which a sequence of pictures can be quickly recorded. His gun can take 12 pictures a second. Marey's work contributes not only to the study of physiology and locomotion, but his new technology advances cinematography as well and promotes the development of the motion picture.

1881   Joseph Maria Ludwig Eder (1855-1944), Austrian photochemist, and Giuseppe Pizzighelli (1849-1912) invent gelatin silver chloride paper that is sensitive enough for printing by artificial light. It is later (1893) marketed by the Nepera Chemical Company of New York as "Velox" paper, and it revolutionizes the process of making photographic prints.

1881   First three-color halftone photographic reproduction is achieved by Frederic Eugene Ives (1856-1937), American printer and inventor. Using his new gelatin method, he uses three lined screens, each of which separates out a particular primary color, and overlays the images to create a color reproduction. It is based on his earlier halftone process.

(See also 1878)

1882   Alphonse Bertillon (1853-1914), French anthropologist and criminologist, first obtains a patent for his system of photographing criminals for the purpose of identification.

1883   Thomas Alva Edison (1847-1931), American inventor, discovers a phenomenon that occurs in an incandescent lamp. Known as the "Edison effect," an electric current can be made to jump a gap between metal plates. He notes this but does not pursue it, and it later becomes a crucial part of the radio electron tube.

(See also November 16, 1904)

FEBRUARY 1, 1884   First volume of *A New English Dictionary on Historical Principles* is published in England. Better known as the *Oxford English Dictionary*, this first edition is completed in 1933.

1884 George Eastman (1854-1932), American inventor, patents his photographic "film" in which the emulsion or gel is smeared on the paper. This makes photography much easier and simpler as it does away with the heavy and clumsy glass.

1884 Paul Gottlieb Nipkow (1860-1940), German engineer, invents the "Nipkow disk," a cardboard disk with a series of square holes, situated in a spiral. When used with a photoelectric cell and spun, the disk is able to scan areas of lightness and darkness and convert that information into an electrical signal. This mechanical scanning disk becomes basic to early (pre-electronic scanning) television.

(See also October 2, 1925)

1884 First practical fountain pen is patented by Lewis Edson Waterman (1837-1901), American inventor. He devises a pen that uses capillary action to control ink flow. This early pen has to be filled with an eye dropper, but he develops self-filling pens by the turn of the century.

1885 First Linotype composing machine (automated typesetting machine) is patented by Ottmar Mergenthaler (1854-1899), German-American inventor. By discarding altogether the idea of manually setting pre-made letters, he builds a machine that a typesetter uses like a typewriter. His machine has a keyboard that moves precast copper letter and punctuation molds into a line, over which it pours a quick-cooling alloy that becomes a finished line of type. Typesetting is not only considerably speeded up, but the publishing industry begins to boom with its new machine.

1885 Charles Sumner Tainter (1854-1940), American inventor, invents the dictaphone which is a machine for recording dictation. Based on the principle of the phonograph, this machine eventually becomes an integral part of the modern office.

1886 Alternating current is used for the first recorded time as a commercial lighting system in the U.S.

1886 Coaxial cables are first used for telephone lines by R.S. Waring. These consist of an insulated tube of conducting material that contains a central conductor at its core. They prove able to reduce interference from electric power lines and other sources.

MAY 4, 1887 Emile Berliner (1851-1929), German-American inventor, first applies for a patent for what he calls a "Gramophone" to distinguish it from American inventor Thomas Alva Edison's (1847-1931) phonograph. His early model uses a flat-disk record made of zinc with a wax coating instead of a cylinder. In making this flat record he devises an etching process that results in a permanent master recording. This will make mechanical reproduction and mass duplication of records eventually possible. Berliner continues to improve his flat disk method and by 1904, his flat phonograph record in which the needle vibrates from side to side replaces Edison's cylinder method.

1887 Tolbert Lanston (1844-1913), American inventor, patents the first monotype typesetting machine. This mechanized rival of the Linotype composing machine also forms type and composes, and together they revolutionize and dominate typesetting until the mid-20th century. The Linotype is used especially for newspapers and the monotype for quality printing, usually for books.

1887 Ludovic Lazarus Zamenhof (1859-1917) of Poland first introduces the universal language "Esperanto." It uses root words derived from and similar to sources in European languages and develops an international following.

1887 Flash powder for photographs is invented in Germany. Called "Blitzlitchpulver," its mag-

Using the linotype to compose lines of text at a newspaper publishing house.

nesium emits a cloud of acrid white smoke when ignited, and its intense light creates harsh tonal contrasts.

1888 Heinrich Hertz (1857-1894), German physicist, conducts the classic experimental proof of the propagation of electro-magnetic waves and is the first to broadcast and receive what come to be known as radio waves. When Guglielmo Marconi (1874-1937), Italian electrical engineer, devises a practical means of using these "Hertzian waves" as a form of wireless communication, they come to be called radio waves, with "radio" being a shorter version of radiotelegraphy (telegraphy by radiation as opposed to telegraphy by electric currents).

(See also 1895)

1888 First Kodak camera is introduced by American inventor George Eastman (1854-1932). With a portable camera that weighs only 2 lbs (1 km), the user takes the pictures and returns the camera to Rochester, New York. Developed photographs soon are returned, along with a freshly loaded camera. Eastman's motto is, "You press the button—we do the rest." In 1889, he abandons paper film and turns to

celluloid, a tougher material with which he makes flexible, transparent film. Along with his new camera, photography for a mass public has arrived.

1888 First modern shorthand method is offered by John Robert Gregg (1867-1948), American educator, whose *Light-Line Phonography* introduces his phonetic system of shorthand. His method distinguishes similar-sounding consonants by length rather than shading and its script flows much like longhand. His system becomes dominant in the United States.

1888 John J. Loud, American inventor, patents the first precursor to the modern ballpoint pen. His revolutionary approach replaces the pen's steel nib with a rotating ball that is constantly bathed in ink from a reservoir. It leaks constantly and is not practical.

(See also 1938)

NOVEMBER 1889 First nickelodeon is introduced by the Automatic Phonograph Company and placed in the Palais Royal Saloon in New York.

1889 William   Friese-Greene   (1855-1921),

The Eastman Dry Plate and Film Company.

English inventor, pioneers photographing motion and is the first to substitute sensitized celluloid ribbon film for the glass plate base for photographic emulsions. He also obtains the first patent for an intermittent-motion device for shielding film movement between exposures by means of a rotating shutter. The motion pictures he demonstrates this year are probably the first to be actually projected.

1889 Edouard Branly (1844-1940), French physicist, invents the coherer for the reception of wireless telegraphic waves and in so doing, discovers the principle of wireless telegraphy. He discovers that metal filings in a container coalesce or cohere in the presence of Hertzian or radio waves, indicating that such a "coherer" can detect electromagnetic radiation. While not a true antenna, his coherer is the first radio wave detector. In 1894, English physicist Oliver Joseph Lodge (1851-1940) improves Branly's detector and names it a "coherer." He uses it to send wireless messages in Morse Code.

1889 First sound movie camera, the kinetograph, is invented by American inventor Thomas Alva Edison (1847-1931). It uses perforated 35mm film and leads to the development of the kinetoscope which shows the film shot by the kinetograph.

1889 First International Congress of Photography is held.

1890 First folding camera is introduced by George Eastman (1854-1932), American inventor. In 1898, he makes this folding camera pocket-size.

1890 Flexography is first patented in England. This letterpress process of relief printing on nonabsorbent surfaces uses flexible rubber plates mounted on cylinders. It is later called aniline printing.
(See also 1926)

1891 Gabriel Lippmann (1845-1921), French physicist, produces the first direct, permanent color images with his highly original method of color photography. He applies a light-sensitive emulsion to a glass plate on the side opposite to the camera lens. This is based on the principle that light rays can overlap to create a multihued "interference." Although his technique is impractical and has no commercial applications, he is awarded the Nobel Prize in physics in 1908 for having devised the first direct process of color photography.

1891 First telephoto lenses for cameras are patented by Thomas Rudolf Dallmeyer (1859-1906) of England. It is also said to be independently invented by A. Miethe and A. Steinheil, both of Germany.

1892 First practical automatic telephone switching system is designed by American undertaker Almon B. Strowger. His design uses pushbuttons instead of an operator and evolves into the dial system.

1892 First practical typewriter that allows the typist to see the words as they are being typed is patented by Thomas Oliver.

APRIL 14, 1894 Thomas Alva Edison (1847-1931), American inventor, first displays his peep-show Kinetoscopes in New York. Designed as an arcade machine, it is an early form of motion picture projector. The Kinetoscope is a cabinet into which a person can look through a peephole and watch pictures move. An electric motor moves a filmstrip to an eyepiece, where a slotted disk expose the images to the viewer at the rate of approximately 40 frames per second. Edison uses the new 35mm, perforated "film" and makes the crucial discovery that the images can be advanced by sprocketed wheels inserted into holes along the sides of the film. This allows the pictures on the film to move before a flashing light at a carefully regulated speed. These demonstrations serve to stimulate research on the screen projection of motion pictures as well as to entertain.

DECEMBER 28, 1895 First public cinema performance is arranged by Auguste Lumière (1862-1954) and his brother Louis (1864-1948), both French inventors, who stage a 20-minute program of ten films in front of a paying audience. Using their invention called a "cinematographe," they project still images on a screen separated by moments of blackness. The images linger on the viewers' retinas, and they perceive a moving image. Their apparatus can both take pictures and project them. This event is considered by many to be the first genuine movie.

1895 Guglielmo Marconi (1874-1937), Italian electrical engineer, first succeeds in sending wireless Morse code signals a distance of more than 1.5 mi (2.4 km) at his father's estate at Pontecchio, near Bologna, Italy. This is considered to be the birth of wireless. In September of this year, he becomes convinced that radio has enormous potential when he is able to transmit and receive signals from over a hill that blocks his line of sight. Most believed at this time that radio waves, like light waves, travelled in straight lines and were thus limited in distance to the horizon.

1895 *The Illustrated London News* becomes the first British newspaper to be printed by rotogravure. This printing system based on the transfer of fluid ink from depressions in a printing cylinder to the paper makes the era of photojournalism possible. An adaptation to rotary cylinder presses of the photogravure process invented in 1879 by Karel Klic (1841-1926) of Czechoslovakia, it uses a special screen which divides up the subject into tiny square cells. This screen pattern is even over the whole plate but the cells vary in depth and give a variety of ink densities. The printing surface is a copper cylinder or a thin copper plate which is prepared flat and then drawn around a cylinder. It proves most efficient for long printing runs and becomes the process preferred for illustrated magazines with large circulations.

1895 Alexander Stepanovich Popov (1859-1905), Russian physicist, adds a wire to the coherer invented by French physicist Edouard Branly (1844-1940), which boosts both reception and transmission and invents the first modern antenna. His addition of this antenna shows that it is possible to receive "Hertzian waves" that were transmitted from a distance of 260 ft (79 m). The following year he publishes a paper suggesting that his apparatus "may be used for the transmission of signals over a distance." Although Popov transmits a ship-to-shore message nearly 6 mi (9.6 km) in 1898, he spends most of his research time studying the physics of thunderstorms.

JULY 10, 1896 Electricity is used in papermaking for the first time in the United States at the Cliff Paper Company in Niagara Falls, New York.

c. 1896 Charles Pathé (1863-1957), French inventor, first separates the Lumiere brothers' "cinematographe" into a camera and an independent projector. By 1904, he has refined the camera to shoot at variable speeds, thus creating the movie camera on which modern models are based.

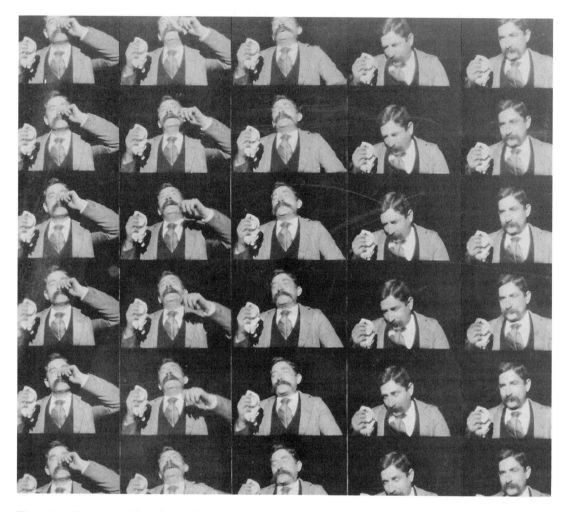

The earliest kinetoscope film submitted for a copyright was of this sneeze, recorded by Thomas Edison.

1896 First daily newspaper to reach a circulation of one million copies is the *Daily Mail* in England.

1896 First film-producing company is founded by Georges Méliès (1861-1938), French filmmaker, after viewing the Lumière brothers' film of December 28, 1895. He makes over 400 films from 1899 to 1913 and is the first to film fictional narratives. As a former magician, he discovers and exploits tricks that a camera can perform, such as slow motion, stop motion, fade out, dissolve, superimposition, and double exposure. Despite his innovations, he never moves his camera for close-ups or long shots. Unable to compete as the industry grows, he dies in poverty.

JULY 1897 The Wireless Telegraph and Signal Company, Ltd. is incorporated in England as the first known commercial organization of its kind. In 1900, the name changes to Marconi Wireless Telegraph Company, Ltd.

1897 Karl Ferdinand Braun (1850-1918), German physicist, constructs the first cathode-ray oscilloscope capable of scanning with an electron beam. He is able to do this upon discovering that electrons can be controlled by magnetism, and that their journey can be traced on a fluorescent screen. The oscilloscope proves to be an important laboratory tool and the precursor to the television tube.

1897 First black shellac-based gramophone records are made. They will eventually be mass-produced by the Deutsche Gramophone Gesellschaft and replace both cylinders and metal disk records.

c. 1897 Frederick G. Creed (1871-1957), Scottish telegrapher and inventor, experiments with the typewriter and the telegraph, searching for ways to send printed messages without using Morse Code, and is credited with producing the first teleprinter. Although the British Post Office buys his machines, they do not catch on. In 1907, however, American inventor Charles L. Krumm designs what is the prototype of the modern teleprinter or teletype machine. It is later refined and marketed in the United States and in 1931, a teletype exchange is inaugurated. News wire services prove is heaviest users.

JUNE 3, 1898 First paid radio message is sent from the Needles, Isle of Wight Station.

1898 First magnetic sound recorder is invented by Valdemar Poulsen (1869-1942), Danish engineer. Called the "Telegraphon," it is a device that records and reproduces sounds by residual magnetization of a steel wire. Thus, in contrast to the phonograph, the sounds are not recorded mechanically on a cylinder or disk but electro-mechanically on wire. He attempts to market his product in the United States as an office dictation apparatus and an automatic telephone message recorder, but is hampered by poor management as well as the system's bulkiness and deficiencies. Poulsen's work nonetheless paves the way for the development of modern recording tape.

APRIL 28, 1899 English steamer *R. F. Mathews*

collides with a lightship and sends the first wireless call for assistance.

SEPTEMBER 1899 This month's cover of the *Saturday Evening Post* is printed in three colors for the first time.

FEBRUARY 28, 1900 German ship *S.S. Kaiser Wilhelm der Grosse* is the first sea-going passenger vessel equipped with wireless service.

JUNE 1900 Nikola Tesla (1856-1943), Croatian-American electrical engineer, writes an article in *Century Magazine* in which he first describes the principle of radar as "a wave reflected from afar [by which you can] determine the relative position or course of a moving object." (See also 1935)

1900 Joseph Poliakoff of Russia invents the principle of the sound track for true talking cinema. This consists of a track which is read optically by a selenium cell. It is well ahead of its time and is not adopted for some time. The modern sound track system works by a modulated light beam that forms a sound track along the edge of the film. When the film is projected, the sound is simultaneously reconstituted through a valve-amplifier system similar to that used on radios.

1900 George Eastman (1854-1932), American inventor, introduces the first "Brownie" camera. His company, Kodak, sells it for one dollar and a roll of film for fifteen cents.

DECEMBER 12, 1901 Guglielmo Marconi (1874-1937), Italian electrical engineer, sends the first wireless signals across the Atlantic Ocean. He sends a radio signal of the letter S (three dots) from the southwest tip of England to Newfoundland, using balloons to lift his antennae. This date is recognized as marking the invention of radio. Marconi's success in sending a radio signal 2,137 mi (3,419 km) across the Atlantic Ocean proves the experts wrong who said that radio waves travelled in a straight line and that their distance was therefore limited to

Kodak Brownie camera.

the horizon. Marconi's successful accomplishment assures the future of radio as an important form of communication. Within a year, regular radio signals are crossing the Atlantic.

1901   Photographic postcards are first printed in England by the Rotary Photographic Company.

JULY 1902   Guglielmo Marconi (1874-1937), Italian electrical engineer, visits St. Petersburg, Russia and demonstrates his wireless apparatus to Russian officials. He establishes wireless communications between Russia and England for the first time.

1902   First telephone conversation via long-distance underground cable is conducted between New York and Philadelphia (90 mi [145 km]).

1902   German expression for radio, "funk," is first coined by Adolph Karl Heinrich Slaby (1849-1913), German physicist, who is known a the "German Marconi."

1903   First high-frequency alternator is built by the General Electric Company using the specifications of Canadian-American physicist Reginald Aubrey Fessenden (1866-1932). Fessenden's goal is to send actual voice mes-

sages through the air, and his high-frequency alternator (a generator that produces alternating current) allows him to produce a continuous radio wave in place of Marconi's intermittent spark-generated pulse. This continuous wave can then be modulated to encode the "shape" of a voice or musical sound.

(See also December 24, 1906)

AUGUST 4, 1903   First International Conference for Radio Telegraphy is held in Berlin, Germany.

(See also 1906)

1903   First German radio engineering development company, Gesellschaft fur drahtlose Telegraphie m.b.H. System Telefunken is established.

1903   First daily newspaper produced at sea is achieved onboard the *S.S. Campania* using dispatches supplied by wireless.

NOVEMBER 16, 1904   John Ambrose Fleming (1849-1945), English engineer, applies for a patent for his invention of the first electron tube. His electronic, two-element vacuum tube is the ancestor of all electronic tubes and is the practical application of the physical phenomenon known as the Edison effect. He calls it a thermionic valve because it controls the flow of electricity just as a valve controls the flow of water. In the United States, this device is called a vacuum tube. It is put to use detecting, modifying, and amplifying electromagnetic radio waves.

1904   Arthur Korn (1870-1945), German physicist, achieves the first telegraphic transmission of a photographic image using a circuit between Nuremberg and Munich. His transmission method breaks down photographs into components and reconstructs them at the receiving end.

1904   First commercially successful method of creating a color photograph on a single plate is

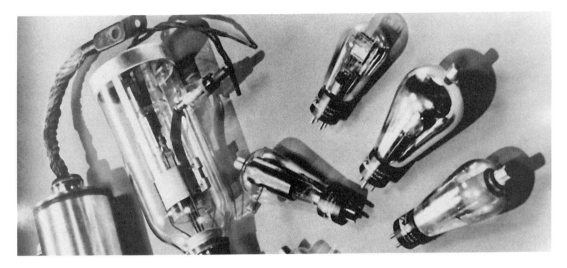

Several variations on the electron tube, used by industries to power machinery.

patented by Auguste Lumière (1862-1954) and his brother Louis (1864-1948), both French inventors. Their "autochrome" process involves covering a photographic plate with small grains of potato starch dyed different colors and then applying a thin film of panchromatic emulsion (emulsion that is equally sensitive to all colors). After an exposure is made through the dyed starch grains, the plate produces a transparency composed of small dots of color that the eye perceives as mixed colors because they are so tiny. This method produces transparencies, not photographs, that have to be projected on a screen or seen through a hand-held viewer.

1904   *The Daily Mirror* is founded in London. It is the first daily newspaper illustrated exclusively with photographs.

MAY 4, 1905   United States Circuit Court settles a dispute between the Marconi and the DeForest Telegraph Companies stating that "Marconi was the first to describe and the first to achieve the transmission of definite and intelligible signals by means of Hertzian waves...."

1905   Albert Einstein (1879-1955), German-Swiss-American physicist, first announces his theory of the photoelectric effect in which light falling upon certain metals is found to stimulate the emission of electrons. This will become the fundamental principle of the modern television camera and will define the manner in which the camera can turn a picture into electricity.

JANUARY 18, 1906   Lee DeForest (1873-1961), American inventor, first patents a device he calls an "audion." This device is an improvement on English engineer John Ambrose Fleming's (1849-1945) electron tube invention of 1904. By adding a third element (a wire grid) to Fleming's diode, he can amplify weak signals far more effectively. His audion or triode becomes an amplifier as well as a rectifier and forms the basis of all radio tubes and other electronic devices as well. It makes radio and television broadcasting possible.

DECEMBER 24, 1906   First successful radio broadcast in the United States is made by Canadian-American physicist Reginald Aubrey Fessenden (1866-1932). This is recognized as marking the beginning of AM radio, as Fessenden's system of Amplitude Modulation (AM) makes practical radio possible. He builds a transmitting station on the Massachusetts coast and sends out speech and music which wireless operators pick up.

(See also 1903)

1906 First International Radio Conference is held in Berlin, Germany.

1907 Boris Rosing, a lecturer at the Technological Institute, St. Petersburg, Russia, first introduces the idea of using a cathode ray tube as a means of reproducing a television picture. Known as "Rosing's Apparatus," it is named the "electric eye" by Rosing. Although this system uses an electronic receiver, it still has a mechanical camera. Rosing's pupil is Russian-American physicist Vladimir Kosma Zworykin (1889-1982), who goes on to invent the iconoscope.

(See also 1923)

JUNE 1908 A.A. Campbell-Swinton, Scottish engineer, first suggests the use of a cathode ray tube as both the transmitter (camera) and receiver. He dose not make available the details of how his proposal would work until 1911, but even then his ideas and plans are far beyond the technology of his time. Still, this is the first description of the modern, all-electronic television system and eventually forms the basis of modern television.

1908 International Radio Telegraphic Conference is held at Berlin, Germany and first proposes the code "SOS" as the wireless distress call.

1909 Motion picture industry first adopts 35-mm film as a standard and it is accepted internationally.

1910 First successful multiplexing of telephone systems is achieved. This allows the simultaneous electronic transmission of two or more messages in one or both directions over a single transmission path.

1911 First photograph taken from an airplane is made by James H. Hare (1856-1946), English-American camera designer and news photographer. Aerial photography will prove to have many wide-ranging and very valuable applications.

AUGUST 13, 1912 First domestic United States law enacted for the control of radio is passed. This act requires radio operators and transmitting stations to be licensed.

1912 Irving Langmuir (1881-1957), German-American chemist, experiments with a gas-filled, tungsten-filament, incandescent lamp. This is the first known application of unalloyed tungsten. As a strong metal with a high melting point, it is also brittle at room temperature. It is made ductile by mechanical working at high temperatures and is then drawn into a very fine wire. It proves to be excellent material for a lamp filament.

1912 Hans Fischer (1881-1945), German biochemist, first proposes the breakthrough concept that color photography could be achieved chemically rather than optically. Although he does not produce color film, his concept is the basis for modern color film.

1912 Corona introduces its first portable manual typewriter.

OCTOBER 29, 1913 Edwin Howard Armstrong (1890-1954), American electrical engineer, files a patent application for a "regenerative circuit" that can considerably amplify radio signals. While working with a vacuum tube, he discovers that feeding the tube's current back into itself greatly enhances its sensitivity. With this, he is able to amplify distant radio signals enough to hear them without earphones. Twenty years later, a court rules that this was first invented by American chemist Irving Langmuir (1881-1957) in 1912. In fact, although De Forest had invented a telephone circuit that

Early typewriter.

had the potential for radio amplification, Armstrong had invented his circuit specifically for the radio. Most now credit him with the invention.

NOVEMBER 24, 1913   First practical trials with wireless apparatus on trains are made.

SEPTEMBER 1914   First air photographs of World War I taken over enemy positions are made by the newly constituted Royal Flying Corps at the Battle of the Aisne.

1914   East coast of the United States is linked to the West coast by telephone for the first time. The system uses seven repeater stations that amplify its signals.

(See also January 25, 1915)

JANUARY 25, 1915   First transcontinental telephone line opens officially between New York and San Francisco.

AUGUST 1915   David Sarnoff (1891-1971), Russian-American business executive of the Marconi Wireless Telegraph Company, first proposes the idea of a "radio music box" or

commercially marketed radio receiver. Although it takes several years for this commercial idea to gain acceptance, Sarnoff becomes general manager of the newly formed Radio Corporation of America (RCA), and in 1921 is able to demonstrate the market potential of radio when he broadcasts the Dempsey-Carpentier boxing match. Within three years, RCA sells more than $80 million worth of receiving sets.

1915   First known transatlantic radiotelephone conversation takes place between France and the United States. This wireless system eventually expands to cover the globe and is even extended to passenger ships.

1918   Edwin Howard Armstrong (1890-1954), American electrical engineer, invents the superheterodyne circuit. Also called the variable frequency receiver, this system simplifies the search for stations by using a single button. Without the superheterodyne, it was very difficult to adjust the receiver properly to the desired frequency. This improved radio receiver also amplifies weak signals and becomes the basis for all such modern devices.

OCTOBER 17, 1919   Radio Corporation of America (RCA) is first formed and acquires the business and property of the Marconi Wireless Telegraph Company of America during its first month in business.

NOVEMBER 2, 1920   Radio station KDKA in Pittsburgh, Pennsylvania, is first station to transmit results of a presidential election (the Harding-Cox returns).

1920   Radio station KDKA in Pittsburgh, Pennsylvania, becomes the first radio station to broadcast regularly scheduled programs. This is done on an experimental basis, since the station is not licensed until November 7, 1921.

(See also September 15, 1921)

1920   First commercial radio link to a land telephone line begins operating between Santa Catalina Island and the California mainland.

1920 Westinghouse Electric Company sells the first "radio receivers" to the public. This device receives the radio frequency wave from the transmitter via the receiving antenna and converts it to sound.

SEPTEMBER 15, 1921 First radio station to be issued a regular broadcasting license in the United States is station WBZ in Springfield, Massachusetts.

MAY 21, 1922 *The New York Times* newspaper first devotes a Sunday page to radio news and programs.

NOVEMBER 17, 1922 First American president to use radio to reach the people is Warren Gamaliel Harding (1865-1923), who speaks from a Madison Square Garden broadcast.

NOVEMBER 26, 1922 First color motion picture is shown in New York. Introduced by the Technicolor Corporation, this is a new two-color method for joining two separate positive prints. By 1928, Technicolor improves the process, making it possible to combine all primary colors, resulting in life-like tones and hues.

1922 First ship-to-shore two-way radio conversation is established between a New Jersey station and the *S. S. America* 400 mi (644 km) at sea.

DECEMBER 15, 1922 British Broadcasting Company (BBC) is formed.

1923 Vladimir Kosma Zworykin (1889-1982), Russian-American physicist, invents the iconoscope, a transmission device that becomes the precursor to the modern television camera, and (in 1925) the kinescope, a reception device almost identical to what becomes modern tele-

vision tubes. Both are completely electronic rather than mechanical. Within the iconoscope's tube is a plate covered with thousands of tiny droplets of cesium-silver, each acting as a microscopic photoelectric cell. As light passes through a lens and falls upon the droplets, each produces a small electric charge, with each droplet's charge being different from the others. These charges are held by the droplets until they are scanned by an electron gun within the iconoscope which converts them into an electronic signal that can be sent to a receiver. The kinescope receives this signal and scans the image back onto a fluorescent screen, converting it back into a visual image. Zworykin soon joins the Radio Corporation of America (RCA) which improves and refines his system.

1923 John Logie Baird (1888-1946), Scottish electrical engineer, first transmits silhouettes or "shadowgraphs" by television using his invention, photomechanical television. Despite initial success and popularity, his system produces an image of very poor resolution and will give way to an electronic system.

(See also October 2, 1925)

1923 National Association of Broadcasters (NAB) is first established in Chicago, Illinois.

FEBRUARY 12, 1924 First network sponsored radio program is the *Eveready Hour* sponsored by the National Carradiobon Company.

FEBRUARY 22, 1924 First presidential radio talk delivered from the White House is made by President John Calvin Coolidge (1872-1933).

JULY 6, 1924 First transatlantic radiophoto—an image of American jurist and statesman Charles Evans Hughes (1862-1948)—is transmitted by RCA (Radio Corporation of America) from New York to London and back again. On April 30, 1926, RCA inaugurates transatlantic radiophoto service. A news image is

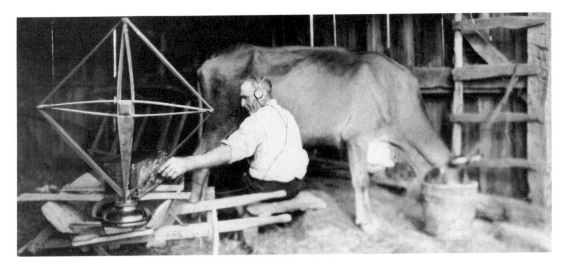

Farmer listening to one of the first radio broadcasts.

received and published in the next day's *New York Times*.

1924 First portable radio is introduced by the Zenith Radio Corporation. It has a built-in loop antenna and a horn-type speaker and sells for $200. It weighs nearly 15 lbs (6.8 kg).

1924 Spiral-bound notebooks are introduced for the first time.

OCTOBER 2, 1925 First to broadcast a live television transmission is Scottish electrical engineer John Logie Baird (1888-1946). His invention is photomechanical television rather than all-electronic television. This system uses a Nipkow disk, which is a cardboard disk with a series of square holes cut in a spiral. When coupled with a photoelectric cell and spun, the Nipkow disk is able to scan areas of lightness and darkness and convert that information into an electrical signal. By using a second disk synchronized with the first, this signal is able to be retranslated into a primitive visual image which is then sent via electromagnetic waves. This mechanical design produces an image of very poor resolution—an inherent flaw—and is eventually replaced by the all-electronic

cathode ray tube design. Still, it causes a great sensation, especially in 1927 when Baird sends a television signal from London to Glasgow.

(See also 1928)

1925 Gregory Breit, Russian-American physicist, and Merle Antony Tuve (1901-1982), American research physicist, conduct ionosphere studies and are the first to develop the radio transmission technique known as radio-pulse. This technique measures the time elapsed between the sending and reception of a pulse of radio energy. This is a major step toward the development of radar.

(See also 1935)

1925 Paul Viertkotter of Germany invents the first flashbulb for cameras. By encasing magnesium wire in glass, artificial illumination is made safe and smoke-free. Foil-filled lamps appear in 1929 which, like the wire bulbs, are set off by batteries.

1925 First practical use of short-wave radio is made by the United States Navy's *S.S. Perry* on a trip to the Arctic.

1925 First Leica camera is made and marketed

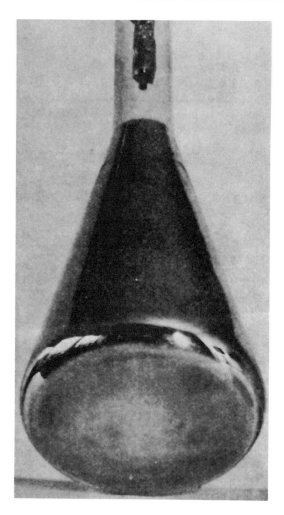

Cathode ray tube.

SEPTEMBER 9, 1926 National Broadcasting Company (NBC) is first organized by RCA (Radio Corporation of America) "to provide the best programs available for broadcasting in the United States."

1926 Aniline printing is first introduced. Used mainly for color printing on plastics and hard-surface materials, it uses very fluid or thin inks colored by aniline dyes which are transferred from flexible rubber plates to a fast-moving web on a rotary press. Later used also on metal foils, fabrics, and corrugated boards, it becomes known as Flexography in 1952.

JANUARY 7, 1927 Philo Taylor Farnsworth (1906-1971), American engineer and inventor, files his first patent application for an electronic television system that uses an electronic camera he calls the "image dissector." While in high school, Farnsworth draws a schematic for an electronic television. His image dissector tube works by transforming the picture into photoelectrons which are moved by magnetic coils in front of a fixed electrical probe that does the scanning. This patent is a legitimate rival to the system put forth by Russian-American physicist Vladimir Kosma Zworykin (1889-1982), and Farnsworth's company and Zworykin's backer (RCA) eventually reach a cross-licensing agreement.

APRIL 5, 1927 Columbia Phonograph Broadcasting System, Inc is first formed. Before the end of the year, it changes its name to the Columbia Broadcasting System (CBS).

OCTOBER 6, 1927 The first talking motion picture, "The Jazz Singer," starring Al Jolson (1886-1950), is produced by Warner Brothers. This is the first time an actor delivers spoken dialogue from the screen. Until Warner's great success, all other major movie producers considered sound an expensive and impractical novelty. By 1930, 95% of major new films have sound.

(See also July 28, 1928)

by the Leitz Company of Germany. This 35-mm, roll-film camera is small and easy to handle, and its fast film and rapid film-advancement mechanism transform photojournalism.

MAY 1926 First wireless messages from the North Pole are sent from the dirigible *Norge*.

AUGUST 1926 First sound film with music is "Don Juan," produced by Warner Brothers. It does not have dialogue.

(See also October 6, 1927)

1927   Transatlantic telephone service is inaugurated between New York and London.

1927   Harold Stephen Black (1898-1983), American electrical engineer, invents a negative-feedback amplifier that has the ability to filter out any amount of distortion. In trying to reduce amplifier distortion, he discovers that subtracting the amplitude of the output signal from that of the input signal causes the two to cancel each other out, leaving only the distortion. This distortion, in turn, can then be amplified, fed back into the system, and used to cancel out the original distortion. He then finds that feeding the output signal back into the system negatively or out-of-phase allows for the removal of almost any amount of distortion. This is a fundamental discovery as it makes possible the development of broadband transcontinental and transoceanic communications systems.

APRIL 1928   First municipal police radio system is put into operation by the city of Detroit, Michigan.

JULY 28, 1928   First all-talking picture, "Lights of New York," makes its debut.

DECEMBER 23, 1928   First permanent, coast-to-coast radio network is established by National Broadcasting Company (NBC).

1928   National Broadcasting Company (NBC) receives its first television station construction permit.

1928   First transatlantic television is achieved by John Logie Baird (1888-1946), Scottish electrical engineer, who sends a signal from London to New York.

(See also October 2, 1925)

1929   Fritz Pfleumer, German engineer, files the first audiotape patent for magnetic recording tape. His workable idea is to cover a strip of paper with a magnetizable iron powder. It is eventually developed into a tape made by bonding a thin coating of oxide to strips of film. Magnetic recording tape is an improvement over records since it is easier to use, store, and edit.

(See also 1935)

1930   First twin-lens 35-mm Rolleiflex camera is made in Germany.

1931   First photo-electric (light) meter for photographers is built in the United States.

OCTOBER 3, 1932   *The Times* of London first uses the typeface known as "Times New Roman." It proves very popular and is made in many varying sizes.

MARCH 12, 1933   American President Franklin Delano Roosevelt (1882-1945) makes the first of his "fireside chats" to the American people. He is the first national leader to use the radio medium comfortably and regularly to explain his programs and to garner popular support.

1933   First talking machine to employ electronics is credited to an inventor named Dudley. His "Voice Operation Demonstrator" or Voder can imitate human speech as its operator uses a keyboard.

1933   First stereophonic recording is made by Alan Dower Blumlein, English engineer. He achieves this by making the record contain two sets of information. He records two sound tracks within a single groove and replays them through one stylus to two separate amplifying systems.

1933   First panchromatic roll film is introduced in the United States. Made with an emulsion that is equally sensitive to all colors, it becomes available in a convenient roll.

1933   Edwin Howard Armstrong (1890-1954), American electrical engineer, invents FM (Frequency Modulation) radio. He devises a method of successfully modulating the frequency of the wavelengths rather than the amplitude, and his system produces clear sound that is immune to electromagnetic interference. It does not become popular until the late 1950s.

JULY 11, 1934   United States Federal Communications Commission is first established. This agency is created to regulate interstate and foreign communications by radio and television broadcasting and eventually telephone, telegraph, and cable television operation, as well as two-way radio and satellite communications.

1935   Leopold Godowsky and Leopold Mannes, both American amateur chemists and musicians, invent a practicable, dye-color film for the Eastman Kodak Company that is called Kodachrome. It is first released this year as movie film and as sheet film in 1938. In 1941 it is available as a negative Kodacolor roll film.

1935   First practical radar equipment for the detection of aircraft is developed and patented by Robert Alexander Watson-Watt, English physicist. Using the principle of echolocation, he eventually is able to display radio wave information on a cathode-ray screen. To find an object with echolocation, a signal is transmitted. Since the velocity of radio waves is, like that of light, a constant, it is easy to calculate the distance to an object by measuring the time it takes for a radio signal to travel that distance. When one of the transmitted signals returns to the source, the back-and-forth time is recorded, as well as the direction from which it is reflected. This information is collated and displayed on a cathode ray screen, thus pinpointing the exact position of the object. By 1938, Britain builds a network of five radio or radar stations on its eastern border that proves to be a decisive factor in the coming war with Germany.

(See also 1943)

1935   First commercial tape recorder using the new magnetized plastic tape is marketed by the German electronics firm AEG. It sells this new

device as the Magnetophon, a record/playback device that uses audiotape.

1935   First softcover reprint or paperback book is introduced by Allen Lane (1902-1970), English publisher and founder of Penguin Books, Ltd. His high-quality, low-price books create a major new reading audience. In 1939, Pocket Books begins publishing small, reasonably priced paperback books in the United States and begins a major revolution in American publishing.

1935   First commercially successful, all-electric organ is invented by Laurens Hammond of the United States. He produces a combination of signal-generating electronic modules that can be linked together in different combinations. His organ starts an entire industry.

1936   First transmission of regularly scheduled television programs from London takes place. The British Broadcasting Company (BBC) adopts the mechanical television system invented by Scottish electrical engineer John Logie Baird (1888-1946), but soon abandons Baird's system and goes with an all-electronic system.

1937   Chester Floyd Carlson (1906-1968), American inventor, takes out the first patent on what he initially calls "electro-photocopy," soon to be renamed "xerography" after the Greek for "dry writing." It is the first copying method to be based on the properties of selenium. He regards copying with photographic methods too slow, and he concentrates on using selenium which is photoconductive. Because selenium will only hold an electric charge in the dark, the charge remains exclusively on the areas where the dark sections of the image are reflected. Coating a drum with powder made of carbon and thermoplastic resin, Carlson rolls an electrically charged sheet of paper along the drum, which releases the image onto the paper where it is then fused into a permanent impression by infrared light.

(See also October 23, 1948)

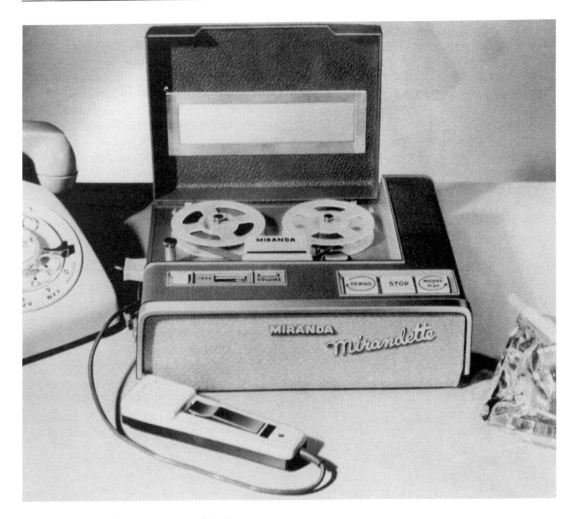

Audio recorder used to accompany a slide show.

1937 First 25-mm single lens reflex camera is the German-made Exakta.

OCTOBER 20, 1938 Radio Corporation of America (RCA) announces that its first television sets will be offered for sale in April 1939 at the World's Fair in New York.

OCTOBER 22, 1938 Chester Floyd Carlson (1906-1968), American inventor, makes the first crude but clear electrostatic copy of the date and site of his achievement, "10-22-38 Astoria." After years of searching for a backer, Carlson finally sells his photocopier to the Haloid Corporation.

(See also 1945)

1938 First workable ballpoint pen is developed by Ladislao Biro of Hungary and his chemist brother Georg. They begin producing their rolling-ball pens commercially in 1943 in Argentina, and they soon appear in both England and America. In 1944, they sell marketing rights to the Shaeffer Company of the United States which launches a heavy advertising campaign. They are eventually outdone by the Bic ballpoint pen.

(See also 1963)

1939 First electronic flash for cameras is developed by Harold Eugene Edgerton, American electrical engineer.

1939 First practical microwave transmitter is invented by John Turton Randall, English biophysicist, and Henry Albert Howard Boot, English physicist. Called a cavity magnetron, this device is capable of generating extremely high frequencies and short bursts of very high power. It becomes an important source of power for radar systems (especially for the British during the Battle of Britain) and later in microwave ovens.

1941 Regularly scheduled television first begins in the United States with transmissions from the Empire State Building in New York. On July 7 of this year, commercial television operations begin officially with the licensing of 21 stations.

1943 Word "radar" becomes popularly known as the military first lifts restrictions on the name. It was coined by S.M. Tucker of the United States Navy and is an acronym for RAdio Detection And Ranging.

(See also 1935)

AUGUST 29, 1945 RCA Victor first offers vinyl plastic records to the public.

1945 Arthur Charles Clarke, English writer, publishes an article in *Wireless World* titled "Extra-Terrestrial Relays" in which he is the first to propose using satellites hovering in stationary orbits at 22,000 mi (35,200 km) above the earth to beam television and radio throughout the world. This entirely theoretical idea sounds like science fiction at the time, but eventually becomes reality.

(See also April 1965)

1945 Harry S. Truman (1884-1972) becomes the first American president to appear on a television network broadcast.

1945 Battelle Memorial Institute, a research organization in Columbus, Ohio, first agrees to support American inventor Chester Floyd Carlson (1906-1968) in his attempt to perfect his invention of xerography. In 1947, the Haloid Company of Rochester, New York buys the rights to produce Carlson's new copier. In 1958, this firm becomes Haloid Xerox, and in 1961, the Xerox Corporation.

(See also October 23, 1948)

1946 First really effective photo-composing machine, the Intertype Fotosetter, is installed at the Government Printing Office in Washington, D.C. Operating on the basic principles of a linotype machine, it has a keyboard that controls the selection of various negatives of the various characters instead of the matrices used by the hot-metal machine. It also uses a camera that replaces the casting mechanism.

1947 Dennis Gabor (1900-1979), Hungarian-English physicist and engineer, first develops the basic concept of holography. He develops the idea as a means to improve the poor resolving properties of electron microscopes. He continues to work on the idea of a hologram (a word he coins, meaning whole or complete picture), and his efforts in this realm of three-dimensional photography eventually find numerous applications with the development of a coherent light source, the laser, in 1960. Once the laser is developed, the hologram is able to produce a true three-dimensional picture.

JUNE 30, 1948 Bell Telephone Laboratories first demonstrates the transistor, a solid-state device for amplifying, controlling, and generating electrical signals that will eventually replace the conventional electron radio tubes. Among its advantages as a replacement for the bulky, power-hungry vacuum tubes are its small size, light weight, mechanical ruggedness, long life,

Radar system used by Air Force.

and low power demands. It revolutionizes radio design.

OCTOBER 23, 1948 New word "xerography," from the Greek word "xeros," meaning dry writing, first appears in *The New York Times*.

1948 First instant-print camera is introduced by Edwin Herbert Land, American inventor. His invention eliminates several steps in traditional film development and delivers a finished print. The key is the film (rather than the camera) which features a pod of chemicals at its edge. When the film is squeezed between the camera's two rollers, the processing fluid in the pod spreads over the picture between the exposed negative and positive sheets. The image is then transferred to the print paper in contact with the film negative. In 1962, Land's Polacolor film makes instant color photography available. In 1972, Polaroid SX-70 color film is introduced. With this new product, the entire development process is done within the camera, eliminating the pulling and peeling of the older system. Processing time is reduced to 10 seconds.

1948   Columbia Records first introduces a long-playing (LP), 12 in (30 cm) record made of vinylite that rotates at 33 1/3 rpm. Created by Hungarian-American inventor Peter Carl Goldmark (1906-1977), this vinyl record has grooves cut much closer together (250 grooves per inch rather than the 80 grooves on a 78 rpm) and is played on a slower turntable. RCA soon releases its competitive version as a 7 in (18 cm), 45 rpm record, but it never reaches the popularity of the LP which has a much greater capacity.

1948   First cable television systems in the United States begin to appear. They are used primarily to deliver a clear signal to rural communities via cables fed directly into homes. Using coaxial cable which consists of an insulated tube of conducting material that contains a central conductor at its core, the thick, layered cable allows transmission of a wide band of frequencies and rejects interference from most normal sources. Apart from dramatically improved reception, cable television holds enormous potential since it can be bidirectional, theoretically enabling viewers to participate in polls from their homes. In the mid-1960s, new technology allows up to twelve channels per cable, and the number continues to rise.

(See also 1975)

OCTOBER 10, 1949   First regular television program to be seen simultaneously in color (in Washington, D.C.) and in black-and-white (network-wide) is the puppet show "Kukla, Fran and Ollie."

1949   First modern, high-speed phototypesetting machine is developed by Rene Higonnet and Louis Moyroud, French inventors, who revolutionize phototypesetting with their invention of the Lumitype. In these machines, the letter matrices are on a disk revolving at a steady high speed and a high-speed lens catches the disk at exactly the right point in its rotation and photographs the correct letter directly. An early version can produce 28,000 characters an hour. It is soon improved to 80,000 characters per hour.

SEPTEMBER 1, 1950   First Mexican television station, XH-TV, opens in Mexico City.

1950   Vidicon camera tube is first used. This television camera is the first camera device to employ the phenomenon of photoconductivity (in which a light pattern is stored on the surface of a photoconductor and the photoconductor is then scanned by an electron beam). This is a very much improved version of the older iconoscope and the improved orthicon tube. It is more sensitive and more efficient.

1950   First machine to be able to successfully recognize speech is developed by Bell Telephone Laboratories. It can distinguish ten spoken numbers from a series of acoustic signals.

(See also 1972)

1951   Cinerama is first attributed to Fred Waller (1886-1954), American photographer. In this process, three synchronized movie projectors each project one-third of the picture on a wide, curving screen. It gives viewers a greater sense of reality than does a flat screen, since it approximates the entire field of human vision. As its costs rise, however, it is abandoned in the 1960s.

1951   Coast-to-coast network television first begins in the United States.

1952   Linotype Machinery Ltd., London, introduces Jubilee, the first original new text face since Times Roman in 1932. It is designed to meet the needs of new high-speed rotary letterpresses.

1952   Three-dimensional (3-D) motion pictures first begin to appear in the United States. The first 3-D film is *Bwana Devil*, and viewers must wear special plastic polarizing glasses to get the true 3-D effect.

1952   Minoan Linear B script (early evidence of

Audience wearing special glasses to view a three-dimensional (3-D) film.

writing in the Aegean area) is first deciphered by Michel George Francis Ventris (1922-1956), English architect and cryptographer, who shows it to be Greek in its oldest form.

1954   First transistor radio to appear on the market is the Regency. Although not a commercial success, it introduces the portable radio to the consumer market and spurs the mass production of transistors.

1954   For the first time, the number of radios in the world exceed the number of copies of newspapers printed daily.

1955   Narinder S. Kapany, Indian physicist, first introduces fiber optics. By surrounding glass fiber with cladding, he takes advantage of the phenomenon of total internal reflection, meaning that there is very little loss of intensity over a great distance. Optical fibers prove to be a revolutionary tool in such fields as medicine and telecommunications.

(See also 1970)

SEPTEMBER 25, 1956   First two-strand transatlantic telephone cable between Scotland and Newfoundland is put into operation. Prior to this, cable telephone connections between the United States and Europe were possible only by short-wave radio telephone.

NOVEMBER 30, 1956   First retransmission of a recorded television program is made using videotape. This allows a California television station to rebroadcast that day's showing of "Douglas Edwards and the News."

DECEMBER 18, 1958   Era of active satellite communications begins as the United States launches *SCORE* (Signal Communications by Orbiting Relay Equipment) into Earth orbit. The 8,750 lb (3,977 kg) relay satellite beams a Christmas message from American President Dwight David Eisenhower (1890-1969) on December 19, 1958. This is the first voice broadcast from space.

(See also July 10, 1962)

1958   First videotape recorders are installed in American television studios. Using a system called Ampex, it has ¹/₂ in (1 cm) wide tape moving at a speed of 200 in (508 cm) per second. Recording is done on three tracks, with two for storing the video signals and one for sound.

Fiber optics converter.

1958 Stereophonic records first become commercially available.

1959 First commercial Xerox copier is introduced.

AUGUST 12, 1960 First passive communications satellite, *Echo 1*, is launched into orbit by the United States. It is a large, foil-covered balloon that inflates in space and off which signals are bounced.

1960 Halogen lamp is first introduced. This incandescent lamp filled with halogen gas pro-

duces a very bright light as the gas regenerates the filament, making it glow at a higher temperature.

1961 International Business Machines Corporation (IBM) introduces its first Selectric typewriter whose rotating ball types the characters while the carriage stays put.

JULY 10, 1962 The first known active communications satellite, *Telstar 1*, is launched by the United States. It is built to transmit high-definition television pictures across the Atlantic Ocean.

(See also July 11, 1962)

JULY 11, 1962 First television programs via an active communications satellite, *Telstar 1*, are exchanged between the United States and Europe. Now silent, the 160 lb (73 kg) satellite is expected to stay in orbit some 200 years.

1962 First practical light-emitting diode (LED) is invented by Nick Holonyak, Jr., of the United States. This is a semiconductor device in which electrons subject to an applied voltage subsequently emit radiation as they fall to lower energy levels. Made of gallium arsenide, his LED is the first semiconductor laser to operate in the visible spectrum. They are now used in calculators, electronic gadgets and toys, and in many household devices.

1963 First felt-tip pen is introduced by Pentel, a Japanese firm. It uses a soft, absorbent tip fed by free-flowing, often colorful ink.

(See also 1979)

1963 Eastman Kodak of Rochester, New York, first introduces the Kodak Instamatic Camera. Its drop-in film cartridge greatly simplifies loading, and it has a built-in, pop-up flash.

1963 The first audiocassette is introduced by the Philips Company of the Netherlands. This new format makes the unwieldy and complicat-

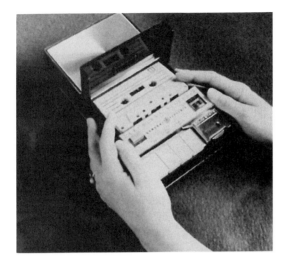

The portable cartridge tape recorder from General Electric.

ed system of recording on magnetic tape easy and economical. The cassette format makes inserting, advancing, and rewinding a tape fast and easy, and becomes very popular.

1963 First "hot line" or direct telephone link between the White House and the Kremlin is established by Washington and Moscow. It is intended to permit the leaders of both countries to be able to speak directly and immediately to each other in times of crisis.

1964 Intelsat (International Telecommunications Satellite Consortium) is first established by the United States and 11 other countries to set up and develop a global communications satellite system.

(See also April 1965)

APRIL 1965 First commercial telecommunications satellite is believe to be *Early Bird*, which is launched by the United States and placed into a geostationary orbit by Intelsat. It can relay 240 telephone conversations simultaneously.

1965 First electronic telephone exchange in the United States is installed in New Jersey.

1965 Soviet Union launches its first domestic communications satellite, *Molniya 1*.

1965 First home videotape recorder is introduced by Sony Corporation of Japan.

1966 First telephone-telegraph cable between North and South America is laid between Venezuela and the United States Virgin Islands.

MARCH 1, 1967 Overseas direct dialing by telephone from New York to London and Paris first begins.

1967 Battery-operated cordless telephones are first developed in the United States. These mobile phones use cellular radio technology and become extremely popular in the 1980s.

1968 First transistorized cable, handling 720 telephone calls simultaneously, is laid between Florida and the Virgin Islands.

1968 Mini-cams are first used by the CBS television network to cover the Democratic and Republican national conventions. These portable cameras allow on-the-scene television coverage and will revolutionize television news coverage and reporting.

JULY 20, 1969 First radio and television transmissions from the surface of the Moon are received on Earth. The activities of the crew of the United States *Apollo 11* spacecraft are witnessed by millions as mankind first sets foot on the Moon.

1970 First long optical-fiber for telecommunications is developed by Corning Glass in the United States. An optical fiber is a filament composed of two types of glass that operates on the principle of total internal reflection. A light beam directed at one end of the filament is sealed in and light loss is prevented by a continuous series of total reflections along the inside lateral walls of the filament. By using digital transmission, telecommunications sys-

Video tape recorder with monitor and T.V. camera.

tems can carry information farther over a smaller cable than its old, copper wire predecessor. Optical fibers are also immune to electromagnetic interference.

(See also 1976)

1972 First speech-recognition system, VIP 100, is introduced by Threshold Technologies. It can only recognize a limited number of words when spoken singly.

1972 First home video-cassette recorder is introduced by Philips.

1972 First word-processing system, Lexitron, becomes commercially available.

1972 Video telephone systems are first introduced by Bell Systems in the United States.

1973 Color duplication photocopiers are first introduced.

1974 First international fax standard is set by the United Nations. It allows facsimile messages to be transmitted at about one page in six minutes.

1974   International Business Machines Corporation (IBM) first introduces its Memory typewriter that can store up to 50 pages of type on a magnetic medium and play it back at a rate of 150 words per minute.

1974   Arthur Fry, American chemist, first develops a glue that allows paper to stick to a surface but also allows it to be easily removed. It is later marketed as Post-It Notes.

1975   First nationwide pay-per-view cable television station, Home Box Office (HBO), begins service. Taking advantage of coaxial cable technology and later digital video compression, the number of television channels rises to 150.

1975   International Business Machines Corporation (IBM) first introduces the laser printer which prints comparably to the quality obtained by photocomposition.

1976   First optical-fiber cable trunk for telecommunications is installed.

1976   Japan Victor Company (JVC) first introduces the video home system (VHS) format for videotape. Its cassette uses a 3/4 in (2 cm) tape that proves popular and eventually becomes the standard system over its rival format, Beta.

1976   International Business Machines Corporation (IBM) first introduces the ink-jet printer.

1976   Raymond Kurzweil of the United States greatly develops speech-synthesis technology to aid the blind by producing the first machine that can read printed text and give a speech-synthesized output. The Kurzweil Reading Machine uses a computer scanner to convert text to sound. By placing printed material face down on a glass plate, the reader activates an optical character recognition system that converts the letters to speech for a verbal recreation of the text.

1977   Fiberoptic cable gets its first commercial use in an AT&T communications network. This cable with a hair-thin glass core and reflective coating can carry voice, video, and data signals as pulses of light. Optical fibers operate on the principle of total internal reflection (by which the beam of light is bent so much that it reflects back completely into the medium from which it originates). Optical fibers achieve this by using a layer of material (cladding) that prevents any light loss. Fiber optics have applications in medicine and especially in the field of communications. By using digital transmissions, telecommunications systems carry more information farther over a smaller cable system than its copper wire predecessor. Optical fiber system also have reduced weight and are immune to electromagnetic interference.

1978   First popular word processing software for personal computers is Wordstar developed by John Barnaby of the United States. This eventually gives way in popularity to WordPerfect and Microsoft Word. Desktop publishing soon becomes a reality.

1978   First telephone link using optical fiber technology in Europe begins between two cities in England.

1978   First automatically focusing camera is introduced by Konica.

1979   First erasable ballpoint pen, called the Eraser Mate, is introduced by the Gillette Company of the United States.

1979   First commercial network of cellular telephones is created in Tokyo, Japan.

1980   First small, portable tape player called the "Walkman" is introduced by Sony Corporation.

1981   Compact Disks (CD) first enter the commercial market as technical standards for compact disk recording and playback are established and accepted worldwide. These 4.8 in (120 mm) diameter plastic disks use tiny pits that are read by a laser to reproduce sound. Unlike traditional "analog" sound recording on records which loses its fidelity over time with use, CDs use "digital" recording. Storing the recorded information as a binary code allows it to be replayed by using a laser beam which never touches the playing surface. In 1982, CBS/Sony and Philips both introduce the first CD players.

1983 First regular cellular telephone system in the United States begins operations.

1984 CD-ROMs are first introduced by Sony and Philips. This Compact Disk Read-Only Memory system is an optical disk that can store very large amounts of data.

1985 Seiko-Epson of Japan first introduces a TV with a 2-inch (5 cm) screen that is a liquid-crystal display made with a polycrystalline silicon.

1986 First digital audio tape (DAT) recorders are made in Japan. Using a code of binary numbers, these new recorders make it possible for anyone to record, store, and play back sound with greater clarity and integrity.

1987 Kodak first introduces a disposable camera called the Fling. The entire (unopened) camera is returned for film processing and a new one is purchased to take more pictures.

1988 First transatlantic optical fiber cable is laid. It carries 37,800 voice channels.

1990 Improved facsimile (fax) machines offering color images become available.

1991 Matasushita of Japan first introduces a video recorder that is programmable by voice commands.

1992 Digital cellular phone systems are first introduced in the United States. This system triples user capacity and makes for better sound quality.

MARCH 1994 Entirely new communications satellite system is announced by Teledesic Corporation. Formed by American billionaires Bill Gates of Microsoft and Craig McCaw of McCaw Cellular, this joint effort plans to blanket the Earth with 840 communications satellites placed in low Earth orbit. Theoretically, if each region of the globe has a satellite passing over it at all times, transmitters could aim straight up, minimizing distortion. Since all that would be needed is a battery-powered station and an 18 inch (46 cm) antenna, even the poorest country can be part of the system. Skeptics cite the immense costs and technical problems to be solved in building and launching such a total system.

1995 Results of radiocarbon dating of cave paintings discovered in December 1994 near Avignon, France indicate that they are the oldest wall paintings ever found, being dated between 30,300 and 32,400 years old. These over 300 paintings of bears, mammoths, wooly rhinos, and other Ice Age animals are done in red, yellow, and black pigment derived from charcoal and iron ores. They rival in beauty the drawings in the famous Lascaux cave which are only about 17,000 years old. Although their purpose will probably always remain a mystery, the existence of such exquisite drawings suggests that artistic ability may have arisen very early in human development.

1996 World's smallest and lightest cellular phone is first produced by Motorola. This first really "pocket" cellular phone, called StarTAC, is about the size of a pager when closed. It opens to full size for use and weighs 3.1 oz (88 g). Its lithium-ion battery powers 2-3 hours of talk time and 22 hours of standby. It presently costs between $1,000 and $2,000.

Computers

c. 6000 B.C. Earliest example of a notched or tally stick is the Ishango bone found on the shore of Lake Edward in Zaire (Congo) and dated to this time. It is believed that this small, notched bone with a piece of quartz stuck in one end is a possible record of some activity based on the lunar cycle. If the Ishango bone is in fact a numerical record of something and not just a decorative or ritualistic piece, then traces of this type of recording system can be found as early as 30,000 B.C.

c. 2400 B.C. Babylonians first use an abacus and also are able to approximate pi as 3.125. In its earliest form, the abacus may have been merely a row of shallow grooves or lines traced in the ground, with pebbles or stones used as counters. The rows stand for units of tens or hundreds. The Semitic word for dust (abaq) is thought to be the root of the modern word abacus. Around 600 B.C., the Greeks call the flat surface or sand tray upon which they draw their calculating lines an "abax." This early mechanical computing device uses columns and counters. The Romans call this table an abacus. Following its Babylonian origins, the abacus makes its way to India and enters China around 1200 B.C., probably brought there by Arab traders.

c. 1200 The first minted "jetons" appear in Italy. These coin-like counters are used on an abacus table and become a very popular way of keeping count. Since they are cast, thrown, or pushed on an abacus table, they are named by the French after their verb to throw (jeter). Commoners have jetons made of copper while wealthier individuals have their jetons struck in silver with their coat of arms.

1617 "Napier's bones" are first described by Scottish mathematician John Napier (1550-1617) in his book *Rabdologia*. Following his invention of logarithmic tables, he invents several automatic calculating machines that use these exponential expressions. He calls one of these a "rabdologiae" in which he places his logarithmic tables on wooden cylinders, the surfaces of which contain numbers. By turning the correct cylinders (containing the digits 0-9) and adding or subtracting the numbers which appear, he obtains the correct result. Napier uses an ancient numerical scheme known as Arabian lattice and lays out his version of it on this set of four-sided rods. Essentially a multiplication table cut up into movable columns, it becomes fashionable and popular and is known also as numbering rods, speaking rods, and multiplying rulers. Napier's tool simplifies routine calculations and spurs others who use his invention to create their own more elaborate calculating devices. It is called Napier's bones because more elaborate versions use ivory or bone to replace the original wood cylinders.

1622   Slide rule is invented by William Oughtred (1575-1660), an English mathematician. Using the newly-invented logarithmic tables, he builds a device consisting of two sliding rulers marked with graduated scales representing logarithms. The user can calculate mechanically by sliding one ruler against the other, thus performing rough but rapid multiplication and division. This is the first linear or straight slide rule. His student, Richard Delamaine (also called Delamaine the Elder), independently invents a circular slide rule of his own design. Slide rules prove to be simple, versatile instruments used primarily by engineers. They perform reasonably accurate calculations quickly, and remain in use until replaced by the pocket calculator.

(See also 1971)

SEPTEMBER 20, 1623   First truly mechanical calculating machine or computer is built by German mathematician and astronomer Wilhelm Schickard (1592-1635). This day he writes a letter to his friend, German astronomer Johann Kepler (1571-1630), describing a wooden calculating machine he has built that can automatically perform all four mathematical functions (addition, subtraction, multiplication, and division). Starting with Napier's calculating device, he places it on cylinders that can be selected by the turn of a dial. Results are generated by turning large knobs, and the answers are displayed in small windows. He uses accumulator wheels to perform six-digit calculations. Because of plague, war, and a fire, Schickard's "calculator-clock" machines are destroyed and remain undiscovered, having no historical influence or impact.

1642   Calculating machine called "La Pascaline" is invented by French mathematician and physi-

cist Blaise Pascal (1623-1662). The 19-year old Pascal constructs a calculating machine that can add and subtract by means of cogged wheels, so as to give his father a much-needed accountant's tool. He uses a mechanism whose principle is similar to the modern odometer in that it contains a row of wheels with numbers on them connected by gears. The cogs and gears connecting the wheels are arranged so that one complete revolution of the dial furthest to the right causes the adjacent dial to turn only 1/10 of a revolution, and so on. Called "La Pascaline," it is conceptually more ambitious than the machine invented in 1623 by German mathematician and astronomer Wilhelm Schickard (1592-1635), but it does not work as well. Still, it becomes the first such device that is manufactured and sold for general use, and Pascal later presents one to Queen Christina of Sweden (1626-1689) in 1652. It becomes well-known and demonstrates publicly that an apparently intellectual activity like doing arithmetic can be performed by a machine.

1672   Samuel Morland (1625-1695), English inventor and diplomat, publishes *The Description and Use of Two Arithmetic Machines* in which he describes his simple adding machines and a more complex model that uses Napier's logarithms. He also designs a device that can count money, making this the first commercially sold calculator in Europe. His more complicated machine multiplies and divides and consists of a flat brass plate with a perforated, hinged gate, a series of circular disks with numbers engraved at the edges, and several semi-circular pins upon which the flat disks can be placed. The machine comes with 30 disks for ordinary multiplication and another five disks for finding square and cube roots.

1673   Idea of stored information is first introduced by German mathematician Gottfried Wilhelm Leibniz (1646-1716). His Leibniz calculator, also called the "stepped reckoner," is the most advanced machine of its time. Its crucial device is a special gear known as the

This calculating machine, named "La Pascaline," was designed for accountants by Blaise Pascal.

Leibniz wheel that acts as a mechanical multiplier. This gear is a metal cylinder with nine horizontal rows of teeth. The first row runs 1/10 the length of the cylinder; the second row, 2/10, etc. The Reckoner has eight of these stepped wheels all linked to a central shaft. Although highly sophisticated and far ahead of its time, it probably does not work well because of the limitations of machinery and the difficulty of producing small physical tolerances. Still, its stepped-drum gear remains the only working solution to the problems posed by calculating machines until about 1875.

MARCH 1679   First to recognize the importance of the binary system of notation (making use of only 1 and 0) is Gottfried Wilhelm Leibniz (1646-1716), a German mathematician and philosopher. He is also the first to develop a generalized exposition of the binary system of notation. This system becomes the basis of modern computer systems. Its importance to information theory and computer technology derives mainly from its convenience in representing systems that have a two-state nature—such as "on-off," "open-closed," or "go-no go." The modern information theory word "bit" is a shortened version of *bi*nary digi*t*.

1801   Jacquard loom is invented by French inventor Joseph Marie Jacquard (1752-1834). His automatic loom can perform the two critical textile-weaving functions of lifting the individual warp threads and "remembering" the pattern being weaved. It uses a system of hooks and needles to do the first and a collection of perforated cards to accomplish the second. His new device revolutionizes the textile industry but also shows the potential of punched cards. As the fabric is woven, the hooks are held stationary by the surface of the card, but when a punched hole is encountered, a hook is allowed to pass through to lift its thread. By stringing together a large number of cards, Jacquard can create an intricate pattern. He patents his invention in 1804, and its existence proves of great importance to the development of computers. His punched cards are the direct forerunners of the input and storage mediums used on early computers.

(See also 1834)

1822   Charles Babbage (1792-1871), English mathematician, partially builds the first prototype of his "difference engine." A difference engine is a machine capable of both storing a

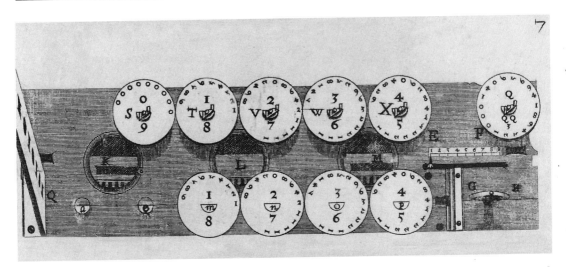

The disk version of Napier's Bones, a multiplying machine built by Samuel Morland.

series of numbers and performing additions with them. This small working model is a six-digit calculator made of toothed wheels and run by a hand crank. It calculates tables using the method of constant differences and prints out the results. This shows that his plans for a larger, steam-powered machine may work.

(See also 1834)

1834 First expression of the idea of an all-purpose programmable computer is attempted by Charles Babbage (1792-1871), English mathematician. His idea of an "Analytical Engine" can be considered the forerunner of the modern electronic computer. The plans for this programmable, automatic machine to replace his 1823, partially-built "Difference Engine" are exceedingly ambitious. Babbage envisions the device as having the capability of performing any mathematical operation, and being programmed by punched cards. It would also have a memory element in which to store numbers as well as other characteristics typical of a modern computer. However, the detailed blueprint plans call for a device the size of a locomotive. When the British government refuses to give him any further financial support,

he is forced to stop work after building only a few isolated components of the massive machine. Babbage is so far ahead of his time that he is severely limited by the state of technology which provides him only with mechanical means to achieve what we now know to be an electronic goal.

1843 First popular treatment of the basic principles of mechanical computing is published by Augusta Ada Byron, Countess of Lovelace and daughter of English poet Lord Byron (1788-1824). She translates an article on English mathematician Charles Babbage's (1792-1871) "Analytical Engine," written by Italian mathematician and military engineer Luigi F. Menabrea (1809-1896). As an amateur mathematician, she is able to add many pages of explanatory notes, and her translation, which is published in a popular scientific journal, is an astute analysis of the machine and of Babbage's ideas.

1847 Boolean algebra is first described by English mathematician and logician George Boole (1815-1864) in his pamphlet *The Mathematical Analysis of Logic*. In this and his 1854 work, *An Investigation of the Laws of Thought*, Boole establishes what is now called symbolic logic by

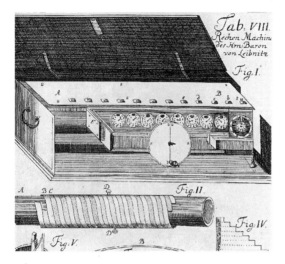

Gottfried Leibniz's calculator, which performed the four fundamental operations of arithmetic.

devising a system that can convert logic into mathematical symbols. He sets out logical axioms and chooses symbols and operations similar to algebraic notation to express those logical forms and perform operations. By wedding this to the binary number system (operations in Boolean algebra are AND, OR, and NOT, with objects either sharing or not sharing particular properties), he works out a way of reducing human reasoning to a series of yes or no choices. A century later, computer scientists are able to use Boolean algebra to work out programs of logical commands that are communicated to a computer through binary circuit operations.

(See also 1948)

1850 Modern logarithmic slide rule is invented by French mathematician Amedée Mannheim (1831-1906) and becomes the standardized calculator of choice. As a 19-year old artillery officer, Mannheim designs a very simple slide rule, adding a movable cursor. Although it is now considered an integral part of a slide rule, this is the first time that a movable cursor is combined with simple sliding logarithmic scales. This important addition enables fairly complex

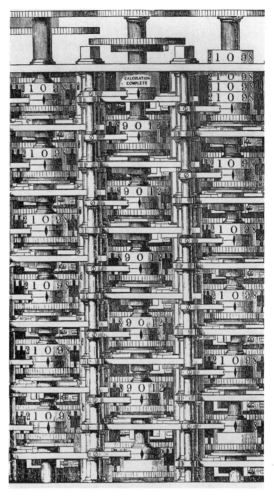

Charles Babbage's difference engine.

operations to be easily carried out. This version dominates mechanical calculation for the next century.

(See also 1971)

OCTOBER 1853 First printing calculator is designed by Pehr Georg Scheutz (1785-1873), Swedish publisher, and his son, Eduard. This year they build a difference engine that they call a tabulating machine. It is able to operate to the fourth order of difference, process 15-digit numbers, and print out results rounded off to eight digits. For its time, it is very fast and accurate and offers the first concrete demonstration of the enormous mathematical potential of machines. Openly based on a Babbage design, the machine wins a gold medal at the 1855 Paris Exhibition. Scheutz eventually sells a machine to an American and another to the British government, for whom it computes 600 different types of tables.

1873 Mechanical tide predictor is invented by Scottish mathematician and physicist William Thomson, Lord Kelvin (1824-1907). He constructs an elaborate machine with several pairs of lower gears that cause a series of upper gears to turn a proportionate amount. The upper gears then move rods up and down according to other gear ratios which in turn, raise or lower a pulley, resulting in a pen moving up or down on a paper chart. Tide computation works as long as one knows what gear ratios to use. This machine takes advantage of the earlier invention of the ball and disk integrator which is eventually used in several computing machines of this period.

1878 First direct multiplication machine is built by Ramon Verea (1838-1899), Spanish-American inventor. This year he receives a patent for a calculator that can perform direct multiplication (rather than repeatedly adding numbers), as well as division. His device is able to "look up" the product of two numbers on a pair of cylinders and then multiplies the inner gears accordingly. After receiving his patent, Verea states that he has no commercial ambitions for his machine, but built it only to demonstrate "that a Spaniard can invent as well as an American."

1884 First data processor is built by American inventor Herman Hollerith (1860-1929). This year he files the first of a series of patents for an electromechanical system that counts and sorts punched cards. His cards contain statistics which are run through a sorter which groups them into specific categories and then through a

tabulator, which counts the perforations and displays the totals. Although it does not calculate or compute, his machine can collate and add, making it a real data processor. It is now possible to count, collate, and analyze large amounts of information. He powers his machine with electricity.

(See also July 1890)

1884 First adding machine with a numeric keyboard and a built-in printer is constructed by American inventor William Seward Burroughs (1855-1898). Although he founds the American Arithmometer Company the next year and goes into production, his design cannot stand up to regular use. In 1892, he patents his redesigned and rebuilt keyboard calculator. This new machine prints out each figure entered, as well as the final results. It becomes a highly successful device that eventually outsells all other calculators on the market. Really an adder-subtractor, it can also multiply and divide by way of repeated additions and subtractions.

July 1890 First mechanized national census is achieved as American inventor Herman Hollerith (1860-1929) puts his electronic sorting and tabulating machines to work on the United States census. His reader machine has metal pins that pass through holes punched in cards the size of dollar bills, momentarily closing electric circuits. The resulting pulses advance counters assigned to details such as income and family size. His sorter machine can be programmed to pigeonhole cards according to patterns of holes, which is an important aid in analyzing census statistics. Hollerith's automatic sensing makes it possible to classify and count the data for the United States census of 1890 in one-third the time it had taken with the handwritten tally sheets used in the 1880 census. This marks the real beginning of modern data processing as his system is adopted worldwide. In 1911, Hollerith joins two companies to form the Computing-Tabulating-

Illustrations of Herman Hollerith's electronic tabulating and sorting machine.

Recording Company, which later becomes the International Business Machines Corporation (IBM).

1893 First practical multiplication-table calculator and the first to be commercially successful is "The Millionaire." Built by Otto Steiger, a Swiss engineer, it contains a mechanism that represents numerical products by racks of different lengths. His durable and practical calculator is immediately welcomed in businesses and universities throughout the world, and by 1935, over 4,500 machines are sold.

1897 Russia performs its first census using the electrical sorting and counting machines designed by the American inventor Herman Hollerith (1860-1929).

(See also July 1890)

NOVEMBER 16, 1904 First electron tube is invented by John Ambrose Fleming (1849-1945), an English physicist. After experimenting with the Edison effect (which is the passage of electricity from a hot filament to a cold plate with an evacuated tube), and finding that it is due to the newly discovered electrons that are boiling off the hot filament, he applies for a patent on this date for what is variously known as the Fleming valve, diode, kenotron, and thermionic tube. He calls this device a diode vacuum tube and observes that the diode acts like a valve since it turns on for current in one direction and off for current in the other. His valve or tube sees immediate application as rectifiers in radios and later in early electronic computers.

(See also 1906)

1906 Lee DeForest (1873-1961), American inventor, inserts a third element to Fleming's diode vacuum tube and invents the triode vacuum tube. This allows the electrical signal to be amplified for the first time, making radios and other electronic equipment (like computers) practical by magnifying weak signals without distortion. The triode becomes the basis of the familiar radio tube, making radio a practical, effective medium. Further, its binary function (on and off capability) makes this tube invaluable to the designers of early electronic digital computers.

(See also 1904)

1911 Leonardo Torres y Quevedo (1852-1936), Spanish mathematician and engineer, demonstrates his design for the first chess-playing device that is completely automatic. It employs a Boolean network of electromagnetic relays and is one of the earliest attempts to imitate complex human thought processes.

1914 Thomas John Watson (1874-1956), American business executive, becomes head of the Computing-Tabulating-Recording Company and is one of the first to see the enormous potential of punch card machines. CTR leases instead of sells its punch card equipment, and once large companies install their machines, it is impossible to return to the old methods. CTR also sells the punch cards to their customers (who are required to buy from CTR). Watson instills his vision in his salesmen and engi-

neers, and in 1924, founds the International Business Machines Corporation (IBM).

1918 American Telephone and Telegraph (AT&T) establishes the first commercial service of a five-channel phone line between Baltimore and Pittsburgh. Their technique of frequency-division multiplexing which permits a single communications link to carry multiple conversations or streams of data simultaneously will eventually prove indispensable for transferring data between computers.

1919 William Henry Eccles, English physicist, and F. W. Jordan publish one of the first papers on flip-flop circuits (multiple change of poles) and demonstrate how to connect two triodes in a circuit which has two stable states. This is an essential advance toward an electronic computer.

1920 Word "robot" is first coined by Karel Čapek (1890-1938), Bohemian novelist and playwright, in his play *R.U.R. (Rossum's Universal Robots)*. In this story, a humanlike machine is created that is more precise and reliable than humans themselves, and threatens all with extinction. The machine's name, "robot," is taken from the Czech word "robota" for servitude or forced labor, and becomes an international word for an automaton. The term "robotics" is first used by American science fiction writer Isaac Asimov to mean the study of the construction, maintenance, and use of robots. Technical advances in computers and cybernetics lead to the increased use of industrial robots, and future robots may be equipped with microprocessors and specialized sensors.

1923 Moore School of Electrical Engineering is first established at the University of Pennsylvania. Founded with a bequest from the wealthy Philadelphia industrialist Albert Fitler Moore, it becomes instrumental in the early development of the all-electronic digital computer.

JANUARY 1, 1925 Bell Telephone Laboratories is first established formally as a consolidation of various research activities within American Telephone and Telegraph (AT&T). Bell Labs becomes an important player in the early development of computer systems.

1928 First cathode ray tube (CRT) is patented by Vladimir Kosma Zworykin (1889-1982), Russian-American physicist. It proves to be vitally important to the development of the computer, besides its primary significance to television. These same, large bottle-shaped CRTs that will be used as picture tubes for television sets and in radar to display data will be used in the early electronic computers. They will function as memory tubes with "guns" in the bases of the tubes, shooting positively and negatively charged electrons at the faces of the tubes, thus storing bits in the form of charge spots that are visible to the eye.

1928 First to apply Hollerith machines to the physical sciences is Leslie John Comrie (1893-1950), New Zealand-English mathematician and astronomer, who prepares tables of the moon's position. This marks an important transition point, with punch card equipment being used for advanced scientific purposes and not only for statistics and business.

1928 Eighty-column punch card is first adopted by International Business Machines (IBM), effectively setting the standard that lasts for 60 years. Data or programs are recorded as patterns of holes at the intersections of 80 columns and 12 rows. They are retrieved in one of two ways: by shining a light through the openings or by penetrating them with fine bristles. This is very much the same method that the census-tabulating equipment devised by American inventor Herman Hollerith (1860-1929) functioned nearly 40 years earlier.

1930 First analog computer is built by Vannevar Bush (1890-1974), American electrical engineer. Called a differential analyzer, his purely mechanical machine is designed to speed the solution of complicated differential equations which take months to work out manually. Re-

sembling a huge erector set, this difficult-to-use, semiautomatic calculator has impressive computational power. Still, Bush realizes that analog devices are not well-suited for accurate, versatile computing, and he later builds a much faster electromechanical version that uses vacuum tubes.

1931 First *Nautical Almanac* is printed that uses calculations obtained by machines.

1932 Bell Laboratories uses a computer-like device to reproduce music for the first time.

1933 Totalisator or "tote machine" that uses electromagnetic relays is installed at the first racetrack in the United States. Invented by an American electrical engineer, Henry L. Straus, this machine that calculates betting odds and displays payoffs to bettors was used at English racetracks before making its appearance at Arlington Park near Chicago, Illinois. Consisting of many relays, it is really a special-purpose electromechanical calculator.

1933 First to use an automatic calculator for problems in atomic theory is English physicist Douglas Rayner Hartree (1897-1958). He uses a differential analyzer made by American electrical engineer Vannevar Bush (1890-1974).

1934 Harold Locke Hazen, American engineer, publishes his classic paper, "Theory of Servo-Mechanisms," which is the first to treat the notion of automatic control systematically. A servomechanism is an automatic device that is used to correct the performance of a mechanism by means of an error-sensing feedback. A year earlier, Hazen's design of a light-sensitive servomechanism began the development of devices that can use an instrument's output to control a process automatically.

1934 Konrad Zuse, German engineer, makes the important discovery that an automatic calculator requires only three basic units: a control, a memory, and a calculator for the arithmetic. He discovers this after realizing that once one has the instructions coded for the control mechanism, there is no longer any need for any calculating-plan forms, and the process becomes only a series of boxes acting like a memory. He soon puts this discovery into practice.

(See also 1938)

1935   IBM 601 multiplying card punch first begins production. These machines form the backbone of most scientific and commercial calculation until the advent of the electronic computer.

1937   Concept of a "Turing machine" is first proposed by English mathematician Alan Mathison Turing (1912-1954) in his seminal paper titled "On Computable Numbers." Turing offers the specifications for a hypothetical universal computer (Turing machine) that can solve all problems that are computable. His idea that a machine can carry out any solvable mathematical computation by following a series of instructions contributes mightily to the development of the theoretical basis of the computer.

1937   George Robert Stibitz, American mathematician, experiments with having relays do calculations at Bell Laboratories, which begins a relay computer project. His use of cast-off telephone relays, old batteries, flashlight bulbs, and strips cut from a tin can produces the first electrical binary adder. Stibitz is one of the first to realize that simple relays—on-off electromechanical switches—could also be used to perform symbolic logic. This leads to his design of the Complex Number Calculator.

(See also January 8, 1940)

FEBRUARY 13, 1938   First American crossbar central office (connecting two telephones in direct dialing) opens in Brooklyn, New York. Crossbars and relays are at this time being used by Bell Laboratories in a computer project.

1938   First binary calculating machine is built by

Konrad Zuse, German engineer. This year he completes construction of the Z1, an entirely mechanical device whose binary instructions are punched on movie film. For memory, the Z1 uses slotted metal plates with pin positions in the slots representing 0 or 1. This early machine leads Zuse to build others.

(See also December 5, 1941)

1938   First fully electronic analog computer is built by George A. Philbrick of the United States. He substitutes electronics for the mechanical shafts of American electrical engineer Vannevar Bush (1890-1974) and invents what he calls "Polyphemus, The Computing Cyclops" to take the drudgery out of differential equations. Since the results are displayed on the circular screen of an oscilloscope, he names his creation after the one-eyed character of Greek mythology.

JANUARY 8, 1940   Complex Number Calculator built at Bell Labs becomes operational. This machine looks like a large desktop calculator and is the first machine to service more than one terminal, as well as the first to be used from a remote location (via telephone). Composed mainly of standard telephone relays, it can add, subtract, multiply, and divide complex numbers. On September 11, 1940, the Calculator is demonstrated before the American Mathematical Association via a remote terminal. This is the first example of remote job entry (RJE).

DECEMBER 5, 1941   First fully operational, general-purpose calculator with automatic control of its operations is built by Konrad Zuse, German engineer. Called the Z3, his machine features floating point arithmetic (for varying decimals) and uses 2,600 relays. It also has some memory, storing 64 twenty-two bit numbers and is fairly fast. Because of the war, Zuse works in isolation of any developments going on anywhere else, and is ignorant of the earlier work of English mathematician Charles Babbage (1792-1871) as well. His Z3 is destroyed during an air raid in 1944.

1941   First flight simulator is perfected by Helmut Hoelzer of Germany. During the development of the V-2 rocket, he builds a simulator using an analog computer that can generate voltages imitating those sent to the rocket's guidance system as a result of such external influences as wind velocity. The guidance system in turn produces flight-control voltages that can be analyzed to determine if they would have corrected the rocket's course.

1942   John Vincent Atanasoff, American physicist, and Clifford E. Berry construct a prototype of what many regard to be the first electronic digital computer. It is also the first electronic computing device to use binary arithmetic. They use vacuum tubes (for its logic) and capacitators (for its memory) arranged on a drum. It is fed information on punch cards. Although it never becomes operational, a federal judge in 1974 voids the patent granted to ENIAC (Electronic Numerical Integrator and Computor) and declares Atanasoff and Berry the inventors of the modern computer.

JANUARY 1943   Harvard Mark I becomes operational at IBM's laboratory in New York. It is 51 ft (16 m) long and contains 3,304 electromechanical switches able to add or subtract 23 digit numbers in three-tenths of a second. It is also the first American calculating device to offer the convenience of loading instructions from punched paper tape and the first digital computer in the United States to work from a program and produce reliable results. Also called the Automatic Sequence Control Calculator (ASCC), it is designed and built by a team led by American mathematician Howard Hathaway Aiken (1900-1973). It is succeeded by the Harvard Mark II which becomes operational July 1947. Although larger and more sophisticated than Mark, it becomes outmoded by the capabilities of ENIAC (Electronic Numerical Integrator and Computer).

1943   William Bradford Shockley (1910-1989),

American physicist, devises the acoustic delay line which is the first type of memory fast enough to keep up with a vacuum-tube electronic computer. The delay line briefly stores data by converting it from electrical impulses to much slower sound waves passing through a tube of mercury. Improved versions of the delay line can hold as many as 1,000 bits of data for increments of 1/1000 of a second, until the computer is ready for it.

1943   Project Whirlwind, one of the most innovative and influential computer projects in computer history, is first begun at the Massachusetts Institute of Technology under United States Navy sponsorship. It begins as a feasibility study for a general-purpose flight trainer or simulator. It will evolve into a project to design a digital computer following a 1945 decision to convert from analog electronics to digital electronics.

(See also March 1951)

SEPTEMBER 1944   John von Neumann (1903-1957), Hungarian-American mathematician, visits the Moore School of Engineering for the first time and inspects ENIAC (Electronic Numerical Integrator and Computor). He later will give the computer its first real job which has to do with a large and complex calculation of the feasibility of a proposed design for the H-bomb.

1945   First full description of the basic details of a stored-program (memory) computer is offered by John von Neumann (1903-1957), Hungarian-American mathematician. In his now-famous memorandum, "First Draft on a Report on the EDVAC," he incorporates ideas from other computer giants and proposes his basic scheme for computers. His plan calls for a central arithmetic logic unit, a central control unit, provisions for input and output, and most important, a memory for storing programs. A key to his stored-program concept is his suggestion that codes be treated as numerals that

The Mark I, created by a team lead by Howard Aiken, is regarded as the first general-purpose computer.

can be stored electronically just as data numerals are stored, thus eliminating special instruction wiring. This and his co-authored paper of 1946 which offers comprehensive designs for a parallel stored-program computer are a substantial departure from anything yet proposed and are considered a landmark in the history of the computer sciences.

(See also 1946)

FEBRUARY 16, 1946 ENIAC (Electronic Numerical Integrator and Computor) is dedicated and becomes fully operational. Until the Atanasoff-Berry legal ruling in 1974, this is considered the first electronic digital computer. Despite the ruling that gives American physicist John Atanasoff priority, this is still the first large, general-purpose electronic computer. Using vacuum tubes rather than electromechanical switches, it is twice as large as the Harvard Mark I, but also a thousand times faster. Still, its creators realize it is obsolete, lacking internal storage of programs. Since its programs are wired into the circuitry, to change programs requires a technician to unplug and replug hundreds of wires, a task that can take two days. ENIAC is used to solve H-bomb problems

and then begins steady work calculating artillery-aiming tables for the army.

(See also 1942)

1946 Most influential paper in the history of computer science first becomes available as what is known as the "Princeton Reports." Written by Hungarian-American mathematician John von Neumann (1903-1957) in collaboration with American colleagues Arthur Walter Burks and Herman Heine Goldstine, this paper details what comes to be known as "the von Neumann machine," and provides the foundation for essentially all future computer development. Titled "Preliminary Discussion of the Logical Design of an Electronic Computing Instrument," it defines the stored-program concept and uses "flow diagrams" now known as flow charts.

1946 First high-level language for computers is invented by Konrad Zuse, German engineer. Called "Plankalkul," it can be used for both numerical and non-numerical problems. He formulates its syntax in part by writing a chess-playing program. Although it promises to simplify the writing of software by giving the programmer easier instructions to interpret and

The first electronic computer, ENIAC.

input (compared to long strings of ones and zeros), it is not really known until many years later.

1947　First transistor is invented by American physicists William Bradford Shockley (1910-1989), John Bardeen (1908-1992), and Walter Houser Brattain (1902-1987). While investigating semiconducting materials as replacements for vacuum tubes, this team also seeks to control both the direction and the amplification of an electric current. They conduct several ground-breaking experiments and mount a germanium oxide semiconductor between two fine metal wires which serve as metallic contacts through which the current flows. They succeed in creating the first "point-contact resistor" or "transfer resistor", which becomes known as the transistor and starts the semiconductor revolution. Improved transistors would eventually replace the big, hot, power-hungry vacuum tubes used in ENIAC and other early computers. This is one of the most important developments in electronics in the twentieth century.

1947　First practical, truly random-access memory (RAM) is invented by English electrical

engineer Frederic Calland Williams. His Williams tube makes possible the retrieval of any item of stored data or program and comes is used in first-generation digital computers throughout the world. To built it , he modifies a cathode ray tube (CRT), and an electron beam inside the tube scans successive lines across its face, "painting" dots and dashes of phosphorescent electrical charge on the screen to represent binary ones and zeros.

JANUARY 1948　First random-access, magnetic storage device for computers is built by Andrew Donald Booth, English physicist. His magnetic drum storage system holds far more data than Williams tubes but is very slow. The drum registers information as magnetic pulses in tracks around a metal cylinder by using a row of read/write heads to both record and recover data. Since it is too slow to be used as memory, it finds applications as a storage device.

JANUARY 1948　SSEC (Selective Sequence Electronic Calculator) is first introduced by International Business Machines Corporation (IBM). A hybrid machine with more than 20,000 relays and 12,500 vacuum tubes, it can handle large problems more rapidly than any of its predecessors. Operation sequence is controlled by means of instructions stored in a variety of memories (vacuum tubes and relays as well as punched paper tape). Although it is operational for only four years, it produces moon-position tables that are used in 1969 to plot the course of *Apollo 9* to the Moon.

1948 "Cybernetics" is first coined by American mathematician Norbert Wiener (1894-1964). This year he publishes his book, *Cybernetics: or, Control and Communication in the Animal and the Machine*, which founds the science of cybernetics (derived from the Greek word for "steersman"). Interdisciplinary in nature and based on common relationships between men and machines, it is very popular in its time and is a major influence on later artificial intelligence research. He contends in his book that any machine with pretensions to intelligence must be able to learn or modify its behavior in response to feedback from its environment. Cybernetics is eventually used in control theory, automation theory, and computer programs.

1948 Information theory is first detailed by Claude Elwood Shannon, American mathematician. He publishes *The Mathematical Theory of Communications* in which he attempts to show communications engineers how to encode data so that it can be checked for accuracy following transmission between computers. His solution first offers details on information theory and sheds light on the problem of conveying information so that it is not misinterpreted. This becomes a matter central to the field of artificial intelligence. Shannon's great achievement is his ability to apply the binary nature of Boolean algebra (symbolic logic) to relay circuits which are either open or closed.

(See also 1847)

MAY 6, 1949 First practical, fully electronic, stored-program computer goes into operation. Designed by English mathematician, physicist, and engineer Maurice Vincent Wilkes, EDSAC (Electronic Delay Storage Automatic Computer) is built at the University of Cambridge (England) using readily available technology. His machine has a mercury delay line memory that can hold 512 34-bit words. For programming, however, a library of short programs called subroutines are used that are stored on punched paper tapes.

SEPTEMBER 1949 BINAC (Binary Automatic Computer), the first stored-program computer built in the United States, is tested and delivered to Northrop Aviation. Built by American electrical engineers John Presper Eckert, Jr., (1913-1995) and John William Mauchley (1907-1980), it is intended for use on a guided missile program, and is relatively small and fast. It borrows its technology from tape recorders, and like most computers of this era, is not very reliable. To obtain accuracy, BINAC consists of two complete computers, both performing the same calculations and comparing the results.

1949 "Short code" is invented by American electrical engineer John William Mauchley (1907-1980), who attempts to simplify the writing of equations for BINAC (Binary Automatic Computer). In order to represent mathematical symbols like the plus sign or parentheses, his Short Code substitutes pairs of digits in place of the long strings of zeros and ones that computers normally read. This enables algebraic equations to be written in terms that bear a one-to-one correspondence to the original equations. This method constitutes a major step toward high-level programming languages.

1950 First commercially produced computer in the United States to offer magnetic drum-storage technology is the ERA 1101. Built by Engineering Research Associates and originally known as the Atlas, the computer has a capacity of two million bits, offering a large amount of slow memory at relatively low cost.

1950 First practical stored-program computer to operate routinely in the United States in the SEAC. Built by the National Bureau of Standards, SEAC (Standards Eastern Automatic Computer) is based on the EDVAC and uses mercury delay lines for storage. It begins input and output with punched paper tape and later switches to magnetic tape and a Williams tube memory.

1950   First criterion for judging artificial intelligence is offered by English mathematician and logician Alan Mathison Turing (1912-1954). In his paper "Computing Machinery and Intelligence," Turing explores the issue of whether or not a computer can "think." He describes a test by which one can determine whether a computer is "intelligent." He argues that if a person cannot tell whether a problem is being solved by a computer or another person, then the computer should be judged "intelligent." Turing holds that human thought is computation, and from that perspective, intelligent machines are possible.

MARCH 31, 1951   United States Census Bureau takes delivery of the first UNIVAC (Universal Automatic Computer). Invented by American electrical engineers John William Mauchley (1907-1980) and John Presper Eckert, Jr. (1913-1995), it is the first commercial computer and is sold by Remington Rand, the first large-scale seller of computers. It is also the first widely available commercial computer to use stored programs. Mauchley and Eckert design a real computer system or a family of related machines that gives businesses a data-processing system that suits their growing needs. A year after the first UNIVAC installation it is used to predict the results of the 1952 presidential election.

MARCH 1951   Project Whirlwind first becomes operational under American electrical engineer Jay Wright Forrester at the Massachusetts Institute of Technology. Designed for high speed and power and intended for air traffic control, it is the largest digital computer of its time doing real-time computing (it occupies the largest

Early UNIVAC mainframe computer.

floor-area of all the early computers). It processes data and presents the results almost instantaneously on a video screen. An operator can then direct subsequent actions by touching a light gun to the screen. It is intended to perform a broad range of tasks that require direct interaction between operator and machine.

1951   First language compiler is produced by Grace Brewster Murray Hopper (1906-1992), American mathematician, who develops the concept of the compiler programming languages (also known as assembly language or assembler) for the new UNIVAC computer. She also coins the word "compiler" for her A-O and later A-2 programming system. She calls it a compiler because the software combines or compiles instruction sequences to create a program. Until this time, no adequate software was available to translate a programmer's high-level, problem-oriented language into machine-readable instructions. This invention of the first automatic coding system in the United States leads to Flow-Matic, a more advanced and commercially viable English-language compiler.

(See also 1955)

1951 Australia completes the construction of its first stored-program computer, CSIRAC.

1951 EDVAC (Electronic Discrete Variable Automatic Computer) first begins operation. This long-delayed successor to ENIAC encounters many problems in its development and has already seen the British EDSAC (Electronic Delay Storage Automatic Computer) become the first fully functional stored-program computer. Designed for the United States Army, it becomes the first computer that we now call a "von Neumann machine." Designed by Hungarian-American mathematician John von Neumann (1903-1957), American mathematician Herman Goldstine, and American engineer Arthur Burks, EDVAC keeps both its program and its data in its memory which is its most advanced feature.

1951 First computer built specifically for business use, LEO I becomes fully operational. Built in England, LEO (Lyons Electronic Office) is able to calculate taxes and payrolls, as well as calculate the optimum mix for their brands of tea. Its speed is suggested by the fact that a complete calculation of an employee's wages and deductions takes about 1.5 seconds, as compared to about 8 minutes by an experienced human clerk. It remains in service for well over a decade.

1951 Junction transistor is invented by American physicist William Bradford Shockley (1910-1989). Simpler to build and more reliable than the old, point-contact method, this is a much more versatile type of transistor and is much easier to mass produce. In Shockley's junction, or bipolar, transistor, tiny amounts of elements such as phosphorous or boron are added to the semiconductor to alter its electrical properties. Although it will be several years until the junction transistor replaces vacuum tubes, it will eventually revolutionize electronics.

1951 The Computer Group of the Institute of Radio Engineers is first established. In 1972, it

is renamed the IEEE (Institute of Electrical and Electronics Engineers) Computer Society.

MAY 1952 International Business Machines Corporation (IBM) announces the IBM 701 as its first full-function, fully-electronic digital computer. It is a parallel binary machine with 4,000 tubes that relies primarily on punch cards for input and output, but can also operate high-speed printers and store data on magnetic tape. A total of 19 are bought by aircraft companies, universities, atomic research laboratories, large corporations, and the United States Navy, at a monthly rental fee of $15,000.

1952 Integrated circuit idea is first proposed publicly by G.W.A. Dummer, English electrical engineer. An expert in radar, he suggests at a Washington, DC symposium that transistors, resistors, and other components could be combined into a single semiconductor block he calls "wireless electrical circuits" that would be smaller, faster, and more reliable than separately wired parts. His concept of the miniaturization of one solid unit will begin a revolution even more far-reaching than the invention of the transistor.

(See also September 1958)

1952 First fault-tolerant computer is built by Czech electrical engineer Antonin Svoboda (1907-1980), who develops the SAPO in Czechoslovakia. Fault-tolerant computing is the art of building computing systems that continue to operate satisfactorily in the presence of faults (hardware or software). Two general approaches to fault recovery are available: one is fault masking, using redundant systems; another is dynamic recovery, involving automated self-repair. The SAPO uses the first method. It is also the first large, automatic digital computer built in Czechoslovakia.

1952 Autocode, the first working high-level language, is developed by Alick Glennie, English nuclear researcher. Programming languages are described as low-level, high-level, higher-level, English-like, and natural in the order of their distance from machine language. The early and popular high-level languages, like FORTRAN and COBOL, are special-purpose languages useful for certain types of processing. They allow a programmer to write a computer program with relative speed, accuracy, and ease.

1953   Magnetic core memory makes its first appearance and significantly improves the performance of the Whirlwind computer. Developed by American electrical engineer Jay Wright Forrester, it will prove to be the first real system to offer a basis for a large-scale memory at reasonable cost. Forrester uses ceramic ferrite shaped into small, spiral rings arranged on a grid of wires. Four years after conceiving the idea, Forrester's new memory system is able to double operating speed, quadruple input data rate, and reduce memory bank maintenance from four hours a day to two hours per week.

1953   Maurice Vincent Wilkes, English mathematician, first recommends the use of microprogramming. This innovative idea allows frequently used machine-level commands to be housed in the computer rather than loaded each time. Initially used for engineering convenience, it is now a reality in every computer.

1953   First truly practical compilers and high-level languages are invented by J. Halcombe Laning and Niel Zierler of the United States for Project Whirlwind. Their system uses ordinary words such as "stop" and "print" as well as equations in their natural form. It also performs most housekeeping functions automatically. However, so much machine time is required to translate them back into machine language, that it is rarely used.

1953   First commercial computer to use interrupts is the ERA 1103 computer, built by Engineering Research Associates. The capability to interrupt a program, now a feature of modern computer systems, permits a computer to respond quickly to exceptional events. This

machine is geared to scientific and engineering customers, and in 1954 replaces its Williams tube memory, becoming the first commercial computer to have a magnetic core memory.

1953   Speedcoding is invented by American programmer John Backus. Backus writes this early high-level language for the IBM 701 computer. Although its programs occupy more memory and require 15 times as much processing time as the software code in the 701's assembly language, Speedcoding has long-term advantages since it is also able to trim several weeks from a programming assignment. This new language will lead directly to Backus' development of FORTRAN (Formula Translation) software.

(See also 1957)

DECEMBER 1954   First mass-produced computer, the IBM 650 magnetic drum calculator, becomes available. Although IBM is unsure that there will be any market at all for this inexpensive, practical, and reliable medium-size model, it goes ahead and builds 250 of them. Easy to program and operate, with its magnetic data-storage drum allowing fast access, it proves a runaway success with 450 selling in its inaugural year. It also incorporates error-checking procedures that halt the machine if data is incorrectly read from the drum. It becomes a sturdy workhorse—the Model T of the computer industry—with 1,500 being manufactured before phaseout in 1969.

1954   Antonin Svoboda (1907-1980), Czech electrical engineer, builds the Soviet Union's first fault-tolerant computer.

(See also 1952)

1954   First commercially available silicon transistor is produced by an obscure electronics firm named Texas Instruments. Also called a junction transistor, it improves the original germanium version of the device. Because it is made of silicon, it offers much greater reliability at high temperatures and can drop the price of a transistor from $15.00 to $2.50. Texas Instruments puts them in a popular, rugged portable radio.

1955 First heuristic computer program is designed by American behavioral scientist Herbert Alexander Simon and American computer scientist Allen Newell (1927-1992). They unveil their "Logic Theorist," a software package able to apply rules of reasoning and to prove theorems in symbolic logic. Heuristic programming is required whenever an algorithmic or strictly quantitative solution is not available. This program is considered a milestone in the field of artificial intelligence.

1955 First business-oriented high-level language, "Flow-matic," is developed by Grace Brewster Murray Hopper (1906-1992), American mathematician. Also called B-O, it becomes one of the main inputs in influencing the early development of a common business language.

(See also 1959)

1956 TX-O mainframe is one of the first general-purpose, programmable computers to be built with transistors. Spawned by the Whirlwind Project, it is designed at the Massachusetts Institute of Technology's Lincoln Lab. This experimental computer hosts many imaginative feats of programming by students and computer enthusiasts.

1956 International Business Machines Corporation (IBM) begins the era of magnetic-disk storage with its RAMAC 305 computer which has the first disk drive, called the IBM 305 disk file. The storage component of the Random Access Method of Accounting and Control (RAMAC) can hold 5 million bytes of data. It consists of 50 magnetically coated metal platters, stacked one atop the other and rotated by a common drive shaft.

1956 First operating system is developed cooperatively by General Motors and North American Aviation. An operating system is the software (programs and data) that initiates the interaction of the electronic and electromechanical components of a computer so that they make up a useful system for carrying out calculations. Called the GM-NAA I/O system, this system uses a group of supervisory commands to syncopate and accelerate the processing of individual jobs in an IBM 704 mainframe by performing each task in three phases: input, execution, and output. Because the system enables a number of jobs to be carried out in a continuous sequence, it popularizes the technique known as batch processing.

1956 First commercial computers to use magnetic core storage are the IBM 704 and the UNIVAC 1103.

1956 Term "artificial intelligence" is first coined by American mathematician John McCarthy. Defined to mean the creation of artifacts that exhibit intelligent behavior, artificial intelligence (AI) is a name given by McCarthy for the proposal to obtain funding for a conference on the subject. Organized by McCarthy and his colleague, American electrical engineer Marvin Lee Minsky, the conference held in the summer at Dartmouth College marks the birth of the field of AI.

1956 Information Processing Languages (IPLs) is first developed by American computer scientist Allen Newell (1927-1992) and colleagues. This family of computer languages opens new frontiers for artificial intelligence. Besides permitting the use of list processing (in which data is represented as lists of words, phrases, and other symbols), the languages make possible a technique called associative memory. This allows programmers to store concepts in a computer's memory as they believe the human brain may be organized.

1957 FORTRAN (Formula Translation) is first introduced by International Business Machines Corporation (IBM). As the programming language most widely used by scientists and engineers in the 1970s, FORTRAN enables a computer to perform repetitive tasks from a single set of instructions. Created under the leadership of American programmer John Backus, FORTRAN soon becomes the standard high-level programming language for technical projects.

1957 First "light-pen" word processing system, the TX-2, is developed at the Massachusetts Institute of Technology. Foreshadowing the era of word processing, this computer system is equipped with unique text-handling utilities and offers a system that allows software writers to "type" words into their programs by pointing a light pen at characters on a computer-generated display of the alphabet. The text handler also provides for tab and margin settings, erasing characters, and moving words or entire paragraphs.

1957 AI Lab is first established at the Massachusetts Institute of Technology. Created by the founders of artificial intelligence (AI), American mathematician John McCarthy and American electrical engineer Marvin Lee Minsky, this laboratory will be the birthplace of McCarthy's LISP programming language (1960) as well as the site of Minsky's highly influential description of how the human brain might organize knowledge.

1957 System Dynamics software is invented by American electrical engineer Jay Wright Forrester. This new computer-modeling strategy is quite different from its predecessors, for it does not predict the future of a complex system by utilizing its historical performance. Rather, it is able to predict by building a system model and then testing its accuracy against historical fact.

1957 Control Data Corporation is first established. Co-founded by William C. Norris, this Minnesota-based corporation is the first computer company to be publicly financed. Initial capitalization is accomplished through the sale of 600,000 shares of common stock priced at $1 per share.

JULY 1958 First SAGE (Semi-Automatic Ground Environment) center becomes operational. This U.S. air defense system incorporates the first production-model computer with built-in graphics. It also is the first large-scale communications network to serve computers, linking hundreds of radar stations to regional command centers. Over the next five years, 26 other centers are built with each containing a SAGE computer. Consisting of 55,000 tubes—the most of any computer before or since—and weighing 250 tons, a SAGE computer can run 50 monitors, track 400 airplanes, and communicate with other information sources to give a single, integrated picture of what is going on in the sky.

SEPTEMBER 1958 First working integrated circuit is built by Jack St. Clair Kilby, American electrical engineer. He proves that resistors, capacitors and tiny transistors can be made simultaneously on a single piece of semiconductor material. In 1959, Kilby's patent adds a crucial new element by making the connections directly on the insulating wire of the semiconductor chip, thus using no wires at all. The miniaturization achieved by this one solid unit will lead eventually to the mass-production of computer chips.

(See also 1959)

1958 Input/output capability is first introduced commercially on the IBM 790. These small, special-purpose processors work independently from a computer's central processing unit (CPU) to speedily store and retrieve data. These processors can relieve the CPU of most input/output tasks.

1958 Daniel Delbert McCracken of the United States writes the first FORTRAN (Formula Translation) textbook.

(See also 1957)

1958 First planar transistor is developed by Jean Hoerni, Swiss-American physicist. His planar process for manufacturing transistors is more resistant to damage and contamination than original integrated circuits and therefore produces a more reliable product. He diffuses the mesa or tiny, round plateau into the wafer itself, effectively embedding the transistor's various parts chemically into a piece of silicon, and the result is a completely flat transistor without any protruding parts. His design also relies on thin coatings of silicon dioxide. Hoerni's planar process is a major technical breakthrough and leads directly to a commercially feasible integrated chip.

1958 International Business Machines Corporation (IBM) introduces its second-generation computers, the IBM 7000-series mainframes. These mainframes are the company's first transistorized computers and are five times faster than their vacuum-tube predecessors. The 7090 becomes the workhorses of the scientific computing of the early 1960's, and IBM sells or leases more than 400 of these $3 million machines.

1958 Concept of block structure is first introduced in the high-level computer language ALGOL. Devised by a committee of computer scientists from Europe and the United States (working for the International Federation for Information Processing), ALGOL is a general-purpose programming language that is suitable for communicating algorithms, executing them efficiently on different computers, and teaching computer science. Block structure is the technique of dividing programs into self-contained units. It becomes very popular in Europe but not in America where FORTRAN (Formula Translation) is well-established.

JANUARY 1, 1959 Notion of time-sharing is first proposed by American mathematician John McCarthy. As a technique of organizing a computer so that several users can interact with it simultaneously, it proves to be an excellent way of efficiently using expensive equipment. Time-sharing software enables a mainframe computer to run several programs at once,

attending to each for a few milliseconds in a cyclic manner. Taking advantage of a mainframe's greatest asset, speed, this technique allows several terminals to be linked to the mainframe which then executes a small portion of each user's program so quickly that the human user is unaware that sharing is taking place.

(See also 1961)

JUNE 1959 John Backus, American programmer, first describes the Backus Normal Form (BNF). This set of rules and definitions provides much-needed order to what is becoming a babel of computer languages. This standard method of describing a programming language's grammar is later renamed Backus-Naur Form to give due credit to Danish astronomer Peter Naur.

1959 First major description of COBOL (Common Business-Oriented Language) is completed. Written and developed by American mathematician Grace Brewster Murray Hopper (1906-1992), COBOL uses word commands (English language words, phrases, and statements) instead of mathematical symbols, and significantly simplifies the training and knowledge needed to run the program. It excels at the most common types of data processing for business and becomes available the following year. This language becomes the most widely used programming language for commercial applications in the 1970's and contributes to the widespread use of computers in business, governmental, and science applications.

1959 First international conference on information processing is held in Paris, France, and sponsored by the United Nations Educational, Scientific, and Cultural Organization (UNESCO).

1959 First practical integrated circuit is designed by Robert Norton Noyce (1927-1990), American physicist. Using the basic plan of American electrical engineer Jack St. Clair Kilby and incorporating Swiss-American physicist

Jean Hoerni's planar transistor, Noyce uses a flat transistor to replace connecting wires as well as conducting channels printed directly on the silicon surface. The first "chip" is born and will result in a miniaturized, mass-produced product that is reliable, cheap, and very powerful. It becomes the basic building block of the next generation of computers and eventually makes the development of the microprocessor and the personal computer possible. These early chips contain only a few individual circuits, but will eventually carry more than a million.

1959 First major programming language for commercial applications, Commercial Translator, becomes available from IBM.

1959 First computer to use a paged virtual memory is the ATLAS computer of the University of Manchester (England). Designed by Thomas M. Kilburn of England, this method of providing very large amounts of memory has a real core memory of only 16,000 words and a drum of an additional 9,600 words that automatically swaps information with the core memory as required.

1959 ERMA (Electronic Recording Method of Accounting) is developed as a computerized bookkeeping system that allows computers to read numbers for the first time. In addition to magnetic drum and magnetic tape storage, a check sorter, and a high speed printer, it also includes a special scanner to read account numbers preprinted on checks in magnetic ink.

1959 Computer-Assisted Manufacturing (CAM) is first exhibited at the Servomechanisms Laboratory of the Massachusetts Institute of Technology. In this demonstration project, a milling machine is operated by means of instructions written in a language called APT (Automatically Programmed Tools). This computer-controlled machine produces an aluminum ashtray for each attendee.

1960 LISP, the first computer language de-

signed expressly for writing programs in the field of artificial intelligence, becomes available. Created by American mathematician John McCarthy, LISP (short for list processing) offers programmers great flexibility in organizing their software and is a great help in trying to get computers to mimic the workings of the human mind.

1960 LARC (Livermore Advanced Research Computer), one of the first supercomputers, becomes operational at the Lawrence Livermore Laboratory in California. Designed by Sperry Rand and built with 60,000 transistors, this machine provides unprecedented speed for nuclear weapons research, being nearly 100 times faster than any of its contemporaries. Instead of binary numbers, LARC uses decimal numbers in its computations and is the largest computer of its type ever built.

1960 First Control Data Corporation computer, the CDC 1604, is produced. Designed by American computer scientist Seymour R. Cray, who uses the new transistor technology, it is one of the most powerful computers of its time and is priced at just under $1 million. This machine establishes Control Data as a major supplier of large computers.

1960 First commercial modem designed specifically for converting digital computer data to analog signals for transmission across a long-distance telephone network becomes available. Called DATAPHONE, this high speed data transmission system is the beginning of what will become a world-wide linking capability among computer systems. The word "modem" is actually an acronym for Modulator/Demodulator which describes the process of changing computer signals (digital) into telephone signals (analog) and vice versa.

1960 First neural network, the MARK I Perceptron, is invented by Frank Rosenblatt, American psychologist. This machine has a 400-cell photoelectric eye and can recognize letters of the alphabet. Its circuitry is intended to mimic the learning processes of the human brain. Although his machine proves to have limited pattern-recognition capabilities, it attracts wide attention to the subject of artificial intelligence.

1960 Digital Equipment Corporation produces its first computer, the PDP-1, the precursor to the minicomputer. Selling for only $120,000 (compared to over a million dollars for a mainframe), the Programmed Data Processor model 1 weighs 250 lbs (114 kg), has a CRT graphics display, requires no air conditioning, and can be operated by one person. It is the brainchild of American electrical engineer Kenneth H. Olsen, who believes that there is a need for small, rugged, inexpensive real-time machines that do not require housing in a computer center and a staff of trained operators. He is one of the few engineers to foresee the need for small computers. This leads directly to the first minicomputer.

(See also 1965)

1960 First IBM 1401 computer is introduced. This economy model is a stripped-down version of other IBM computers, but as a transistor-based machine, it provides data-processing capabilities to small companies that had previously been unable to afford a computer. Renting for as low as $2,500 a month, it is by far the most popular computer of its day and IBM leases more than 15,000 of these workhorses.

1960 MICR (Magnetic Ink Character Recognition) is first adopted by the United States banking industry. The banking industry becomes the primary user of magnetic ink. Having its checks and deposit slips printed with this ink allows them to be scanned through MICR readers which translate the characters into electronic signals.

1961 First industrial robot, Unimate, begins work at a General Motors plant in New Jersey. This one-armed automaton obeys step-by-step commands stored on a magnetic drum and works with pieces of die-cast metal using a specialized gripper.

1961 First major IBM product to be designed with transistor electronics is the IBM 7030 or Stretch. One of the earliest of the supercomputers, it uses quick-switching transistors and a new technique called multiprogramming. Intended to be two orders of magnitude in terms of performance above existing computer technology, it becomes the most powerful computer of its time. Although a technical success, it proves a financial disaster due to cost overruns and schedule delays.

1961 First computer time-sharing system in the United States is demonstrated at the Massachusetts Institute of Technology. American mathematician John McCarthy installs the Compatible Time-Sharing System (CTSS) that is run on an IBM 709 computer. The CTSS handles processing requests from keyboard-equipped terminals in milliseconds, proving the worth of McCarthy's time-saving concept.

(See also January 1, 1959)

1961 American Federation of Information Processing Societies (AFIPS) is first established. This is a national federation of professional societies that is founded to represent the member societies on an international level and for the advancement and diffusion of knowledge of the information processing sciences.

FEBRUARY 20, 1962 Computers play their first real role in manned space flight when an IBM 7090 continuously calculates the position and speed of American astronaut John Herschel Glenn, Jr.'s *Friendship 7* capsule. Computers will be crucial to the United States space program, and it is safe to say that there would have been no space program without them.

1962 MOS (metal-oxide semiconductor) is first developed by the Radio Corporation of America (RCA). Combining 16 MOS transistors into a single integrated circuit quickens the pace of electronic integration. MOS transistors are smaller, require less power, and are manufactured easier than standard transistors.

1962 Superconductivity switch is first patented by Brian D. Josephson of England. Called a Josephson junction, this device consists of a thin insulating layer sandwiched between two superconducting films. Although this switch works 10 times faster than an ordinary transistor and has low power requirements, it generates little interest outside the laboratory because it must operate at a temperature near absolute zero.

1962 First interactive computer game called "Spacewar" becomes publicly available. Featuring shoot'em up graphics that inspire many future video games, it allows dueling players to fire at one another's spaceships using an early version of a joystick. They must also steer away from a black hole.

1962 First departments of "Computer Science" are established at Stanford and Purdue universities. They will soon become a popular undergraduate major in scores of universities.

1962 First computerized stock quotation system, QUOTRON, is introduced. Using a Control Data computer, it becomes popular with stockbrokers and signals the end of the traditional paper "ticker tape" method. By 1966, the New York Stock Exchange completes the automation of its basic trading functions.

1962 First graphic-oriented minicomputer is the LINC (Laboratory Instrumentation Computer). Designed and built by Wesley Clark of the United States, it is also the first computer able to process data from laboratory experiments in real time. Initially used in biomedical studies, it is inexpensive and small.

1962 First commercial computer to use silicon

chips is the SDS 910 produced by the California company called SDS.

1962 ATLAS computer at the University of Manchester (England) is the first to employ Thomas M. Kilburn's idea for a two-level memory. This technique of virtual memory allows a computer to use its storage capacity as an extension of memory, permitting it to run outsize software and to switch rapidly between multiple programs.

1963 First universal standard for computers, ASCII (American Standard Code for Information Interchange), is developed. This seven-bit code (actually 128 unique strings of seven ones and zeros) is developed to allow computers from different manufacturers to exchange data. Each code sequence stands for a letter of the English alphabet, one of the Arabic numerals, one of an assortment of punctuation marks and symbols, and a special function. Produced by a joint government-industry team, it creates a standard code usable on any computer, regardless of manufacturer. Used on most personal computers today, it has been adopted as a standard by the United States government.

1963 First artificial intelligence (AI) text is *Computers and Thought*. Edited by American computer scientist Edward Albert Feigenbaum and Julian Feldman, this collection of 21 papers is now considered a classic in the AI field. Significantly, it reprints the 1950 paper of English mathematician and logician Alan Mathison Turing (1912-1954) in which the "Turing test" was introduced.

1963 Sketchpad software is created by American electrical engineer Ivan Edward Sutherland. He publishes this interactive, real-time computer drawing system as his doctoral thesis. A designer or engineer who knows nothing about programming can, using Sutherland's software and a simple light pen, draw geometric figures on-screen and manipulate them to solve complex engineering problems.

APRIL 7, 1964 Term "computer architecture" is first introduced by International Business Machines Corporation (IBM) to describe its family of six mutually compatible computers and 40 peripherals that work together. This IBM System/360 is the first byte/addressable machine and also contains primitive integrated circuits. Designed to replace all earlier IBM computers, the 360 series is a huge gamble ($5 billion) that pays off. Many 360 features become standards in large segments of the computer industry, such as the use of eight-bit bytes for the representation of characters and the use of nine-track tapes, as well as multiple-spindle disk systems with removable disk packs. The 360 series is tremendously successful, and within two years orders for the 360s reach 1,000 units a month.

1964 Term "word processing" is first stated by International Business Machines Corporation (IBM). It uses this phrase as a means of marketing its new product, a typewriter that can record words on magnetic tape. This recording ability means that revisions can be made and text can be recycled, permitting an unlimited number of what appear to be hand-typed, personalized letters to be created.

(See also 1979)

1964 On-line transaction processing first becomes available in IBM's SABRE reservation system. SABRE (Semi-Automated Business Research Environment) links 2,000 terminals in 65 cities via telephone lines to a pair of IBM 7,090 mainframes and processes airplane seat reservations for American Airlines. The system has a 500 MGB magnetic disk memory and delivers data on any flight in less than three seconds. It is the largest commercial real-time data processing network of its time.

1964 First real, modern supercomputer is built. Control Data Corporation produces its CDC 6600 supercomputer, the fastest in the world at this time. It is three times faster than its IBM Stretch competitor and is the first RISC (Re-

duced Instruction Set Computer) computer, possessing a simplified data flow and instruction set. It uses 10 smaller computers (called peripheral processors) funneling data to a central processing unit (CPU) performing 3 million instructions per second. It is only surpassed in speed by its own successor in 1968.

1964 International Business Machines Corporation (IBM) team develops FORMAC (Formula Manipulation Compiler), the first computer algebra program. Developed to do non-numeric mathematics (i.e., formal algebraic manipulation), it is an extension of FORTRAN (Formula Translation) and also is the first to be used fairly widely on a practical basis for mathematical problems needing formal algebraic manipulation.

1964 Computer-Aided Design (CAD) gets its first real start with a joint project between General Motors and IBM called DAC-1 (Design Augmented by Computers). Built to help design automobiles, it features an interactive mode that revolutionizes computer graphics. Auto designers need only to touch an electronic stylus to the contours of a car displayed on a computer screen, and they can enlarge, rotate, or alter the shape of certain parts or the entire car. The first car with any DAC-designed parts is the 1965 Cadillac whose trunk lid is designed with this new system.

1964 BASIC (Beginners All-purpose Symbolic Instruction Code) is created by American mathematician Thomas Eugene Kurtz and John George Kemeny. This easy-to-learn programming language is derived from FORTRAN (Formula Translation) and is written to help their Dartmouth College students program the school's mainframe. BASIC soon becomes hugely popular among personal computer users since it is not machine language and has some advanced features despite its very simple language.

1964 PL/I or Programming Language One is first introduced. The product of a joint effort between International Business Machines Corporation (IBM) and its clients, this universal language incorporates the best features of FORTRAN, COBOL, and ALGOL and can perform both analytical tasks for scientists and data processing jobs for business.

1965 First diagnostic expert system program, DENDRAL, is devised by Edward Albert Feigenbaum, American computer scientist. Expert systems are programs designed to apply the accumulated expertise of human specialists. This program applies a battery of "if-then" rules about chemistry and physics to identify the molecular structure of organic compounds.

1965 First commercially successful minicomputer, the PDP-8, is produced by Digital Equipment Corporation. This $18,000 computer built with integrated circuits can fit on a desktop and proves perfect for small businesses. About the size of a refrigerator, it can run only one program at a time and contains only 4K words of memory. But it also costs only a fraction the price of a mainframe, meaning that customers who had formerly relied on large computers can now afford to switch to their own small, efficient ones. This combination of speed, small size, and reasonable cost allows it to win a place in thousands of manufacturing plants, small businesses, and scientific laboratories.

1965 First general-purpose digital computer to fly in space is a 19 in (48 cm) IBM machine that fits in the United States space capsule *Gemini 3*.

1965 First object-oriented language, Simula, is created by Kristen Nygaard and Ole-Johan Dahl, Norwegian computer programmers. A programming language is said to be "object-oriented" if it supports objects as a language feature, and if the objects are required to belong to classes that can be modified through inheritance (which is a mechanism for sharing code and behavior). With the goal of creating a computer language adept at simulating complex phenomena, Nygaard and Dahl invent a new type of language that groups data and instructions into modular building blocks called objects, with each representing one facet of a system intended for simulation.

1965 First floppy disks are developed by the International Business Machines Corporation (IBM). Also called mini disks or flexible disks, they are initially made as round 8 in (20 cm) disks for use in the IBM System 30 series minicomputers. These are soon modified to a 5.25 in (13 cm) size (with capacities of either 360 kilobyte or 1.2 kilobytes), and then to 3.5 in (9 cm) disks available in 720 KB and 1.44 MB capacities. The newer 3.5 in (9 cm) model is encased in hard plastic, and since it is no longer flexible, becomes known as a diskette. Floppy disks are used to load programs into a computer, to exchange data between computers, and to back up data stored on a hard disks. On early systems that have no hard drive, the floppy also contains the operating system as well as application programs. Floppy disks are made by depositing a metallic oxide material on a mylar substrate. The oxide coating is ferro-magnetic and responds to the magnetic fields generated by the heads in the disk drive. Floppy disks have made it extremely easy and inexpensive to share information and programs.

1967 First commercial computer to use a cache (auxiliary) memory is the IBM 360/85. As a storage subsystem, cache memory is faster than main memory and holds operating instructions and data likely to be needed, therefore speeding up operations.

1967 Fairchild Semiconductor Corporation is the first to put more than 1,000 transistors onto a single, random-access memory (RAM) chip. Using four transistors per bit, the integrated circuit has a capacity of 256 bits. Although it is 10 times smaller and much faster than an equivalent amount of magnetic core memory, it is slow to become a commercial success because of its steep price.

1967 United States Internal Revenue Service completes the computerization of federal income tax processing throughout the country.

OCTOBER 1968 Term "software engineering" is first coined at a NATO (North Atlantic Treaty Organization) conference. To this point, software had mostly developed as the result of creative individuals working largely on their own. By the time of this conference held in Garmish, Germany, many university professors feel the need for increasing the discipline and precision of creating software, and they consequently work to foster a movement that will regularize, define, and automate its practice. Software engineering can be defined as that form of engineering that applies the principles of computer science and mathematics to achieving cost-effective solutions to software problems. Software is defined as "computer programs, procedures, and possibly associated documentation and data pertaining to the operation of a computer system."

1968 CMOS (Complementary MOS) technology, a variation of the highly efficient Metal-Oxide Semiconductor transistor, first becomes available. In a CMOS chip, semiconductor material is used to produce two varieties of transistors, called n-type and p-type. Older MOS chips had only one variety. Incorporating both kinds reduces their power consumption but also makes them expensive to manufacture. CMOS chips will eventually come down in cost as manufacturing techniques improve, and they will make possible the portable, battery-powered personal computers of the future.

1968 Apollo Guidance Computer (AGC) is the first navigational device built for NASA (National Aeronautics and Space Administration) to include integrated circuits and magnetic core-rope memory for its permanent program. This guidance system will successfully steer *Apollo 11* to a safe manned Moon landing on July 20, 1969.

1968 First 1K (1.024-bit) RAM (random-access memory) chip is introduced by the Intel Corporation. The first of its kind, it achieves a significant breakthrough since now it is possible to replace magnetic cores that take up so much space with a tiny integrated circuit. Magnetic core memory, the mainstay of the computer industry since the mid-1950s, will now begin to be phased out.

1968 First phased-array radar comes on line at Eglin Air Force Base in Florida. This powerful new system uses a computer to coordinate the transmission of radar signals from thousands of stationary antennas arranged on a flat surface and angled skyward. This system allows for the tracking of multiple targets by precisely controlling the timing of the signals and creating a narrow but powerful radar beam that sweeps across the sky many times a second.

1968 First prototype mouse is demonstrated by its inventor, American electrical engineer Douglas Carl Engelbart. The product of years of research and actually developed as early as 1965, this first mouse is part of a keyboard console that has five piano-like keys that Engelbart calls a chord keyset. The Engelbart mouse uses a wooden housing with wheels placed at right angles to track cursor movement. An improved version of this concept is first popularized in 1984 by Apple Computer who introduces it on its Macintosh personal computer. Today's mouse is a small hand-held interactive input device that, when rolled over a flat surface, controls placement of the cursor on a computer's terminal display screen. Single or multiple clicks select a course of action from a menu or an icon. Having a roughly oval shape and being connected to the computer by a wire that is suggestive of a tail, is believed to be the reason it is called a mouse. IBM-PC and PC-compatibles eventually adopt the mouse.

(See also January 24, 1984)

JUNE 1969 First "unbundling" of software is introduced by International Business Machines

Corporation (IBM). Prior to this major decision, software was not sold separately, and the hardware customer had to buy a "bundle" consisting of hardware plus all available software. The software companies argue that the hardware manufacturers are actually selling software to their customers and including its cost in the price of the hardware, and they charge unfair competition. This decision by IBM to sell its software separately allows customers to purchase only as much software as they need, and it also permits them to purchase from other suppliers. The birth of the commercial software industry arrives.

1969 First packet-switched network, called ARPANET, is established to link computers in research facilities that work for the United States Department of Defense. ARPANET, short for ARPA (Advanced Research Projects Agency) Network, is designed to provide effective and efficient communications between heterogeneous host computers so that hardware, software, and data resources can be shared conveniently and economically by a wide community of users. This packet-switch system replaces the telephone's inefficient circuit-switching and divides information into small parcels (packets) which are eventually reassembled. This cross-country linking of computers lays the foundation for computer networking and what eventually becomes the Internet Protocol.

(See also 1986)

1969 First fully mobile robot is built at the Stanford Research Institute. This wheeled device uses a video camera to see its environment and possesses obstacle detectors and sensors to measure distance. Despite showing some problem-solving ability, machine vision, and other forms of artificial intelligence, it is still a very primitive device.

1969 Data General Corporation introduces its Nova minicomputer. It is one of the first commercial machines to use medium-scale inte-

grated circuits (MSI). These new computer circuits are smaller, faster, and contain more complex functions. The Nova costs $8,000 and is one of the least expensive and most successful of the early minis. In 1971, its Super Nova series uses semiconductor memory.

1969 First time-sharing operating system, Multics, appears in preliminary form at the Massachusetts Institute of Technology. Headed by Fernando Corbato, Multics is written in high-level language and proves to be the most powerful and flexible of the general interactive systems.

1969 Translation of a language by machine first becomes practical with SYSTRAN. Developed by Peter Toma for the United States Air Force, the earliest version translates Russian into English. Later editions translate other language pairs. SYSTRAN employs linguistic rules and a huge database linking words and idioms. Its best use is on technical literature, and its results are not perfect but certainly useable.

1969 RS-232-C standard for serial communications is first agreed upon by a United States industry group. The implementation of this common standard permits computers and peripheral devices like printers and modems to transfer information serially, one bit at a time. It is this important standard that defines connector pin assignments and signal descriptions.

1969 Unix Operating System is first developed by American programmers Kenneth Thompson and Dennis Ritchie. Combining many of the time-sharing and file-management features offered by Multics, but in a simpler, more consistent way, it becomes especially popular with scientists and proves to be one of the most influential systems in computing history. Unix provides a hierarchical file system in which directories hold files and other (sub)directories. The resultant file tree is shared by all users, making it straight-forward to name and find files. It also provides a single, uniform way of accessing a file's contents. It proves so successful because it is the first operating system to focus entirely on interactive use; because it is made available to universities at almost no cost; and because it is the first system to run on machines of vastly differing architectures.

1970 First Automated Teller Machine (ATM) is installed at a bank in Valdosta, Georgia, providing on-line computerized banking services. An ATM is an unattended computer terminal-type device that offers most of the services available from a teller. This machine operates when presented with a magnetized card and a special personal identification number (PIN). Among the services it offers are cash withdrawals, transfer of funds between accounts, deposits, and account balance inquiry. Although now commonplace, they are initially rejected because of their erratic performance.

1970 Computers monitor freight trains for the first time as TRAIN (TeleRail Automated Information Network) is introduced to United States rail. This system continuously updates the location of every freight car in the country, totaling more than one million. In 1975, TRAIN II also is able to record their cargo and destinations.

1970 Computer language Pascal is first developed by Swiss computer scientist Niklaus Wirth. This procedure-oriented language facilitates an approach to writing software called structured programming in which a program is divided into small, logically arranged tasks that are both easy to write and to understand. Pascal is also important for its spin-off languages: concurrent Pascal, object-oriented Pascal, Modula-2, and Ada.

1971 First rule-based expert system, MYCIN, is designed by American medical information scientist Edward Hance Shortliffe. This pioneering expert system is programmed with 500 "if-then" medical rules, and helps doctors diagnose blood infections and prescribe antibiotics. The MYCIN system contains the first

integrated architecture for interactive consultation between an expert user and an Expert System (ES), and is the first to apply ES to medical diagnosis. Its success leads to the creation of "knowledge shells" which are problem-solving programs empty of specific advice that can be filled in by human experts in any field.

1971 First microprocessor, the Intel 4004, is designed by American engineer Marcian Edward Hoff, Jr. Confronting the difficult and expensive dilemma of having to design a custom chip for every new, complex component and device that is invented, Hoff develops a general-purpose chip that can perform any logical task, much like the central processor of a computer. His custom chip fits 2,250 transistors on a sliver of silicon the size of the head of a tack and inaugurates the era of very large scale integration. Hoff's new idea is literally a programmable processor on a chip, expandable if necessary and effecting major cuts in manufacturing costs. It has all the functions of a computer's central processing unit.

1971 First mass-produced pocket calculators become available in the United States. Developed by American electrical engineer Jack St. Clair Kilby and fellow Texas Instrument engineers Jerry Merryman and James Van Tassel, the battery-powered Pocketronic has tiny logic circuits, a small, power-efficient keyboard, and a new thermal printer. Although weighing 2.5 lb (1.1 kg) and costing $150, it proves appealing to average consumers as well as professionals. In 1972, Hewlett-Packard introduces the HP-35, the first hand-held scientific calculator with an LED (light-emitting diode) display. Pocket calculators soon are able to perform such a wide range of functions that they not only reach a mass market but make the slide rule obsolete. In 1976, Keuffel & Esser present their last slide rule to the Smithsonian.

JANUARY 1972 Hewlett-Packard announces its first hand-held calculator, the "electric slide rule" called HP-35. This marks the beginning of the end of the traditional "slipstick" used by scientists and engineers. This miniaturized calculator takes only five seconds to solve a navigation problem that normally requires five minutes of slide-rule work. It is an instant success. (See 1974)

APRIL 1972 First 8-bit microprocessor, the Intel 8008, is introduced. Powerful enough to run a minicomputer, this 8-bit chip can for the first time handle all the letters of the alphabet (upper and lower case), all 10 numerals, punctuation marks, and many other symbols. Despite its power, it has many technical drawbacks and is soon replaced by the much more efficient and powerful Intel 8080 microprocessor.

1972 First software program to integrate grammar rules, word definitions, and logical reasoning is designed by American mathematician Terry Allen Winograd. He creates SHRDLU, an experimental artificial intelligence program that controls an arm playing with blocks on a table. The program executes commands and can respond correctly to hypothetical questions.

1972 First video game, "Pong," is introduced by Nolan Bushnell, founder of the Atari Corporation. His simple video game is a coin-operated version of table tennis and becomes a financial success. Bushnell eventually sells his company for $15 million.

1972 First "windows" environment is introduced by Alan Kay of the United States. He develops the "Smalltalk" software system which allows partitioning of a screen to display files, menus of commands, and icons. Together with its manipulation by a mouse, it becomes an essential part and one of the most attractive features of the new Macintosh personal computer.

(See January 24, 1984)

1972 ILLIAC IV, designed at the University of Illinois and built by Burroughs, makes parallel processing possible. Parallel processing is the use of multiple resources or concurrency to increase throughput, increase fault-tolerance, or reduce the time needed to solve particular problems. It is the only route to reach the highest levels of compute performance. As the first computer to abandon the traditional one step at a time method of operation, ILLIAC IV has 64 processors operating simultaneously on separate parts of a problem. It is ahead of its time but also very difficult to program.

1972 First relational database is developed by Edgar F. Codd of the United States. Codd's relational model provides a simple and intuitive method for defining a database, storing and updating data in it, and submitting queries of arbitrary complexity to it. Such databases make facts easy to find by storing them in tables with rows of entries crossed by columns. An intersection between a column and a row gives a characteristic for an entry, and the same column appearing in several tables serves to link them, revealing relationships. This database method eventually gives rise to thousands of electronic libraries.

1972 Computerized axial tomography is invented by Godfrey Newbold Hounsfield of England. This new technique, called a CAT scan, records multiple images electronically as an x-ray camera circles the body. A computer then converts the array of pictures into a three-dimensional image of a cross section. The CAT scanner not only provides an accurate view of the body's internal structure, but can also identify areas of diseased tissue.

(See also 1977)

1972 Prolog, the most widely used logic programming language, is created by Alain Colmerauer of France. As a nonprocedural language that provides sets of facts and relations between them, Prolog (Programming in Logic) offers an alternative to writing step-by-step instructions for a computer. Prolog is used in artificial intelligence and facilitates the writing of expert systems which emulate the human powers of deduction.

1973 National Aeronautics and Space Administration (NASA) uses two off-the-shelf IBM computers for its Apollo Telescope Mount (ATM). This is NASA's first use of computers not designed specifically for the space program.

1973 First international computer chess tournament is held.

1973 Ethernet links minicomputers and becomes the first local area network (LAN). Invented by Robert Metcalfe of the United States, it permits machines to share software, data, and peripherals. It is eventually extended to personal computers.

1973 Universal Product Code (UPC) first appears to automate supermarket operations. Each product has a unique code of wide and narrow bars that are read by a laser scanner which informs a central computer. The computer signals the price to the cashier and deducts the item from inventory. Developed by Xerox, the UPC eventually becomes standard on nearly every item bought and sold and is also used to keep track of all kinds of products.

1973 Winchester disk is first introduced by International Business Machines Corporation (IBM). This new high-capacity, high-speed rotational storage device redefines computer storage. Housed in a component called the IBM 3340, this hard disk consists of a rigid aluminum alloy disk that is coated with a magnetic oxide material. Since they are rigid, they can be spun much faster than a floppy (up to 3,600 rpm). This unit contains four disks with read/write heads that actually float above the surface of the disk on a cushion of air. It has a capacity of 30 megabytes of data.

JULY 1974 *Radio Electronics* publishes an article that describes what is essentially the first home computer. Jonathan Titus of the United States describes his Mark 8, which uses an Intel 8008 chip, and tells how to build a "personal computer." This product is essentially a computer kit that, because it is so difficult to build,

maintain, and operate, compels its early users to band together into clubs or "users' groups." The first of these, the Homebrew Club, is founded in March 1975 in the Santa Clara Valley south of San Francisco (now known as "Silicon Valley").

(See also January 1975)

1974 Activity inside the brain is revealed for the first time through a new technique called Positron Emission Tomography (PET). Controlled by a computer, this research technique shows concentrations of radioactively tagged substances which indicate areas of chemical activity.

1974 Xerox PARC creates one of the earliest computer workstations. Called Alto, it provides near-mainframe speed and contains several features that eventually find their way into personal computers.

1974 Intel Corporation introduces the Intel 8080 microprocessor which is five times faster than the Intel 8008. With its 64K (64,000 bytes) memory, it can handle four times as much as its predecessor and becomes the chip of choice. It makes dozens of new products possible and becomes the basis for the first personal computers.

1974 Texas Instruments introduces the TMS 1000, the first integrated circuit to include a microprocessor, permanent read-only memory (ROM), and input/output circuitry. Designed to handle data in four bits at a time, it is not capable of further memory expansion but proves ideal for certain uses. It becomes a popular controller chip and is used on devices such as microwave ovens and hand-held video games.

1975 First commercial array processor, the AP-120B, is introduced. An array processor consists of many separate processing elements, each capable of carrying out basic arithmetic operations under the control of a master program. This add-on system can transform an ordinary minicomputer into a lightning-fast calculator. It achieves its amazing speed by performing different operations on many pieces of data simultaneously. It is especially well suited for repetitive, number-intensive tasks like weather analysis or the processing of wind tunnel data.

JANUARY 1975 Cover of *Popular Mechanics* displays the Altair 8800 microcomputer, which marks the first major public awareness of the concept of the personal computer. Offered by a small New Mexico company called MTIS (Micro Instrumentation and Telemetry Systems), this product offers a complete computer kit. Although the system has no software, small memory, uses switches instead of a keyboard, and has no monitor, it completely sells out and thousands of orders pour in during this initial year of its offering.

AUGUST 1975 First commercial packet-switching network, Telenet, is created by Larry Roberts of the United States. A civilian counterpart to ARPANET (whose network architecture it follows), it offers extras beyond linking customers in different cities. Telenet is the first VAN (Value-Added Network) and provides such services as error checking and electronic mail.

SEPTEMBER 1975 First issue of *Byte* magazine appears. Founded explicitly to discuss personal computing and microcomputers, it is followed by a half dozen more periodicals that address the same topic.

1975 First fault-tolerant computer, the Tandem-16, is produced by Tandem Computers. Eventually renamed the NonStop System, it operates on a buddy system of multiple processors and can therefore run during repair or expansion as well as during data errors or hardware failures. Each processor has the other's application program in its memory, and if one fails, its "buddy" automatically takes over. It becomes widely used for money-producing Automated Teller Machines (ATM).

1975 First digital computer built into an un-

manned spacecraft is the Viking Command Computer Subsystem (VCCS) that NASA puts into its Viking probes to Mars. This system has dual components for reliability, using two power supplies, two central processors, and two memories. NASA uses redundant systems throughout its spacecraft as much as possible.

1975 Air traffic control first becomes fully computerized in the United States. An IBM 9020 mainframe and backup is located in each of the 20-route control centers around the country and automates the radar tracking of aircraft. Prior to this automated system, tracking was done manually using wooden markers and handwritten data strips.

1975 William Henry Gates III, Harvard University freshman, and his friend, Paul Allen, develop the computer language BASIC, which is the first software produced for a personal computer. Their BASIC interpreter is an easy-to-use, high-level language rather than machine code, and enables Altair's users to write programs. When Altair's producer, MITS, buys the software, Gates drops out of school and with Allen, establishes the Microsoft Corporation in 1977. Created to market their new program, Microsoft goes on to become the largest developer of personal computer software in the United States, supplying programs for Apple, Commodore, Tandy Radio Shack, and IBM. As president and CEO of Microsoft, Gates is the nation's youngest billionaire. When he takes Microsoft public in 1986, he is only 31 years old.

1975 Laser printer is first introduced commercially by International Business Machines Corporation (IBM). The quality of computer printouts improves radically with this new device. Essentially a photocopying machine with a microprocessor-controlled laser as a light source, it combines text with finely detailed graphics on the same page, and is so fast, it is able to produce 215 pages per minute.

1976 Development of the first optical character

reader enables the Kurzweil Reading Machine to turn print into spoken words for the sightless. This computer is equipped with a voice synthesizer that uses more than a thousand pronunciation rules and can scan a page in less than one minute. It is the first computer that can read aloud.

1976   United States publishing firm Van Nostrand Reinhold issues the first edition of its *Encyclopedia of Computer Science*.

1976   First commercially successful vector processor, the CRAY-1, becomes available. This computer is able to calculate many numbers simultaneously and can perform 100 million arithmetical operations per second. Part of its great speed comes from its C-shape design which reduces wire length and signal travelling time. It is the fastest computer of its time.

1976   First word processing software for personal computers is "Electric Pencil," developed by American programmer Michael Shrayer. He designs a program to prepare manuals that accompany other software he devises. It quickly becomes very popular.

1976   First standard for packet-switching is agreed on and approved by the International Telecommunications Union. With the approval of the X.25 standard for packet-switching (the technique of dividing files into small parcels for transmission over a network), computer-to-computer communication is simplified and in turn, soon spurs the development of data banks and electronic mail.

1976   First disk operating system for home computers, CP/M, is designed by American computer scientist Gary A. Kildall. His Control Program for Microcomputers introduces software compatibility and makes it possible for one version of a program to run on a variety of computers built around 8-bit microprocessors. With its principle feature the ability to write and read information to and from a disk drive, CP/M quickly becomes the most popular

operating system and dominates the market through the 1970s.

1976   Apple Computer is first established by American computer innovators Stephen G. Wozniak and Steven P. Jobs. They begin the company in the garage of Jobs' parents' home in Cupertino, California, when they receive an order for 50 Apple I computers. They incorporate on January 1, 1977, and it becomes the fastest growing company in American history. Apple becomes a major power in personal computing and sets the early pace for ease-of-use and graphical interfaces.

(See also 1977)

AUGUST 1977   First desktop computer is the TRS-80 microcomputer. Developed by the Tandy Corporation, it leads the shift from do-it-yourself kits to preassembled, ready-to-use computers. As a completely assembled system ready to be used, this new microcomputer appeals to a much wider audience than simply hobbiests and tinkerers.

1977   First personal computer designed for a mass market is the Commodore PET (Personal Electronic Transactor). Simple to operate, it has a monitor and cassette deck for storing programs and data, and comes fully-assembled. With a memory limited to 12K, it sells for only $595.

1977   Godfrey Newbold Hounsfield of England makes Magnetic Resonance Imaging (MRI) practicable and gives physicians their first detailed pictures of the body's soft tissues. MRI uses a computer to control an intense electromagnetic field and radio waves to produce images that do not expose the patient to the hazards of an x ray. Knowing the frequencies at which certain atoms in the body will spin, the computer searches for certain ones (such as those that might indicate cancer cells). Once the radio waves are turned off, the atoms emit pulses of absorbed energy, and the computer reads these pulses and uses them to draw a three-dimensional image of the scanned area.

1977 Apple II computer is first introduced by Apple Computer, Inc. and becomes an instant success. Described as the Volkswagen of personal computers, it sells for only $1,195 (without a monitor) and is designed originally to enliven video games. Hooked up to a television, it provides brilliant color graphics. It eventually appeals to a wide range of users and is an exceptional educational tool.

1977 Charles Babbage Institute for the History of Information Processing (CBI) is first established at the University of Minnesota. Its purpose is to support the study of the history of information processing and to be a clearinghouse for information about research resources related to its history, as well as a repository for archival material.

1978 OSI (Open Systems Interconnection) is first formulated by the International Organization for Standardization in Switzerland. OSI sets the standard or protocol that facilitates communication between computers of different makes. Agreement on a set of communication standards does away with any need for computers to have facilities for translating foreign protocols. The term "open" is used to mean only the freedom from any technical barriers.

1978 5-1/4 floppy diskette is first adopted by Apple Computer and Tandy Radio Shack whose new computers have disk drives for this size. This smaller version of the original IBM 8-in (20 cm) floppy becomes the standard medium for personal computer software. Pioneered by Shugart Associates, the 5-1/4 floppy becomes the standard for the 1980s.

1979 First popular word processing program for microcomputers, Wordstar is created by John Barnaby, American programmer. Introduced by MicroPro International, this word processing package can display characters as fast as

they are typed and line-for-line as they will appear when printed. A word processor is a device that can create, manipulate, store, revise, and output text. It becomes a giant in the industry.

1979 Ada programming language is first introduced. As a very powerful language for military computers, it is developed for the United States and NATO (North Atlantic Treaty Organization) by an international team led by the French computer scientist Jean Ichbiah. Named after Lady Lovelace (Augusta Ada Byron) who was the popularizer of the work of English mathematician Charles Babbage (1792-1871), it is designed specifically to reduce software development and maintenance costs, especially for large, constantly changing programs with long lifetimes. It soon replaces a jumble of programming languages and becomes the standard for American military software.

1979 First electronic spreadsheet, VisiCalc, is developed for Apple II by American electrical engineer Daniel Bricklin and American computer programmer Robert Frankston. As an electronic version of the financial-planning aid called a spreadsheet, VisiCalc (for Visible Calculator) automates the recalculation of a spreadsheet's conclusions after a change in its assumptions. The program is so successful that people buy it first and then buy whatever hardware (the Apple II) will run it. VisiCalc is the first example of "the software tail that wags the hardware dog," and is the program that makes a business machine of the personal computer.

1980 First optical data-storage disk called a WORM (Write Once, Read Many) is offered by Dutch electronics firm, Philips Corporation. This method stores information via indelible laser marks on a thin metal layer and holds 1.3 gigabytes of information (almost 60 times the capacity of a typical microcomputer disk of this time). This new method is best suited to storing information that might expand but not change (since it cannot be erased or overwritten because the laser makes permanent marks). Although optical disks are more expensive and slower than magnetic disks, the advantages are their extremely large capacity and increased security. They are specially suited to huge databases (like libraries and banks) where speed is not necessary. Eventually, erasable optical disks are introduced.

(See also 1982)

1980   First hard disk for microcomputers is introduced by Seagate Technology. Designed to fit in the space occupied by the standard floppy disk drive, this new hard disk can hold 30 times the storage capacity of a floppy. It marks the beginning of the great expansion of the personal computer's storage capacity.

1980   "Pac-Man," one of the most popular video games of the 1980s, is first introduced. This phenomenally successful game represents a maze through which a player can manipulate the Pac-Man face by using a joystick. Pac-Man has to eat all the cookies in the maze while avoiding little ghosts that try to trap him. Pac-Man is partly responsible for the huge growth in video game income which reaches a peak of $7 billion in 1982.

AUGUST 12, 1981   First IBM PC is introduced. After building its first microcomputer in only one year using parts and systems from outside suppliers, the computer giant greatly spurs the growth of the personal computer market with this new machine. IBM decides to challenge the Apple Corporation and targets its new machine not at individuals, homes, and schools (Apple's traditional customers) but at large corporations. It eventually dominates the business microcomputer market.

1981   First portable computer, the Osborne 1, becomes available. Invented by American journalist and entrepreneur Adam Osborne, this new machine weighs 24 lb (11 kg), uses CP/M operating software, and comes complete with monitor and disk drives. Although it produces

a very popular product, the Osborne Computer Corporation goes bankrupt in September 1983.

1981   MS-DOS is first introduced by the Microsoft Corporation. Adopted by IBM as the operating system for its newly introduced personal computer, it soon emerges as the dominant operating system (eclipsing CP/M) and spurs a boom in software development.

1981   First electronic telephone directory accessible over telephone lines using a computer is introduced in France. Called Teletel, this marks the beginning of telecomputing, as the system grows in six years from only 600 connected homes to 3 million subscribers using 6,000 services of all kinds.

1981   DBASE II is first made available for microcomputers by American engineer Wayne Ratliff. This program enables microcomputers to organize large files of information, like payrolls and inventories, and brings the power of database-management software to personal computers.

MAY 1982   First popular workstation is introduced by SUN Microsystems Inc. of California. The high performance and low cost of this system firmly establishes SUN as a leading workstation company.

1982   First erasable optical disk is introduced by Philips Corporation, a Dutch electronics firm. This prototype has the potential of adding the versatility of purely magnetic storage disks to the huge storage capacity of optical techniques. This new device combines special materials, a laser beam, and magnetism to add information to the disk, wipe it clean, and then rewrite it with new data in place of the old.

(See also 1980)

1982   Lotus 1-2-3 is first introduced. Designed by entrepreneur Mitchell D. Kapor to be used with the new IBM PC, this spreadsheet program brings the most successful software product of all time to the microcomputer. Blending the spreadsheet capability of VisiCalc with graphics and data retrieval, this product proves wildly successful and sells almost half a million copies in only a few months.

FEBRUARY 1983 First fiber optics telephone line between cities is opened by AT&T between Washington, D.C., and New York.

1983 Turbo Pascal is first created by Phillipe Kahn of France. His new compiler is a fast, cheap program that can be used on an IBM PC and its compatibles and is capable of producing compact programs. It becomes an instant success as well as the standard version of Pascal.

1983 First computerized musical keyboard is invented by Raymond Kurzweil of the United States. Called the Kurzweil 250, it is able to reproduce the sounds of 30 different musical instruments. It achieves this by storing these various complex musical tones in the form of mathematical models.

1983 First talking computers are produced by Digital Equipment Corporation. The company's DECtalk is a device that converts text to speech using a 10,000-word dictionary and several hundred rules of punctuation. Callers enter their questions with touch-tone phone codes and then hear DECtalk read the proper answers. It is used primarily by businesses to handle initial telephone inquiries.

1983 First VHSIC (Very High Speed Integrated Circuits) are introduced by TRW. Built using new techniques that allow circuit elements to be made smaller and placed closer together than ever, these new chips are capable of 25 million switching operations per second.

1983 First of the so-called "PC clones," the Compaq Portable Computer is introduced by Compaq Computer Corporation. Called a clone because it runs on the same software as the IBM PC, these machines take advantage of the temporary shortage of the highly-demanded IBM PC and become enormously popular.

Customers who are unable to buy a PC purchase over $100 million worth of Compaq's 28 lb (13 kg) personal computers in the year following their introduction.

JANUARY 24, 1984 Macintosh computer is first introduced by Apple Computer, Inc. Based on Motorola's 68000 microprocessor, this new "friendly" personal computer popularizes the use of a mouse-driven graphical user interface (GUI) that is easy to learn. Apple's advertising campaign successfully compares its simplicity and understandability to the difficult and often unforgiving IBM-style PCs, and more than 100,000 Macintosh are sold within six months. The Macintosh user interface, with icons, windows, pull-down menus, and the mouse pointing device sets new standards for ease of use for the personal computer industry.

1984 First public alarm over computer viruses is raised by Fred Cohen of the United States. Speaking at a conference on computer security, he warns against viruses or small programs that make copies of themselves and can infect a computer tied into a network. These destructive programs are called viruses because they can infiltrate from the outside (via a network connection) and harm data and programs.

1984 Gallium arsenide is first used as the basis for high-speed integrated circuits. Long-known as being able to conduct electrons three to six times faster than silicon circuits while consuming less power, gallium arsenide chips do not become standard until new manufacturing techniques make them practical.

1984 First computerized navigation system for cars and trucks is introduced. Called ETAK, this system displays a video map on a dashboard-mounted screen and works using an electronic compass and a computer-controlled map. It also uses dead reckoning from a known starting point to estimate a vehicle's location, and has sensors that count wheel revolutions to record distance traveled.

1984 Three-inch (7 cm) diskette is first adopted by Apple and Hewlett-Packard. Their acceptance of this new, sturdy diskette with a rigid plastic jacket leads to its industry-wide acceptance. Developed by the Sony Corporation, it is smaller than its 5 in (13 cm) predecessor but can store much more data. It uses an access door that springs shut to protect its magnetic surface when removed from its drive. It soon becomes the standard and is called a diskette to distinguish it from its flexible "floppy" predecessor.

1985 Compact disks first become available for computers. Called CD-ROM (Compact Disk-Read Only Memory), this optical storage medium can hold 550 megabytes of data. For example, *Grolier's Electronic Encyclopedia*, composed of nine million words, occupies less than one-fifth of one disk. They become a very popular method of reproducing text, graphics, and sound from the same disk. Whether used for text or music, they are prepared by encoding digital data onto a master disk. A laser burns small holes or pits into a thin metallic film sandwiched between a plastic substrate and a protective plastic or glass film. The master can be used to make copies. When a CD player or a computer reads data from a disk, it uses a photodetector or diode to catch low-power laser light reflected off the disk's surface.

1985 "High Sierra" format for CD-ROMs is first adopted by computer and electronics firms. This format for storing data and directories on CD-ROMs brings standardization and therefore compatibility to CD-ROM formats. Where previously each compact disk publisher had devised its own storage plan, resulting in no one machine being able to read all of them, the new format will ensure that any computer or CD player can read any CD-ROM disk.

1985 Intel 80386 microprocessor is first introduced. This 32-bit chip packs 275,000 transistors and boosts personal computer speed and power to near-mainframe levels. Able to perform as many as four million operations per second and handle up to four gigabytes (billion bytes) of memory, it allows an entire personal computer to be built within 10 square inches. It is also compatible with Intel's earlier processors and can run almost any software written for them.

1985 PageMaker desktop publishing program is first introduced by Aldus and provides the essential features to make desktop publishing a reality. Desktop publishing refers to the creation and printing of high-quality documents by using a self-contained computer system with software that allows page images to be viewed and edited on the screen before printing. Pagemaker combines text from a word processor with images from a scanner, and provides for altering layouts, colors, and type styles. The results are then fed to a laser printer. The Apple version of this software is called PostScript.

1985 First megabit memory chips are introduced by International Business Machines Corporation (IBM). Small enough to pass through the eye of a needle, these chips store one million bits of data each, which is four times as much as their largest-capacity predecessor, and are used on IBM's powerful 3090 mainframe computers.

1986 Compaq Computer Corporation introduces the Deskpro 386, the first 32-bit personal computer built around Intel's 80386 chip. This model is powerful enough to perform demanding graphics and other computer-assisted design, engineering, and manufacturing tasks that were formerly accomplished only by minicomputers. Within two years, this model is outmoded by another Deskpro that is twice as fast.

1987 Apple Computer introduces its Macintosh II, the first Mac to include sockets for add-on circuit boards. This feature is standard on nearly all other personal computers, and Macintosh seeks to better compete. Already possessing a faster microprocessor than its rival, as well as a math coprocessor (a chip tailored for speed in number crunching), these new sockets on the Mac will allow it to be fitted with additional memory or to be connected to a variety of peripherals.

1987 IBM Personal System/2 is first introduced. In this new generation of personal computers, IBM uses a new operating system called OS/2 which allows the use of a mouse to manipulate data and programs. The most powerful of its four models uses an Intel 80386 microprocessor which brings this machine closer to the capabilities and features of Apple's new Macintosh.

1988 Hypercard is first developed by William Atkinson of the United States. Created for Apple's new Macintosh personal computer, this program enables the user to browse through information by association, similar to the human mind, and provides quick, natural access to large amounts of data. The availability of Hypercard underscores the potential of and points the way toward the future development of what is called hypermedia software which could combine literature with music, art, and film.

1988 American computer innovator and Apple Computer's co-founder Steven Jobs develops the Next computer. This innovative PC is the first personal computer to incorporate a drive for an optical storage disk. It also simplifies programming via object-oriented languages and is equipped with voice recognition. Switching the computer's emphasis from hardware to software, Jobs's new machine is innovative but not very popular.

1989 Megabarrier for logic chips is first broken by Intel Corporation which produces its million-transistor i860 microprocessor. Like its megabit memory chip, this nearly quadruples the number of transistors it holds.

1990 International Business Machines Corpo-

ration (IBM) makes window environments available on its PCs for the first time.

1991 "Clipboard" computers first become available. Not much bigger than an actual clipboard, these new computers replace the keyboard with a liquid crystal display (LCD) screen and an electronic stylus. Users input data by printing individual letters directly on the screen. Powerful software is able to translate the user's handwriting.

1992 "Flash memory chip" is first introduced by Intel Corporation. This is a plug-in memory card based on a chip that keeps its memory without power. Invented by Toshiba in 1985, this much-improved version does everything that a hard disk can do without any of the mechanical disadvantages and shortcomings. The new flash chip has all the flexibility of a RAM chip but is really more like a ROM chip. Its circuitry allows one to replace chunks of information held in its transistors, much as one might store changes to data or programs on a hard drive. Laptops and notebook computers are perfect users of this new technology.

1992 Apple Computer introduces its first portable computers. The Powerbook laptop can run both Mac and IBM-PC software, but it is expensive.

1992 Apple Computer introduces its first pocket-size portable called "Newton." A computerized executive organizer, it is equipped with built-in wireless communications ability.

1993 First prototype fully "optical" computer is produced by American electrical engineers Vincent Heuring and Harry Jordan. Although it is unable to perform complex calculations and is as large as a compact car, this "proof of principle" machine demonstrates that it is possible to build a computer that stores its information in the form of light. In this machine, pulses of laser light fly through 2.5 mi (4 km) of coiled fiber-optic cable, with each pulse representing a one in binary code while darkness represents a zero. Since today's electronic computers face a built-in speed limitation, computers based on optics can operate at the speed of light. Such optics-based computers hold the promise of being hundreds or even thousands of times more powerful than today's machines. This prototype is described by some as the optical Model T.

1993 Palm-size computers are first introduced by several firms. Called a Personal Digital Assistant (PDA), these hand-held PCs have no keyboard, and users write directly on its plastic screen with a special pen. The software converts the handwriting to type.

1993 First Pentium chip is introduced by Intel, the world's dominant computer chip maker, as part of its effort to make it the new industry standard. It replaces the less powerful Intel 486 semiconductor model. During 1994, however, a flaw is discovered in the new chip that can produce errors in certain high-level mathematical calculations. On December 20, 1994, Intel announces that it will replace any of the approximately four million Pentium chips it has sold.

1994 Computer beats a human world champion at chess for the first time. Russian grandmaster,

Garry Kasparov, is outwitted by "Genius 2," a computer program designed by English physicist Richard Lang. Before deciding on a move, the computer program examines each of the roughly 36 possibilities in an average game situation. For each of these 36 "branches," it looks 16 moves ahead. Lang's computer uses a speedy Intel Pentium microprocessor that can execute 166 million instructions per second.

1996 New software called Java is first deployed. This object-oriented programming language allows Web developers to create and automatically deliver working programs, not just files, to any computer on the network. It is unlike other languages popular with Internet developers in that it is not restricted to any particular type of computer chip or operating program. A programmer can thus create an application using Java, embed it in a Web page, and any computer that visits the page with a Java-based browser can operate the program. Although still very new, Java is regarded as one of the most important computer software developments since Mosaic (the Internet browser that opened up the World Wide Web to a general audience). It also may fundamentally alter the way software is created, sold, and delivered.

Earth Sciences

c. 5000 B.C. Primitive quarrying and mining first begin with the Neolithic Revolution. Characterized by the use of stone tools shaped by polishing or grinding, this age follows the Paleolithic Age or age of chipped-stone tools, and is followed by the Bronze Age, where metal tools are first used.

c. 600 B.C. First known geological observation is made by Greek philosopher Thales of Miletus (624-546 B.C.), who describes the effect of streams. He is also the first to ask what the universe is made of and to try to answer this question without seeking a supernatural explanation. He says water is the fundamental or primary element from which everything is made.

c. 550 B.C. First person known to study fossils is Greek philosopher Anaximander (610-c.546 B.C.). He theorizes that they are evidence of previous life forms. He also is the first to draw a map of the entire earth as he knows it. He states that life originated in the water and that humans developed from fish.

c. 525 B.C. First to teach that the earth is spherical is Greek philosopher Pythagoras (c.582-497 B.C.). He also says the earth's core is fire and that running water sculpts its surface.

c. 450 B.C. First to state that the earth moves

through space is Greek philosopher and mystic Philolaus.

c. 400 B.C. First mention of the lost continent of Atlantis is made by Greek philosopher Plato (c.4227-347 B.C.) in his dialogue *Timaeus*. This moralistic tale mentions a submerged continent on the other side of the Pillar of Hercules (Gibraltar). Although written as part of a fictional story, the legend of Atlantis persists and is believed by many modern geologists. It is known that around 1400 B.C., an island in the Atlantic Ocean did explode volcanically, possibly inspiring Plato's tale.

c. 300 B.C. First Greek to observe and describe real tides is Greek geographer and explorer Pytheas. He speculates correctly that they are influenced by the moon. As an explorer, he sails westward from Greece beyond Gibraltar and up the northwestern coast of Europe, possibly reaching Norway.

(See also c.700)

c. 240 B.C. First to correctly estimate the earth's size is Greek astronomer Eratosthenes (c.276-c.196 B.C.). He also makes a map of the known world that includes details from the Caspian Sea to Ethiopia, and from the British Isles to Ceylon. Knowing that on the summer solstice, the sun is directly overhead in Syene, while at the same time it is seven degrees from zenith in Alexandria, he guesses that the difference can only be due to the curvature of the earth. Knowing the exact distance between Syene and Alexandria, he then calculates the diameter of the earth by assuming it is a sphere with an equal curvature on its entire surface. His final circumference calculation of about 25,000 mi (40,225 km) is almost correct. Centuries later, most academics discount his calculation, considering it to be far too large, since it implies that most of the earth is covered by water. He also suggests that the earth's seas are all connected, and that the overall effects of water, earthquakes, volcanoes, and fluctuations of the sea are insignificant in proportion to the overall size of the earth.

c. 20 B.C. First suggestion of the existence of unknown continents is made by Greek geographer Strabo (63 B.C.-19 A.D.). He also discusses the land-forming activity of rivers, recognizes the long-dormant Vesuvius as a volcano, and notes that earthquakes are less prominent when volcanoes are active. He is regarded as the father of modern theories that volcanos make mountains, and establishes the notion that volcanoes act as safety valves for pent-up subterranean pressures. He notes the presence of marine fossils in the desert and states that the sea had once covered certain portions of the land. He also teaches that entire continents can be moved over time and that probably all islands were at one time connected to larger continents.

c. 44 A.D. First division of the earth into zones is made by Roman geographer Pomponius Mela. He uses five zones to divide the earth: North Frigid, North Temperate, Torrid, South Temperate, and South Frigid. Although these divisions are generally believed today, Mela considers the Torrid Zone to be uninhabitable and impassable because of its burning heat.

132 First record of any kind of seismometer is that invented by Chinese astronomer Chang

Heng (78-139). None of its details or how it functions is known.

c. 700 First medieval statement that the tides are governed by the phases of the moon is made by English scholar Bede (673-735). Also called the Venerable Bede, he states correctly that high tide does not occur at the same time everywhere, and that tide tables have to be prepared separately for every port. (See also c.300 B.C.)

c. 900 First to divide all substances into the grand classification of animal, vegetable, and mineral is Persian physician and alchemist Rhazes (c.860-c.930). He also subclassifies minerals into metals, volatile liquids (spirits), stones, salts, and others.

914 Al-Masudi, Arab historian and geographer, records accurate principles on evaporation and the causes of ocean salinity. He is the first in the East to combine history and scientific geography in a major work. He is known as the "Herodotus of the Arabs."

c. 1175 First mention in China of fossils is by Chu Hsi, whose book *Chu Tsi Shu Chieh Yao* states that fossils were once living organisms.

c. 1335 First known recording of tides for the purpose of prediction of floods is made by English scholar Richard of Wallingford (c.1292-1336). He later builds a mechanical clock that indicates low and high tide.

c. 1350 Giacomo Dondi (1298-1359), Italian physician and astronomer, writes his *Tractatus de Cause Salsedinis* in which he discusses the salinity of the oceans as well as the phenomenon of the tides. He suggests that the combined effects of the sun, moon, and planets have an effect on the earth's tides. He is also the first to recommend the extraction of salts from mineral waters for medicinal purposes.

c. 1357 First version of isostasy is offered by German philosopher and mathematician Albert of Saxony (c.1316-1390). This geological notion implies that rock formations slowly find their natural level by sinking or rising according to their densities.

(See also 1889)

1473 First published use of the word "geology" is found in *Philobiblon*, written by medieval writer Richard de Bury (1287-1345) more than a century before. This book uses the word "geology" to describe the science or law of the earth as opposed to "theology" which is the science of the divine. Geology comes from the Greek words "ge" for earth and "logos" for reason.

1502 First Westerner since Greek times to scientifically classify stones and minerals is Italian physician Camillus Leonardus (also known as Camillo Leonardi). In his lapidary titled *Speculum Lapidum*, he states that the identification of stones depends on their hardness, porosity, gravity, and density. In 1505, his *De Mineralibus* classifies minerals according to their physical characteristics.

1505 First printed book on earthquakes is *De Terremotu et Pestilentia* by Filippo Beroaldo (1453-1505) of Italy. This short work on earthquakes also contains some information on the plague.

1513 First written description of the Gulf Stream in the North Atlantic is given by Spanish explorer Juan Ponce de Leon (c.1460-1521).

1546 First handbook on mineralogy is written by Georgius Agricola (1494-1555), German mineralogist. In his *De Natura Fossilium*, he classifies minerals and other earth materials in terms of their geometrical form (spheres, cones, and plates). This is the first mineral classification that is really empirically based, and it offers a comprehensive system. In this work, he is also the first to distinguish between "simple" substances and "compounds." It is because of this

An illustration from *De Terremotu*, the first book on earthquakes.

book that Agricola is considered the father of mineralogy.

1546 Origin of mountains is first considered by German mineralogist Georgius Agricola (1494-1555) in his *De Ortu et Causis Subterreaneorum*. He argues that there are many factors in mountain building, but the main mechanism is erosion caused by moving water, with mountains forming along the banks of ever-deepening river beds. In this work he also describes his ideas on the origin of ore deposits in veins, and correctly attributes them to deposition from aqueous solution.

1556 First to base the study of geology on observation as opposed to speculation is German mineralogist Georgius Agricola (1494-1555). His opus, *De Re Metallica*, is published this year, one year after his death. This monumental work summarizes every aspect of mining and metallurgical processes of his time. It has 273 woodcut illustrations that show machinery and processes, and it takes him 25 years to complete. As the product of a lifetime of practical experience, it contains little hearsay (although it does offer the concept that varying

Georgius Agricola illustrates the technical aspects of mining.

proportions of earth, air, fire, and water account for the differences in the external characteristics of minerals). It is a popular and definitive work, and earns Agricola the title "father of mineralogy."

1561   First printed book in Europe dealing solely with the origin of mountains is *De Montium Origine* written by Valerio Faenzi of Italy.

1565   First catalog of a geological collection based on Agricola's classification system is the

Calculorum qui in Corpore ac Membris Hominum by German geologist Johannes Kentmann (1518-1574).

(See also 1546)

1565   First illustrations of fossils are published by Swiss naturalist Konrad von Gesner (1516-1565), in his *De Rerum Fossilium*. He does not, however, consider them to be the remnants of past living organisms, but discusses them along with other diverse things found in the soil (minerals, ores, stalactites, and prehistoric stone tools). He also classifies minerals into 15 categories.

1570   First systematic observation on the range of strata is made by British geologist George Owen (1552-1613). He makes the observation that masses of minerals found in the earth are not simply thrown together in a haphazard manner but have their own regular order. Owen also traces bands of carboniferous limestone for some distance. He is often called the father of English geology.

1580   First work in English to deal solely with earthquakes is *Concerning the ... Effects of All Earthquakes,* written by Thomas Twyne (1543-1613) of England.

1580   First in France to maintain that fossil shells and fossil fish are the remains of animals that once lived in the sea is French natural philosopher and potter Bernard Palissy (c.1510-1589). He also investigates the formation of underground springs and explains volcanoes and earthquakes as the result of underground fires.

1590   First known survey of the New World and its relation to the old is *Historia Natural y Moral de las Indias,* written by Spanish theologian and explorer José de Acosta (1539-1600). He speculates that Indians populated the New World by migrating over a land bridge that formerly connected Asia and North America.

1603   First to use the word "geology" is Italian

naturalist Ulisse Aldrovandi (1522-1605). He uses it to suggest the materials or the rocks that make up the subject of the science.

1609 Anselmus Boethius de Boodt (c.1550-1632), Belgian physician, publishes his *Gemmarum et Lapidum Historia* in which he makes one of the first attempts at a systematic description of minerals. He also classifies minerals, speculates on the nature of coral, and argues that gems have no magical powers. He distinguishes five degrees of hardness in stones.

1617 First method of determining distances by trigonometric triangulation is offered by Willebrord van Roijen Snell (1591-1626), Dutch mathematician. His work, *Eratosthenes Batavius*, published this year, founds the modern art of mapmaking.

1619 First reference to high temperatures within the earth is made by French physician Jean Baptiste Morin (1583-1656) in his *Nova Mundi Sublunaris Anatomia*. He also proposes that there are three layers of earth which correspond in reverse order to the three recognized regions of air.

1657 First book to use the word "geology" in the modern sense—as the science of the earth—is *Geologica Norvegica* by Norwegian geologist Mikkel Pederson Escholt (c.1610-1669). It also offers a clear and near-modern view of earthquakes and ranks as the first scientific treatise printed in Norway.

(See also 1690)

1663 Isaac Vossius (1618-1689), Dutch scholar, publishes *De Motu Marium et Ventorum,* in which he first states the correct theory that the overall ocean circulation of the North Atlantic region is essentially clockwise.

1667 First law of crystallography is stated by Danish anatomist and geologist Nicolaus Steno (1638-1686). This principle states that the crystals of a specific substance have fixed characteristic angles at which the faces, however

distorted they may be, always meet. He also argues that the earth is layered with the petrified remains of once-living organisms; that fossils are not the "practice-creations" of God; and that "tongue stones" or "glossopetrae" are really shark's teeth. He also describes rock stratification and states that tilted strata were at one time horizontal. In his 1669 book *De Solido* he distinguishes between marine and land fossils, states that some belong to extinct forms, and suggests that they might be used to stratigraphic ends.

1674 One of the first contributions to the establishment of hydrology as a quantitative science is made by French hydrologist Pierre Perrault (c.1611-1680). He does field work on the Seine River and shows conclusively that precipitation is more than adequate to sustain the flow of rivers. He publishes his *De L'Origine des Fontaines* anonymously this year.

1680 First distinction between igneous rock and aqueous rock is made by German philosopher and mathematician Gottfried Wilhelm Leibniz (1646-1716). He states his theory in *Protogaea*, saying that certain rocks had cooled from a state of fusion (igneous), and others had been formed by the action of water into more or less stratified masses.

1683 First map to show geologic features is contained in an article by English zoologist Martin Lister (1638-1712), published by the *Philosophical Transactions* of the Royal Society. He recognizes that the geographical distribution of different strata can be represented on a map.

1688 First to suggest that fossils might be useful in revealing the historical past of the earth is English physicist Robert Hooke (1635-1703). In this work on earthquakes, he states that mollusk fossils deserve to be regarded as historical objects, like old coins, that can tell us about the past. He also concludes from the types of fossils he finds that the climate of

England once had been much warmer. He attributes earthquakes and volcanoes to the existence of fires within the earth.

1690   First published reference to geology in England is in the title of a book by English cleric Erasmus Warren. His work, *Geologia; or, A Discourse Concerning the Earth Before the Deluge*, is of little scientific value.

1691   Earth's water cycle is first detailed by English astronomer Edmond Halley (1656-1742). He publishes a paper this year describing how the sun evaporates sea water, which rises to mountain heights and is condensed into rain, which then falls and penetrates into the earth to emerge in springs and rivers.

1691   First book on artesian wells is *De Fontium Mutinensium* written by Italian physician Bernardino Ramazzini (1633-1714). Artesian wells are those dug into springs from which water flows under natural pressure without pumping. The term is derived from the French province of Artois where naturally flowing wells were drilled as early as the 12th century. Ramazzini's book contains diagrams and also explains how the wells originated.

1696   First to suggest a literal interpretation of the six "days" of creation is English mathematician and Anglican priest William Whiston (1667-1752) in his book, *A New Theory of the Earth*. In this work he attempts to explain Biblical accounts scientifically, and thus states that the great flood can be attributed to a passing comet. He also suggests that the earth is much older than most believe. Despite his religious orientation, he is one of the strongest advocates of the need to harmonize science and religion.

1699   First major fossil catalog is written by Welsh botanist and geologist Edward Lhuyd (1660-1709). His eight-volume *Lithophylacii Britanici Ichnographia* is a landmark work that lists and classifies over 1,700 types of fossils.

1703   First instrument built to study an earthquake's shocks is designed by Jean de Hautefeuille (1647-1724), French physicist. He builds a rudimentary seismoscope that uses drops of mercury. Although fairly crude, it is still considered an early instrument.

1705   First mastodon bones discovered in North America are found in Albany, New York, by American cleric and writer Cotton Mather (1663-1728). Mastodons are an extinct type of elephant that were distributed throughout the world and became extinct relatively recently. Their remains are quite common and often well-preserved.

1715   First definitive statement on the origin of springs is made by Italian zoologist Antonio Vallisnieri (1661-1730). He says that spring water gushes from mountains because rain and snow that had penetrated the ground earlier was now resurfacing.

1719   First to suggest a theory of stratified rock formations is English geologist John Strachey (1671-1743) in his *A Curious Description of the Strata Observ'd in the Coal-Mines of Mendip*. This leads to his 1727 book, *Observations on the Different Strata of Earths and Minerals*, in which he states that there is a relation between surface features and rock structure. This is one of the earliest attempts at establishing a stratigraphical succession.

1722   First scientific study of ferrous metals and the processes for working them is written by French physicist René Antoine Ferchault de Reaumur (1683-1757). His book, *L'Art de Convertir Le Fer Forge En Acier et L'Art D'Adoucir Le Fer Fondu*, demonstrates the importance of carbon to steel and helps the French iron and steel industries to improve their production. He also shows that certain "tourquoises" found in southern France are in fact fossil teeth of extinct animals.

1725   Luigi Ferdinando Marsigli (1658-1730), Italian soldier and oceanographer also called

Marsili, publishes his *Histoire Physique de la Mer*, the first complete book dealing with the sea. He also is the first to use a naturalist's dredge.

c. 1735 First to classify minerals according to their crystalline forms is Swedish botanist Carolus Linnaeus (1707-1778).

1735 First to find permafrost is German explorer Johann Georg Gmelin (1709-1755), who discovers it during an expedition to Siberia. Permafrost is earth, part or all of which is permanently frozen. It is estimated that nearly 20% of the earth's surface is permafrost.

1738 First measurements confirming the earth is flattened at the poles are offered by French mathematician Pierre Louis Moreau de Maupertuis (1698-1759). This year he publishes his *Sur La Figure de La Terre* in which he states the results of his expedition to Lapland in 1736 to measure the curvature of the earth. His calculations show that the earth is an oblate spheroid.

1743 First geological map is made by English physician and geologist Christopher Packe (1686-1749). He produces "A New Philosophico-chrorographical Chart of East Kent," which maps the geology of East Kent, England. He initiates the use of hachuring (shading) by parallel lines drawn in the direction of a slope to show the pattern of valleys. Hills and valleys are easily distinguishable as are the ranges of chalk, slate, and clay.

1744 First national geographic survey is conducted by Cesar Francois Cassini de Thury (1714-1784), French astronomer and geographer and a member of the famous Cassini family of astronomers. He leads the triangulation of France, which is a technique for the precise determination of distances and angles used for road building and other civil engineering construction. As a result of this geodesic survey, he begins to prepare a great topographic map of France that results in the first modern survey map. This is eventually completed by his son, French mathematician and geographer Jacques Domenique Cassini (1748-1845) and serves as the basis for the *Atlas National* of 1791.

1745 First (bituminous) coal mine in North America is opened in Richmond, Virginia. This deposit was known as early as 1673.

1746 First true geologic or mineralogic map that ignores political boundaries is made by French geologist Jean Etienne Guettard (1715-1786). He writes an article in which he declares that the mineralogical formations on the opposing French and English coasts are identical, indicating they were once part of the same geological system. This article also contains his map that ignores political boundaries and shows how the formations or great "bandes" of rock appear to end at the English and French coasts but are in fact connected under the Channel. Later, Guettard makes a major geologic discovery when he recognizes volcanic rock in Central France, far removed from any active volcano.

1746 First use of the word "geognosy" is made by German chemist and mineralogist Johann Heinrich Pott (1692-1777). He uses the word, which is derived from the Greek word *gnosis* for knowledge, in his book *Lithogeognosia*. It is later popularized by the German geologist Abraham Gottlob Werner (1750-1817), who uses it to describe his scientific study of the earth and to distinguish it from the geology of his time, which he says is not based on factual knowledge.

1747 First system of natural classification of minerals is offered by Johan Gottschalk Wallerius (1709-1785), Swedish mineralogist. His *Mineralogia* becomes the first major treatise on this subject. He treats the chemical properties of minerals as well as their external characteristics.

1749 First attempt to measure the gravity of the earth with a plumb line is made by French mathematician and hydrographer Pierre Bouguer (1698-1758). This year he publishes *La Figure*

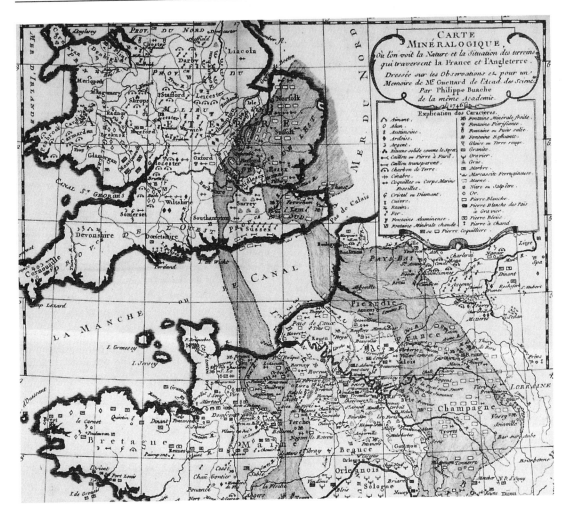

Early map showing the locations of mineral resources.

de la Terre in which he describes an observation made high in the Andes mountains. Now called the Bouguer correction, it states that the lessening of the pull of gravity found at higher altitudes is partly compensated by the gravitational attraction of the intervening rock.

1752 Modern statistical geography is first established by Anton Friedrich Busching (1724-1793), German geographer, who publishes the first of 11 volumes of his *Neue Beschreibung*. Six of these volumes focus on European geography. Altogether they help develop a scientific basis for the study of geography by stressing statistics over descriptive writing.

1755 First work on American geomorphology is written by American cartographer Lewis Evans (1700-1756). This year he publishes *An Analysis of a General Map of the Middle British Colonies in America* which, in map and text, describes the geology of eastern North America. Geomorphology is concerned with the earth's surface features, and specifically the problem of the origin of landforms and the forces that act upon them.

1756   First major classifier of mountains is German mineralogist Johann Gottlob Lehmann (1719-1767). In his *Versuche einer Geschichte von Flotz-Gebrungen*, he classifies mountains according to three types, of which strata play an important role, and also offers a theory of their origins. This work lays the foundation of stratigraphical classification and spurs local geology or the investigation of specific sites.

1758   Blowpipe technique for analyzing ores is first introduced by Axel Fredrik Cronstedt (1722-1765), Swedish mineralogist. The mouth blowpipe is a conical brass tube curved at the small end and terminating in a round orifice. By placing a substance in the flame of a blowpipe, information is obtained about its general nature by the changes it undergoes in the flame and the residue that remains after burning. Cronstedt thus systematizes and improves the technique of observing color changes as a means of making a chemical analysis. The blowpipe remains a useful tool to the skilled user for over a century, and is eventually replaced by the system of spectral analysis. This same year, Cronstedt begins the classification of minerals not only according to their appearance but also according to their chemical structure—an important development.

1759   Threefold classification of rocks of the earth's crust is first established by Italian mining engineer Giovanni Arduino (1714-1795). He originates the classification of primary, secondary, and tertiary rocks. Primary or "Primitive" includes the schists and associated mass found at the core of mountains and containing no organic remains. Secondary is comprised of limestones, marls, shales and other stratified sedimentary materials, many of which are full of fossils. Tertiary is made up of generally looser detritus, derived from the disintegration of other rocks, and sometimes full of the remains of plants and animals. He also adds a fourth (Quaternary or Volcanic) that consists of lavas and tuff accumulated by repeated erup-

tions and inundations of the sea. His categories are readily accepted.

1760   First to suggest that earthquakes emit shock waves is English astronomer John Michell (c.1724-1793). He publishes a long article this year in which he offers his theories on earthquakes, suggesting that one can calculate the center of a quake by noting the time at which the motions were felt. His work is ahead of its time as he states that earthquakes are partly tremulous, but they also move the earth partly by waves or vibrations which succeed one another. He also says earthquakes uplift the strata as well as fracture and rearrange it. He considers all the known ideas about earthquakes and evaluates each according to known facts. Because of his pioneering work on vibrations and waves, he is called the father of seismology.

1763   First to state that much of the rock of France is volcanic in origin is French geologist Nicolas Desmarest (1725-1815). He writes an article in the *Encyclopedie* in which he states that the dark, fine-grained basalt rock in France is the result of volcanic activity. The idea that it and much of Europe was built on the remains of once-active volcanoes stirs a great controversy. Proponents of this idea become known as "Vulcanists" and they are opposed by scientists like the German geologist Abraham Gottlob Werner (1750-1817), who leads the "Neptunist" school. This camp argues that all rocks are formed only by water or the actions of water.

1763   First work on what becomes the Mason-Dixon Line begins. English astronomers Charles Mason (1730-1787) and Jeremiah Dixon are hired to begin a survey of the Pennsylvania-Maryland area in order to settle a dispute. This will take five years and will result in the creation of the Mason-Dixon Line which becomes famous in American history as the boundary between the free states and the slave states.

1765   Bergakademie Freiberg is first established

in Germany. This illustrious German mining school becomes well-known for its faculty and its famous alumni.

1766 First national geological survey is begun by Jean Etienne Guettard (1715-1786), French geologist, and Antoine Laurent Lavoisier (1743-1794), French chemist. This year they begin what is for them an 11-year project sponsored by the Minister of Mining, who plans a geological survey of the entire nation, resulting in 214 quadrangle maps. After 11 years, they have 16 quadrangle sheets completed and another dozen or so in various stages of completion. It is later taken up by others. Although never completely finished, their maps are important not only as the first of their kind but because they also contain stratigraphic information in columns placed in their margins.

1768 James Cook (1728-1779), English navigator, begins the first of three ocean voyages during which he is the first to take the subsurface temperature of the ocean.

1769 Benjamin Franklin (1706-1790), American statesman and scientist, develops the first published chart of the Gulf Stream. He uses temperature measurements and observation of water color to track its course. He suggests that ships going to Europe stay in its current, and that those returning to America avoid it.

1772 Antoine Laurent Lavoisier (1743-1794), French chemist, makes the first quantitative analysis of seawater.

1774 Abraham Gottlob Werner (1750-1817), German geologist, publishes his *Über die äussern Kennzeichen der Fossilien* in which he first establishes a new language and new methods for classifying minerals. Placing little emphasis on crystal shape and chemical composition, he bases his classification on the external characteristics of minerals (such as color, external shape, fracture, transparency, hardness, specific weight, etc.). He devotes his entire career to the principle that all strata were laid down as sedi-

ment by the action of water (Neptunist school), and stridently resists the opposing school (Vulcanists) that argues for the action of heat and volcanoes.

1775 First determination of the density of the earth using a plumb line is made by English astronomer Nevil Maskelyne (1732-1811). Choosing a roughly symmetrical, small mountain, he sets up observatories on its south and north slopes and uses a plumb line, determining the angle between it and fixed stars. After several mathematical calculations and corrections, he finds the density to be 4.7. One of the variables he factors in is the density of what he calls "common stone" found beneath the mountain, but since he does not attempt to determine what type of rock it is, his calculations are not accurate.

1775 First colored geologic map is produced by Friedrich Gottlieb Glaser (1749-1804) of Germany in his *Versuch einer Mineralogischen Beschreibung der Gefürstetn Graffschaft Henneberg*. It shows a part of Saxony.

1779 First person to climb mountains in order to investigate them scientifically is Swiss physicist Horace Benedict de Saussure (1740-1799). This year he publishes the first part of his four-volume *Voyages dans les Alpes* in which he summarizes his findings and also regularly uses the word "geology." Until Saussure, the high regions of the Alps were regarded fearfully as "montagnes maudit"—desolate and dangerous tracts. Saussure not only provides accurate observations on mountains and glaciers, but he is the first to also describe their beauty.

1780 First to show in map form the relative time relationships between stratigraphic formations and the fossils found there is Jean Louis Giraud Soulavie (1752-1813), French geologist. In his *Histoire Naturelle de la France Méridionale*, he distinguishes five kinds of sedimentary strata, characterizing each by the types of fossils found there. He observes that some species have changed over time and others have disappeared entirely. He also infers that the time period for all of this to occur must have been extremely long. Pointing the way to stratigraphic correlation by fossils, he is considered the founder of stratigraphical paleontology.

1784 Chemical composition as a means of classifying minerals is first introduced by Irish geologist Richard Kirwan (1733-1812) in his *Elements of Mineralogy*.

1784 Mathematical theory of crystal structure is first offered by French mineralogist René Just Haüy (1743-1822). This year he publishes his *Essai d'une Theorie sur la Structure des Crystaux* in which he joins mathematics and mineralogy and lays the foundation for a new way of understanding crystals. He offers his law of crystal symmetry which says that crystals are the result of the stacking together of tiny, identical units. This new revelation about crystal structure lays the groundwork for the science of crystallography.

(See also 1801)

MARCH 1785 Uniformitarianism is first proposed by Scottish geologist James Hutton (1726-1797) in a paper, "Concerning the System of the Earth," that he presents to the Royal Society of Edinburgh. It is out of this brief paper that geology has its beginnings as a real science and an organized field of study. Hutton's theory of "uniformitarianism" contains one of the most fundamental principles of geology—that natural processes have been constantly at work shaping the earth over enormously long periods of time. He says that the prime force behind this shaping is the internal heat of the earth. His theory of geologic gradualism gets its name because it views the geologic evolution of the earth as a slow, continuous transformation that is only rarely marked by sudden, catastrophic change. These ideas meet considerable resistance, but Hutton eventually prevails and is known as the father of geology.

(See also 1830)

1786 Mont Blanc, in southeast France on the Italian border and the highest point of the Alps, is first climbed by Jacques Balmat and Michel-Gabriel Paccard. The notion of mountains as inaccessible, accursed places is beginning to change, and they eventually become subjects of scientific investigation.

1787 First book on the geology of America is written by Johann David Schopf (1752-1800), German physician and geologist. After serving as a Hessian surgeon, he tours the new United States from Rhode Island to the Carolinas and publishes *Beiträge zur Mineralogischen Kenntnisz der Ostlichen Theils von Nordamerika* in Erlangen, Germany. In it he records his comprehensive observations about American landforms, records the kind of rock he discovers and examines its mineral composition, and makes the best possible effort to explain what he sees.

1787 First systematic treatise on geologic formations is written by Abraham Gottlob Werner (1750-1817), German geologist. In his *Kurze Klassifikation*, he classifies rock types according to his system of sequential precipitation and deposition from the waters of a universal ocean. It is here that he says that basalt rock is aqueous (formed by the action of water) and not volcanic in origin (formed by the action of heat). Unlike Scottish geologist James Hutton (1726-1797), who recognizes the important geologic role of heat and volcanic action (and is therefore called a Vulcanist), Werner believes incorrectly that virtually all strata had been laid down as sediment by the action of water (and is called a Neptunist).

1790 James Rennell (1742-1830), English geographer, begins his scientific study of winds and currents, and makes the first comprehensive study of the Atlantic Ocean currents.

1792 First to study volcanic rock experimentally is Lazzaro Spallanzani (1729-1799), Italian naturalist and biologist. This year he publishes

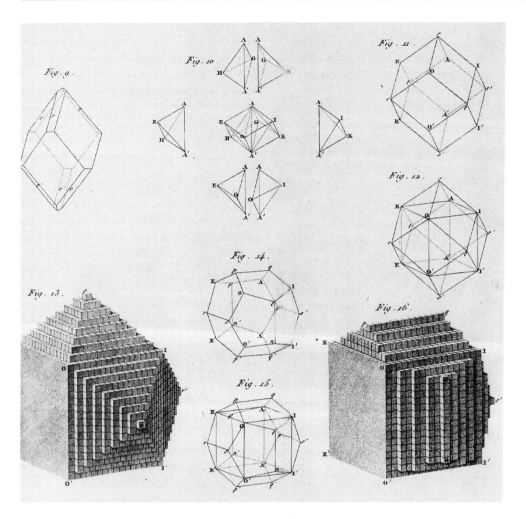

The discipline of crystallography was founded by René Just Haüy.

the first of his six-volume work, *Viaggi alle due Sicilie e in Alcune Parti dell'Appenine*, in which he describes his observations of Vesuvius, Stromboli, Vulcano, and Etna. In order to observe properly, he suffers burns while measuring the flow of red-hot lava and by actually descending into a volcano. He is also overcome by gas at Etna. His studies earn him the status of pioneer in volcanology.

1796 First to extend a system of classification to fossils is French anatomist Georges Cuvier (1769-1832). He notes that although many

fossils are different from living animals, they nonetheless belong to one of the four phyla he had established (Vertebrata, Mullousca, Articulata, and Radiata). Despite his awareness that the deeper the fossil is found, the older the rock (and the more the fossil differs from the life forms he knows), he refuses to make any judgments about an evolutionary theory, suggesting rather that the world was subjected to several catastrophic events.

1798 First accurate determination of the earth's mean density is made by English chemist and

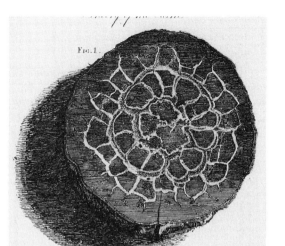

James Hutton uses this slice of iron-stone to illustrate his theory of uniformitarianism.

physicist Henry Cavendish (1731-1810). Aware that the earth's mass can be calculated if the value of the gravitational constant is known, he devises an experiment to determine the gravitational constant for smaller bodies, from which he will extrapolate for the entire earth. Using a suspended beam with two lead balls attached to a torsion balance, he measures the period of oscillation and is able to calculate the gravitational force between the two balls. Using this value, he is able for the first time to measure the earth's mass and also arrives at a mean density figure of 5.48 times that of water. This is very nearly correct.

1798 Experimental geology is first founded by Scottish geologist and chemist James Hall (1761-1832). This year he begins a series of papers in which he provides experimental evidence for the geological theories of his colleague, Scottish geologist James Hutton (1726-1797). By melting minerals and cooling them at a controlled rate, he proves that he can obtain different kinds of rocks. He continues to experiment extensively with igneous rock and shows that they are produced by intense heat. He is also known as the founder of geochemistry.

1799 First geologic map based on scientific principles is begun by English geologist William Smith (1769-1839). After extensive observations of his native land, Smith discovers that each successive layer of strata has its own characteristic forms of fossils and that with study, one can identify a stratum by its fossil content. He then plans to make his discovery known by making a map that uses different colors to indicate the succession of sedimentary beds or groups of beds. In 1815 he eventually publishes his landmark map, *A Delineation of the Strata of England and Wales with Part of Scotland*, which lays down the principles of paleontological stratigraphy. This geologic map, roughly 6 x 9 ft (2 x 3 m), is the first attempt to represent on a large scale the geological relations of such an extensive tract of ground. Its color key makes it easily understood.

1801 Science of crystallography is first founded by French mineralogist René Just Haüy (1743-1822). He publishes his five-volume masterpiece, *Traité de Minéralogie*, in which he explains that if crystalline forms are identical (or different), then that identity or difference is found in its chemical composition. Crystallography will become highly significant to physics with the eventual development of x-ray techniques.

1803 First color map of the eastern region of the United States is made by French traveler Constantin François Chasseboeuf de Volney (1757-1820). His map is part of a broad geological description of this area, which he divides into several regions of granitic rocks, sandstones, calcareous rocks, marine sands, and river deposits. The map itself is part of a book he publishes, in which the colors used are explained. Volney gives a copy to his American friend William Maclure (1763-1840)), and it inspires him to produce a geological map of his own.

(See also 1809)

1803 Friedrich Wilhelm Heinrich Alexander von Humboldt (1769-1859), German naturalist, writes an article in which he first states that the Atlantic Ocean is a valley dug out by a torrential current. He also studies the oceanic current off the west coast of South America that is originally named after him but is now called the Peru Current.

NOVEMBER 13, 1807 Geological Society of London is founded, becoming the first association of its kind in the world. Its first president is English geographer and geologist George Bellas Greenough (1778-1855), who becomes primarily responsible for its success. By April, 1811, the society has 200 members.

(See also 1811)

1809 First frequently reprinted and widely distributed geological map of the eastern coast of the United States is made by William Maclure (1763-1840), American geologist. Although his map is crude and appears to have been done hastily, it is widely influential and stands as a major symbol of the beginning of the geological exploration of the American continent.

1809 Principles of scientific paleontology are first offered by English geologist William Martin (1767-1810). In his work *Outlines of an Attempt to Establish a Knowledge of Extraneous Fossils on Scientific Principles* he offers the sound beginnings of a science of past plants and animals based on fossil evidences.

1811 Geological Society of London publishes its first *Transactions* containing 18 articles.

1811 First major shift of the Mississippi River occurs during a series of earthquakes centered in New Madrid, Missouri. Continuing into 1812, these major quakes alter much of the midwest topography and are considered the strongest in United States history (estimated at 8.0 on the Richter scale) because they affect such a large area. The New England earthquakes of 1727 and 1755 were not nearly as severe.

1811 Color geologic map is first used to dem-

onstrate how fossils can be used to accurately determine the geological chronology of a particular area. French mineralogist Alexandre Brongniart (1770-1847) and French anatomist Georges Cuvier (1769-1832) publish their *Essai sur la Géographie Minéralogique des Environs de Paris* in which they offer a table of stratigraphical succession linking fossils and geological chronology. This work places paleontology on an accurate, scientific basis. It is based largely on the work of Brongniart.

1812 First accurate reflection goniometer capable of measuring the angles between faces of small crystals is designed by English chemist William Hyde Wollaston (1766-1828). His optical goniometer measures the interfacial angles of small crystals by reflecting a beam of light from them. It is built to check the known laws of crystal morphology, some of which he is able to correct. His device greatly advances mineralogical research.

1812 Catastrophism is first proposed by French anatomist, Georges Cuvier (1769-1832). He publishes his *Recherches sur les Ossemens Fossiles* in which he not only reconstructs many fossil vertebrates, but also states his geological theory that the earth has had a long, peaceful history punctuated by sudden physical changes such as a flood. He argues that after such a total event, life is created anew on the earth and the remaining fossil record then contains only the remnants of the ages before the flood. He states that the last catastrophe was the Flood described in Genesis.

1812 Mohs hardness scale is first proposed by German mineralogist Friedrich Mohs (1773-1839). He devises a decimal scale to measure the resistance of a smooth surface to scratching or abrasion. It is measured against the result of scratching its surface with a substance whose hardness is known. The Mohs scale is composed of 10 minerals that have been given values. In the order of softness to hardness, they are: 1. talc, 2. gypsum, 3. calcite, 4. fluorite, 5.

apatite, 6. orthoclase, 7. quartz, 8. topaz, 9. corundum, 10. diamond. If a mineral is scratched by topaz but not by quartz, its Mohs hardness is 7.5-8.

1813   First successful English textbook on geology is written by English geologist Robert Bakewell (1768-1843). Many editions of his *Introduction to Geology* is published in the United States after it is edited by American chemist Benjamin Silliman (1779-1864).

1815   First British mineralogist to follow Haüy's classification system is English physician and chemist John Kidd (1775-1851). This year he publishes his *Geological Essay*, which embraces the laws of crystallography as proposed by French mineralogist René Just Haüy (1743-1822).

1816   First broad compilation of American mineral resources is published by American geologist Parker Cleaveland (1780-1858). His *Elementary Treatise on Mineralogy and Geology* becomes especially popular in Europe and earns him international recognition. In 1818 he becomes the first American member of the Geological Society of London.

1816   First to propose that glaciers once covered most of Europe is Swiss civil engineer Ignatz Venetz (1788-1859). His article on glaciers and their movement influences German-Swiss mining engineer Johann von Charpentier (1786-1855) who, in turn, persuades Swiss-American naturalist Jean Louis Rodolphe Agassiz (1707-1773) of this fact.

(See also 1834)

1818   First geologic analysis of American rock is published by American physician John White Webster (1793-1850).

1818   First deep-sea study of the ocean floor is conducted by English explorer John Ross (1777-1856). He leads an expedition into Baffin Bay (in the Atlantic Ocean between west Greenland and the Canadian Arctic archipelago) and succeeds in bringing up muddy sediments and

worms using a "clamp" he designed himself. His findings are ignored. Later he makes other contributions to oceanography during his search for the Northwest Passage, the non-existent North American waterway linking the Atlantic and Pacific oceans.

1818   *American Journal of Science* is first established by American geologist Benjamin Silliman (1779-1864). Silliman uses this journal to introduce Britain's most advanced geological ideas to America. He also cultivates students who later have a profound effect on American geology. It is through this journal that Silliman becomes the best-known American geologist.

1819   First successful French textbook on geology is written by Jean Francois d'Aubuissons de Voisins (1769-1843), French geologist. His book, *Traité de Géognosie*, uses the word *geognosy* as employed by his mentor, German geologist Abraham Gottlob Werner (1750-1817), to mean the arrangement and classification of minerals according to their age.

1819   First "Reader in Geology" at Oxford University is English geologist William Buckland (1784-1856). Although already a professor of mineralogy at Oxford, he is appointed to a teaching position in geology. This is the first such appointment in a major English university and contributes to the public recognition of geology as a science. Buckland has the distinction of publishing in 1836 the last major geological textbook to be written from the diluvialist (Great Flood) point of view. It is titled *Geology and Mineralogy Considered with Reference to Natural Theology*.

1821   Britain's first official survey geologist, John MacCulloch (1773-1835), Scottish chemist and geologist, publishes his work on the classification of rocks. He does much of his geological work while in government service.

1822   First recognized dinosaur fossil is discovered by English physician Gideon Algernon Mantell (1790-1852) and his wife Mary Ann

Mantell while searching for fossils around the quarries of Sussex. Mantell's wife discovers large fossil teeth remains of a dinosaur, and by the time of his death, Mantell discovers four out of the five genera of dinosaurs known during his time. From his work, the dinosaurian reptiles known as *Iguanodon*, *Hylaeosaurus*, *Pelorosaurus*, and *Regnosaurus* are unearthed and documented. Mantell's fossil collecting takes over his home (which swarms with sightseers) and ruins his medical practice. His wife eventually takes the children and moves out.

1822 Cretaceous Period is named by the Belgian geologist Jean Julien d'Omalius d'Halloy (1783-1875). As part of the Mesozoic Era, it is the interval of geological time that began about 136 million years ago, following the Jurassic Period and preceding the Tertiary Period. Its name refers to chalk (*creta* in Latin) which was the characteristic rock formed in most of Europe during this time. It was during this period that the earth assumed many of its present features.

1823 First consistent classification of rocks is offered by Karl Caesar von Leonhardt (1779-1862), German mineralogist and geologist, in his *Characteristik der Felsarten*.

1826 First geologic map of Germany is prepared by German geologist Christian Leopold von Buch (1774-1853). This is a magnificently detailed map composed of 42 sheets. Buch's extensive travels and wide geological investigations eventually lead him to break with his mentor, German geologist Abraham Gottlob Werner (1750-1817) who believes the Neptunist theory, after he visits the Auvergne region of France and realizes that this area is obviously the result of volcanic activity.

1827 First extensive study of the nature and effects of volcanic action is made by George Julius Poulett Scrope (1797-1876), English geologist, in his *Memoir on the Geology of Central France*. His astute observations on the age

of these volcanoes in the Auvergne region of France play a major role in deposing the Neptunist theory that all rocks were formed by sedimentation from the oceans. He also states simply that volcanoes merely build up in time as rock and ashes are ejected from a volcanic vent, and that if this activity continues for a long time, the volcano becomes very large. Scrope's writing has a major influence on Scottish geologist Charles Lyell (1797-1875), who travels to Auvergne to see for himself.

1827 First to distinguish minerals from rocks is Alexandre Brongniart (1776-1847), French mineralogist, in his *Classification et Caracteres Mineralogiques des Roches Homogenes et Heterogenes*. A mineral is crystalline and inorganic with certain specific physical characteristics. Rock is made up of one or more minerals and is divided into one of three types according to its mode of origin.

1829 Polarizing microscope is invented by English physicist William Nicol (1768-1851). Using plane-polarized light (which vibrates in only one plane, as opposed to ordinary light which vibrates in all directions), this new device illuminates the specimen which appears bright against a dark background. Nicol also develops a method of making thin sections of minerals and fossil wood for examination on his new microscope. It is not until the work of English geologist and chemist Henry Clifton Sorby (1826-1908), however, that microscopic petrography actually comes into being.
(See also 1851)

1829 Jurassic Period is named by Alexandre Brongniart (1770-1847) and German naturalist Friedrich Wilhelm Heinrich Alexander von Humboldt (1769-1859). They designate this part of the Mesozoic Era between the Triassic and Cretaceous Periods after the Jura Mountains between France and Switzerland (because of the chalk sequence in its strata).

1830 Uniformitarianism is first successfully popularized by Charles Lyell (1797-1875), Scottish geologist, in his classic work, *The Principles of Geology*. Written in three volumes, this landmark work is very well received and serves to validate the gradualist ideas of Scottish geologist James Hutton (1726-1797) concerning what the proper time scale of geology should be. He eloquently and persuasively documents his argument that there exists a continuity of life and a natural process in the physical world. He also estimates the age of some of the oldest fossil-bearing rocks at the startling age of 240 million years. He achieves his goal of writing a book "to establish the principle of reasoning" in geology, and dismisses theological explanations of natural phenomena that are unscientific and unsupported. This work greatly influences English naturalist Charles Robert Darwin (1809-1882), who said, "I always feel as if my books came half out of Lyell's brain."

(See also March 1785)

1830 *Jarbuch für Mineralogie, Geognosie, und Petrefaktenkunds* is first established by German geologist Heinrich Georg Bronn (1800-1862) in Heidelberg. He hopes this publication will speed up and bring together the scientific communication of geological and paleontological information.

1833 First major book published by a state survey in the United States is written by American geologist Edward Hitchcock (1793-1864). This work, *Final Report on the Geology of Massachusetts*, also includes the first geological map of an American state.

1834 First to propose the idea of the extensive movement of glaciers is Johann von Charpentier (1786-1855), German-Swiss mining engineer. He suggests that the huge erratic boulders found scattered throughout the Alpine regions have been brought there by ancient glaciers. He precedes Swiss-American naturalist Jean Louis Rodolphe Agassiz (1807-1873) in this notion of glacial geology and actually takes him into the field to demonstrate that the valley glaciers

in the Alps once extended to the lower levels, indicating that the glaciers of recent past geologic ages were more extensive than at present. (See also 1840)

1834 Triassic Period is first named by German geologist Friedrich August von Alberti (1795-1878). As the earliest period of the Mesozoic Era, it is named by Alberti after the "Trias," which is a sequence of strata in central Germany. The name refers to the threefold division of the strata of rocks of marine origin.

1835 Ordnance Geological Survey is first established in Britain. It later becomes the Geological Survey of Great Britain.

1836 Cambrian geologic period is first named by Adam Sedgwick (1785-1873), English geologist. During his investigations of ancient Welsh rocks, he gives the name Cambrian (from the ancient name for Wales) to the oldest fossil-bearing rocks. The Cambrian geological period of time is thus considered to be from 570 to 500 million years ago.

1837 Human paleontology is first founded by French paleontologist and archeologist Edouard Armand Isidore Hippolyte Lartet (1801-1871). This year he discovers the jawbone fossil of a primate and turns his focus solely to the study of the origins of humankind. In 1852, he finds evidence that humans coexisted with now-extinct animals. He also discovers some of the earliest known art, and goes on to show that the Stone Age was comprised of a succession of phases in human culture.

1838 Geochemistry is first coined by Christian Friedrich Schonbein (1799-1868), German physical chemist. This year he publishes an article in which he first uses the word geochemistry to describe the study of the chemical and physical properties of rock formations.

1838 Geological term *facies* is first used by Swiss geologist Amanz Gressly (1814-1865), who applies it to sedimentary rock in his article,

"Observations Geologiques sur le Jura Soleurois." He uses the word to mean the sum total of the organic and inorganic characteristics of a sedimentary formation.

1838 Micropaleontology is first founded by German biologist Christian Gottfried Ehrenberg (1795-1876). He founds the study of fossil microorganisms after discovering the microscopic fossil organism content of various geological formations and realizing that such single-cell fossils comprise certain rock layers.

1839 Charles Lyell (1797-1875), Scottish geologist, first identifies the Pleistocene Period in the history of the earth. Spanning a time period from about 2.5 million to 10,000 years ago, this epoch includes several advances and retreats of ice sheets and glaciers. Pleistocene geology is somewhat synonymous with glacial geology.

1839 Silurian geologic period is first named by Roderick Impey Murchison (1792-1871), Scottish geologist, in his 800-page work *The Silurian System*. The period is below those which English geologist William Smith (1769-1839) had already ordered, and it establishes for the first time the stratigraphic sequence of early Paleozoic rocks. This enables geologists to accurately trace the earth's history backward across an increasingly long span of time. Murchison names his system after the Silures, an old British tribe that was present during Roman times.

1839 Devonian Period of geologic time is first named by English geologists Adam Sedgwick (1785-1873) and Roderick Impey Murchison (1792-1871). After researching deposits in Devon and Cornwall, they choose this name to distinguish the rocks of the Paleozoic Era between the Silurian and Carboniferous periods.

1839 "Miller indices" for crystals are first introduced by English mineralogist and crystallographer William Hallowes Miller (1801-1880). In his book *A Treatise on Crystallography*, he offers a system of indexing that uses three

numbers to describe each type of crystal face. Also called Millerian indices, these very accurate crystallographic symbols are still in use today.

1840 Theory of the Ice Age is first put forth by Jean Louis Rodolphe Agassiz (1807-1873), Swiss-American geologist, in his *Etudes sur les Glaciers*. After spending five summers among the glaciers of his native Switzerland, he states his theory of glacier movement, demonstrating through experimentation the fact that glaciers move. He also explains the transporting power of glaciers which, in turn, explains much about modern geological configurations. His ideas would not be accepted for another 25 years.

1841 Permian Period is first named by English geologist Roderick Impey Murchison (1792-1871). This is the only name not based on a type of rock found in Europe. By invitation from the Czar of Russia, Murchison studies in Russia and recognizes a sequence of beds that prove younger than rocks of the Carboniferous period he knows to exist in England. He names them Permian (after the Russian district of Perm where they are found) to distinguish them from the rocks of the Carboniferous Period.

1841 First geologic map of France is completed. Begun by French geologists Armand Petit Dufrenoy (1792-1857) and Jean Baptiste Armand Louis Léonce Elie de Beaumont (1798-1874) in 1825, this great geologic map is titled *Carte Géologique de la France au 1:500,000*. It is inspired by and modeled after the pioneering map of England and Wales done in 1815 by English geologist William Smith (1769-1839).

1841 Edward Forbes (1815-1854), English naturalist, pioneers the use of a dredge in the scientific study of shallow water. One of the first to consider the sea as an entity and to divide the ocean into natural zones on a scientific basis, he is regarded—along with American oceanographer, Matthew Fontaine Maury

(1806-1873)—as the co-founder of oceanography.

(See also 1855)

1842 Richard Owen (1804-1892), English zoologist, first coins the word dinosaur (Greek for terrible lizard) and refers to the gigantic proportions of these beasts.

1843 William Edmond Logan (1798-1875), British-Canadian geologist and the first director of the Geological Survey of Canada, publishes his research on the geological formations of North America. His work makes him a pioneer in Pre-Cambrian geology. He is also the first to determine that coal beds are formed in place, as well as the first to discover reptile remains of the Carboniferous System (345-280 million years ago).

1844 Friedrich Wilhelm Heinrich Alexander von Humboldt (1769-1859), German naturalist, publishes the first of his five-volume work, *Kosmos*. This ambitious project is considered the first reasonably accurate encyclopedia of geography and geology, and many regard it as the founding text of the science of geophysics.

1846 Robert Mallet (1810-1881), English engineer, publishes an article on the dynamics of earthquakes. During his career, he compiles the first modern catalog of recorded earthquakes and the first seismic map of the earth. He also coins the word seismology. His earthquake catalog records all known quakes from 1606 B.C. to 1850. After the massive earthquake in Naples, Italy in 1857, he conducts his own experiments by exploding gunpowder and measuring the rate at which shock travels through different kinds of material.

1846 Biogeography is founded by English naturalist Edward Forbes (1815-1854). This year he publishes a paper in which he details how most of the plants and animals of his homeland migrated there from the Continent during one of three separate episodes of the glacial epoch.

This pioneering work becomes the starting point of biogeography—the study of the distribution and dispersal of plants and animals throughout the world.

1849 First to use the word petrography to describe and systematically classify rocks is Karl Friedrich Naumann (1797-1873), German mineralogist, in his *Lehrbuch der Geognosie*.

1851 Microscopical petrography is first practiced by Henry Clifton Sorby (1826-1908), English geologist and chemist. Convinced of the value of the microscope as a tool in all sciences, he prepares thin sections of rocks (1/1000th of an inch) and gives the first description of their microscopical structure. His subsequent findings establish the value of this new technique, and he becomes known as the father of microscopical petrography (the study of rocks in thin section, based on the optical properties of constituent mineral grains).

1852 Edward Sabine (1788-1883), English physicist, first demonstrates that the frequency of disturbances in the earth's magnetic field parallels the rise and fall of the sunspot cycle of the sun. This also provides the first demonstrative sun-earth link that is not related to either the sun's radiation of light or its gravitation.

1854 Richard Owen (1804-1892), English zoologist, prepares the first full-sized reconstruction of dinosaurs for display at the Crystal Palace in London. Although quite inaccurate, these large exhibition models generate enormous interest in dinosaurs that still remains unabated.

1855 Matthew Fontaine Maury (1806-1873), American oceanographer, publishes *The Physical Geography of the Sea*, the first textbook in oceanography. With Edward Forbes (See 1841), the co-founder of oceanography, Maury spends years studying the physical and mechanical aspects of the science of the sea. He also publishes the first chart of the depths of the Atlantic Ocean.

1855   First major synthesis of the theories of ore deposition is put forth by Carl Bernhard von Cotta (1808-1879), German geologist. In his book, *Die Erzlagerstatten*, he also clearly expresses the modern theory that molten materials which poured from volcanoes or formed within the crust of the earth may be crystallized into rock at any geological epoch and therefore are not a good indicator of age.

1866   Othniel Charles Marsh (1831-1899), American paleontologist, is appointed professor of vertebrate paleontology at Yale University, the first such professor in the United States. He makes many major fossil discoveries in the American West and discovers the pterodactyls, the flying lizards of the Cretaceous Period.

1866   Gabriel Auguste Daubree (1814-1896), French geologist, first suggests that since the alloy nickel-iron is a common component of meteorites (and by inference of other planets as well), it could be that the earth's core is also formed of that alloy. He becomes an excellent experimental geologist and carries out experiments on the formation and fracturing of rocks and the chemical action of underground water on limestone.

1867   Comitato Geologico d'Italia is first established in Italy by Italian crystallographer and statesman Quintino Sella (1827-1884). It is founded to promote that country's geological survey.

1868   Service Géologique de France is first established by French geologist Jean Baptiste Armand Louis Léonce Elie de Beaumont (1798-1874) in order to promote a geological survey of France.

1870   First complete remains of the dinosaurs of the Cretaceous Period are discovered by American paleontologist Edward Drinker Cope (1840-1897). This begins his lifelong writings about the rich fossil fields of the American West. His discovery establishes that the Age of Mammals began much earlier than most thought.

1872   Anton Dohrn (1840-1909), German zoologist, founds the first marine biological center in Naples, Italy and spurs the creation of oceanographical institutes in other countries.

1873   Alexis Anastay Julien (1840-1919), American geologist, and C. E. Wright publish the first microscopic investigation of rocks in the United States.

1874   Paleocene Epoch of geologic time is first proposed by Alsatian botanist Wilhelm Philipp Schimper (1808-1880). Beginning approximately 60 million years ago, it is a part of the early Tertiary Period that precedes the Eocene Epoch and follows the Cretaceous Period. In North America, this epoch is characterized by a general warming trend in climatic conditions, the absence of dinosaurs, and the expansion and evolution of mammals.

1875   Results of the first river-borne exploration of the Colorado River are published by American geologist John Wesley Powell (1834-1902). His work, *Exploration of the Colorado River of the West and Its Tributaries*, originates and formalizes a number of concepts that become part of the standard working vocabulary of geology.

1876   First comprehensive work on the petrography of American rocks is completed by Ferdinand Zirkel (1838-1912), German geologist. It is published as Volume VI of the United States Geological Survey. Petrography is the description and systematic classification of rocks. This same year, George Wesson Hawes (1848-1882) of the United States publishes an article, which becomes the first American memoir on the petrography of rocks.

1876   Mineralogical Society of Great Britain and Ireland is first established.

1877   Thomas Chrowder Chamberlin (1843-1928), American geologist, publishes his *Geology of Wisconsin*. He is one of the first to propose that there was not only one Ice Age but actually several.

1878   International Union of Geological Sciences is founded and holds its first session in Paris, France. Its purpose is to promote discussion and comparison of results of geological research from various countries.

1878   Albert Heim (1849-1937), Swiss geologist, publishes his *Untersuchung Über den Mechanismus der Gebirgsbildung*. He is the first genuine European geological artist. His talent lay in his power to describe accurately the most complex geological structures and to then illustrate them in brilliant drawings. His studies of the Swiss Alps greatly advances knowledge of the dynamics of mountain building and of glacial effects on topography and geology.

1879   Geological Survey of the United States of America is first established under the direction of American geologist Clarence Rivers King (1842-1901).

1879   Ordovician System of geological strata is first established by English geologist Charles Lapworth (1842-1920). He proposes that a complex series of strata exist between the Cambrian System and the Silurian System that is in fact a separate system with its own geologic time period. His suggestion is accepted almost immediately by the geological community.

1880   First precise seismograph is invented by English geologist John Milne (1850-1913). His device senses motion with a horizontal pendulum whose movement is recorded on a drum, usually by a pen. He uses his device to record several earthquakes in Japan and soon establishes a chain of seismological stations around the world, marking the beginning of modern seismology.

1880   Seismological Society of Japan is first established.

1881   First textbook in geophysics is written by

Osmond Fisher (1817-1914), English geologist. In his *Physics of the Earth's Crust*, he states that the earth's fluid interior was subjected to convection currents rising beneath the oceans and falling beneath the continents. This modern view is mostly ignored during his life. He also proposes that a catastrophic event had wrenched the moon from the earth's Pacific basin, and that land masses drifted slowly toward the cavity to fill it.

AUGUST 26, 1883   First of a series of violent explosions occurs on the insular volcano of Krakatoa. The ensuing eruption on this Indonesian island is one of the most catastrophic in history. Explosions on August 27, 1883, are heard 2,200 mi (3,540 km) away in Australia and propel ash to a height of 50 mi (80.5 km). A series of tsunamis or tidal waves are triggered, the greatest of which is 120 ft (36.6 m) high and kills some 36,000 people on the islands of Java and Sumatra. All life on what remains of Krakatoa is buried under a thick layer of sterile ash, and plant and animal life does not regenerate for five years.

1884   First microscopic study of Japanese rocks is made by Bunjiro Koto (1856-1935), Japanese geologist. He also studies earthquakes in Japan.

1885   Rayleigh (seismic) waves are first proposed by John William Strutt Rayleigh (1842-1919), English physicist. These are a type of surface wave caused by an earthquake that travels on the free surface of an elastic solid. Their motion is a combination of longitudinal and vertical vibration that gives an elliptical motion to the rock particles. Of all the seismic waves, they have the strongest effect on a seismograph.

1885   Gondwanaland is first named by Austrian geologist Eduard Suess (1831-1914). This year he publishes the beginning of his massive, four-volume work, *Das Antlitz der Erde*, in which he refers to this hypothetical, former supercontinent in the Southern Hemisphere that he postulates broke up and gradually drifted apart during the early Mesozoic Era. He says its pieces formed Africa, Antarctica, Australia, India, and South America. He argues that the fossils of a certain fern unite all these places, and chooses the name for this original supercontinent from a region in India known for this fern called Gondwana, the land of the Gond peoples.

1886 First catalog of the crystal forms of all known minerals is prepared by Viktor Moritz Goldschmidt (1888-1947), Swiss-Norwegian geochemist. Using a goniometer, he compiles and synthesizes data on crystal morphologies and publishes his three-volume work, *Index der Kristallformen der Mineralien.*

1888 First seismograph in the United States is installed at the Lick Observatory on Mt. Hamilton, near San Jose, California, which opens this year.

1889 Principle of isostasy is first established by American geologist Clarence Dutton (1841-1912). He proposes the principle and the name to describe the equilibrium between the lighter (surface) and denser (underground) parts of the earth's crust. His principle states that the level of the earth's crust is determined by its density, resulting in the lighter materials buoying up and forming mountains and plateaus, and the denser sinking to form basins. When this balance (isostasy) is disturbed by erosion or other forces, a compensating movement can occur.

1893 First to use the term geochronology is American geologist Henry Shaler Williams (1847-1918), who publishes an article in which he uses the term to describe the geologic time scale.

1894 First comprehensive textbook of geomorphology is *Morphologie der Erdoberfläche*, written by German geologist Albrecht Penck (1858-1945). Geomorphology is concerned with the earth's surface features and the physical, chemi-

cal, and biological processes that act upon them. Landforms of all types are of particular interest, as are the determination of quantitative relations between them and the forces that act upon them.

1899 Thomas George Bonney (1833-1923), English geologist, discovers diamonds in eclogite rocks in South Africa and first proposes the theory that eclogite is the parent rock of diamonds. This small group of igneous and metamorphic rocks contain garnet (red) and green pyroxene. Diamonds are often found as enclosures of the rock's garnet, leading to Bonney's theory (which has yet to be disproved).

1901 First comprehensive interpretation of the Alps as a whole is offered by Maurice Lugeon (1870-1953), Swiss geologist. He extends to the Swiss Alps the interpretation of recumbent folding phenomena. This says that geologically older layers are found atop more recent ones. Lugeon's work pioneers the development of modern Alpine geology.

1905 Daniel Moreau Barringer (1860-1929), American mining engineer and geologist, first proposes that the well-known, mile-wide crater in Arizona is not an extinct volcano but was formed by the impact of a large meteorite. Although this idea is derided, most modern evidence seems to indicate he is correct.

1905 First in-depth studies of the chemistry of marine deposits are conducted by Jacobus Henricus Van't Hoff (1852-1911), German physical chemist, who publishes his *Zur Bildung der Ozeanischen Salzablagerungen.*

1906 First clear evidence that the earth has a central core is provided by Richard Dixon Oldham (1858-1934), Irish geologist. Studying the waves that reach a seismic instrument from a distant earthquake, he focuses on the two waves (compression and shear waves) that follow a path deep through the earth's interior. When he finds that at a certain point, the shear wave is deflected, he realizes that it is because shear waves cannot pass through a liquid. This is proof that the earth has a central core of dense, molten fluid. Oldham becomes a pioneer in the application of seismology to the study of the interior of the earth.

1907 First method of dating the earth is suggested by Bertram Borden Boltwood (1870-1927), American chemist and physicist. Basing his ideas on the notion that lead is always found in uranium minerals, and that lead might be the final, stable product of uranium disintegration, he suggests that a method of determining the age of the earth's crust might be possible. This method would be based on the quantity of lead found in uranium ores which would then be directly related to the known rate of uranium disintegration. His idea that the earth has a uranium clock in it is proven correct when radioactive dating is eventually perfected.

(See also 1929)

1908 Frank Wigglesworth Clarke (1847-1931), American chemist and geophysicist, publishes *The Data of Geochemistry* in which he unifies all the chemical analyses of rocks carried out in the laboratories of the United States Geological Survey. Clarke is also the first to estimate the average chemical composition of the earth's crust. He estimates that igneous rock accounts for about 95% of the volume of the earth's crust; shales, 4%; sandstones, 0.75%; and limestones, 0.25%.

1909 First attempt at geological thermometry is made by Esper Signius Larsen (1879-1961), American petrologist, and Frederic Eugene Wright (1877-1953), American geologist, who publish a joint paper this year. In one of the earliest systematic efforts to establish criteria for geological thermometry, they use quartz as a geological thermometer because its presence and properties are known to be related to the temperature under which it was formed.

1909 Mohorovicic discontinuity is discovered by Croatian geologist Andrija Mohorovičić (1857-1936). He studies earthquake patterns and deduces that the earth possesses a layered structure. Based on his analysis of different types of wave speeds and arrival times, he is able to calculate the depth of the boundary where material changes from the earth's crust to its mantle. He states that the separation between these top two layers is not gradual but sharp, and lies from 10-40 mi (16-64 km) below the surface. This separation is called the Mohorovicic discontinuity.

1910 Paleogeography is pioneered by American paleontologist Charles Schuchert (1858-1942). This year he publishes an article containing his principles of paleogeography—the study of the distribution of lands and seas in the geological past.

1911 Modern geochemistry is founded by Swiss-Norwegian geochemist Viktor Moritz Goldschmidt (1888-1947). This year he publishes his *Die Kontaktmetamorphose im Kristiania-Gebiete*, in which he makes extensive studies of the alteration of rocks by heat and offers fundamental advances in correlating the mineralogical and chemical composition of metamorphic rocks. Now recognized as the founder of modern geochemistry and inorganic crystal geochemistry, he was placed in a concentration camp by the Nazis and managed to escape to England via Sweden in 1942.

(See also 1923)

1911 Alexander Karl Behm (1880-1952), German physicist, discovers the principles of echo-sounding. His experiments in an aquarium illustrate that it is possible to measure the sea's depth by timing the echo of an underwater explosion.

(See also 1917)

1913 Friedrich Johann Karl Becke (1855-1931), German mineralogist, offers the first comprehensive theory of metamorphic rocks. These rocks (whether igneous, sedimentary, or metamorphic) are the result of the alteration of preexisting rock, usually by elevated temperatures and pressures.

1913 Gutenberg discontinuity is discovered by German-American geologist Beno Gutenberg (1889-1960). He is the first to explain satisfactorily the existence of the shadow zone inside the earth where earthquake waves are not felt. He postulates the existence of a liquid core at the center of the earth that has a radius of about 2,100 mi (3,379 km). This zone forms a band that circles the earth at a fixed distance from the epicenter of the earthquake, and waves hitting it are refracted away from this shadow zone. He knows that this band is liquid because transverse waves do not penetrate it. The sharp boundary between the core and the rocky mantle above it is later called the Gutenberg discontinuity.

1914 Geologic terms lithosphere and asthenosphere are first introduced by American geologist Joseph Barrell (1869-1919). Studying the crust-mantle interface, he describes the lithosphere as the crust and part of the mantle that exhibit the same characteristics of rigidity. The lithosphere is approximately 43-60 mi (69-96.5 km) beneath the surface. He calls the asthenosphere that less rigid region from 60-435 mi (96.5-700 km) down that separates the lithosphere from the lower mantle or mesosphere.

1915 Continental drift theory is first proposed by German geologist Alfred Lothar Wegener (1880-1930). He publishes his *Die Entstehung der Kontinente und Ozeane* which offers geology the revolutionary theory of continents in motion. He argues that the continents were originally part of a single, original landmass he calls Pangea, which slowly separated and gradually drifted apart. Although others noticed the puzzle-like fit of the continents and some even speculated that they had come apart in a sudden, catastrophic splitting, no one ever offered the startling idea of a gradual, continuous drift.

Wegener is at a loss, however, to offer a convincing mechanical explanation to account for this movement. Current plate tectonics has its origins in Wegener's theory.

1916 Planetesimal hypothesis is first offered by American geologist Thomas Chrowder Chamberlin (1843-1928). In his book *The Origin of the Earth*, he argues that a star once passed near the sun, pulling matter from it that later cooled and condensed into small fragments (planetesimals) which then coalesced and formed the planets and the earth.

1917 Paul Langevin (1872-1946), French physicist, first succeeds in using sound waves or an acoustical echo as an underwater detector. His system uses piezoelectricity to create ultrasonic waves. This employs the principle that certain sound vibrations can cause an electrical effect. By World War II, the system called sonar (Sound Navigation and Ranging) is perfected. Modern sonar, which employs a transducer (a device that converts energy from one form to another), is used for mapping ocean bottoms for fish or wreck locations as well as for submarine detection.

1922 Aleksandr Evgenevich Fersman (1883-1945), Russian geochemist and mineralogist, publishes his *Geokhimia Rossii* which focuses for the first time on regional geochemistry and establishes him as one of the founders of geochemistry. Geochemistry is the measurement and study of the distribution and migration of the chemical elements and isotopes in the earth and its components. In 1929 Fersman formulates the concept of geochemical migration of the elements.

1922 First reliable data on the concentration of elements within the earth is produced by American geochemist and petrologist Henry Stephens Washington (1867-1934). From 1896-1906 he conducts petrological studies in Asia Minor, southern Europe, Brazil, and the United States. He also determines average crustal

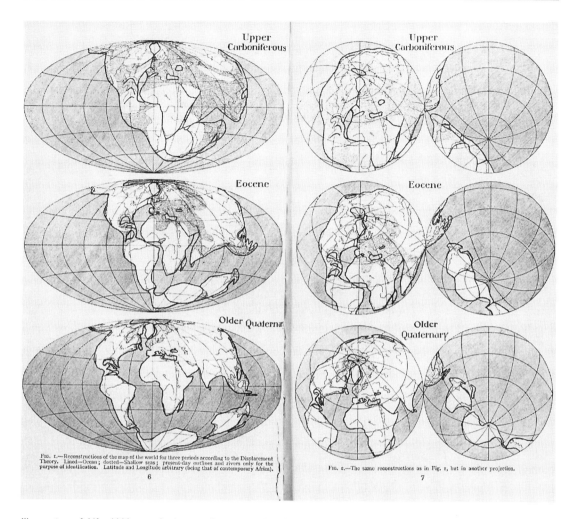

Fig. 1.—Reconstructions of the map of the world for three periods according to the Displacement Theory. Lined—Ocean; dotted—Shallow seas; present-day outlines and rivers only for the purpose of identification. Latitude and Longitude arbitrary (being that of contemporary Africa).

Fig. 2.—The same reconstructions as in Fig. 1, but in another projection.

Illustration of Alfred Wegener's theory of continental drift.

densities and makes chemical studies of meteorites and igneous rocks.

1923 Inorganic crystal chemistry is founded by Swiss-Norwegian geochemist Viktor Moritz Goldschmidt (1888-1947). This year he publishes volume one of an eight-volume study, *Geochemische Verteilungsgesetze der Elemente*, that lays the foundation of inorganic chemistry. By working out the chemical consequences of the properties of the elements and using the new discoveries of atomic and ionic sizes, he is the first to predict in what kind of minerals which elements would appear. This transforms mineralogy from a purely descriptive science to one capable of prediction.

(See also 1911)

1923 Felix Andries Vening Meinesz (1887-1966), Dutch geophysicist, first develops a method of measuring gravity in ocean basins by means of a device carried in a submarine. This contributes to a better understanding of the earth's crust beneath the ocean.

1924 First petroleum deposit discovered by using the seismic refraction method is made at Orchard Dome, Texas. This discovery validates the usefulness of seismic surveys in subsurface geology. The technique is based on determinations of the time interval that elapses between the start of a sound wave and its arrival at a detector in a certain spot. Different materials with differing densities have unique effects on seismic waves, and a special signature is given off by each one.

1929 First calculation of mineral age from lead isotopes is achieved by American geologist Clarence Norman Fenner (1870-1949) and American chemist and geophysicist Charles Snowden Piggot (1892-1973). Using mass spectroscopy enables them to determine the age of rocks by understanding the earth's radioactivity. A mass spectrometer allows them to apply the ideas of American chemist and physicist Bertram Borden Boltwood (1870-1927), who stated that the quantity of lead in a mineral was directly related to uranium disintegration, and that knowledge of the rate of disintegration would provide science with a geologic clock.

(See also 1907)

1935 First seismic measurements obtained in open seas are made by American geologist William Maurice Ewing (1906-1974). He studies the structure of the earth's crust and mantle in the Atlantic Ocean basins and in the Mediterranean and Norwegian seas, and proposes that earthquakes are associated with the central oceanic rifts that encircle the globe. Ewing's work initiates the seismic study of the ocean floor by the use of explosives.

(See also 1939)

1935 Richter scale is first introduced by American geophysicist Charles Francis Richter (1900-1985). This new scale becomes a widely used method of measuring the magnitude of earthquakes. Richter uses a torsion pendulum seismograph that registers earth movement on a scale of numbers that is arranged so that each increase of one unit represents a tenfold in-

crease in the magnitude of the earthquake. The strongest earthquakes recorded have reached 8.9 on the Richter scale.

1936 First core samples of the ocean bottom are obtained by American chemist and geophysicist Charles Snowden Piggot (1892-1973). At this time he is an investigator at the Geophysical Laboratory of the Carnegie Institution, and he measures the radioactivity of the rocks he finds with his deep-sea drilling.

1937 Bruno Herman Max Sander (1884-1979), Austrian geologist, invents a microscopic technique which he applies to igneous, sedimentary, and metamorphic rocks to correlate the internal optical structure of crystals with their history of deposition or metamorphism. He also is regarded as the founder of the study of structural petrology, which is the study of deformed rocks and their tectonic history.

1939 First deep-sea photographs are taken by American geologist William Maurice Ewing (1906-1974). Combining the photos with collected seismic data, he suggests that sea floor spreading may be worldwide and episodic in nature. During his career Ewing makes fundamental contributions to an understanding of the ocean floor and brings this field into the 20th century by using ultrasound reflection, gravity measurements, and core sampling.

(See also 1935)

1940 J-B tables are first proposed by English astronomer Harold Jeffreys (1891-1989) and New Zealand mathematician Keith Edward Bullen (1906-1976). These are accepted as the standard tables of travel times for earthquake waves.

1947 Carbon-14 dating technique is first developed by American chemist Willard Frank Libby (1908-1980). Using the knowledge that the isotope carbon-14 is found in all carbon-containing products of life, and that after an organism dies it produces no more carbon-14 (which then begins to breakdown and undergo radioactive decay), he realizes that he has discovered a practical dating tool. Using a Geiger counter with unprecedented accuracy, he is able to roughly date the age of an ancient piece of organic matter by measuring how much carbon-14 is left in it (since he knows exactly the rate at which it breaks down). The carbon-14 dating method enables geologists to learn a great deal about the earth's recent history.

1948   First complete survey of American mineral resources is completed. The United States Bureau of Mines and the Geological Survey cooperate in issuing this first overall appraisal of "the Nation's mineral wealth—its magnitude and degree of accuracy." This work is titled *Mineral Resources of the United States*.

1956   Mid-Oceanic Ridge is discovered by American oceanographer and geologist Bruce Charles Heezen (1924-1977), and William Maurice Ewing (1906-1974), American geographer. Using sonar, they are able to measure and record this gigantic, undersea mountain chain that encircles the earth and runs continuously for 40,000 mi (64,360 km). They postulate the Heezen-Ewing theory which states that an enormous central rift (at certain points double the size of the Grand Canyon) exists along the ridge and may be subject to underwater earthquakes. Their discovery sparks further undersea investigations.

1956   First comprehensive stratigraphical picture of the Jurassic Age is offered by William Joscelyn Arkell (1904-1958), English paleontologist, in his *Jurassic Geology of the World*. It also stimulates further development of a stratigraphical classification applicable to all geological systems and periods.

1961   New stratigraphic term, the megagroup, is first proposed by American geologists David H. Swann and H. B. Williman. They define it as a rock-stratigraphic unit containing lithologically similar formations and ranking

above a group. It is primarily a rock unit, and its upper and lower limiting surfaces may cross and recross faunal or floral zones or unconformities.

1962   Theory of sea floor spreading is first proposed by American geologist Harry Hammond Hess (1906-1969). This year he presents evidence for his theory, stating that the sea floors crack open along the crest of the mid-ocean ridges and that new sea floors form there, spreading apart on either side of the crest. This new theory becomes central to the new science of plate tectonics and helps explain why the continents may have pulled apart.

1963   First list of biogenic minerals is prepared by German-American paleontologist Heinz Adolf Lowenstam. Biogenic sediments are sedimentary materials that are wholly or partly formed through biological activities in the sea. Shells or broken fragments of shells that accumulate on the sea floor are biogenic, as are the white sands associated with coral reefs (since they consist primarily of calcium carbonate and are derived from the coral).

1964   Theory of the geomagnetic dynamo is first stated by English geophysicist Edward Crisp Bullard (1907-1980). He studies the earth's crust and internal constitution and bases his new theory on convective motion within the earth's core. This idea enables him to conduct important studies on the radioactive heat generation within the earth, and to learn more about its thermal history.

c. 1965   First mathematically computerized program to analyze a system of related variables in hydrology, geomorphology, and geology is produced by Donald Robert Coates, American geologist.

1965   First to use the term plates geologically is Canadian geologist John Tuzo Wilson (1908-1993). He uses this simple word to describe the rigid sectors—separated by mid-ocean rifts, major mountain chains, and shear zones—that make up the earth's surface and which move with respect to one another. He also offers a model of their behavior in terms of plane geometry and establishes global patterns of faulting and the structure of the continents .

1965 Photogeology is first practiced seriously by American geologist Harold Rollin Wanless (1898-1970). This year he publishes his *Aerial Stereo Photographs,* in which he uses aerial photographs to interpret the exposed geological features on the earth's surface. This early use is eventually followed by aircraft using infrared film and later earth-orbiting, remote sensing satellites that reveal a wealth of extremely varied geological information.

1966 First metamorphic facies map of Europe and the entire Soviet Union is prepared by N. L. Dobretzov of Russia and colleagues. Facies means any distinctive type rock, broadly corresponding to a certain environment or mode of deposition. Metamorphic rocks are those derived from preexisting rocks by mineralogical, chemical, and/or structural changes in response mainly to changes in temperature and pressure.

1967 International Committee on the History of Geology holds its first meeting at Erevan, Armenia.

1967 Plate tectonics is first used as a geologic term by D.P. Mackenzie of England and R.L. Parker of the United States. This concept states that the earth's surface is divided into at least six major and seven minor plates or large rock segments that are moving in respect to one another. It compares the continents to huge icebergs that are carried along by the currents underneath them. Plate tectonics helps explain continental drift as well as earthquakes, sea floor spreading, and the phenomenon of magnetic reversal.

JULY 24, 1969 First samples of moon rocks are brought back to earth. United States spacecraft *Apollo 11* splashes down on earth after its

historic eight-day voyage to the moon, and crew members return with lunar samples taken from the Sea of Tranquility. Later radiation tests show the age of these rocks to be between 2,700 and 4,700 million years. The oldest earth rocks to date are about 4,500 million years. Further study of these moon rocks suggest that its geology is simpler than that of the earth. It has a crust of about 40 mi (65 km) thick and is composed chiefly of anorthsite, a relatively rare, feldspar-like rock. Its mantle is believed to consist of denser rocks that are rich in iron and magnesium.

JULY 23, 1972 First earth resources technology satellite (*ERTS 1*) is launched by the United States. This infrared camera-in-space works on the principle that all living and non-living objects on earth emit, absorb, and reflect—in their own distinctive fashion—electromagnetic radiation. This satellite and its many follow-ons scan the earth from space and provide photgraphic information that proves valuable to such diverse fields as agriculture, geology, cartography, urban planning, and oceanography among others. The satellite is later renamed *Landsat 1*.

1974 First deep sea vents are discovered by American oceanographers. These hydrothermal vents or "hot spots" are found on the floor of some of the ocean's deepest waters where no light reaches. They are milky-blue plumes of hot water pouring out of chimney-like openings in the ocean floor. They also find indications of some forms of marine life that appear to flourish in absolute darkness.

(See also 1986)

SEPTEMBER 3, 1976 First pictures of the surface of Mars are transmitted to Earth. United States unmanned spacecraft *Viking 2* touches down on Mars and transmits images of its rocky surface and planetary geology. It also shows a deeply pitted surface covered by very fine soil. Surface temperatures range from -114°F to

-23°F (-73°C to -5°C). Analysis of the photos reveal that massive flows of water have at one time flowed across the Martian surface.

JUNE 26, 1978 First satellite dedicated to the gathering of oceanic data, *Seasat 1* is launched by the United States. Using remote sensing technology, its mission is to measure global ocean dynamics and physical characteristics. Despite a power loss that limits its life to only 99 days, it achieves 80% of its objectives. It obtains data on ocean surface wind speeds, wave heights, rain, surface temperature, and swell lengths.

1978 First attempt to mine the deep ocean floor for manganese nodules begins in the central Pacific Ocean. It is under the management of an international consortium of mining companies (United States, Canada, West Germany, and Japan). An estimated trillion and a half tons of these small, black/brown, friable lumps are believed to be on the Pacific Ocean floor alone, and could be an important source of industrial metals.

1980 Asteroids and dinosaur extinction is first linked. American physicist Luis Walter Alvarez (1911-1988) and colleagues discover an unusually high concentration of iridium at a sedimentary layer that marks the Cretaceous-Tertiary boundary. Since iridium is rare on earth but occurs in asteroids and meteorites, and since this time frame coincides with that of the disappearance of the dinosaurs, he speculates that a giant asteroid collided with earth, causing a prolonged dust-blackout and mass extinctions.

1983 Plate tectonics and climate are first linked by Robert A. Berner, American geochemist, Antonio C. Lasaga, Cuban-American chemical physicist, and Robert Minard Garrel (1916-1988), American geologist. They publish a computer model indicating that the carbon dioxide content of the atmosphere is highly sensitive to changes in the rate of sea floor spreading and continental area, indicating that plate tectonics is a major factor in the world's climate.

1986 First documented life forms on deep sea vents are described by American oceanographer Laverne Duane Kulm and colleagues. They publish the first detailed report of the colonies of clams and tube worms, together with mineral precipitates, that are associated with the venting of fluids at ocean depths of about 6,562 ft (2,000 m) in the subduction zone where the Pacific Plate is sinking below Oregon. They thrive in absolute darkness where no food is available by feeding off the mineral-rich water that is vented. This demonstrates that fluids and gases emerge not only from continents but also from the ocean floor.

(See also 1974)

1989 World's oldest known rocks are discovered. Rocks that are 3.96 billion years old from near Canada's Great Slave Lake replace rocks from west Greenland, which are about 100 million years younger. Their age is determined through measurements of uranium and lead contents of the mineral zircon in granite rocks. This discovery supports the idea that even older rocks, an original crust, must have existed because granites are formed from preexisting rock.

(See also 1992)

1991 First demonstrated link between periodic changes in Earth's orbit and its longer climate cycles is obtained by American geologists Dennis Kent and Paul Olsen. They extract 4 mi (6.4 km) deep sample cores obtained from an area beneath New Jersey. Besides being able to learn more about fluctuating climates 200 million years ago, they find that many climate changes mirror periodic changes in Earth's rotation and orbit. This information has implications for astronomers, since the orbits of other planets determine the Earth's longer climate cycles, the cycles themselves can indicate how these planets were behaving 200 million years ago.

1992 Oldest earth's crust is discovered by R. Maas and colleagues in western Australia. It is dated between 3.9 and 4.2 billion years.

1992 National Geologic Mapping Act first becomes law in the United States. This act will expedite production of a geologic map database for the United States that can be applied to land use management assessment, conservation of natural resources, groundwater management, and environmental protection. It is expected to reverse the drastic curtailment in geologic map production that has taken place since the 1970s.

1993 Atmospheric rivers of water vapor are first identified by American climatologist Reginald Newell, who creates a precise vapor-flow map using balloon measurements and satellite readings. He finds these rivers of water vapor all over the globe, the longest of which runs about 4,000 mi (6,436 km). They are generally 150 mi (241 km) wide and only 1 mi (1.6 km) deep, with 350 million pounds of vapor flowing past a given spot each second. Although these rivers are transient, there are at least a few of them in the atmosphere at any given point in time. Newell believes the rivers form when two air masses, thick with evaporated water from the oceans, collide. The air at the edge has nowhere to go but up, and as it rises it cools. As it cools, its capacity to hold water vapor diminishes, and the filaments of water-saturated air that results are the rivers. Clouds eventually develop and rain results. These rivers often get sucked into a low-pressure system and can feed directly into a cyclone or hurricane.

JUNE 1994 Largest deep earthquake ever recorded is first sensed this month by seismologists in the United States. The center of this quake occurs 400 mi (644 km) below the surface of the earth where two of Earth's 60 mi (96 km) thick surface plates are colliding and one is being subducted (thrust downward) into the deeper, hotter mantle. Although not felt at the surface, the earthquake creates reverberations that are recorded for months. By analyzing the

oscillations, researchers are able to calculate the density and elasticity of the rock in different parts of the mantle. Knowledge of these properties is essential in understanding the slow mantle flow that drives the continents around the Earth's surface.

DECEMBER 1994 New hypothesis on dinosaur extinction is first offered by an international team of Kevin O. Pope, American geologist, Kevin H. Baines, American atmospheric physicist, Adriana C. Ocampo, American planetary geologist, and Boris A. Ivanov of Russia. They suggest a new hypothesis explaining how the mass extinctions by asteroid impact may have come about 65 million years ago. They find that the impact site is unusually rich in sulfur, and suggest that the vaporization of more than 100 billion tons of it may have filled the atmosphere with life-killing sulfur dioxide and sulfuric acid for decades.

1995 New type of tectonic plate boundary is discovered in the Indian Ocean. Until now, the 12 known plates were classified as one of three types: mid-ocean ridges where plates diverge; deep-sea trenches where two plates converge and one dives under the other; and transform faults where plates slide past each other (like the San Andreas fault). This new type of plate boundary is both convergent and divergent at the same time. Scientists find that the Indo-Australian plate has actually been breaking apart for several million years. The Australian plate is pivoting counterclockwise around a point 600 mi (965 km) south of the tip on India, pushing into the Indian plate to the east, and pulling away from it to the west.

JANUARY 1996 First evidence for deep recycling of the earth's crust is obtained by scientists from South Africa and Scotland. They find a speck of staurolite encased in a diamond, suggesting that at least part of the earth's crust plunged down through a subduction zone and into the mantle, then to be carried back to the surface. Scientists have long suspected that over billions of years of geologic time, much of the earth's surface gets pushed down into the planet's hot interior, undergoing great transformations before reappearing at the surface. It seems likely that the staurolite crystal, which is a common mineral of clay sediments and is considered a blemish when found in a diamond, formed within metamorphosed crustal rocks which were later carried into the mantle where diamonds are formed.

Energy, Power Systems, & Weaponry

c. 50,000 B.C.   Sometime between 50,000 B.C. and 30,000 B.C., the bow and arrow is introduced. This is probably the first composite mechanism invented by humans. The bow consists of a stave made of wood that is bent and held in tension by a string or cord. The arrow is a wooden shaft tipped by a pointed stone, usually flint. It enables the energy stored over a comparatively long period (drawing the bow) to be rapidly released. Although used later as a military weapon, the bow and arrow's primary importance to early humans is as a hunting weapon.

c. 1000 B.C.   Crossbows are first used in China. This mechanized bow and arrow system increases the user's ability to draw back the bow. It is designed to catch the bowstring in a hook when drawn back, and the arrow is laid in a groove before shooting. The fully mechanized crossbow does not appear in Europe until the 10th century. This system employs a winch controlled by a wheel with notches and uses a ratchet to turn the wheel. It is a fearsome weapon that shoots heavy bolts that are shorter than arrows and equipped with heavier points. It is outlawed for use against Christians by the Lateran Council in 1139, but remains the su-

preme hand missile weapon until the introduction of firearms.

(See also April 1139)

c. 500 B.C.   First known use of a toothed gearwheel occurs about this time. Gears operate in pairs to transmit and modify rotary motion and torque or turning force. The teeth of one gear engage the teeth on a mating gear, and when attached to a rotating shaft can power such devices as a pulley or a waterwheel.

c. 500 B.C.   Rotary mills driven by donkeys or slaves first appear in the Mediterranean world. Used to grind grain to make flour for baking, they use a grinder shaped like an hourglass that rests on a fixed lower stone shaped like a cone. The grain is poured in from above, and the top stone is turned by pushing or pulling a long wooden pole attached to the upper stone.

c. 400 B.C.   Catapults are first used by the Chinese in warfare. This ancient military device operates by the sudden release of tension on wooden beams or twisted cords of rope or other fibers. This artillery device can forcefully propel stones and other large projectiles for some distance. It is known by the Greeks at about this same time, but it is the later Romans who make improvements to its basic design and who perfect its use in battle.

c. 300 B.C. First to draw a line of latitude from east to west across a map is Greek geographer and disciple of Aristotle, Dicaearchus (c.355-c.285 B.C.). He places an orientation line on a world map, running from the Greek island of Rhodes through the Spanish port of Gibraltar. This line marks the fact that, on any given day, all points on it see the noonday sun at an equal angle from the zenith.

c. 250 B.C. Archimedes (c.287-212 B.C.), Greek mathematician and engineer, invents the Archimedean screw for raising water. This pump is essentially a metal pipe made in a helix or corkscrew shape that draws water upwards as it revolves. This hollow, helical cylinder can scoop up water from a lake or pond and deliver it to a higher level when it is rotated. It is also used to remove water from the hold of a ship.

c. 100 B.C. Wheel-of-pots driven by river currents first appears in Egypt as a method of drawing water. An earlier system of a pulley and chain-of-pots was reportedly used to water the Hanging Gardens of Babylon around 700 B.C.

c. 100 B.C. Romans first introduce chain mail armor, consisting of interlinked metal rings sewn directly to fabric or leather. This form of body armor for soldiers prevails until medieval times when armorers improve on this early version and fabricate mail independent of cloth or leather in the form of both a shirt and a coat. Chain mail forms the main armor of western Europe until the 14th century when it is replaced by plate armor.

c. 85 B.C. Antipater of Thessalonica (northern Greece) makes the earliest known reference to the use of a windmill. A windmill harnesses natural wind energy to do work, and it becomes one of the prime movers that replaces animal muscle as a power source. This windmill is a primitive, horizontal type in which the paddle

wheel revolves in a horizontal plane. These very early models do not use any gearing. The windmill reaches Europe via the Arabs near the end of the 12th century in the form of "post mills" with sails to catch the wind.

c. 75 A.D. Pliny (23-79), Roman scholar, is the first to record the use of oil as an illuminant. Although the burning of grease or oil is known by very early humans, the Greeks introduce sophisticated lamps with handles, spouts, and nozzles used for receiving oil and holding wicks. Roman metal workers also create elaborate bronze and iron oil lamps.

c. 650 Blades of Damascus first come into demand as far away as Europe. These weapons are made following the ancient Indian method. Starting with round ingots—weighing about 2.2 lbs (1 kg) and measuring about 5 in (13 cm) in diameter and .5 in (13 mm) thick—these are hammered together in one or more directions to make blades that are then forged, cooled, and tempered. The process, which involves repeated heating and hammering together of strips of iron, results in the addition of small amounts of carbon. The finished sword blade is exceptionally hard and has a characteristic streaked appearance because of the varying carbon content of the iron strips. It becomes a fearsome weapon.

670 Callinicus of Egypt invents Greek fire. This mixture of inflammable petroleum, potassium nitrate, and possibly quicklime is able to burn on water and is used in naval battles to burn wooden boats. The Byzantine Greeks launch the deadly compound from tubes mounted on the prows of their ships, and in 673 they effectively destroy the Arab fleet attacking Constantinople.

c. 950 Black powder, or gunpowder, is believed to be invented in China at this time, although some Chinese texts make the first reference to fireworks as early as 600. A mixture of saltpeter (potassium nitrate), sulfur, and charcoal, the powder burns rapidly when ignited and can propel missiles when exploded in a confined space. As the first explosive invented, it comes to Europe around the 13th century and is used in firearms. It is not used for peaceful purposes, like mining or road building, until the 17th century. It remains the only available explosive material until the mid-19th-century discovery of nitroglycerine.

1086 *Domesday Book* first records the existence of 5,624 mills in England. All of them appear to be watermills of the primitive horizontal, rather than the vertical, type.

APRIL 1139 Lateran Council, under Catholic Pope Innocent II, first prohibits the use of the crossbow in war except against infidels or non-Christians. This law has little effect on actual practice.

(See also c. 1000 B.C.)

1242 Roger Bacon (c. 1220-1292), English scholar, publishes his *De Mirabile Potestate Artis et Naturae* in which he first gives the ingredients for making gunpowder (saltpeter, charcoal, and sulphur). This knowledge may have reached the West from the Orient during the Mongol invasion earlier that century.

1288 First known gun is made in China. It is more like a hand cannon that uses black powder, or gunpowder, and is small enough to be used by one person. The Chinese initially use bamboo to house the explosion and the projectile, but soon switch to a metal tube.

(See also 1313)

c. 1290 Longbow first comes into use in England. With an effective range of 240 yds (219 m), it proves to be the dominant English weapon into the 16th century. Probably of Welsh origin, it plays an important role in the battles of Crecy, Poitiers, and Agincourt. Averaging 5 ft (1.5m) in length and requiring a 60-90 lb (27-41 kg) pull as well as proper training to use, it proves to be a devastating weapon against the

slow-firing crossbows and even slower early firearms.

1313 First Western account of a gun or firearm is a German description of an "iron pot" or "vase." These primitive guns use black powder or gunpowder and are made of wooden or iron staves bound together with hoops. They are more like hand cannons than small arms. Some attribute this invention to the German monk, Berthold Schwarz.

1324 Cannons are first recorded in use in the West at the siege of Metz in northeastern France. These early cannons are made of bronze or welded wrought-iron strips and shoot cast-iron balls. Gunpowder cannons play no real part in this battle or any other for another century, but large cannons eventually take the place of the siege engines that were more like ancient catapults.

1326 Recognized as the first Western illustration of a firearm or gun is a drawing in *De Officiis Regnum* by Walter de Milemete of England. It appears to be a small cannon that fires arrows.

c. 1375 Earliest known use of the word "petroleum" is found in the Wardrobe Account of Edward III (1312-1377), king of England. The written entry states: "Delivered to the King in his chamber at Calais: 8 lbs of petroleum."

1421 Earliest known reference to a fuse for hollow, explosive shells is made. By this time a slow-burning fuse made of soft cord soaked in nitre and diluted alcohol and then dried is used instead of the old, red-hot wire that is pushed directly into the powder charge.

c. 1425 First "matchlock" small arms are introduced. These are hand-carried long guns that ignite the powder charge by lighting a match when the trigger is pulled. With this "serpentine" fuse method, a shooter can stand and take aim, knowing that firing depends only on pressing a lever. This is the first major small-arms invention, and it allows a soldier to concentrate on his target. It also allows him to hit a moving target.

(See also 1515)

1435 First hand grenades begin to appear in Europe. Although called grenades, these are really hand-thrown gunpowder bombs, which are effective because of their blast and not because of shrapnel action.

c. 1450 The Dutch invent a type of windmill called the "wipmolen." Its top can be turned so that its sails can face and catch the prevailing wind.

c. 1475 Foresights for aiming a hand-held firearm first come into use. This allows a soldier to increase his accuracy by enabling him to take better aim.

1515 Matchlock firearm is first replaced by the wheellock or firelock musket. With this gun, the charge or powder is ignited by sparks produced when a serrated wheel is spun against iron pyrite. This eliminates the need for a burning fuse which is not only dangerous, but eliminates the element of surprise in battle. It also can be loaded and primed at leisure and then fired immediately. They are not as reliable as the matchlock in firing, and must be kept very clean. They are also expensive and more likely to need repairs.

1522 The Spanish effectively use the arquebus in Italy. Also called the harquebus, it is the first weapon fired from the shoulder. This is a muzzle-loaded, smoothbore matchlock gun that uses a serpentine fuse, and it is the first handgun to have a stock shaped to fit the shoulder. Its name is derived from the German words for "hooked gun."

(See also c.1550)

1525 Snaphaunce lock is invented by the Dutch. This is an early flintlock that uses a piece of flint on an S-shaped cock that produces a spark when pivoted against a piece of steel. The spark fires the powder in the pan which in turn ignites the charge in the gun chamber.

(See also 1550)

1525 Rifled gun barrels are invented. These spiral cuts or grooves in a gun's barrel make the bullet leave with a spinning action that provides greater accuracy and stability to its flight. The invention of rifling has been attributed to a German gunsmith named Kutter or Kotter.

(See also 1631)

1550 The Spanish invent Miquelet. This improved flintlock uses a spring mechanism to fire the powder in the pan rather than igniting it by manual means.

c. 1550 First muskets begin to appear. This smoothbore shoulder gun evolves from the arquebus, but is larger and more awkward to use. Its advantage is its increased range and more destructive bullet. It is able to penetrate the finest armor of its time, and the infantry using a musket has more effective missile power against armor than any soldiers before their time. These early shoulder muskets are heavy and must be supported by a forked rest.

1603 Coke is discovered in England. This solid, charcoal-like substance is produced by heating coal to very high temperatures and is later used in foundries (foundry coke) and for heating purposes (oven coke).

(See also 1709)

1609 First attempt to harness tidal power is made at the Bay of Fundy in Canada. Several small mills are powered by this especially high and fast-running tide.

1630 Modern form of the flintlock gun is invented. Its new ignition system using flint against steel replaces the matchlock and the wheellock. It has a frizzen (striker) and pan cover that are made in one piece. Cheaper and requiring less maintenance than a matchlock, it is militarily practical and becomes the firearm of choice. It is widely used until early in the 19th century.

1631 First military use of weapons with rifled (spiral grooves) barrels is made by the Landgrave of Hesse who equips his troops with rifled carbines. Rifling improves the trueness of a bullet's flight and consequently helps a shooter's aim.

1642 City of Boston in the American colonies builds a small aqueduct to bring water into this city of 2,500. This is the first city water supply built in the colonies.

1659 Natural gas is discovered in England. Usually a mixture of methane and ethane, both of which are gaseous under atmospheric conditions, it usually occurs geologically in association with petroleum.

1663 Edward Somerset Worcester (1601-1667), English inventor, builds what many consider to be the first steam engine. His apparatus can lift water by using steam power. This machine that contains alternating vessels filled with cold and hot water can shoot a stream of water 40 ft (12 m) into the air. The British Parliament grants him a patent this year.

1670 Gas distilled from coal is first discovered. Heating bituminous coal in the absence of air (called destructive distillation) results in a gas very useful for burning, especially for heating purposes. Coal tar and coke are also obtained as byproducts of this process.

1678 Christiaan Huygens (1629-1695), Dutch physicist and astronomer, first proposes a theoretical gunpowder engine in which a piston is raised and lowered to work a cylinder. He works on this idea of creating a vacuum in a metal cylinder by the explosion of gunpowder. The piston would then be pushed by air pressure and would provide power. He actually

builds small prototypes, but finds using gunpowder is impractical as well as dangerous.

1687 Denis Papin (1647-1712), French physicist, publishes a work that describes a machine with a piston that moves under the effect of steam. It works by means of a piston sliding into a vertical cylinder at whose base is water. The water is heated and the piston is pushed up by the force of the steam. When the steam condenses, the piston moves down and lifts a weight by means of a pulley. This is one of the first atmospheric engines.

1698 Thomas Savery (c.1650-1715), English engineer, patents the first steam engine. Called the Miner's Friend, it is a monocylinder suction pump that drains water from mines and is based on principles established in 1687 by French physicist, Denis Papin (1647-1712). It is slow, noisy, expensive, and especially dangerous since it uses steam under high pressure.

1709 Abraham Darby (c.1678-1717), English ironmaster, first successfully smelts iron ore with coke. Coke is the name for the solid residue of mostly carbon that remains after certain types of bituminous coals are heated to a high temperature. Darby finds that coke can be used in a larger furnace than charcoal and can therefore create a hotter fire (which produces iron at a faster rate).

1710 First porcelain factory in Europe opens in Germany at Meissen, near Dresden. Porcelain is vitrified pottery, or pottery that is changed into glass by heat, and which has a white, fine-grained body that is usually translucent. The word is derived from "porcellana," a word used by Italian explorer Marco Polo (1254-1324) to describe the pottery he saw in China. In 1800, the first steam engine in Germany used for power is installed in the royal porcelain factory in Berlin.

1712 Thomas Newcomen (1663-1729), English engineer, devises an improved version of the Savery steam engine. Unlike the machine built in 1698 by English engineer Thomas Savery (c.1650-1715), it does not use steam under high pressure but rather allows air pressure to do the work. This is the first really practical steam engine as well as the first to use a piston and cylinder. The use of air pressure rather than the force of high-pressure steam in the engine's cycle makes this machine an atmospheric engine. It remains in use for nearly fifty years in England and Europe, pumping out flooded coal and tin mines.

1722 First Newcomen steam engine built outside of England is installed at Königsberg, Germany.

(See also 1712)

1723 Jacob Leupold (1674-1727), German mechanic, publishes the beginning of a nine-volume work, *Theatrum Machinarum Generale*, that is the first systematic treatise on mechanical engineering.

c. 1740 Benjamin Huntsman (1704-1776), English inventor, invents the crucible or cast process of making steel. His steel is more uniform in composition and more free from impurities than any steel previously produced. He does not patent his process which he attempts to keep secret, but it is eventually copied by others. Huntsman heats small pieces of carbon steel in a fire-clay crucible placed in a coke fire. Since the fire reaches temperatures of 2,900°F (1,593°C), he is able to melt steel for the first time, thus producing a highly desirable, homogenous metal of uniform composition.

1749 Benjamin Franklin (1706-1790), American statesman and inventor, first installs a lightning rod on his home in Philadelphia, Pennsylvania to test his theory that a tall metallic rod will attract lightning. A lightning rod works because it diverts the electric current from the nonconducting parts of a structure, allowing it to follow the path of least resistance and pass harmlessly through the rod and into the ground via cables.

1751 Nicolas Focq of France invents the iron-planing machine. This device marks the beginnings of the production of machine tools. Operated by a hand-crank, his machine allows the cutting tool to travel across the surface to be planed.

(See also 1817)

1752 John Smeaton (1724-1792), English civil engineer, invents the fantail for windmills. This device keeps the main sails facing the wind at all times. As the first person to investigate scientifically the design of windmill sails and waterwheels, Smeaton proposes five sails instead of the traditional four and also introduces cast iron into millwork.

1754 First pump powered by a water wheel in the American colonies begins operations in Reading, Pennsylvania.

1755 First atmospheric steam engine in the American colonies is a Newcomen machine sent from England.

(See also 1712)

1765 First known example of a deliberately redundant technology is a steam engine built by William Brown that has four steam boilers. Three operate constantly and a fourth is kept in reserve in case it is needed. As a general principle that can be applied to any operating system, redundancy is simply the provision of more essential sub-systems than are actually needed so as to provide back-up in case of failure.

1765 James Watt (1736-1819), Scottish engineer, invents the separate condenser for the steam engine. By introducing a separate chamber to Newcomen's steam engine, he produces a machine not only much more efficient in its use of fuel but also significantly quicker in the work it performs. In the Newcomen engine, a single steam chamber is cooled to condense the steam and produce a vacuum, which must then be heated up again. This is highly wasteful. Watt's second, separate chamber can be kept constant-

Benjamin Franklin experimenting with lightning.

ly cold while the other is kept hot. It is much more efficient and accomplishes a great deal more work since it does not take long pauses to constantly re-heat the chamber.

1769 James Watt (1736-1819), Scottish engineer, obtains his first steam engine patent which includes his separate condenser, the steam jacket, and the closed-top cylinder. It is Watt who first makes the steam engine into a "prime mover" and more than just a pump. Because of Watt, steam engines powered by burning coal can deliver large quantities of energy to any

needed spot. This means that the location of large factories need no longer be confined to a source of moving water. It also heralds the dawn of the Industrial Revolution.

c. 1771 First waterwheel-driven cotton mill is built by Richard Arkwright (1732-1792), English inventor. His mill powers a device that can spin thread by mechanically reproducing the motions ordinarily made by the human hand. His machinery eventually replaces many other manual steps of textile manufacture, and he soon converts his power source to steam. This can also be considered the first factory since it uses a single energy source to power scores of machines.

1775 John Wilkinson (1728-1808), English industrialist, builds a boring machine for drilling holes in metal. The first modern machine tool, this new creation can drill precise, round holes whose accuracy is essential for boring cylinders for the newly invented steam engines. Precise holes mean less leakage. He also becomes a master working with iron, building the first iron furnace c. 1748 and the first iron bridge in 1779. Fittingly, he chooses to be buried in a cast-iron coffin of his own design.

1781 James Watt (1736-1819), Scottish engineer, invents the sun-and-planet gear which adapts the back-and-forth action of a steam engine to a rotary motion. It is this mechanical attachment that converts the rocking motion of a piston into a useful rotary movement that is most responsible for making the steam engine the primary power source of the Industrial Revolution.

1782 Jonathan Carter Hornblower (1753-1815), English engineer, builds the first compound steam engine to work. His invention of the double-beat valve allows him to build the first reciprocating compound steam engine (one with two cylinders). When Scottish engineer James Watt (1736-1819) claims patent infringement and wins, Hornblower loses the opportunity to develop his concept further. It is ignored for another generation.

1782 James Watt (1736-1819), Scottish engineer, first patents the parallel-motion, double-acting steam engine that admits steam to both sides of the piston. This is the first steam engine in which the piston both pushes and pulls, and it is a major improvement in steam engine technology.

1783 James Watt (1736-1819), Scottish engineer, first introduces the "horsepower" as a unit of work. Seeking a means to measure the power

of his steam engine, he tests a strong horse and decides it can raise a 150-pound weight nearly 4 ft (1.2 m) a second. He then defines one "horsepower" as 550 foot-pounds per second (33,000 foot-pounds per minute). This unit of power is still used.

1784 James Watt (1736-1819), Scottish engineer, is the first to use steam to heat a building as he heats his office.

1784 Henry Cort (1740-1800), English inventor, first patents his "puddling process" in which the carbon in iron is separated by stirring molten pig iron in a reverberating furnace (one in which the flames and hot gases swirling above the metal provide heat so that the metal does not come in contact with fuel). This process combined with his iron bar breakthrough significantly increases iron production, and Cort becomes known as the Father of the Iron Trade.

1784 Henry Shrapnel (1761-1842), English artillery expert, invents "spherical case" shot. This spherical projectile is filled with bullets and is discharged by a fuse. As the projectile moves toward its target, it explodes and sends bullets flying in all directions. Later, the term "shrapnel" comes to be used to designate a shell that splinters.

1785 First steam-powered spinning mill is built at Papplewick in England.

1786 John Rennie (1761-1821), Scottish engineer, invents a clutch to regulate the machinery used in a flour mill. This important device provides a convenient means of starting and stopping a machine and proves to be important to all driving motors.

1788 James Watt (1736-1819), Scottish engineer, invents a centrifugal governor that automatically controls an engine's output of steam, never allowing it to grow too large or get too small. Watt's governor consists of two metal

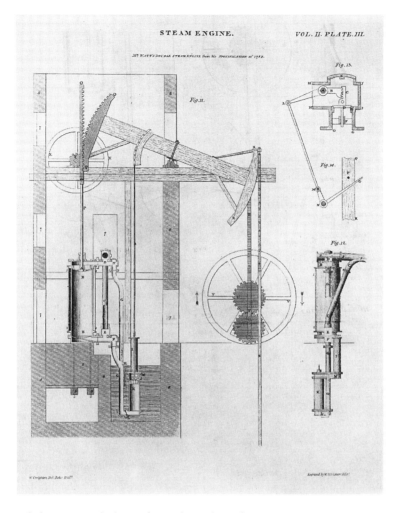

James Watt's sun-and-planet gear, which transforms the rocking of a piston into rotary motion, propels the steam engine forward as the primary power source of the Industrial Revolution.

spheres mounted on a vertical rod that is spun by the engine's output of steam. The faster the rod spins, the farther the two spheres are thrown outward by centrifugal force. The farther they are thrown out, the more they choke off the steam outlet. As the steam output decreases, the slower the spheres rotate, therefore reopening the steam outlet and reversing the process. This could be considered the first practical example of automation, since it is a device that controls a process by means of the variations in the process itself.

1789 First cotton factory to be powered by steam opens in Manchester, England.

1790 James Watt (1736-1819), Scottish engineer, invents a pressure gauge or indicator that shows the pressure of the steam in engine cylinders.

1792 William Murdock (1754-1839), Scottish inventor, illuminates part of his house with coal-gas. He is the first to realize that gas is a more convenient energy source than coal and that it can be piped and easily controlled. He goes on to develop methods for manufacturing, storing, and purifying coal-gas.

(See also 1802)

1794 Philip Vaughn of England first receives a patent for a radial ball bearing to be used in conjunction with a carriage axle. His specifications are too exacting for the machine tools of his time, and widespread use of ball bearings does not take place until the 1860s, when they are used by bicycle makers. Modern ball bearings are hardened steel balls that roll easily and translate a sliding action into a rolling action.

1795 Oliver Evans (1755-1819), American inventor, publishes *The Young Millwright and Miller's Guide* in which he details his automation of a flour mill, inventing the grain elevator, conveyor, drill, hopper boy, and descender. His mechanized flour mill becomes the first in the United States to be powered by steam.

1796 Joseph Bramah (1748-1814), English engineer and inventor, first builds a practical hydraulic press that operates by a high-pressure plunger pump. In 1812, he patents a hydraulic jack. These practical applications of hydraulic principles open a major new source of power to the manufacturers and builders of the Industrial Revolution.

1799 William Murdock (1754-1839), Scottish inventor, first patents his long D slide valve for steam engines. This injects and removes steam alternately from each end of the cylinder and is a major improvement in steam engine design and performance.

1799 Philippe Lebon (1767-1804), French chemist and engineer, first patents a method of distilling gas from wood and uses it in a lighting fixture he calls a "Thermolamp." Although the French government declines to finance his plans for a large-scale lighting system, he contributes greatly to the development of coal-gas lighting and is considered a pioneer of gas lighting.

1801 First high-pressure steam engine built in the United States is made by Oliver Evans (1755-1819), American inventor. He builds a high-pressure steam engine with a cylindrical shell and a cylindrical single flue. It does not use a condenser like the Watt engine, but is simpler and more compact. Because it vents steam to the atmosphere, it requires large amounts of water.

(See also 1804)

1801 Philippe Lebon (1767-1804), French chemist and engineer, uses his gas "Thermolamp" to illuminate the Hotel Seignelay in Paris. This is the first time an entire building is lit with gas.

1802 Matthew Murray (1765-1826), English engineer, first uses a cast-iron "bedplate" or foundation underneath steam engines, making them somewhat portable and easier to place and use. Until this, steam engines had specially built masonry structures supporting them.

1802 Richard Trevithick (1771-1833), English engineer, patents and builds a high-pressure, non-condensing, direct-acting portable steam engine with a return flue boiler. This is the first full-size steam engine to use high-pressure steam. His powerful engines prove that pressures higher than the atmosphere's can be safely used to power machinery.

1802 William Murdock (1754-1839), Scottish inventor, first installs gas lighting at the main Birmingham factory of Watt and Boulton. Gas lighting eventually transforms the way people spend their evenings as well as making factories able to operate for longer hours.

1804 Oliver Evans (1755-1819), American inventor, first builds a double-acting vertical, high-pressure steam engine that uses a steam pressure of three atmospheres. He is unsuccessful in marketing it, however.

1805 First percussion lock for guns is created by Scottish inventor Alexander John Forsyth (1769-1843). It is based on the detonation of an

Gas lighting on a street in Scranton, Pennsylvania.

explosive when struck sharply and is the prede-cessor of the percussion cap. This method takes advantage of the explosive property of potassi-um chlorate and fulminate of mercury, both of which detonate when struck with a sharp blow. (See also 1814)

DECEMBER 1806 First gas mains laid in a public street are lead pipes laid from Haymarket to St. James Street in London, England.

JANUARY 28, 1807 Frederick Albert Winsor (1763-1830), English inventor, conducts the first successful demonstration of street lighting as he lights Pall Mall in London, England, using coal gas.

APRIL 30, 1812 First gas company comes into being as the British Parliament grants a charter to the London and Westminster Gas Light & Coke Company.

1814 Joshua Shaw (1777-1860), English-American artist, makes the first percussion cap for a bullet. He fits a small copper cap contain-ing a tiny charge of fulminate of mercury over a metal nipple. When the gun hammer hits the cap it explodes, sending a jet of flame down an open channel in the nipple, and firing the main charge. Percussion weapons have the advantage of speed and sureness of fire over the old flintlocks. They are also simpler, cheaper, and easier to maintain. Still, they do not replace flintlocks for another 30 years.

1815 Samuel Clegg (1781-1861), English inventor, lays the first gas mains. He uses cast iron pipe. The gas is used for lighting purposes.

1816 First interior commercial use of gas in the United States is in the New Theater in Philadelphia, Pennsylvania, where gas lights are installed.

1816 Robert Stirling (1790-1878), Scottish clergyman and inventor, patents his safe, small, and reliable engine. He builds a steam engine that has no pressurized boiler and produces an external-combustion engine that alternately heats, compresses, and cools a gas to produce energy. Although quiet and efficient, it is expensive to operate and does not replace the steam engine.

FEBRUARY 5, 1817 First gas company in the United States, the Gas Light Company of Baltimore, is incorporated.

1817 Richard Roberts (1789-1864), English inventor, builds a small planing machine for smoothing metal. This is the first improvement of a metal-planing lathe since Focq, and it significantly advances the precise manufacture of the specialized machines.

(See also 1751)

1821 First electric motor is invented by Michael Faraday (1791-1867), English physicist and chemist. In 1820, Danish physicist Hans Christian Ørsted (1777-1851) discovers that an electric current can produce a magnetic field. Faraday uses this discovery and builds a device consisting of a hinged wire or needle, a magnet,

and a chemical battery. When the current is turned on, a magnetic field is set up in the wire, and it begins to spin around the magnet. This demonstrates in a simple way the principle of an electric motor.

1823 J. Norton first develops the elongated bullet which is far superior aerodynamically to a round ball. It also sustains its velocity in flight. The introduction of rifling (the helical grooves cut into a rifle's barrel) makes possible the change from round lead balls to this elongated shape called cylindroconoidal.

1824 Claude Burdin (1778-1873), French mining engineer, first coins the term "turbine" to describe a power source that derives from the movement of a fluid or a gas. He takes it from the Latin "turbo" for spinning top.

1827 First modern water turbine is built by Benoit Fourneyron (1801-1867), French engineer and inventor. His reaction turbine can produce about six horsepower. It uses a radial outward flow with water being directed through stationary guide vanes at the center into curved blades of an outer runner. This use of two wheels greatly improves its efficiency since the inner, fixed wheel has gates that channel the water to the blades of the outer rotating wheel, both of which are submerged.

1827 Jacob Perkins (1766-1849), American inventor, first builds a horizontal steam engine with a piston that acts as the exhaust valve. It is later successfully developed as the uniflow engine. In this type of steam engine, the steam enters through admission valves at the ends of the cylinder and escapes through exhaust ports uncovered by the piston. This type of single-purpose design eliminates the repeated heating and cooling that occurs in other, older designs.

1828 Johann Nikolaus von Dreyse (1787-1867), German inventor, obtains the first patent on a muzzle-loading rifle. This is one attempt to find a method that would allow rifled guns to be loaded as easily as smoothbores can be

Illustration showing the major technological accomplishments of the nineteenth century, the steam press, steam-powered locomotive, steamboat, and electric telegraph.

loaded with sperical balls. This solution does not prove practical.

(See also 1836)

1828 Marc Seguin (1786-1875), French engineer, first patents his multiple fire-tube boiler that proves to be a marked advance over the water-tube boiler used in traditional steam engines. Fire-tube boilers get their name because the hot gases circulate through tubes to heat the water surrounding the tubes. These boilers, though limited in maximum pressure and capacity, are very reliable. It is this efficient engine that English inventor George Stephenson (1781-1848) uses to power his "Rocket" locomotive in 1829.

1828 James Beaumont Neilson (1792-1865), English inventor, first patents his new blast furnace which uses a hot-air blast instead of cold air for the smelting of iron. This important advance triples the output of iron per ton of coal, doubles its yield per unit amount of blast, and allows a profitable recovery of iron to be made from lower grade ores.

1831 Jacob Perkins (1766-1849), American

inventor, first patents a high-pressure model of a heat circulator and invents a radiator for use with a hot-water central heating system.

1832 Hippolyte Pixii (1808-1835), French engineer, uses Faraday's discovery of electromagnetic induction to build the first practical generator. His hand-driven machine produces an alternating current and has stationary coils around which a field magnet revolves. This first generator produces alternating current (AC) which Pixii later is able to convert into direct current (DC).

1836 John Frederic Daniell (1790-1845), English chemist, invents a new zinc-copper battery, the Daniell cell. By placing a barrier between the zinc and copper, he stops the formation of hydrogen which impairs its function, making the current in older, voltaic batteries decline rapidly. His improvement makes this the first reliable source of electric current.

1836 Samuel Colt (1814-1862), American inventor, first patents his revolver or repeating pistol. He uses a pawl mechanism to rotate the cylinder that holds the bullets, and the cylinder is locked into place by cocking the hammer. His rapid-fire weapon does not become accepted until after 1846 when the United States realizes its value during the Mexican-American War. Colt becomes the largest arms manufacturer in the world by 1855.

1836 Johann Nikolaus von Dreyse (1787-1867), German inventor, first patents the "needle" rifle with a bolt breech-loading mechanism. This gun is loaded through the rear of the barrel. This design proves to be the preferred method of loading both large and small modern guns.

FEBRUARY 25, 1837 First patent for an electric motor is granted to American inventor Thomas Davenport (1802-1851). His patent is titled "Improvement in Propelling Machinery by Magnetism and Electro-magnetism" and describes a motor driven by a galvanic battery.

1839 William Robert Grove (1811-1896), English physicist, makes the first fuel cell when he discovers that he can reverse electrolysis (and combine hydrogen and oxygen) to produce an electric current. A fuel cell is a type of battery that converts chemical energy into electrical energy (direct current).

1840 Gas meters are first used. The concept of selling gas by the measure, instead of a flat rate, originates with the Chartered Gas Light Company of London.

1845 Charles Wheatstone (1802-1875), English physicist, first develops an improved version of the dynamo as he builds an electric generator that induces electricity in coils by using an electromagnet powered by a battery. It provides a much steadier current than the versions of his time.

1849 George Henry Corliss (1817-1888), American inventor, first patents a more efficient and economical system of valves and governors to control steam and exhaust valves. He gains fame by building a 700-ton (636-metric ton), 1,400-1,600 horsepower Corliss steam engine that powers all the exhibits at the Centennial Exhibition at Philadelphia, Pennsylvania in 1876. His engine runs continuously for six months.

1849 Claude Etienne Minie (1804-1879), French Army captain, invents a bullet with a flat base and a pointed nose called the "Minie Ball." Shaped much like modern bullets, it has greater accuracy and range and can be loaded and fired quickly. The Minie system uses a hollow-base bullet made of soft lead that is smaller than the barrel, allowing it to easily slide down the barrel while loading. The sudden inflammation of the propellant charge of black powder expands the bullet bottom into the grooves of the rifle bore. Highly accurate at 200 yds (183 m), these bullets are lethal to 1,000 yds (914 m) and effectively eliminate horse cavalry from battlefields.

1849 James Bicheno Francis (1815-1892), English-American hydraulic engineer, and Uriah Atherton Boyden (1804-1879), American engineer and inventor, first develop the inward-flow turbine for generating power. Called the Francis type, it has an outside casing with guide vanes that direct water against all the spirally curved vanes on the inner rotor simultaneously.

c. 1850 James Young (1811-1883), Scottish chemist, first makes kerosene or coal oil which he distills from coal and oil shale. Kerosene soon becomes the most popular fuel for oil lamps. The discovery of huge quantities of crude oil in America in 1859 makes all other sources of kerosene obsolete, and by the end of the century, kerosene is the chief product of American oil refineries. Because it belongs to the family of hydrocarbons called alkanes or paraffins, it is sometimes referred to as paraffin oil.

1850 William George Armstrong (1810-1900), English engineer, invents a hydraulic accumulator or actuator that makes it possible for

The Corliss steam engine, a symbol of the nineteenth-century belief in progress.

hydraulic machinery to be run some distance away from a direct water source. His machine is basically a large, water-filled cylinder with a piston that can raise water pressure within the cylinder and in supply pipes to 600 lbs (272 kg) per square inch. This allows machinery such as hoists, capstans, turntables, and dock gates to be worked in almost any situations.

1851 William Kelly (1811-1888), American inventor, is the first to develop the pneumatic conversion process for making steel. He discovers this process, in which air is blown through molten pig iron to oxidize and remove unwanted impurities, before English metallurgist Henry Bessemer (1813-1898), but does not patent it immediately. (See 1856)

1851 Gustav Adolphe Hirn (1815-1890), French physicist, experiments with superheated steam and is the first to demonstrate, by investigating the heat balance of a steam engine, the economic advantages of superheating. Superheated steam is water vapor at a temperature higher than the boiling point of water at a particular pressure. It use permits more efficient operation of devices that convert heat into mechanical work.

(See also 1866)

1851 William Thomson (1824-1907), Scottish mathematician and physicist also called Lord Kelvin, first describes the concept of the heat pump. Today's air conditioners apply this concept (that heat is absorbed by the expansion of a working gas and given off at a higher temperature in a condenser). Although he does not build the pump, Kelvin describes the concept whereby a warm body can be made warmer and a cool body cooler.

(See also 1927)

1852 Mineral lubricating oils for engines are first introduced. They are made from oil shale in Scotland.

1853 Alexandre Edmond Becquerel (1820-1891), French physicist, builds the first thermionic device, converting heat directly into electricity by means of thermionic emission or the ejection of electrons from a heated surface. He finds that a measurable electric current can be produced by a potential of a few volts in the air if the air is heated between hot platinum electrodes. He also later invents an actinometer, a device that can measure the intensity of light by observing the strength of an electric current produced between two metal plates. This work leads to the eventual development of the photoelectric cell.

1854 William Halliday of the United States revolutionizes the entire concept of wind power with his invention of a windmill tower topped by a light wheel fitted with an automatic mechanism that breaks its speed in high winds or when a tank of water is full. Easy to build, this type of windmill can generate enough electricity to power a small house and becomes commonplace in the settling of the West.

1855 William George Armstrong (1810-1900), English engineer, first introduces his breech-

loading, rifled gun with a screw breech lock. It is adopted by the British Navy in 1859 and marks the beginning of modern artillery.

1856 Henry Bessemer (1813-1898), English metallurgist, first introduces his converter for the cheap manufacture of steel. His "blast furnace" adds oxygen directly and not only burns off the carbon but raises the temperature as well. By using ordinary oxygen, he accomplishes this without spending any additional money on fuel. He is eventually able to sell high-grade steel at one-tenth the going price, beginning the era of cheap steel.

1857 Horace Smith and Daniel B. Wesson of the United States introduce the first metallic cartridge revolver. Their new revolver has rim-fire copper cartridges that can be loaded quickly and easily from the rear of the weapon. It is not until the Smith-Wesson patent expires in 1872 that new revolvers appear.

1858 Germain Sommeiller (1815-1871), Italian-French engineer, invents a rock drill powered by compressed air. This device forces air into a closed container in order to build up pressure which is then released to power a drill. His pneumatic drill is used to tunnel under Mont Cenis between Switzerland and France and build the first major mountain tunnel.

JULY 1, 1859 First issue of the *American Gas Light Journal* is issued. It is the first publication to serve the gas industry and record its progress.

AUGUST 27, 1859 Edwin L. Drake (1819-1880), retired American railroad conductor, drills the first successful oil well in Titusville, Pennsylvania. He is hired by the Seneca Oil Company. Drake uses the percussion drilling method and a cable-tool. This is also the first well deliberately drilled for the purpose of finding oil.

1859 Gaston Plante (1834-1899), French physicist, invents the first storage battery. Also called an accumulator, it has lead plates immersed in sulfuric acid and is completely rechargeable. To charge his battery, Plante uses a hand-driven generator, demonstrating the transformation of mechanical energy into electrical energy, and finally into chemical energy in the battery. It is essentially the same battery used in today's automobiles.

JUNE 1, 1860 One of the earliest books on petroleum, *The Wonder of the Nineteenth Century! Rock Oil in Pennsylvania and Elsewhere*, is written by Thomas A. Gale of the United States.

DECEMBER 21, 1860 First recorded "dry hole" in the United States is a 500-ft (152-m) well drilled on the Pennsylvania shore of Lake Erie that finds neither oil nor gas.

1860 First successful repeating rifle is invented by Christopher Spencer of the United States. This rifle has a magazine in the buttstock which contains seven cartridges that are fed into the chamber in succession by means of a trigger-guard operating lever. Despite its advances, this early repeating system is not adopted for military use.

1860 B. Tyler Henry of the United States designs the first Henry rifle. It carries 15 cartridges in a tube under the barrel and can mechanically feed them into the breech to be fired instead of being manually fed one at a time. His system forms the basis for the successful Winchester repeating rifle that becomes popular around 1870.

1860 Joseph Wilson Swan (1828-1914), English physicist and chemist, invents a primitive electric light using a filament of carbonized paper in an evacuated glass bulb. Lack of a good vacuum prevents it from working very well, but it is essentially this design that American inventor Thomas Alva Edison (1847-1931) later uses.

(See also October 21, 1879)

NOVEMBER 1861 First shipload of petroleum to cross the Atlantic is loaded at Philadelphia and embarks for London. This marks the beginning of regular traffic in full-cargo shipments of oil.

1862 First practical application of a generator is made when a lighthouse in the Straits of Dover uses an arc lamp powered by an electric generator.

1862 First practical machine gun is invented by Richard Jordan Gatling (1818-1903), American inventor. His Gatling gun fires 250 rounds per minute and consists of ten breach-loading rifle barrels that are cranked by hand around a central axis. It makes its military debut at Butler's siege of Petersburg, Virginia, during the American Civil War. The ten-barrel model, firing 1,000 rounds per minute and covering a range of 2,400 yds (2.2 km), is a deciding factor in the battle of Santiago, Cuba, during the Spanish-American War of 1898.

(See also 1884)

1863 First anti-pollution legislation aimed at the oil industry is passed by the Pennsylvania legislature to prevent the running of tar and distillery refuse into the creeks in the producing districts.

1863 First oil pipeline is a 7 mi (11.3 km) section that Samuel Van Syckel of the United States builds between his oil production plant and a railway station in Pennsylvania.

1866 First oil well in Texas is completed near Oil Spring in Nacogdoches County. It obtains 10 barrels a day at 106 ft (32 m).

1866 Jean Jacques Sulzer-Hirzel (1806-1883), Swiss inventor, and Charles Brown (1827-1905), English inventor, first develop a valve gear that allows the use of superheated steam in an engine. Because superheated steam does not condense inside the cylinder, it increases the engine's efficiency.

1866 Robert Whitehead (1823-1905), English engineer, invents the modern torpedo. His design is a spindle-shaped underwater missile driven by a compressed air engine. It can move at six knots for a few hundred yards while carrying an explosive charge in its head. It can also stay at a set depth by a valve actuated by water pressure and linked to horizontal rudders. In 1876 he improves this weapon by adding a servo-motor to give it a truer course through the water, and in 1896 he adds a gyroscope.

1867   Alfred Nobel (1833-1896), Swedish inventor, first produces the stable, solid explosive called dynamite. After exhaustive experimentation, he finds that the highly volatile nitroglycerine can be safely handled when absorbed by a porous clay called kieselguhr. This special clay soaks up the volatile nitroglycerine without changing its chemical makeup, and the resulting dough-like substance can be made into hard cakes or sticks. Five times more powerful than gunpowder, dynamite revolutionizes the mining industry and soon finds many military applications.

1867   First bolt-action, repeating rifle is developed by Frederic Vetterli of Switzerland. Bolt-action rifles soon become adopted by armies throughout the world.

1867   Zenobe Theophile Gramme (1826-1901), Belgian engineer and inventor, builds the first practical, commercial generator for producing alternating current. Two years later, he achieves the same thing with direct current. In 1873 while a Gramme dynamos is being exhibited at an exhibition in Vienna, Austria, it is accidentally discovered that the device is reversible and can therefore also be used as an electric motor. With these first practical dynamos, he essentially establishes and founds the electrical industry.

1867   George Herman Babcock (1832-1893), American inventor, and his colleague Stephen Wilcox first develop a high pressure steam boiler that separates water from steam and

recycles it back. It becomes the accepted method of steam generation.

1867   Georges Leclanche (1839-1882), French chemist, invents the zinc-carbon battery. His new battery contains a conducting solution (electrolyte) of ammonium chloride, a negative terminal of zinc, and a positive terminal of manganese dioxide. These electric cells with porous containers are a precursor of the dry cell and of today's portable batteries.

1872   First long-distance fuel pipeline is completed in the United States. It carries natural gas a distance of 25 mi (40 km).

1875   Alfred Nobel (1833-1896), Swedish inventor, first introduces blasting gelatine. He mixes nitroglycerine and collodion (a low nitrogen form of guncotton) and finds that they form a gelatinous mass that is water resistant and more powerful than dynamite.

1875   Robert Augustus Chesebrough (1837-1933), American manufacturer, first introduces Petrolatum which becomes known by its product name of Vaseline. This smooth, semisolid blend of mineral oil with waxes crystallized from petroleum becomes useful as a lubricant, carrier, and waterproofing agent in many products.

OCTOBER 21, 1879   First practical incandescent light bulb is produced by Thomas Alva Edison (1847-1931), American inventor. After spending $50,000 in one year's worth of experiments in search of some sort of wire that could be heated to incandescence by an electric current, he finally abandons metal altogether and discovers a material that will warm to white heat in a vacuum without melting, evaporating, or breaking. Using only a filament of scorched cotton thread, his electric light burns for forty continuous hours. The age of the electric light is born.

1880   Pierre Eugène Marcellin Berthelot (1827-1907), French chemist, first introduces the bomb calorimeter for determining the calorific value of fuels. This device makes it possible to determine the actual thermal efficiency of a heat engine.

1881 Thomas Alva Edison (1847-1931), American inventor, builds the first central electric power station at Pearl Street, New York. It has three 125-horsepower steam generators that supply current to 225 homes. He develops an entire system for generating and distributing electricity from a central source. Some regard this accomplishment a greater achievement than his invention of the incandescent light bulb.

1882 First central electric power station in England is installed at Holborn, London. It is designed by American inventor Thomas Alva Edison (1847-1931).

1882 First hydroelectric generating plant in the United States is built at Appleton, Wisconsin.

1883 Stuart Perry (1814-1890), American inventor, builds the first windmill with steel blades.

1884 Charles Algernon Parsons (1854-1931), English engineer, patents and builds the first parallel-flow reaction steam turbine and high speed (DC) generator. It operates on the turbine principle that energy can be derived from the movement of a fluid or gas. His steam turbine consists of a rotor in which several vaned wheels are attached to a shaft. Steam enters and expands, causing the shaft wheels to rotate. Other stationary blades force the steam against those that rotated, making use of as much energy as possible. The steam continues and encounters another set of turbine blades designed to work with the same steam at a slightly lower pressure. His new turbine achieves a speed of 18,000 revolutions per minute compared to the previous maximum of 1,500 rpm. Turbines later become the most widely used method of providing electricity for large-scale processes.

1884 Hiram Stevens Maxim (1840-1916),

Cannon used in the last quarter of the nineteenth century.

American-English inventor, first demonstrates his single-barrel, belt-fed and water-cooled automatic gun. Capable of firing 666 rounds per minute, the automatic gun loads, fires, extracts, and ejects the cartridge all on the momentum of recoil. Operating off the energy from the erupting shell, it renders the hand-cranked Gatling gun obsolete and is adopted by the military establishment of every major country.

1884 Lester Pelton, American engineer, invents the Pelton wheel or impulse turbine. His device involves high-pressure water jets that are directed against the rim of a wheel fitted with large twin buckets. It is most commonly used where a small volume of water falls a great distance.

1884 Nikola Tesla (1856-1943), Croatian-American electrical engineer, invents the electric alternator, an electric generator that produces alternating current (AC). He is convinced that AC has many advantages over DC (direct current which Edison supports), and he is eventually proven correct.

1885 William Stanley (1858-1916), American electrical engineer, first perfects the electric transformer and develops an alternating-current (AC), constant-potential generator. Transformers are used to overcome electrical loss that occurs during its transporting. His work establishes the practicality of alternating-current systems.

c. 1885 First solar cells are made by Charles Fritts, American inventor. His thin wafers are the size of a quarter and are made of light-sensitive metal selenium. A current is generated when sunlight strikes the cells, but his system is extremely inefficient with less than one percent of the light energy actually being converted into electrical energy.

(See also 1954)

1889 Frederick Augustus Abel (1827-1902), English chemist, and James Dewar (1842-1923), Scottish chemist and physicist, first patent cordite, a smokeless powder derived from nitroglycerin and guncotton (nitrocellulose). It is safe to handle and can be shaped and cut into precisely divided sizes. It is used exclusively for firing shells since it is smokeless and allows armies to conceal their battle positions.

1889 John Moses Browning (1855-1926), American inventor, begins work on what becomes the first successful gas-operated machine gun. He discovers a way of using the expanding gases and recoil from exploding ammunition to eject, reload, and fire his weapon automatically. His rapid-fire weapon becomes an enormous success.

(See also 1918)

1890 First hydroelectric plant to generate alternating current (AC) rather than direct current (DC) is built on the Willamette River near Oregon City, Oregon.

(See also 1894)

1891 First long distance high-voltage line for carrying electricity is completed and spans the 120 mi (193 km) between Lauffen and Frankfurt am Main, Germany.

1891 Charles M. Kemp of the United States patents the first commercial solar water heater.

1891 Charles Algernon Parsons (1854-1931), English engineer, builds the first steam turbine engine to be fitted with a condenser.

(See also 1884)

1893 First widely used semi-automatic pistol is designed by Hugo Borchardt of the United States. It is produced in Germany and becomes the ancestor of the Luger.

1894 First dam built specifically to generate power for a hydroelectric plant is built on the Willamette River site in Oregon where a plant already exists.

1895 Charles Gordon Curtis (1860-1953), American inventor, first develops a multistage turbine engine whose blades and disks become progressively larger when the steam expands. It is later adopted by most of the navies in the world.

1897 Rudolf Diesel (1858-1913), German inventor, builds his first commercial Diesel engine. His high compression engine is four times as efficient as a steam engine and does not need a complex ignition system to ignite the fuel. His new internal combustion engine injects fuel into an engine whose piston compresses air in a ratio as high as 25 to 1. Such high compression causes the air to reach temperatures of nearly 538°F (1,000°C), a temperature that is high enough to ignite fuel with any complex spark ignition system. As a large, heavy engine, its early applications are primarily for stationary purposes. It eventually is improved and finds widespread application in the shipping and locomotive industries.

1900 First offshore oil wells are drilled. This new technique allows the search for petroleum to be conducted beneath the deposits of the continental shelf. The drilling is usually conducted from specially designed, fixed platforms.

1900 Thomas Alva Edison (1847-1931), American inventor, first produces a nickel-iron battery. It has a long life and proves to be very durable. This battery can withstand considerable abuse and can be overcharged, over-discharged, or even left idle with little negative effect.

1901 Peter Cooper Hewitt (1861-1921), American inventor, invents the mercury vapor discharge lamp. Based on the knowledge that electricity, when passed through certain gases at very low pressures, discharges a glowing rather than an arcing light, Hewitt invents the first discharge light using mercury vapor in a glass tube. Although it produces a high proportion of ultraviolet light which is invisible to the human eye, it is very successful and eventually leads to the practical fluorescent light.

(See also 1934)

1902 First military tank is built by F. R. Simms of England. This armored vehicle is powered by an internal-combustion engine and carries three guns.

(See also October 1914)

1904 First efforts to harness geothermal power are made in Italy. Power is generated from the underground steam that pours from the "fumaroles" or natural crevices at Larderello in Tuscany, Italy. The following year a 25 kw steam-driven generator is installed.

1906 Tungsten, an exceptionally strong metal, is first used successfully as a lamp filament material and is later employed in many electrical and electronic applications.

1910 Georges Claude (1870-1960), French engineer and chemist, produces the first neon light when he discovers that a glass tube filled with neon gas glows with an eye-catching red color when it is charged with electricity. These will eventually replace ordinary incandescent bulbs in advertising signs.

OCTOBER 1914 E. D. Swinton of England first suggests that tractor tracks be added to an armored vehicle, and the British army begins calling this new offensive vehicle a "tank."

(See also September 1916)

1915 Depth charge bombs are first used by the Allies against German submarines. They consist of canisters filled with explosives and fitted with a valve that is triggered by water pressure against a spring.

SEPTEMBER 1916 British army makes first use of a tank in battle. It gets bogged down in the cratered battlefield, but is used much more effectively the following year at the Battle of Canbrai in northern France.

1918 Louis Schmeisser of Germany invents the first successful submachine gun. It is a fully automatic shoulder weapon that can fire 400 rounds per minute. The effectiveness of this automatic gun shows that a small shoulder arm that fires pistol ammunition can be useful militarily. It has a barrel only 7.9 in (20 cm) long.

(See also 1920)

1918 John Moses Browning (1855-1926), American inventor, first perfects his .30 calibre, gas-operated Browning automatic rifle and it is adopted by the United States Army. Browning machine guns play a later role in World War II and the Korean War.

1920 Viktor Kaplan (1876-1934), German engineer, first patents the Kaplan type reaction turbine engine. Commonly used with low pressure, it has four very large blades on an outer rotor, and the pitch of the blades can be altered while running to get best results from the water pressure.

1920 John Taliaferro Thompson (1860-1940), American inventor, first develops a submachine gun that has a large drum magazine and fires .45 calibre cartridges. Capable of firing 800 rounds per minute, it is first used in combat by the United States Marines in Nicaragua in 1925. Called the "Tommy gun," this easy-to-use automatic weapon becomes associated with the criminal gangland wars of the 1920s and 1930s.

1921 First magnetron is invented by Albert Wallace Hull (1880-1966), English physicist. It is basically a vacuum tube that functions under the joint action of an externally applied magnetic field and the electric field between its anode and cathode. When connected to a resonant line, it can be made to operate as an oscillator. This oscillator is capable of extremely high frequencies and short bursts of very high power and can be used to generate microwaves. It also becomes an important power source in the development of radar systems.

1927 Thomas Graeme Nelson Haldane, English inventor, receives the first patent for a heat pump.

(See also 1851)

1928 First practical adhesive tape called "Scotch Tape" is introduced in the United States by the Minnesota Mining and Manufacture Company.

1934 First practical fluorescent lamp is developed by a team of American scientists at a General Electric laboratory. Based on the mercury vapor discharge lamp, which it improves by reducing its operating voltage, this new light employs the principle that fluorescent chemicals called phosphors will convert ultraviolet light to light that is visible to the human eye.

(See also 1901)

1941 L. D. Goodhue and W. N. Sullivan, both of the United States, invent an aerosol spray or "bug bomb" for applying insecticides.

c. 1942 United States military first introduces a shoulder-type rocket launcher called the bazooka. Developed for attacking tanks and fortified positions, it is a smooth-bore steel tube that is recoilless and delivers a powerful explosive.

JULY 16, 1945 First atomic bomb made of plutonium is successfully tested in a desert area near Almogordo, New Mexico, by the United States. This successful detonation achieves the goal of the secret wartime Manhattan Project.

AUGUST 6, 1945 First atomic bomb is dropped on Hiroshima, Japan, by the United States. This uranium bomb weighs 2,000 tons (1,818 metric tons) and has an explosive energy of 20,000 tons (18,181 metric tons) of TNT. After an equally destructive plutonium bomb is dropped on Nagasaki three days later, Japan surrenders unconditionally, ending World War II.

1945 First nuclear reactor in Canada begins operations at Chalk River, Ontario. It is a small, natural-uranium, heavy water-moderated reactor.

DECEMBER 24, 1946 First nuclear reactor in the Soviet Union begins operations.

AUGUST 11, 1947 Construction begins on the first peacetime nuclear reactor in the United States at Brookhaven, New York.

1948 First nuclear reactor in France is a small experimental facility built at Fontenay-aux-Roses.

1950 First nuclear reactor in England begins operations at Windscale. On October 7, 1957, a major release of radioactivity into the environment occurs here as a "controlled" release goes wrong and destroys the reactor's core. The government attempts to keep the accident a secret.

1952 First experimental breeder reactor (EBR-1) is put into operation by the United States at Arco, Idaho. The reactor generates 1,400 kw of heat and 170 kw of electricity—the first significant amount of electricity produced in this manner.

MAY 5, 1953 First atomic cannon shell is successfully fired by the United States. Although it is used on a fairly small gun 11.7 in (297 mm), tactical nuclear weapons such as this do not prove to be suited to the battlefield and have never been employed.

MARCH 1, 1954 The United States successfully tests its first dropped thermonuclear bomb (an H-bomb) which explodes at Eniwetok Atoll.

JUNE 27, 1954 Soviet Union announces that a 5,000 kw atomic power plant has begun operations inside Russia. This is the first nuclear power plant.

1954 Bell Telephone Laboratories revolutionize solar cell technology by developing the first silicon photovoltaic cell. Researchers' initial design yields a conversion efficiency of 6% and is later improved to 15%.

(See also 1957)

OCTOBER 17, 1956 First large-scale nuclear power plant designed for peaceful, energy-related purposes begins operations at Calder Hall in England.

DECEMBER 1957 First full-scale, commercial nuclear power plant in the United States begins operations at Shippingport, Pennsylvania.

(See also 1989)

1957 First use of silicon solar cells as a power source is achieved as Bell Telephone uses them experimentally to power a telephone repeater station in rural Georgia. The key to these photovoltaic cells is in creating a semiconductor that releases electrons when exposed to radiation within the visible spectrum.

(See also March 17, 1958)

1957 Soviet Union first develops an Intercontinental Ballistic Missile (ICBM) to which the United States will respond, beginning the "arms race" in nuclear weaponry delivery systems. These long-range missiles prove relatively immune to electronic countermeasures since they

use inertial guidance systems. They evolve into highly lethal mass killing systems by the use of multiple, independently-targeted warheads. None has ever been used in warfare.

MARCH 17, 1958 First real application of silicon solar cells is achieved when a solar array is used to provide electricity for the radio transmitter on the American satellite *Vanguard 1*. Solar cells have been used on almost every scientific satellite since.

1958 First large-scale nuclear power plant in the Soviet Union becomes operational at Troitsk, Siberia.

1962 First advanced gas-cooled nuclear reactor that is fueled by enriched uranium is built in England.

1963 First Canadian deuterium uranium (CANDU) nuclear reactor begins operations.

1966 Largest electricity-producing power plant to use tidal power is built on the coast of France. The Rance Tidal Works has a 2,500 ft (762 m) dam at the mouth of the Rance River, and its generators produce 240 megawatts of electric power.

APRIL 1971 Magnetohydrodynamic (MHD) power generators first begin operations in both the Soviet Union and in Germany. This new technology generates electrical power by passing a conducting fluid through a magnetic field, and it is expected to have a higher efficiency rate than traditional electric generators.

MARCH 28, 1979 First major American nuclear accident occurs due to technician error at the Three Mile Island nuclear power plant in Pennsylvania results in a partial meltdown. Some radioactivity escapes but no one is injured. The unit is permanently shut down.

1981 Largest solar power station to date, Solarplant

Shippingport, Pennsylvania, atomic-electric generating station, the first full-scale nuclear power plant exclusively for civilian needs.

One, opens in California. It can generate 1000 megawatts of electricity.

1985 Largest commercial fast breeder reactor to date begins operations in Creys-Manville, France. Called the Superphenix, it generates 1,200 megawatts, but is shut down permanently in 1992 due to constant technical problems.

1986 Largest dam in the world first opens in Venezuela and is able to generate 100,000 megawatts of power.

1986 First fusion reaction caused by a laser occurs for one-billionth of a second at the Lawrence Livermore Laboratory in the United States. Fusion is the nuclear process that powers the sun, and results in an enormous production of energy.

1988 Draper Prize is first established by the National Academy of Engineering in the United States. Named in honor of American aeronautical engineer Charles Stark Draper (1901-1987), it bestows a prize of $375,000 on the

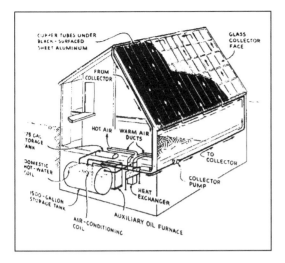

A 1958 diagram of a solar-heated home designed by the Massachusetts Institute of Technology.

winner and is considered the engineering equivalent of the Nobel Prize.

1989 First United States nuclear power plant to be dismantled is the Shippingport Atomic Power Station in Pennsylvania.

(See also December 1957)

1991 Philips Company of The Netherlands first develops a new light bulb that uses electromagnetic induction to excite a gas and produce light. The breakthrough bulb is extremely energy efficient, having no parts to wear out, and will last for up to 60,000 hours.

1992 Largest array of thin-film photovoltaic modules ever built first begins operations in Davis, California. Nearly 10,000 modules produce up to 479 kw and power 124 homes.

1994 First lithium battery with a water-based electrolyte is developed by Canadian physicist Jeff Dahn. Although only a prototype, it can potentially be used to power electric cars in the future. Electric vehicles presently lack a battery that is powerful enough, safe enough, and cheap enough. Although lithium as an electrode material had proven powerful enough, it reacts violently when immersed in water. This entirely new battery is designed to use a water-based electrolyte, thus making it not only environmentally nontoxic but safe and inexpensive.

1996 First 100-megawatt solar power plant is scheduled to open in the Nevada desert this year. A joint venture between the American corporations Enron and Amoco, the solar power plant will cost approximately $150 million and can supply a city of 100,000. It plans to initially sell energy for 5.5 cents a kilowatt-hour, about three cents cheaper on average than electricity generated by oil, coal, or gas. The technology uses a conventional thin-film, silicon-based photovoltaic cell that is able to transform into electricity about 8% of the sunlight that reaches it. The low price will be maintained by tax exemptions and guaranteed governmental purchases.

Mathematics

c. 30,000 B.C. Earliest evidence of Stone Age man's ability to devise a tally system is the notched shinbone of a young wolf found in Czechoslovakia dating to this time. The tally system is the earliest and most immediate technique for visibly expressing the idea of number. It matches the collection to be counted with some easily used set of objects like fingers, stones, or marks on a stick. This 7 in (18 cm) bone found in 1937 has 55 deeply cut notches, more or less equal in length, arranged in groups of five.

c. 8000 B.C. Clay tokens are first used in Mesopotamia to record numbers of animals and plants. The shape of these tokens is used to represent a specific commodity. For example, a cylinder stands for an animal and cones and spheres refer to certain quantities of grain. Each token functions on a one-to-one basis and represents a single commodity. By 6000 B.C., clay tokens have spread throughout the Middle East and are eventually replaced by complex tokens with more elaborate markings.

(See also c. 4000 B.C.)

c. 4000 B.C. Complex clay tokens first appear bearing elaborate markings and having a wide variety of shapes. The profusion of many more shapes and markings is believed to reflect the increasingly complex structure of Sumerian society with its increased number of finished goods. Some speculate that it may also be an indication of the first system of coercive taxation and redistribution.

c. 3500 B.C. First fully developed number system is used in Egypt. This system allows counting to continue indefinitely with only the occasional introduction of a new symbol. The emergence of Egyptian government and administration could not have taken place without such a system of numbers. As a system with a base of 10, it uses special pictographs for each new power of 10 up to 10 million.

c. 2400 B.C. Positional notation is first used by the Sumerians in Mesopotamia. They are the only pre-Grecian people to make even a partial use of such a number system. Positional notation is based on the notion of place value, in which the value of a symbol depends on the position it occupies in the numerical representation. The great advantage of this system (that we still use today) is that it does not require a great many symbols to express numbers, no matter how big or small. An example of this is the number 515. Starting from left to right, the first 5 is in what we recognize as the hundred position, and thus it stands not for five of something but for 500. The 1 is in the ten position. This is so because we have a system with a base of ten. The Sumerian system has a base of 60 instead of 10 and uses cuneiform symbols.

c. 1700 B.C. Oldest mathematical document in existence is an Egyptian papyrus written by a scribe named Ahmose or Ahmes. Dated to this period, it becomes the main source of our knowledge of Egyptian mathematics (along with another papyrus, the Golenischev or Moscow papyrus, dated slightly after this time). As a practical handbook telling how to solve everyday problems, it informs us how the Egyptians counted, reckoned, and measured. It contains 85 problems and shows the use of fractions, the solution of simple equations and progressions, some geometry, and measuring area and volume. No generalizations or rules for solving a particular type of problem are given, however. It is also called the Rhind papyrus after the Scottish collector Alexander Henry Rhind, who purchased it at a Nile resort.

c. 1200 B.C. Oldest extant Chinese mathematical work is the *Chou-pei*. Experts disagree about the actual date of this anonymous publication, mainly because it appears to be the work of several people over different periods. Some consider it to be as late as 200 A.D. Its name seems to refer to the use of the gnomon in studying the circular paths of the heavens, and the book is primarily concerned with astronomical calculations. It does include the use of fractions and the properties of a right angle. It also indicates that the Chinese understand what will come to be known as the Pythagorean theorem of the right triangle.

876 B.C. First known reference to the use of a symbol for zero is made in India. An unknown Indian mathematician suggests that an untouched abacus level be given a special symbol. Until this innovation, a gap was often used in positional notation systems, but it was not

always strictly applied and confusion sometimes resulted.

585 B.C. First to treat geometry in an abstract way is Greek philosopher Thales (624-546 B.C.). Borrowing Egyptian geometry, he converts it into an abstract study by stating that it deals with imaginary lines of zero thickness and perfect straightness. He also is the first to prove a mathematical statement using a regular series of arguments. By stating what is known, he then proceeds in a step-by-step manner to reach the desired proof. In this, he can be said to be the inventor of deductive mathematics.

c. 500 B.C. Pythagoras (c.560-c.480 B.C.), Greek philosopher, formulates the idea that the entire universe rests on numbers and their relationships. He is known for establishing several mathematical principles and discoveries (such as being the first to work out by strict mathematical deduction that the square of the length of the hypotenuse of a right triangle is equal to the sum of the squares of the lengths of its sides). This is still known as the Pythagorean theorem. He also studies sound and finds that the relationship of pitch is correlated with the string length of a musical instrument. The Pythagorean dictum that "all is number" deeply influences the development of classical Greek philosophy and medieval European thought (especially the astrological belief that the number harmony of the universe affects all human actions). Scientifically, however, Pythagorean mathematics also aids in the discovery of Kepler's law which posits the relation between the different orbital radii of the planetary solar system.

c. 500 B.C. Chinese rod numerals first appear. This system uses 18 separate symbols made up of variations of vertical and horizontal lines to represent the digits one through nine and the first nine multiples of ten. Using this stick figure-like system, the Chinese can represent numbers as large as desired or needed. Throughout the history of Chinese mathematics, its

The Muse of Arithmetic standing between a man performing counter reckoning and another using Arabic numerals.

main purpose is to maintain the calendar. As this is a right granted only by the emperor, few outside that select circle know or practice mathematics.

c. 440 B.C. First known textbook on geometry, *Elements*, is written by Hippocrates of Chios, Greek geometer. He arranges the known propositions of geometry in a scientific fashion, and this work may have served as a model for Greek mathematician Euclid's *Elements*. His book is known through the writings of others since no trace of it remains. He is also believed to be the first to support himself openly by accepting payment for teaching mathematics. He should be distinguished from his more famous contemporary, Greek physician Hippocrates of Cos (460-c.370 B.C.).

c. 440 B.C. First known attempt to square the circle is made by Greek philosopher Anaxagoras (c.500-c.428 B.C.). When he is imprisoned for impiety after asserting that the sun is not a deity but rather a huge, red-hot stone, he passes the time trying to solve this problem, which will fascinate mathematicians for the next two thou-

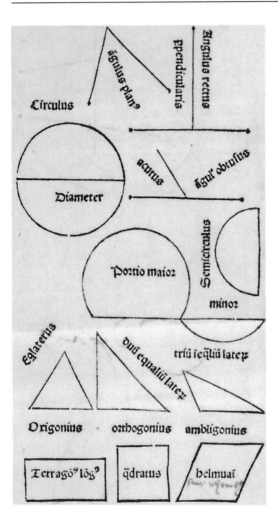

Diagrams from Euclid's *Elements*.

sand years. Also called the quadrature (square) of the circle, this is a theoretical problem that has no real need to be solved in terms of a practical application. Addressing such a mathematical problem indicates how much closer Greek mathematics is to philosophy than to practical affairs. Although his contribution to this problem is not known, no other problem exercises a longer fascination than that of constructing a square equal in area to a given circle.

c. 425 B.C. Quadratrix curve which serves to divide an angle into any number of equal parts

is discovered by Greek mathematician Hippias of Elis. It is the first example of a curve that cannot be drawn by the traditionally required straightedge and compass, but has to be plotted point by point. Hippias actually invents the curve as a device for trisecting angles, and when it is later used in the quadrature of the circle, it gets its name quadratrix. Hippias is also one of the earliest, with Greek geometer Hippocrates of Chios, to teach mathematics for money.

c. 400 B.C. First Greek mathematician to apply mathematics to mechanics is Archytas of Tarentum (428-327 B.C.). A statesman and general as well, he distinguishes harmonic from geometrical and arithmetical progressions and contributes to a theory of acoustics. He also is reputed to be the first to study geometry on a circular cylinder, discovering in the process some of the properties of its oblique section, the ellipse. Further, he devises an ingenious solution to the problem of doubling the cube by means of cylindrical sections.

389 B.C. First university is founded at Academaeus in Athens by Greek philosopher Plato (c.427-c.347 B.C.), who teaches that geometry is a necessary preparation for the study of philosophy. Although not an original mathematical thinker, one of his greatest contributions is the invention of the analytic method or analysis as a method of proof. This logical technique begins with what is given and then proceeds step by step to what is proved. He also suggests that one can reverse the process and begin with the proposition that is to be proved and from it deduce a conclusion that holds. Plato also is the first to make well known the Pythagorean notion of the five possible regular solids (solids with equivalent faces and with all lines and angles formed by those faces equal). Plato is perhaps most important to the history of mathematics for his role as an inspiration to others.

c. 350 B.C. First to consider systematically the geometry of the cone is Menaechmus, Greek mathematician. He shows that ellipses, parabolas, and hyperbolas are all curves produced by the intersection of a cone and a plane. In doing so, he invents what becomes known as conic sections, a subject that will, in a century, raise geometry to its highest point during these ancient times.

c. 270 B.C. Conon of Samos, Greek mathematician and astronomer, succeeds Greek mathematician Euclid (c.325-c.270 B.C.) at Alexandria and becomes the teacher and friend of Greek mathematician and engineer Archimedes (c.287-c.212 B.C.). He may be the first to study the mathematical curve that later becomes known as the "spiral of Archimedes."

c. 250 B.C. Aśoka (272-232 B.C.), king of the Maurya empire in India, erects stone pillars in several major cities that contain the earliest preserved representation of the number symbols we use today.

c. 250 B.C. Archimedes (c.287-c.212 B.C.), Greek mathematician and engineer, first calculates the most accurate arithmetical value for pi (π) to date. He uses this value to figure the areas and volumes of curved surfaces and circular forms. He also uses a system for expressing large numbers that uses an exponential-like method and demonstrates that nothing exists that is too large to be measured. He finds areas and volumes of special curved surfaces and solids and is able to determine the ratio of a cylinder to the sphere it circumscribes. He also devises the first mathematical exposition of the principle of composite movements and is able to calculate square roots by approximation. In addition to the major status he holds in physics, he is considered the greatest mathematical genius of antiquity.

c. 240 B.C. Conchoid curve is discovered by Greek mathematician Nicomedes, who uses it to solve the problem of duplicating or doubling the cube. He also uses it to trisect an angle. Because of its shape, it is named conchoid or "mussel-like." English mathematician and sci-

Page from the first complete edition of Archimedes's works, which were not published in their entirety until 1544.

entist Isaac Newton (1642-1727) uses the conchoid in constructing curves of the third degree.

c. 225 B.C. Theory of conic sections is first introduced by Greek mathematician Apollonius of Perga (c. 261-c. 190 B.C.). In his highly original work titled *Conics*, he produces several new geometrical figures. In particular, he names three of these different curves ellipse, parabola, and hyperbola—none of which was considered

Appolonius of Perga.

by Greek mathematician Euclid (c.325-c.270 B.C.)—-all of which are produced by cutting through a cone at particular angles. His work achieves a special relevance 18 centuries later when German astronomer Johann Kepler (1571-1630) and later English scientist and mathematician Isaac Newton (1642-1727) calculate the elliptical orbits of the planets.

c. 150 B.C. Spiric sections are first described by Greek mathematician Perseus. A spiral is a plane curve that unwinds around a point while moving ever farther from the point. Perseus produces quartics known as spiric sections by cutting the anchor ring or torus by a plane.

6 B.C. Liu Hsin, Chinese scholar and astronomer, writes his treatise on the calendar, *San-t'ungli*, and is the first person known to use decimal fractions. Using the number 10 as the number base, a decimal fraction has a positive integral power of 10 in its denominator and can be written in many ways (e.g., 71/100, 0.71, or .71). Liu Hsin uses it specifically in his calculations of a better estimate for pi.

c. 75 Hero of Alexandria, Greek mathematician and engineer also called Heron, gives algebraic

solutions to first and second degree equations. He also approximates square and cube roots and is the first to give a formula for the area of a triangle as the function of its sides. In optics, he demonstrates that the angle of incidence equals the angle of reflection.

c. 100 First exhaustive work in which arithmetic is treated independently of geometry is *Introductio Arithmetica*, written by Nichomacus of Gerasa, Greek mathematician. Unlike Greek mathematician Euclid (c.325-c.270 B.C.), he illustrates things by using real numbers instead of drawing lines, and his method is inductive instead of deductive. This work also contains the Pythagorean theory of numbers.

c. 100 Menelaus of Alexandria, Greek mathematician, writes his *Sphaerica* in which he gives the first definition of a spherical triangle and proves the theorems on the congruence of spherical triangles. He is also the first to separate trigonometry from stereometry and astronomy.

c. 150 First use of the terms minutes and seconds is made by Greek astronomer Claudius Ptolemy (c.100-c.170), also known as Claudius Ptolemaeus. He subdivides each degree in a circle into 60 "partes minutiae primae," and each of these again into 60 "partes minutiae secundae." He also begins using the Greek letter omicron (since O is the first letter of the Greek word for empty or nothing) to designate a blank in numbers but not in the manner of today's zero. He also offers the first notable value for pi (π) since Greek mathematician and engineer Archimedes (c.287-c.212 B.C.)

c. 250 Diophantus (c.210-c.290), Greek mathematician, writes his *Arithmetica* which is the earliest treatise on algebra now extant. He is also the first Greek to write a significant work on algebra, solving problems using his own symbols and using what would now be called algebraic equations. He becomes best known for Diophantine equations, or equations (indeterminate) that have no single set of solutions.

He is also the first Greek to treat fractions as numbers and is credited with introducing the symbol for minus.

(See also 1585)

c. 390   Theon of Alexandria, the last recorded member of the Museum at Alexandria, publishes Greek mathematician Euclid's *Elements* with his own notes and commentary. Until 1808 when a 10th century Euclid manuscript is discovered that was taken from one predating Theon, this copy is the oldest version of the *Elements*. He is also known as the father of Alexandrian mathematician and philosopher Hypatia (c. 370-415).

(See also c.400)

c. 400   Hypatia of Alexandria (c.370-415), mathematician, writes commentaries on Greek mathematicians Diophantus (c.210-c.290) and Apollonius (c.261-c.190 B.C.). As the recognized head of the Neoplatonist school of philosophy at Alexandria, she is noted as the only notable woman scholar of ancient times and the first woman mentioned in the history of mathematics. Her eloquence, beauty, modesty, and remarkable intellectual gifts attracted a large number of pupils. Her death by stoning and slashing by a fanatical Christian mob leads to the departure of many scholars from Alexandria and marks the beginning of its decline as a major center of ancient learning.

499   Aryabhata (476-c.550), Hindu mathematician and astronomer, writes his *Aryabhatiya,* which contains his significant description of the Indian numerical system. He also works on series, permutations, and arithmetic progressions and is one of the earliest to use algebra. He gives an accurate value for pi.

c. 500   Ammonios, Alexandrian philosopher, is the first to divide mathematics into arithmetic, geometry, astronomy, and music. This division or classification will form the famous "quadrivium" of Western medieval education.

c. 500   Akhmin papyrus is written in Egypt. Written in Greek, it is a manual on practical arithmetic and is the oldest extant treatise on this subject. It is of Byzantine origin and was written by a Christian. It is named after the city on the upper Nile River where it was discovered.

595   First known occurrence of Hindu (later Arabic) numerals is found on a plate made in India during this year.

(See also 662)

625   Wang Hs'iao-t'ung, Chinese mathematician, writes his *Ch'i-ku Suan-ching* in which numerical cubic equations appear for the first time in Chinese mathematics.

662   Earliest known mention of Hindu (later Arabic) numerals outside of India is made by Severus Sebokht, a bishop in Syria. He writes of Hindu computations "which excel the spoken word ... and are done with nine symbols."

773   Positional notation is first discovered by the Arabs when Al-Mansur (709-775), second caliph of the Abbasid dynasty, commissions a translation of the Hindu astronomical treatise *Siddhanta*, which dates to the first half of the sixth century. In translating this document, they first learn that the position of a digit in a column can affect its value (e.g., a 5 in the right hand column means 5, but in the left hand column it means 50). They also learn the use of a zero to indicate an empty column. These numerals and positional notation eventually come to the West as "Arabic numerals," and are used in the early 9th century by Arab mathematician Al-Khwarizmi (c.780-c.850), who produces the first astronomical tables in Arabic.

c. 825   First detailed information on the Hindu (later Arabic) numerical system is found in a treatise written by Arabian mathematician Al-Khwarizmi (c.780-c.850). Centuries later, it is titled *Algoritmi de Numero Indorum* when it is translated from the Arabic into Latin. Several years later, his other work titled *al-Kitab al-*

jabr w'al-muqubala becomes the source for the modern term "algebra" (taken from the Arab word in his title, "al-jabr," meaning transposition).

c. 825   First Arabic translation of Greek mathematician Euclid's *Elements* is made by Arabian mathematician Al-Hajjai Ibn Yusuf (c.786-833). He also translates the work of Greek astronomer Claudius Ptolemy (c.100-c.170).

(See also 1120)

c. 860   Al-Mahani, Persian mathematician and astronomer, is the first to state the Archimedean problem of dividing a sphere into two segments according to a given ratio, in the form of a cubic equation. He does not solve the problem, however, and he grapples with it so long that it becomes known as "Mahani's problem."

(See also c.950)

870   Thabit ibn Qurra (826-901), Arabian mathematician, founds a school of translators and translates most of the major Greek mathematical works. It is mainly because of his work in translating these Greek texts into Arabic that the West eventually comes to know so many of them. Had it not been for his efforts, the body of Greek mathematical works extant today would be much smaller. He was also a first-rate mathematician in his own right, and his treatise on amicable or friendly numbers is the first example of original mathematical work in the Middle East.

c. 900   First to use a table of sines for mathematical computations in astronomy is Arabian astronomer and mathematician al-Battani (c.858-929), also called Albategnius. As perhaps the outstanding Middle Eastern scholar of his century, he greatly improves the astronomical calculations of Greek astronomer Claudius Ptolemy (c.100-c.170) by replacing geometrical methods with trigonometry. He is also the first to prepare a table of cotangents. Out of a 12th century translation of his work comes the word "sine."

c. 950   First Arab to solve the problem of the cubic equation by means of conic sections is Alchazin, Arabian mathematician and astronomer also called Abu Ja'far al-Kazin. It is also called "Mahani's problem."

(See also c.860)

c. 980   Gerbert (c.945-1003), French scholar, helps initiate the rebirth of learning in Europe, especially in mathematics. As one of the first Christians to study in the Moslem schools of Spain, he reintroduces the abacus to the West and uses Arabic numerals (although not a zero). Gerbert is not only a profound scholar but in 999 he is elected pope, taking the name Sylvester II.

c. 1020   Al-karkhi, Arabian mathematician, writes a treatise on algebra considered the best produced by an Arab. He is also the first to work with higher roots.

1120   Adelard of Bath (c.1090-c.1150), English scholar, makes the first Latin translation of Greek mathematician Euclid's *Elements* from an Arabic version. This introduces Euclid to Europe and serves as the major geometry textbook in the West. He also introduces Arab trigonometry to the West and uses Arabic numbers.

(See also c.825)

c. 1134   Plato of Tivoli, Italian translator also called Plato Tiburtinus, translates several Arab mathematical works and is the first to give in Latin the complete solution of a quadratic equation.

1202   First great Western mathematician to advocate the adoption of "Arabic notation" is Italian mathematician Leonardo Fibonacci (c.1170-c.1230), also called Leonardo da Pisa. In his book on algebraic methods, *Liber Abaci*, he explains the use and advantages of Arabic numerals over Roman numerals and makes clear the values of positional notation using a base of 10. He also describes "the nine Indian figures" (since it is rightly a Hindu or Indian invention), as well as the sign for zero. Called zephirum in Arabic, this word and its variants account for our present words cipher and zero. Since Leonardo's father is a Pisan engaged in business in northern Africa, the son studies under a Muslim teacher and travels in Egypt, Syria, and Greece where he becomes very familiar with Hindu-Arabic numerals.

c. 1250   First text to treat trigonometry as a study separate from astronomy is written by Nasir al-Din (1201-1274), Persian mathematician and astronomer also called Nasir Eddin. He also publishes a proof of the parallel postulate.

c. 1270   William of Moerbeke (c.1215-c.1286), Flemish scholar, translates the mathematical works of Greek mathematician and engineer Archimedes (c.287-c.212 B.C.) from Greek into Latin. Although his translations are extremely literal (since he knows little mathematics), they are well received and serve as the basis for the first printed versions in the early 16th century. It is this translation that is available to Renaissance scholars like Italian artist Leonardo da Vinci (1452-1519).

(See also 1543)

c. 1360   Nicole Oresme (c.1323-1382), French mathematician, writes a mathematical work, *De Proportionibus Proportionum,* that contains the first known use of fractional exponents or fractional powers. In another work he locates points by coordinates and foreshadows modern coordinate geometry.

c. 1400   Al-Kashi, Persian mathematician and astronomer, writes a short work on arithmetic and geometry and is the first to use finite decimal fractions. His masterpiece, *Risala al-Muhitiyya,* first explains in detail all arithmetic operations using them.

1435   First formal account of the uses of perspective is given by Italian artist Leone Battista Alberti (1404-1472) in his *Della Pittura.* In

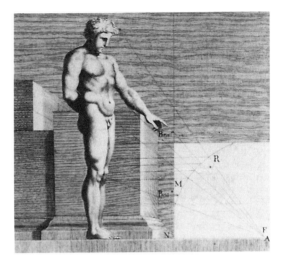

Illustration of mathematical perspective, as explained by Italian artist Leone Battista Alberti.

this treatise on how to graphically give the illusion of three-dimensional space on a flat surface, he mentions his debt to Italian architect Fillippo Brunelleschi (1377-1446) who discovered the mathematical laws of perspective around 1410. Alberti is a forerunner of what is today called "projective geometry."

(See also 1478)

1464   First systematic European exposition of plane and spherical trigonometry considered independently of astronomy is written by Regiomontanus (1436-1476), German astronomer and mathematician also named Johann Muller. His *De Triangulis Omnimodes* treats trigonometry as a theoretical science and introduces the use of algebraic and trigonometric methods into Germany.

1478   Piero Della Francesca (c.1420-1492), Italian painter, writes his *De Prospectiva Pingendi,* which is the first true treatise on perspective. This method of graphically depicting three-dimensional objects and spatial relationships on a two-dimensional plane is based on mathematical laws of perspective discovered by Italian architect Fillippo Brunelleschi (1377-1446)

Mathematicians and painters alike studied perspective.

around 1410. Francesca's actual paintings reflect his intimate knowledge and study of geometry and mathematics. He leaves behind a manuscript on regular solids that is published by Italian mathematician Luca Pacioli (1445-1517), also known as Luca De Borgo, in 1509 as *Divina Proportione*. Francesca's work is a major step in the development of perspective since the time of Italian artist Leone Battista Alberti (1404-1472) and he deals with some very difficult problems of perspective.

(See also 1435)

1478　First dated, printed arithmetic is the *Arte dell'Abbaco* published in Treviso, Italy. Its author is unknown.

MAY 1482　First printed edition of Greek mathematician Euclid's geometry classic is the *Elementa Geometriae,* published in Venice, Italy. The translation into Latin from the Arabic is was completed by Giovanni Campano (also called Jordanus Campanus of Novara) about 200 years prior to publication.

1484　Nicholas Chuquet (1445-1500), French mathematician, writes an arithmetic known as *Triparty en la Science des Nombres.* In this very

advanced work, he discusses rational and irrational numbers, the theory of equations, and the role of the zero symbol. He is also the first to use the radical sign with an index to indicate roots.

1489　Johannes Widmann (c.1460-c.1500), German mathematician, publishes his *Behende und Hubsche Rechenung*, which is the first printed book to contain the + and - symbols. They are not used as symbols of operation but merely to indicate excess and deficiency. Widmann's book is a commercial arithmetic, and his use of the symbols reflects their actual use to indicate a surplus or lack of warehouse measures.

1491　First printed example of the modern process of long division is demonstrated by Filippo Calandri of Italy in his *Pictagoras Arithmetrice Introductor*. It also contains the first illustrated mathematical problems published in Italy.

1494　First printed work on algebra is written by Italian mathematician Luca Pacioli (c.1445-1517). His *Summa de Arithmetica, Geometrica, Proportioni et Proportionalita* also contains the first printed example of a detailed description of the method of double-entry bookkeeping. More influential than it is original, this work actually contains material in four fields: arithmetic, algebra, elementary Euclidian geometry, and double-entry bookkeeping. It is this last aspect with its practical applications and commercial aspects that makes it so popular, and Pacioli is now generally regarded as the father of double-entry bookkeeping.

1511　First mathematician to study the properties of the mechanical curve known as a cycloid is Charles de Bouelles (c.1470-c.1553), French mathematician also called Bouvelles or Bovillus. As a circle rolls along a horizontal plane, a fixed point on its circumference will trace out a graceful curve called a cycloid—a name later given by Italian astronomer and physicist Galileo Galilei (1564-1642). Bouelles also publishes his *Le Livre et de l'Art et Science de*

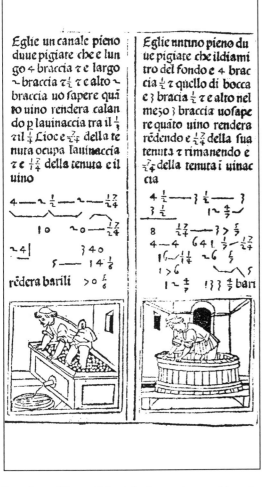

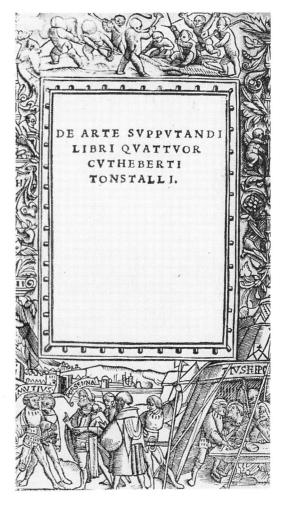

Page from Filippo Calandri's first written arithmetic accompanied by illustrated problems.

Title page of the first separate arithmetic printed in England, compiled by Cuthbert Tunstall.

Géométrie this year, which is the first book on geometry to be published in French.

1522 First separate arithmetic printed in England is *De Arte Supputandi* by English scholar Cuthbert Tunstall (1474-1559).

1522 First work on conic sections to appear in Christian Europe is the *Elementis Conicis*, written by German mathematician Johannes Werner (1468-1528). This work marks the renewal of interest in these particular curves first examined by Greek mathematician Apollonius (c.261-c.190 B.C.).

1525 First German textbook on algebra is *Die Coss* written by German mathematician Christoff Rudolff (c.1500-1545). This work is especially significant as one of the earliest printed works to make use of decimal fractions as well as the modern symbol for roots. He also publishes a book in colloquial German on whole numbers and fractions that is a first of its kind in that country.

1534 Petrus Apianus (1495-1552), German geographer and astronomer also called Peter Bienewitz or Peter von Bennewitz, publishes his *Instrumentum Sinuum sive Primi Mobilis*, which contains a table of sines for every minute of arc. These are the first such tables to be printed.

FEBRUARY 12, 1535 First to work out a general solution of the cubic equation (equations of the third degree) is Italian mathematician Niccolo Tartaglia (c.1499-1557), also called Nicolo Fontana. Ten days after this discovery, he takes part in a public contest and defeats all comers at solving cubic equations. He intends to keep his solution a secret but confides it to Italian mathematician Girolamo Cardano (1501-1576) in 1539. Cardano eventually reveals the secret to all in a book he writes.

(See also 1545)

1543 Niccolo Tartaglia (1499-1557), Italian mathematician also called Nicolo Fontana, edits the Latin translation by Flemish scholar William of Moerbeke (c.1215-c.1286) of the mathematical writings of Greek mathematician and engineer Archimedes (c.287-c.212 B.C.). This is the first printed version of these works, and it is through Tartaglia that they become fully known to the West.

(See also c. 1270)

1545 Girolamo Cardano (1501-1576), Italian mathematician, publishes his *Ars Magna*. In this treatise on algebra, he discloses the discovery of the rule that allows the cubic equation to be solved, despite having pledged his secrecy to its discoverer, Niccolo Tartaglia (c.1499-1557), also called Nicolo Fontana. Although Cardano does give credit to Tartaglia, the method becomes known as "Cardano's rule." This work also contains the first major Western treatment of negative numbers and the first usage of a complex number in a computation. The incident between Cardano and Tartaglia draws attention to the issue of the ethics of secrecy in science, and eventually it becomes a scientific

Niccolo Tartaglia.

tradition that credit goes not to the person who first makes a discovery but to the person who first publishes it.

c. 1550 Rheticus (1514-1576), German mathematician whose given name is Georg Joachim von Lauchen, completes his trigonometric tables. They are the best tables of their time, with sines calculated out to 10 places. He is also the first to define the trigonometric functions as ratios of the sides of a right triangle. Prior to him, trigonometric functions were related to arcs of circles rather than angles.

1556 First work on mathematics printed in the New World is a small commercial compendium by Juan Diez. It is printed in Mexico City.

1557 First English treatise on algebra is written by Robert Recorde (c.1510-1558), English mathematician. In his book *The Whetstone of Witte*, he uses the modern symbol for equality (=) for the first time. He states that he selected this symbol because no two things could be more equal than two parallel lines.

1559 Jean Buteo (1492-c.1565), French mathematician, publishes his *Logistica*, which is one

Mathemetician Girolamo Cardano.

of the earliest works on elementary geometry in French.

1570 First complete English translation of Euclid's *Elements* is made by Henry Billingsley, English mathematician. He is assisted in translating this work from Greek by English mathematician, astrologer, and alchemist John Dee (1527-1608).

1572 Rafaello Bombelli (c.1526-1573), Italian mathematician, publishes a book on algebra in which he points out the reality of the term imaginary roots and lays the foundation for a better knowledge of imaginary quantities. This is the first consistent theory of imaginary numbers to be used. Through his ingenious reasoning and what he calls "a wild thought," Bombelli shows the important role that conjugate imaginary numbers are to play in the future.

1579 Franciscus Vieta (1540-1603), French mathematician also known as Francois Viete, publishes his *Canon Mathematicus seu Ad Triangula*. This is the first book in Western Europe to systematically develop methods for solving plane and spherical triangles using all

six trigonometric functions. He also calculates pi (π) correctly to nine decimal places.

(See also 1591)

1583 Mathematical words tangent and secant are used for the first time by Thomas Finck (1561-1656), Danish mathematician and physician, in his *Geometriae Rotundi*. Since it is Finck who first uses these words for certain trigonometrical ratios, he is credited with their invention.

1585 First comprehensive system of decimal fractions and their systematic, practical applications is published by Simon Stevin (1548-c.1620), Belgian-Dutch mathematician and engineer, in his *De Thiende*. While in no way the inventor of decimal fractions, Stevin is certainly the first systematic user of them, and it becomes his goal to explain their base-ten advantages so they will be more popularly adopted. It is his wish to teach everyone "how to perform with an ease, unheard of, all computations necessary between men by integers without fractions." In this work, he also argues for the decimal system of weights and measures. He is also the first to translate the work of Greek mathematician Diophantus (c.210-c.290) into a modern language.

(See also c.250)

1591 First systematic algebraic notation is introduced by Franciscus Vieta (1540-1603), French mathematician also known as Francois Viete. His book published this year, *Isagoge in Artem Analyticum*, founds modern algebra. Vieta contributes to the development of symbolic algebra, as he is the first to introduce the practice of using letters as symbols (vowels to represent unknown quantities and consonants to represent known ones). Vieta is considered the father of modern algebra, and his book is considered the first work on symbolic algebra. It is also the first that would be recognized by a modern-day student as a book on algebra. Vieta himself does not use the term algebra but prefers the term analysis, and as a result, analysis comes to mean the use of algebraic methods to solve problems. Algebra, of course, is still used to note that branch of mathematics dealing with the rules governing the manipulation of equations.

1594 John Napier (1550-1617), Scottish mathematician, first conceives the notion of obtaining exponential expressions for various numbers and begins work on the complicated formulas for what he eventually calls logarithms. Napier uses the distinction between arithmetic and geometric progression, and it occurs to him that a number can be expressed in exponential form. That is, 4 can be written as 2 to the second power, and 8 can be 2 to the third power. Written in this manner, multiplication is reduced to addition or subtraction of exponents. He works for the next 20 years to obtain the exponential expression of various numbers.

(See also 1614)

c. 1600 First noted mathematician to set foot in the New World is Thomas Harriot (1560-1621), English mathematician and astronomer. He is sent as a surveyor by his patron, Sir Walter Raleigh (1554-1618), to the colony in Virginia in 1585. After returning to England, he introduces the mathematical symbols for "greater than" (>) and "less than" (<). He is also the founder of the English school of algebraists and is the first to factor equations. It is his adoption and regular use of English mathematician Robert Recorde's (c.1510-1558) equality sign (=) that leads to its ultimate acceptance.

1600 Bartholomaeus Pitiscus (1561-1613), German mathematician, publishes his *Trigonometriae; sive, De Dimensione Triangulorum,* which is the first book on trigonometry to bear this title as well as the first time the word itself is used.

c. 1600 Joost Burgi (1552-1632), Swiss mathematician, conceives of the idea of logarithms independently of Scottish mathematician John Napier (1550-1617), but publishes his discovery in 1620, six years after Napier. Most agree

A page from John Napier's book of logarithmic tables.

that Napier had the idea first, and while his approach to logarithms is geometrical, Burgi's is algebraic. Despite these and other slight differences, the essence of the principle of logarithms is in Burgi's work, and most regard him as an independent discoverer who loses credit when Napier publishes first.

(See also 1614)

1613 Francois d'Aguilon (c.1566-1617), Belgian mathematician also called Aguilonius, publishes his *Opticorum* in which he gives the earliest description of stereographic principles and first uses the term stereographic projection. This technique is used for the projection of a sphere on a plane and is essentially the use of geometry with perspective.

1613   Pietro Antonio Cataldi (1552-1626), Italian mathematician, publishes his *Trattato del Mondo Brevissimo di Trovare la Radica Quadra delli Numeri* in which he discusses his invention of the common form of continued fractions. He also develops the first symbolism for continued fractions and shows how to use them to determine square roots.

(See also 1737)

1613   Willebrord van Roijen Snell (1591-1626), Dutch mathematician, succeeds his father as professor of mathematics at the University of Leiden and applies his prodigious mathematical talents to astronomy. He is the first to use trigonometric triangulation to measure distance and the meridian, and thus lays the groundwork for modern mapmaking. He also discovers the law of refraction.

1614   First logarithm tables are published by Scottish mathematician John Napier (1550-1617) in his *Mirifici Logarithmorum Canonis Descriptio*. As the originator of the concept of logarithms as a mathematical device to aid in calculations, he initially calls them artificial numbers, but later coins the term logarithm, meaning number of the ratio. His tables simplify routine calculations to a great degree and reduce multiplication to simple addition or subtraction. The tables quickly become a powerful tool in computations, especially so for astronomy. Most agree with the famous saying that the invention of logarithms, "by shortening the labors doubled the life of the astronomer." Altogether, logarithms are a revolutionary tool for mathematicians. Napier also worked with Belgian-Dutch mathematician and engineer Simon Stevin's (1548-c.1620) decimal fractions and brought them to their present

form by first introducing the use of the decimal point.

(Sere also 1594)

1629   Albert Girard (1595-1632), French mathematician, publishes a work on trigonometry that contains the first use of the abbreviations sin, tan, sec for sine, tangent, and secant. He also is the first to use brackets in algebra and the first to understand the use of negative roots in the solution of geometric problems.

1631   Cross multiplication sign (×) is first used by English mathematician William Oughtred (1575-1660) in his book *Clavis Mathematicae*. Oughtred places particular emphasis on the use of mathematical symbols and uses over 150 of them, but this is the only one that really remains in use.

1635   Paul Guldin (1577-1643), Swiss mathematician, publishes his *Centrobaryca* in which he introduces the centrobaric method. The centrobaric method is a mathematical theorem that gives the relation between volume and area of a revolving figure. It comes to be known as Guldin's theorem, although it was a actually first stated by the 4th century Greek mathematician Pappus in his *Mathematical Collections*.

1635   Francesco Bonaventura Cavalieri (1598-1647), Italian mathematician, publishes his *Geometria Indivisibilibus*, expounding his method of indivisibles which becomes a forerunner of integral calculus. He is also the first Italian writer to recognize and popularize the value of logarithms and is largely responsible for introducing their use as a computational tool in Italy.

c. 1635   Theory of numbers is founded by French mathematician Pierre de Fermat (1601-1665). He single-handedly begins the systematic study of the properties of natural numbers, being the first to carry this study past the stage where Greek mathematician Diophantus (c.210-c.290) had left it. Although he never publishes during his lifetime, he goes on to explore perfect and amicable numbers, figurate numbers, magic squares, Pythagorean triads, divisibility, and above all, prime numbers. His interest in mathematics is always as a pleasureable hobby, and despite his "amateur" status, he is considered as a mathematician of the highest rank who makes many substantive contributions.

1637 René Descartes (1596-1650), French mathematician and philosopher, publishes his *Discours de la Methode*, which contains a 106-page essay titled *La Géométrie*. In this essay, he founds analytic geometry by showing how geometric forms may be systematically studied by analytic or algebraical means. His work is a unique combination of algebra and geometry, enriching both since this new method can solve problems much easier than either one could separately. This application of algebra to geometry paves the way for the eventual development of calculus by English scientist and mathematician Isaac Newton (1642-1727). Descartes is the first to use the letters near the beginning of the alphabet for constants and those near the end for variables. It is because of him that we use x and y in today's algebra. He also introduces the use of exponents and the square root sign. Since Descartes writes this important appendix in an intentionally obscure manner, it is not until 1649 when his friend, the French mathematician Florimond de Beaune (1601-1652), publishes his *Cartesii Geometrium* that Descartes work is really understood. Beaune is able to translate Descartes' obscure rendering into understandable Latin with explanatory notes.

1647 First to use the phrase "method of exhaustion" is Flemish mathematician Gregory St. Vincent (1584-1667), also known as Gregoire de Saint-Vincente, in his *Opus Geometricum Quadraturae Circuli et Sectiones Coni*. The method of exhaustion is traditionally attributed to Greek astronomer and mathematician Eudoxus (c.400-c.347 B.C.), and it is used to prove that the areas of circles are to one another as the

squares of their diameters. It also proves that the volumes of pyramids (of the same height and with triangular bases) are proportional to the areas of their bases. In this book he makes several basic discoveries on conic sections and is also the first to use a geometric series on the paradoxes of Zeno.

1653 First mathematician to make a systematic study of the relations exhibited by the arithmetic triangle is French mathematician and physicist Blaise Pascal (1623-1662) in his *Traité du Triangle Arithmetique*. Although written this year, it is not published until 1665. Pascal is by no means its originator (since the Chinese knew of it centuries before), but is rather the first Westerner to consider it systematically. He provides a method for finding a triangular arrangement of numbers in which each row are binomial coefficients. It is also in this treatise that Pascal introduces mathematical induction into mathematics.

1655 John Wallis (1616-1703), English mathematician, publishes his *Arithmetica Infinitorum* in which he systematizes and extends the work of French mathematician and philosopher René Descartes (1596-1650) and Italian mathematician Francesco Bonaventura Cavalieri (1598-1647). He is also the first to extend the notion of exponents to include negative numbers and fractions. Considered one of the first to write a serious history of mathematics, Wallis is one of the most influential mathematicians of his time, and his work is a precursor to calculus.

(See also 1656)

1656 First to use the sideways figure eight symbol(∞) for infinity is John Wallis (1616-1703), English mathematician. He is also the first to interpret imaginary numbers geometrically.

(See also 1655)

1657 First formal treatise on the mathematical treatment of probability is written by Christiaan Huygens (1629-1695), Dutch physicist and astronomer. His short work, *De Ratiociniis in*

Ludo Aleae (*On Reasoning in Games of Dice*), is based on the ideas brought forth during the Pascal-Fermat correspondence begun in 1654 when the two wrote each other and speculated on the notion of mathematical probability. It appears first as an appendix to Dutch mathematician Frans Van Schooten, Jr.'s (1615-1661) book *Exercitationes Mathematicae*. In this work, Huygens introduces the important concept of mathematical expectation, or as he calls it, "the value (price) of the chance" to win a game. Huygens work remains the only text available on probability theory for the next 50 years.

(See also 1713)

1657 Bernhard Frenicle de Bessy (1602-1675), French mathematician, publishes his *Solutio Duorum Problematum Circa Numeros Cubos et Quadratos* in which he works on Pythagorean numbers and discovers that the number of magic squares increases enormously by writing down 880 magic squares of the order of four. A magic square is a square matrix divided into cells and filled with numbers so that each column, every row, and the two main diagonals can produce the same sum, called the constant. Although they are often used as symbols or charms and thought to have special, magical properties, they are useful to mathematicians who study them as problems in number theory.

1658 First to use three variables in analytic geometry is Jan Hudde (1628-1704), Dutch mathematician. He is also the first to use a letter as a symbol for a negative quantity.

1659 First convergent series of logarithms is given by Pietro Mengoli (1625-1686), Italian mathematician, in his *Geometria Speciosa*.

1659 First to use the symbol for division (÷) is Swiss mathematician and astronomer Johann Heinrich Rahn (1622-1676) in his *Teutsche Algebra*. The symbol is introduced to England in a 1668 translation.

1662 First president of the newly created Royal Society of London for the Improvement of Natural Knowledge is English mathematician William Brouncker (1620-1684). He also is known for demarcating the first infinite series for the area of an equilateral hyperbola and the infinite continued fraction method for pi.

1662 First to apply mathematics to the integration of vital statistics is John Graunt (1620-1674), English statistician. In his *Natural and Political Observations . . . Made Upon the Bills of Mortality*, he becomes the first to draw an extensive set of statistical inferences from a mass of data. He obtains his information from the weekly and yearly statistics published in the Bills of Mortality which tallies the number of burials and other related information in several London parishes. In his book, he reduces 57 years of statistics (1604-1661) to a series of tables from which he draws several conclusions. Among these, he notes that more male babies are born than female, and that women live longer than men. He also concludes that 36 out of every 100 people die by the age of 6 and that hardly anyone lives to be 75. As the first to establish life expectancy and to publish a table of demographic data, he is considered the founder of vital statistics. Graunt also has the honor of being the first tradesman (he is a shopkeeper) admitted to the Royal Society. In fact, he is a charter member, being one of its original 119 fellows.

(See also 1693)

1664 Binomial theorem is discovered by English scientist and mathematician Isaac Newton (1642-1727). Any power of a binomial (consisting of two terms) may be found without performing the successive multiplications by using this formula. This discovery introduces Newton to the study of infinite series and eventually to calculus. Although he discovers (but does not prove) his theorem this year, he does not communicate it until 1676 in two letters to German natural philosopher Henry Oldenburg (c.1615-1677).

1667 First to systematically study the convergent series and to distinguish between a convergent and divergent series is James Gregory (1638-1675), Scottish mathematician and astronomer. Convergence in mathematics is used to describe sequences and functions. A convergent series has a limit of sequences but a divergent series has no limit.

1668 First to calculate the area under a curve using the newly developed analytical geometry is Nicolaus Mercator (c.1620-1687), German-Danish mathematician and engineer whose real name is Kaufmann. His calculations are included in his *Logarithmotechnia,* published this year. Mercator becomes one of the first members of the Royal Society and also designs the fountains at Versailles.

1669 Isaac Barrow (1630-1677), English mathematician, publishes his *Lectiones Opticae et Geometricae.* In it, he introduces the differential triangle and develops a method of tangents that differs from differential calculus in its notation. He is also the first to note the reciprocal relationship between differentiation and integration. Barrow is also remembered for relinquishing his mathematics chair at Cambridge to his illustrious pupil, English scientist and mathematician Isaac Newton (1642-1727). After his resignation, Barrow takes up religious study.

1669 First systematic account of the discovery of fluxional (differential and integral) calculus is contained in a paper written by Isaac Newton (1642-1727), English scientist and mathematician. Entitled "De Analysi per Aequationes Numero Terminorum Infinitas," this paper is circulated privately by Newton and lays the foundation for his later elaboration on calculus. By Newton's time, mathematicians were attempting to solve increasingly complex algebraic problems involving changing quantities of motion, mass, and energy with little success. Newton's differential and integral calculus gives them a new tool capable of revealing simple patterns in complex and subtle natural phe-

nomena. In differential calculus, the rate at which a known quantity changes is found. Integral calculus solves the opposite problem of finding an unknown quantity for which only the rate of change is known. Calculus proves to be an indispensable scientific tool, and without it, it would be impossible to understand the behavior of gravity, the motion of the planets, or such forces as light, electricity, and magnetism. In 1684, German philosopher and mathematician Gottfried Wilhelm Leibniz (1646-1716) publishes his own version of calculus. This eventually provokes a strained controversy between the two men and their adherents over whether Leibniz stole the idea from Newton, but historians now believe that Leibniz's discovery was independent.

(See also 1687)

1678 Ceva's theorem is discovered by Italian mathematician Giovanni Ceva (1647-1734), who gives static and geometric proofs for this theorem concerning lines that intersect at a common point when drawn through the vertices of a triangle. It is contained in his *De Lineis Rectis.* He later writes one of the first works in mathematical economics.

(See also 1711)

1684 Gottfried Wilhelm Leibniz (1646-1716), German mathematician and philosopher, first publishes his account of calculus in the new scientific journal, *Acta Eruditorum.* He discovers it independently of English scientist and mathematician Isaac Newton (1642-1727), although later than him. Newton, however, publishes his discovery later than Leibniz, in 1687. Of the two systems, Leibniz's calculus proves more convenient and flexible because of its well-devised system of notation.

1685 First serious history of mathematics, *De Algebra Tractatus: Historicus & Practicus,* is written by John Wallis (1616-1703), English mathematician.

1687 Isaac Newton (1642-1727), English sci-

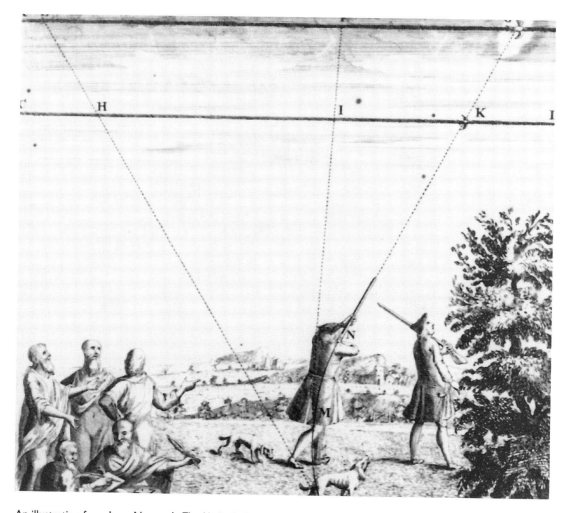

An illustration from Isaac Newton's *The Method of Fluxions and Infinite Series*, a study in calculus.

entist and mathematician, publishes his *Philosophiae Naturalis Principia Mathematica*, which, besides its laws of motion and universal gravitation, contains the first published account of Newton's invention of calculus. This landmark work is generally considered the basis of applied mathematics.

c. 1688 First mathematician to give public lectures on Newtonian theory is Scottish mathematician and astronomer David Gregory (1661-1708). The nephew of the great Scottish mathematician and astronomer James Gregory (1638-

1675), he is a professor of mathematics at Edinburgh until 1691 when he is appointed to Oxford.

1690 Isochrone or isochronous curve along which a body will fall with uniform velocity is discovered by Swiss mathematician Jakob Bernoulli (1654-1705). The equation he calculates for this shows that the required curve is the semicubical parabola. It is in connection with this problem that he encounters Gottfried Wilhelm Leibniz's (1646-1716) calculus. In writing about the isochrone this year, Bernoulli uses the word integral, and a few years later, Leibniz agrees that *calculus integralis* would be a better name than *calculus summatorius* to distinguish it from *calculus differentialis*.

1693 Edmond Halley (1656-1742), English astronomer, compiles the first set of detailed mortality tables. Presently, such tables are used in the insurance business. In his article "Degrees of Mortality of Mankind . . . with an Attempt to Ascertain the Price of Annuities upon Lives," he offers a mathematical formula that essentially originates the science of life statistics.

1696 First textbook on calculus is written by Guillaume Francois Antoine de l'Hospital (1661-1704), French mathematician. His *Analyse des Infiniment Petits* represents the first systematic treatment of calculus and emphasizes the contributions of German philosopher and mathematician Gottfried Wilhelm Leibniz (1646-1716). This work does much to popularize calculus. It also contains the Swiss mathematician Johann Bernoulli's (1667-1748) method of finding the limiting value of a fraction whose numerator and denominator tend simultaneously to zero.

1706 First use of the 16th letter of the Greek alphabet (π or pi) as the symbol for the ratio of the circumference to the diameter of a circle is made by William Jones, English mathematician. In *Synopsis Palmariorum Matheseos* for beginners, Jones correctly gives the circumference-to-diameter ratio to 100 decimal places. It is not until the famous Swiss mathematician Leonhard Euler (1707-1783) uses π that it becomes generally adopted.

(See also 1737)

1707 First Italian work dedicated to pure integral calculus is written by Gabriel Manfredie (1681-1761), Italian mathematician. His *De Constructione Aequationum Differentialium Primi Gradus* also contains a valuable arrangement of the theory of differential equations.

1711 First clear mathematical writings on economics is *De Re Nummaria* by Italian mathematician Giovanni Ceva (1647-1734). Ceva is a professor of mathematics at the University of Mantua and is also employed by the Duke of Mantua. He is also an engineer.

1713 First full-length treatment of the theory of probability is the posthumously published *Ars Conjectandi* by the Swiss mathematician Jakob Bernoulli (1654-1705). In this work he reproduces Dutch physicist and astronomer Christiaan Huygens' (1629-1695) earlier treatise and includes a general theory of permutations and combinations, all accompanied by mathematical theorems. This highly innovative work discusses what becomes known as Bernoullian numbers and the Bernoulli theorem, and analyzes games of chance according to several variables.

1715 First treatise on the calculus of finite differences is the *Methodus Incrementorum Directa et Inversa*, written by Brook Taylor (1685-1731), English mathematician. Taylor's invention of finite differences adds a new branch to mathematics. He applies his new mathematical method to the study of the form of movement of vibrating strings. He is the first to attempt to explain this movement with mechanical principles.

1719 First arithmetic printed in the American colonies is James Hodder's *Arithmetick; or that Necessary Art Made Easy,* published in Boston, Massachusetts. It was originally published in London in 1661.

(See also 1729)

1720 Colin Maclaurin (1698-1746), Scottish mathematician, publishes his *Geometria Organica* in which he discusses the general properties of planar curves and extends the work of English scientist and mathematician Isaac Newton (1642-1727) on conics, cubics, and higher algebraic curves. Some 20 years later, he becomes the first to give a correct theory for distinguishing maxima from minima (the problem of locating the points at which the largest and smallest values of a function occur). Maclaurin is elected professor of mathematics at Aberdeen at the age of 19 and publishes this work two years later.

(See also 1742)

1722   English mathematician Roger Cotes (1682-1716) has his *Harmonia Mensurarum* posthumously published. It is one of the earliest works to recognize the periodicity of the trigonometric functions. It also contains the first publication of the cycles of the tangent and secant functions. It is said that English scientist and mathematician Isaac Newton (1642-1727) remarked on Cotes' early death that, "If Cotes had lived, we might have learned something." Newton owes Cotes a great deal for his criticism and correction of Newton's *Principia*, making a second edition of the text possible.

1728   Daniel Bernoulli (1700-1782), Swiss mathematician, studies the mathematics of oscillations and is the first to suggest the usefulness of resolving a compound motion into the motions of translation and rotation.

1729   First arithmetic book written by an American-born individual is *Arithmetick* by Isaac Greenwood (1701-1745). It is published in Boston, Massachusetts.

1730   First algebra book printed in the American colonies is Pieter Venema's *Arithmetic, or the Art of Ciphering*. It is published in New York.

(See also 1788)

1731   First treatise on solid analytic geometry, *Recherches sur les Courbes à Double Courbure*, is written by French mathematician Alexis Claude Clairaut (1713-1765). This elegant work is written by the prodigious Clairaut at 16, but he waits until he is 18 to publish it. It carries out what French mathematician and philosopher René Descartes (1596-1650) had suggested almost a century before (i.e., their study through

projections on two coordinate planes). Clairaut is elected to the Academy of Sciences by special dispensation this same year.

1733   First to discuss the consequences of denying Euclid's fifth postulate (the parallel axiom) is Girolamo Saccheri (1667-1733), Italian mathematician. In his *Euclides ab Omni Naevo Vindicatus*, he is also the first to suggest the construction of a non-Euclidean geometry independent of that axiom, but he is unable and unwilling to proceed further with his studies. Mathematics will wait nearly a century for this to occur.

(See also February 1829)

1737   Leonhard Euler (1707-1783), Swiss mathematician, offers the first systematic discussion of continued fractions. A continued fraction is the quotient of natural numbers or their negatives such that the denominator contains a quotient the denominator of which contains a quotient, and so on.

(See also 1613)

1737   Leonhard Euler (1707-1783), Swiss mathematician, adopts the 16th letter of the Greek alphabet (π or pi) as the symbol for the ratio of the circumference to the diameter of a circle. Following his adoption and use, especially in his major work *Introductio in Analysin Infinitorum* (1748), it is generally accepted. It is thought that this letter is chosen because it is the first letter of the Greek word for perimeter.

1742   First systematic exposition and defense of Newtonian fluxional calculus is given by Colin Maclaurin (1698-1746), Scottish mathematician. In his *Treatise of Fluxions*, Maclaurin also expands the study of higher plane curves and applies classical geometry to solve physical problems. Despite its contributions, this work ultimately retards British mathematics by espousing Newtonian methods over the superior Leibniz analysis. Maclaurin's countrymen remain for a period of time indifferent to the advances in calculus being made on the European continent.

1744 Leonhard Euler (1707-1783), Swiss mathematician, publishes his *Methodus Inveniendi Lineas Curvas Maximi Minimive Proprietate Gaudentes* in which he systematically treats the calculus of variations and creates the new field of analytical mechanics. It becomes the first textbook of this new field. This highly influential work offers new perspectives for many mathamaticians by applying the calculus of variations to the classical problems of isoperimetric and geodetic curves.

1749 First use of the word statistik is made by Gottfried Achenwall (1719-1772), German statistician, in a work in which he uses statistical methods to describe the constituents and economic conditions of major countries. He is considered one of the founders of statistics. To the practitioner, statistics is the art of making inferences from a body of data, or more generally, the science of making decisions in the face of uncertainty.

1750 First woman university chair of mathematics is Italian mathematician and philosopher Maria Gaetana Agnesi (1718-1799). The daughter of a mathematician, she is a child prodigy who masters Latin, Greek, and Hebrew at an early age and writes a published defense of higher education for women when she is only 9 years old. An expert in differential calculus, she is recognized by Pope Benedict XIV (1675-1758) and appointed to the chair of mathematics and natural philosophy at the University of Bologna. There is some question as to whether she actually assumed this position, but it is known that after her father's death in 1752, she withdrew from all mathematical activity and devoted her life to charitable projects, becoming the director of a home for the sick and poor until her death.

1756 First French version of Isaac Newton's

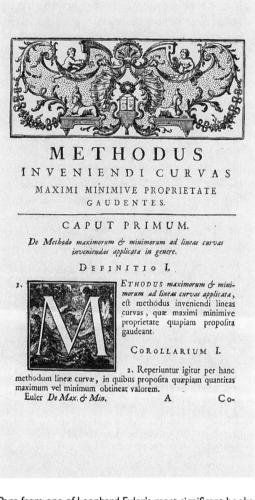

Page from one of Leonhard Euler's most significant books, in which he creates the field of analytical mechanics.

Principia is published. It had been translated some years earlier by the Marquise du Chatelet (1706-1749), French mathematician, physicist, and linguist also named Gabrielle Emilie Tonnelier de Breteuil. His work is now published posthumously.

1758 First French historian of mathematics, Jean Etienne Montucla (1725-1799) publishes his *Histoire des Mathématiques*. A second edition appears in 1799 and a third is completed by French mathematician Joseph Jerome le Francois de Lalande (1732-1807) and published in 1802.

1763 Thomas Bayes (1702-1761), English mathematician, is the first to use probability theory inductively in his *Essay Towards Solving a Problem in the Doctrine of Chances*. It is published posthumously this year. In this work he proposes his theorem of inverse probability whereby probabilities of unknown causes are inferred from observed events. According to this theory, knowing the frequency of an event enables one to calculate the chance of it falling between two degrees of probability, no matter how small (in other words, if it will occur again). This is also called probability inference.

1765 First to prove rigorously that the number for pi (π) is irrational (not expressible as an integer or whole number or as the quotient of two integers) is Johann Heinrich Lambert (1728-1777), German mathematician, in his *Beiträge zur Gebrauch der Mathematik und deren Anwendung*. In this work, he also gives the first systematic treatment of hyperbolic functions and the notation of these functions.

1765 Vincenzo Riccati (1707-1775), Italian mathematician, publishes his *Institutiones Analyticae*, which is the first extensive treatment on the integral calculus. It further introduces and elaborates the use of hyperbolic functions.

1770 Edward Waring (1734-1798), English mathematician, publishes his *Meditationes Algebraicae* in which he states a number theory that becomes known as Waring's problem. He also is the first to use a method for approximating the values of imaginary roots.

1788 First algebra book written by an American-born individual is *A New and Complete System of Arithmetic* by Nicholas Pike (1743-1819). It is published in Newburyport, Massachusetts.

1795 Johann Karl Friedrich Gauss (1777-1855),

German mathematician, discovers the law of quadratic reciprocity. As a prodigious 18-year old, Gauss works independently and is ignorant of the earlier attempts by such greats as Leonhard Euler (1707-1783) and Joseph Louis Lagrange (1736-1813) to prove it. He later elaborates on his discovery and offers several proofs.

(See also 1801)

1797 First useful and consistent graphical representation of complex numbers produced by Caspar Wessel (1745-1818), Norwegian surveyor and cartographer. This technique allows complex numbers to be pictured as corresponding to points on a straight line. His work goes virtually unnoticed until republished in 1897 by the Danish Academy of Sciences.

1797 Johann Karl Friedrich Gauss (1777-1855), German mathematician, writes his doctoral dissertation in which he gives the first wholly satisfactory proof of the fundamental theorem of algebra (that every algebraic equation must have at least one root, real or imaginary; or that every algebraic equation with complex coefficients has complex solutions).

1798 First treatise devoted exclusively to number theory is the two-volume *Essai sur la Theorie des Nombres* by Adrien Marie Legendre (1752-1833), French mathematician. It is also the first to include the method of least squares as a way of calculating orbits. Legendre distinguishes himself further with work on elliptic integrals, which eventually provides mathematical physics with its basic analytic tools.

1799 Gaspard Monge (1746-1818), French mathematician, publishes his *Geometrie Descriptive*, which is the first published exposition of his invention of descriptive geometry. This work revolutionizes engineering design. A skilled draftsman at only 16, Monge develops his own general method of applying geometry to problems of construction and displays his skills and method by working out the proper locations of gun emplacements for a proposed fortress. Typically, such an operation could be performed only by a lengthy, time-consuming process. Monge's new geometrical method works so well and so quickly, that it is classified a military secret until 1795.

(See also 1822)

1800   Louis François Antoine Arbogaste (1759-1803), French mathematician, publishes his *Calcul des Derivations* in which he states the Arbogast method. By using this method, the successive coefficients of a development are derived from one another when the expression is complicated. This work also for the first time separates the symbols of operation from those of quantity.

1801   Johann Karl Friedrich Gauss (1777-1855), German mathematician, publishes his *Disquisitiones Arithmeticae*, which is a work of fundamental importance in the modern theory of numbers. In this classic of mathematical literature, he gives the first proof of the law of quadratic reciprocity and applies this law to special cases of equations in which he is able to bring together algebraic, arithmetic, and geometric ideas. This work is considered to be one of the most brilliant achievements in the history of mathematics, and Gauss is quickly recognized as the greatest mathematician of his time. Eventually, history places him alongside Greek mathematician and engineer Archimedes (c.287-c.212 B.C.) and English scientist and mathematician Isaac Newton (1642-1727) as the greatest mathematicians of all time.

1803   Lazare Nicolas Marguerite Carnot (1753-1823), French mathematician and military engineer, publishes his *Géométrie de Position* in which sensed magnitudes are first systematically employed in synthetic geometry. By means of sensed magnitudes, several separate statements or relations can often be combined into a single inclusive statement or relation, and a single proof can frequently be formulated that would otherwise require the treatment of a number of different cases.

1803   Paolo Ruffini (1765-1822), Italian mathematician, makes the first noteworthy attempt to prove that no algebraic solution of the quintic equation (fifth degree or higher) is possible by means of radicals.

(See also 1824)

1810   First privately established mathematical periodical is begun by Joseph Diaz Gergonne (1771-1859), French mathematician. He founds the journal *Annales de Mathematiques Pures et Appliquées* in France. For its first 15 years, it is the only journal in the world devoted exclusively to mathematics.

1817   Bernhard Bolzano (1781-1848), Czechoslovakian mathematician, publishes his *Rein Analytischer Beweis* in which he pursues a purely arithmetic proof of the location theorem in algebra. Doing this requires a nongeometric approach to the continuity of a curve or function. This is also the first successful attempt to eliminate the use of infinitesimals from differential calculus.

1822   Jean Victor Poncelet (1788-1867), French mathematician, publishes his *Traité des Proprietés Projectives des Figures* which gives great impetus to the study of projective geometry and inaugurates its so-called great period. He is also the first mathematician of stature to argue for the development of projective geometry as a separate branch of mathematics. Concerned with the concrete problem of the graphical description of objects in three dimensions, this book is considered to be the foundation of modern geometry.

(See also 1799)

1822   Foundations of the mathematical theory of elasticity are first laid by French mathematician Augustin Louis Cauchy (1789-1857). It is in Cauchy's mathematical work that the theory of stress in elasticity originates. This is one of Cauchy's minor interests, as he is best known as a pioneer who uses clear and rigorous methods in his analysis and the theory of substitution groups.

1824 First proof of the impossibility of solving the general equation of the fifth degree by algebraic methods is offered by Niels Henrik Abel (1802-1829), Norwegian mathematician. This year he publishes a paper in which he shows the impossibility (after initially attempting to prove the possibility) of solving the general quintic equation by means of radicals. This problem had puzzled mathematicians for two and a half centuries, and his proof is more rigorous than Ruffini's (1803) attempt, demonstrating finally that there can be no general formula (expressed in explicit algebraic operations on the coefficients of a polynomial equation) for the roots of the equation if the degree is greater than four. Abel's work also contains the first rigorous proof of the general binomial equation.

1825 Principle of duality in geometry is given its first clear expression by Joseph Diaz Gergonne (1771-1859), French mathematician. This is a principle whereby one true statement can be obtained from another by merely interchanging two words. For example, in the projective geometry of the plane, the words point and line can be interchanged, giving the dual statements, "Two points determine a line," and "Two lines determine a point." Gergonne becomes involved in a controversy with French mathematician Jean Victor Poncelet (1788-1867) as to the priority of this discovery. It is said that while Poncelet may have entered the field earlier, Gergonne had a deeper grasp of the principle.

1826 First reliable actuarial tables are compiled by English mathematician Charles Babbage (1792-1871). His book, *The Comparative View of the Various Institutions for the Assurance of Lives*, establishes basic mortality and morbidity tables in order to calculate insurance risks and premiums.

1827 First comprehensive volume devoted entirely to differential geometry is written by Johann Karl Friedrich Gauss (1777-1855), German mathematician. His *Disquisitones Generales circa Superfices Curvas* becomes the basic work on the theory of curved surfaces (as opposed to ordinary geometry which is interested in the totality of a given figure diagram or figure). It is also the first systematic study of quadratic differential forms in two variables.

1828 First attempt to formulate a mathematical theory of electricity and magnetism is made by George Green (1793-1841), English mathematician. His *Essay on the Application of Mathematical Analysis to the Theory of Electricity and Magnetism* is the first to apply the concept of potential function to more than gravitational problems, and he introduces it into the mathematical theory of electricity and magnetism. Self-taught and employing highly unusual methods as a mathematician, Green marks the beginning of modern mathematical physics in England.

FEBRUARY 1829 Non-Euclidean geometry is discovered by Nikolai Ivanovich Lobachevski (1793-1856), Russian mathematician. This month, he publishes the first of a series of five papers in the *Kazan University Courier* in which he describes his discovery of non-Euclidean geometry. As early as 1826, he had conceived this self-consistent mathematical system that includes the concept that an indefinite number of lines can be drawn in a plane parallel to a given line through a given point. Lobachevski finds that by denying Euclid's axiom (that through any given point, one and only one line can be drawn, infinitely and in both directions, that is parallel to a given straight line), the results show no contradictions but rather are as self-consistent as Euclid's. His conclusion that there can be other geometries as valid as Euclid's goes against two millenia of mathematical thinking and leads to the modern notion that pure mathematics need not be true or false in the same sense as physics, but need only be self-consistent. This profoundly significant mathematical event leads eventually to the mathematics of curved surfaces and to German-Swiss-American physicist Albert Einstein's (1879-1955) demonstration that the universe is non-Euclidean in structure. At about this same time, Hungarian mathematician János Bolyai (1802-1860) is making the same discovery as Lobachevski.

(See also 1832)

MAY 1829   Group theory or the study of groups is essentially created by Evariste Galois (1811-1832), French mathematician. He attempts to publish a short paper on this work with the Academy of Sciences this year, but it is lost by French mathematician Augustin Louis Cauchy (1789-1857). A resubmission the following year disappears in the papers of French mathematician Joseph Fourier (1768-1830), who dies that year, and a third paper in 1831 is finally dismissed by referee Simeon Denis Poisson (1781-1840), who claims it is incomprehensible. As a political activist and radical, Galois is considered a danger by royalists, and he is killed in a duel on May 31, 1831, at the age of 20. It is unclear whether he was set up by police or not. His papers are eventually published with annotations in 1846 by French mathematician Joseph Liouville (1809-1882). Galois' genius is eventually recognized, and his structural and unifying concepts are remembered by the use of terms such as Galois group, Galois field, and Galois theory—all of which become an integral and accepted part of mathematics. He is the first to use the word group in the technical sense and to apply groups of substitutions to the question of reducibility of algebraic equations.

1829   First statistical breakdown of a national census is given by Lambert Adolphe Jacques Quetelet (1796-1874), Belgian statistician and astronomer. He correlates death with age, sex, occupation, and economic status in the Belgian census, and his work helps develop the uniformity and comparability of statistics. In 1835,

he writes *Sur L'Homme et Le Développpment de Ses Facultés*, which is considered the first true statistical work to be produced. He also contributes the notion of the average man and the concept of vita statistics as well.

1830   Augustus De Morgan (1806-1871), English mathematician and logician, publishes his *Elements of Arithmetic* in which he first introduces the term mathematical induction, describing the logical process traditionally used by mathematicians but never rigorously considered or clarified. He analyzes logic mathematically and studies the logical analysis of the laws, symbols, and operations of mathematics.

1832   Publication of another independent discovery of non-Euclidean geometry is made by Hungarian mathematician János Bolyai (1802-1860). This discovery of a self-consistent mathematical system that includes the concept that an indefinite number of lines can be drawn in a plane parallel to a given line through a given point may have been made by Bolyai as early as 1823, but priority goes to Russian mathematician Nikolai Ivanovich Lobachevski (1793-1856), who first publishes in 1829. Bolyai publishes his discovery this year in a 26-page appendix to his father's semi-philosophical book on elementary mathematics. Bolyai's work is totally independent of Lobachevski's, and when Bolyai finally sees the Russian's work, he thinks it has been copied from his own.

(See also February 1829)

1833   William Rowan Hamilton (1805-1865), Irish mathematician, makes one of the first attempts at analyzing the basis of irrational numbers. His theory states that both rationals and irrationals are based on algebraic number couples. This paper is a pioneer attempt to put algebra on an axiomatic basis like geometry. It eventually leads to his discovery of the quaternions.

(See October 16, 1843)

c. 1836   First attempt to work out a mathematical basis for the astronomical phenomenon known as the ether is French mathematician Augustin Louis Cauchy (1789-1857). Using mathematics, he tries to explain the properties of this supposed solid-but-gas substance that most scientists believe exists in outer space and which both light waves and the earth itself supposedly pass through. His mathematical theory is not satisfactory and the notion of the ether itself is eventually disproved by physicists a generation after Cauchy dies.

1837 Pierre Laurent Wantzel (1814-1848), French mathematician, gives the first rigorous proofs of the impossibility of duplicating the cube or trisecting angles by the use of only a ruler and compasses.

1838 Antoine Augustin Cournot (1801-1877), French mathematician and economist, publishes his *Researches sur Les Principes Mathematiques de la Theorie des Richesses* in which he is the first economist to apply mathematics to the study of economics.

1840 Pierre Charles Alexandre Louis (1787-1872), French physician, pioneers medical statistics, first compiling systematic records of diseases and treatments.

1841 First permanent journal devoted to the interests of mathematics teachers rather than to mathematical research, the *Archiv der Mathematik und Physik*, is founded.

OCTOBER 16, 1843 Algebra of quaternions is discovered by Irish mathematician William Rowan Hamilton (1805-1865). He invents a new branch of mathematics which uses geometric principles to manipulate complex numbers. Hamilton's landmark discovery of quaternions can be described as ordered sets of four ordinary numbers satisfying special laws of addition, multiplication, and equality but freeing algebra from the commutative law of multiplication (which says that a times b is also equal to b times a). A century later, Hamilton's noncommutative algebra becomes an integral component of the quantum theory of atomic structure.

1844 Ferdinand Gotthold Eisenstein (1823-1852), German mathematician, helps develop the theory of complex numbers and also discovers the first covariant used in analysis. An instructor of mathematics at Berlin, he is much admired by the famed German mathematician Johann Karl Friedrich Gauss (1777-1855), who calls him an "epoch-making" mathematician. Unfortunately, Eisenstein dies before he is 30, and is unable to fulfill Gauss's hopes.

1844 Existence of transcendental numbers is first proved by Joseph Liouville (1809-1882), French mathematician. A transcendental number is incapable of being the root of an algebraic equation with rational coefficients (therefore, pi is a transcendental number). It is named by Swiss mathematician Leonhard Euler (1707-1783), who said of such numbers, "They transcend the power of algebraic methods."

1847 First complete and general theory of imaginary points, lines, and planes in projective geometry is offered by Karl Georg Christian von Staudt (1798-1867), German mathematician. His *Geometrie der Lage* finally frees projective geometry of any metrical basis and creates a geometry of position, independent of all measurements, that is complete in itself.

1847 First to introduce the term topology is German mathematician Johann Benedict Listing (1808-1882), in his *Vorstudien zur Topologie*. It is derived from two Greek words, topos meaning place or surface, and logos meaning study. Although this work discusses some topological problems and is the first to use the word itself, it is not the first book devoted to the systematic study of topology as a branch of geometry. This honor falls to French mathematician and physicist Jules Henri Poincaré (1854-1912).

(See also 1895)

OCTOBER 23, 1852 First known document to make reference to the Four-Color Conjecture is a letter from English mathematician and logician Augustus De Morgan (1806-1871) to his friend, Irish mathematician William Rowan Hamilton (1805-1865). The conjecture states that any conceivable map drawn on a plane or on the surface of a sphere can be colored, using only four colors, in such a way that adjacent countries have different colors. Earlier this month, the brothers Francis and Frederick Guthrie pose the problem to De Morgan who cannot find a method to prove it true or false. This coloring problem remains neglected until 1878 when English mathematician Arthur Cayley (1821-1895) presents the problem in the London Mathematical Society's *Proceedings*.

(See also 1976)

1854 Symbolic logic is first founded by English mathematician and logician George Boole (1815-1864). In his epochal book, *An Investigation of the Laws of Thought*, Boole establishes both formal logic and a new algebra (Boolean algebra) and begins the mathematization of logic. Boole maintains that the essential character of mathematics lies in its form rather than its content, and he thus focuses his attention on mathematics as symbol rather than only on number and measurement. Boole's algebra of logic, or Boolean algebra, eventually proves basic to the design of computers.

1858 Charles Hermite (1822-1901), French mathematician, publishes his *Sur la Resolution de l'Equation du Cinquieme degre* in which he offers the first solution of the general equation of the fifth degree (quintic) which he achieves by means of elliptic modular functions.

1865 London Mathematical Society is founded and is the first of its kind. It immediately begins publishing its *Proceedings*. English mathematician and logician Augustus De Morgan (1806-1871) is its first president.

1868 First proof of the relative consistency of non-Euclidean geometry is given by Eugenio Beltrami (1835-1900), Italian mathematician.

His *Saggio di Interpretazione della Geometria Non Euclidea* proposes the first Euclidean model of non-Euclidean (hyperbolic) geometry and opens the way for the complete acceptance of the relative non-contradictoriness of non-Euclidean geometry. Beltrami devises a model he calls a pseudosphere that makes the formerly imaginary geometry of Russian mathematician Nikolai Ivanovich Lobachevski (1793-1856) "real" in a mathematical sense. Further, the geometry on the pseudosphere provides an interpretation in which Euclid's parallel postulate cannot possibly hold, yet in which all his other axioms are true.

1870 First full and clear presentation of Galois theory is offered by Camille Jordan (1838-1922), French mathematician, in his *Traité des Substitutions et des Equations Algebriques*. In this work he is the first to systematically develop the theory of groups given by French mathematician Evariste Galois (1811-1832). He also offers the fundamental concept of class of a substitution group and further proves the constancy of the factors of composition.

1873 Charles Hermite (1822-1901), French mathematician, publishes an article in which he proves that the number e is transcendental. He shows that e cannot be the root of any polynomial equation with integral coefficients. This is the first familiar number that is shown to be transcendental.

(See also 1844)

1874 Set theory is founded by Russian-German mathematician Georg Ferdinand Ludwig Philipp Cantor (1845-1918). He also introduces the mathematically meaningful concept of infinite numbers called transfinite numbers. Set theory has the revolutionary aspect of treating infinite sets as mathematical objects that are on an equal footing with those that can be constructed in a finite number of steps. Although Cantor's ideas are subject to severe criticism during his life, by 1900 his set theory becomes a distinct branch of mathematics.

1877 George William Hill (1838-1914), American astronomer and mathematician, studies the trajectory of the moon's perigee and uses infinite determinants for the first time in analyzing its motion. In his *Researches on Lunar Theory*, he discards the usual mode of procedure in the problem of three bodies (moon, sun, and earth) and uses differential equations instead to achieve a simple, useful solution. This new mathematical technique profoundly changes the entire approach to celestial mechanics from now on.

1878 James Joseph Sylvester (1814-1897), English-American mathematician, becomes the first editor of the *American Journal of Mathematics*. During his career, he founds with English mathematician Arthur Cayley (1821-1895) the theory of algebraic invariants. This becomes essential to the theory of relativity.

1882 Ferdinand Lindemann (1852-1939), German mathematician, first proves that pi (π) is transcendental and that therefore it is impossible to square the circle (quadrature) using a ruler-and-compass construction. A complex number is said to be algebraic if it is a root of some polynomial having rational coefficients. Otherwise it is said to be transcendental. Lindemann's proof shows that pi (π) is not a root of any algebraic equation, and that a segment of length of pi (π) is not able to be constructed by Euclidean tools. This puts to an end the ancient dream of squaring the circle.

1882 First rigorous attempt to remove the hidden assumptions in the work of Greek mathematician Euclid is made by Moritz Pasch (1843-1930), German mathematician, in his *Vorlesungen über Neuere Geometrie*. In this work which contains a system of axioms for descriptive geometry, he becomes the first since Euclid to present elements of geometry as abstract postulates of their relations. He helps lay the foundations of modern geometry.

1888 Contact transformations are discovered by Norwegian mathematician, Marius Sophus Lie

(1842-1899). In his work, *Theorie der Transformationsgruppen*, he details his Lie groups which are based on infinitesimal transformations. These reduce the amount of work required in calculus integration and eventually play an important role in quantum theory.

1890 Giuseppe Peano (1858-1932), Italian mathematician and logician, builds a continuous plane curve that completely fills a square, thus constructing the first space-filling curve. With his invention that seems to defy common sense, Peano shows how thoroughly mathematics can shock the senses. His paradox of a continuous curve that passes through every point of a square is a blow to the geometric preconceptions of what a continuous curve ought to be, and it goes against the geometric intuition of most mathematicians. The significance of such abnormal behavior as continuous functions exhibit is that it teaches mathematicians caution about the deceptive evidence of geometric proofs.

1895 First woman to receive a German doctorate in mathematics through the regular examination process is Grace Emily Chisholm (1868-1944), English mathematician. She attends Gottingen University in Germany because females are not admitted into English graduate schools. She later marries English mathematician William Henry Young (1863-1942).

(See also 1906)

1895 First systematic study of topology as a branch of geometry is published by Jules Henri Poincaré (1854-1912), French mathematician and physicist. His *Analysis Situs* founds combinatorial topology. In this branch of mathematics, algebraic methods are used to study the properties of topological spaces, much as algebra is used to study the relationships of distances and angles in geometric figures. It is primarily concerned with those properties of geometric configurations (as point sets) which are unaltered by elastic deformations (like twisting or stretching). In addition, he introduces the notion of the Poincaré group also known as the first homotropy group or the fundamental group of a complex.

1896   Hermann Minkowski (1864-1909), Russian mathematician, publishes his *Geometrie der Zahlen* in which he devises a geometrical method for number-theory study. This is the first major work on the geometrical theory of numbers. Later, when Minkowski's former pupil, German-Swiss-American physicist Albert Einstein (1879-1955), makes known his special theory of relativity in 1905, it is Minkowski who lays a mathematical foundation for it by using a four-dimensional, space-time form of geometry that he develops. Using this, Einstein is then able to extend his work to include a general theory of relativity.

1897   First official international mathematical congress is held at Zurich, Switzerland. Their purpose is to promote friendly relations, to give reviews of the progress and state of the different branches of mathematics, and to discuss matters of terminology and bibliography. These will be held every four years.

1897   First effective theory of the measure of sets of points is created by French mathematician Félix Edouard Justin Emile Borel (1871-1956). This work helps launch the modern theory of functions of a real variable. He also later develops the first systematic theory of divergent sets.

1899   First book to contain a really satisfactory set of axioms for geometry is *Grundlagen der Geometrie*, written by David Hilbert (1862-1943), German mathematician. Since Euclid's geometry is not rigorous in that it simply assumes many things to be so or to be obvious, Hilbert sets out to give it a logical foundation and to make it self-consistent. In doing this, he defines it in terms that do not require human interpretation to give it form, and thus shifts the foundation of geometry from intuition to logic. His book also founds the formalist school whose thesis is that mathematics is concerned with formal symbolic systems.

1906   First systematic point-set theory in abstract space is created by Maurice René Frechet (1878-1973), French mathematician. His work inaugurates the study of abstract spaces and gives the first definition and theory of abstract sets. This lays important groundwork by describing a method for determining the distance between any two points in abstract space, thereby defining the concept of metric space. Geometry now becomes more generalized as the study of a set of points with some superimposed structure.

1906   First comprehensive textbook on set theory and its applications to function theory is written by William Henry Young (1863-1942) and Grace Chisholm Young (1868-1944), English husband and wife mathematicians. Their book, *The Theory of Set Points*, summarizes and explains the great work of Russian-German mathematician Georg Cantor (1845-1918), who founded set theory and introduced the mathematically meaningful concept of infinite numbers called transfinite numbers.

(See also 1874)

1907   Four-dimensional, space-time form of geometry is first introduced by Russian-German mathematician Hermann Minkowski (1864-1909), who publishes his *Raum und Zeit* this year. In this work he demonstrates that relativity makes it necessary to take time into account as a kind of fourth dimension (treated mathematically somewhat differently from the three spatial dimensions). He argues that neither space nor time exists separately, so that the universe consists of a fused space-time. His work provides German-Swiss-American physicist Albert Einstein (1879-1955) with the mathematical basis for his general theory of relativity.

1909   Edmund Landau (1877-1938), German mathematician, publishes his *Handbuch der Lehre von der Verteilung der Primzahlen*, which is one of the first treatises on the analytical theory of numbers. This work also considers the distribution of prime numbers and prime ideals.

1918 First non-Riemannian geometry is created by Hermann Weyl (1885-1955), German-American mathematician, while attempting to unify the electromagnetic field theory of Scottish mathematician and physicist James Clerk Maxwell (1831-1879) and German-Swiss-American physicist Albert Einstein's (1879-1955) gravitational field theory. He explores the concept of a linear construction in the belief that linking this group of similitudes might result in a unified field theory. The result is the creation of a new branch of mathematics. His work serves as a link between pure mathematics and theoretical physics and eventually supports both quantum mechanics and relativity theory.

1921 First to define games of strategy and to develop a game theory is French mathematician Félix Edourad Emile Borel (1871-1956). He initiates the modern mathematical theory of games by closely analyzing several familiar games.

(See also 1944)

1931 Gödel's theorem is first propounded by Austrian-American mathematician Kurt Gödel (1906-1978). This year he submits a paper whose incompleteness theorem startles the mathematical world. Also called Gödel's proof, it states that all formal systems generate inconsistent or paradoxical mathematical statements that cannot be decided upon using the rules that govern that particular formal system. This means that one has to go beyond formal rules to resolve certain problems of any axiom-based mathematics. Gödel's discovery has great impact, showing that the search for certainty in mathematics cannot and does not exist. It also influences the development of the theory of artificial intelligence.

1932 Fields Medal of the International Mathematical Congress is first established. This hon-

or becomes the equivalent of the Nobel Prize in mathematics.

1934 Alexander Osipovich Gelfond (1906-1968), Russian mathematician, first offers his basic techniques in the study of transcendental numbers (numbers that cannot be expressed as the root or solution of an algebraic equation with rational coefficients). Known as Gelfond's theorem, this work profoundly advances transcendental number theory and the theory of interpolation and approximation of complex-variable functions.

1940 First to name and use a googol is nine-year-old mathematician Milton Sirotta. A googol is the figure 1 followed by 100 zeroes, equal to 10 to the hundredth power. This figure becomes important to the principles of number theory and is supposedly Sirotta's nickname for this number. A googolplex is the word used for the number 10 to the googol power.

APRIL 1943 The first volume of the *Quarterly of Applied Mathematics* appears. It is sponsored by Brown University in Providence, Rhode Island.

1944 New branch of mathematics called game theory is founded by Hungarian-American mathematician John von Neumann (1903-1957). He argues that mathematics developed for physical science is inadequate for economics which involves human action based on choice and chance. He proposes a different mathematical approach more suitable for the social sciences and provides an analysis of strategies that take into account the interdependent choices of two or more players. Game theory is best used for complex decision-making processes.

JUNE 1945 The first Canadian Mathematical Congress is held in Montreal, Quebec.

1950 New branch of analysis called the theory of distribution is first developed by French mathematician Laurent Schwartz. A distribution is a generalization of the classical concept of a function. This new theory has applications in potential theory, spectral theory, partial differential equations, and other aspects of pure and applied analysis.

1951 First separate organizational division of mathematics is created by the National Research Council of the United States National Academy of Sciences. The Council decides that a separate body is needed in order to emphasize and to better represent the growing needs of mathematics.

SEPTEMBER 1951 International Mathematical Union is first established by 15 nations. It is founded to promote cooperation among the world's mathematicians as well as wider dissemination of the results of mathematical research.

1955 Homological algebra is first proposed as a new subject by French mathematician Henri Paul Cartan and Samuel Eilenberg, Polish-American mathematician. This new field of study is a development of abstract algebra that is concerned with valid results for many different kinds of space, and is considered as an invasion of algebraic topology in the domain of pure algebra. It proves to be a powerful cross between abstract algebra and algebraic topology with widely applicable results.

c. 1960 New math is first introduced into the curriculum of American public schools. In comparison with traditional mathematics instruction, this new method places greater emphasis on mastering mathematical concepts, such as set theory, and less on practicing computation skills.

1961 Edward Norton Lorenz, American meteorologist, discovers the first mathematical system containing chaotic behavior. Working on large-scale atmospheric circulation and dynamical and statistical approaches to weather prediction, he finds chaotic behavior in a computer model of how the atmosphere behaves. This work gets the attention of others and leads eventually to an entirely new branch of mathe-

matics, chaotic dynamics. The mathematical equations used in this field are inherently nonlinear, or so complex that they cannot be computed, predicted, or sometimes even defined.

1964 Paul Joseph Cohen, American mathematician, first proves that Cantor's continuum hypothesis is independent of other axioms of set theory and therefore demonstrates that it can be introduced to set theory as a separate axiom.

1974 The value of pi (π) is calculated to 1 million places for the first time by a CDC 7600 computer.

1975 Fractals are first suggested by Polish mathematician Benoit Mandelbrot in his book *Les Objets Fractals*. Based on his conversations with geologists, meteorologists, and cartographers, he develops fractal mathematics whose subject deals with geometric shapes that are self-similar or have a similar appearance at any level of magnification. This new field of mathematics is based on the study of the irregularities in nature, and it finds a place in the field of statistical physics by helping to analyze complex, random-seeming events. There are strong links between the dimensionality of fractal shapes and chaotic motions that contribute later to a new branch of physics called chaos theory.

1976 Kenneth I. Appel, American mathematician, and Wolfgang Haken, German-American mathematician, resolve the century-old Four-Color Conjecture first posed in the 1852 by English mathematician Francis Guthrie. Using a new combination of computer methods and theoretical reasoning, they prove for the first time that the four-color conjecture is true. This conjecture poses the question: can every map drawn on a plane be colored with four colors so that adjacent regions receive different colors? Their proof requires over 2,000 hours of computer and contains several hundred pages of complex detail.

1977 Daniel Quillen, American mathematician,

and A. A. Suslin, Russian mathematician, independently discover two very different proofs of the 20-year-old conjecture concerning the structure of generalized vector spaces. These proofs confirm that all abstract spaces of a certain common type are constructed in direct analogy with two- and three-dimensional Euclidean space.

1983 Gerd Faltings, German-American mathematician, first proves the Mordell conjecture, which suggests a relationship between the number of solutions that equations may have and the geometry of certain surfaces determined by these equations. Louis Joel Mordell (1888-1972) was an American mathematician. Faltings' discovery is an important step toward verifying Fermat's last theorem.

(See also June 1993)

1991 Sphere-packing problem posed by German astronomer Johann Kepler (1571-1630) in 1611 is first solved by American mathemati-

cian Wu-Yi Hsiang. Hsiang's solution has implications for physics, for if it were possible to deduce from quantum mechanics that the atoms in a real crystal must pack together with maximum density, then Hsiang's theorem would prove that the regular atomic structure of crystals is also a consequence of quantum mechanics. Kepler's problem has remained unsolved until now.

JUNE 1993 Proof of Fermat's last theorem is first obtained by English mathematician Andrew John Wiles. The heart of his proof relies upon the Taniyama-Weil conjecture, which is a difficult problem in number theory dealing with the nature of elliptic curves. Wiles produces a 200-page paper which is the result of his seven-year attack on this 325-year-old problem that had been declared unsolvable by generations of mathematicians. Wiles has dreamed of solving Fermat's last theorem since he encountered it in a book when he was 10 years old.

Medicine

c. 3000 B.C. First physician mentioned by name in ancient records is the Egyptian named Sekhet'enanach, who cures a pharaoh's malady and earns a lasting place in medical history.

c. 2900 B.C. Chinese Emperor Fu-Hsi (the first of China's mythical emperors) invents the principles of yin and yang, which becomes the basis of ancient Chinese medicine. These two complementary forces or principles are believed to be part of all aspects of life. For medicine, it is believed that health is found in the perfect equilibrium of these two principles, which continually ebb and flow.

c. 2700 B.C. Acupuncture is invented by the Chinese Emperor Shen-Nung (the second of China's mythical emperors), who also writes the *Pen-Tsao*, considered the first medical herbal text. Acupuncture is an ancient medical technique for relieving pain and is practiced in many parts of the world today. Modern Chinese surgeons use it as a form of anesthesia and insert brass-handled needles into certain points of the body. Although there has not been any full explanation for its effectiveness, the ancient Chinese believed that the needles penetrated one or more of the twelve canals in the body called "chin." These canals did not carry blood but were rather channels for the body's vital

principles. Acupuncture is supposed to stimulate or dampen these principles and therefore put the body back into balance.

c. 1500 B.C. Vedic medicine in India first begins. The word "Veda" means knowledge and the "Vedas" are the books of sacred Hindu knowledge. This period of Indian medicine is one of unquestioned authority in which the supernatural plays a major part. Diseases are considered generally the work of demons and the healing gods must be appealed to. Drugs or herbs are considered ineffective without the assistance of the gods or at least some prayer.

c. 500 B.C. Brahmanic medicine first begins in India. This begins the highest period of Hindu medicine. Brahmans were considered the descendants of the god Brahma and thus were custodians of all knowledge. Indian healers belonged to this class, and three of the greatest names in Hindu medicine—Charaka, Susruta, and Vaghbata—are of this period. They knew some anatomy and physiology, practiced surgery and dissection, and classified diseases.

c. 500 B.C. First recorded dissection of the human body for research purposes is conducted by Greek physician, Alcmaeon. He notes the existence of the optic nerve and the tube connecting the ear and the mouth. He also distinguishes arteries from veins and states that the brain is the center of intellectual activity.

c. 500 B.C. Cataract operations are first performed in India. Their method of couching and puncturing the lens becomes the standard method of removing cataracts from the eye.

437 B.C. What may possibly be the earliest known hospital is built in Ceylon (now Sri Lanka).

c. 400 B.C. Hippocrates (460-370 B.C.), Greek physician, provides the first detailed description of tuberculosis. Also called phthisis, from the Greek meaning to decay and waste away, tuberculosis is known to have been described generally by Assyrian and Babylonian physicians. It is also known that ancient Egyptians suffered from this progressive pulmonary disease, since the remains of several mummies offer indisputable evidence. It is not until the masterful description by Hippocrates, however, that the disease is given real, analytical treatment. He accurately details its characteristic symptoms and distinguishes tuberculosis from other respiratory diseases like pneumonia and pleurisy. He also correctly states that pulmonary tuberculosis is contagious. Known as the "father of medicine," Hippocrates is considered to be the first to treat medicine as a science.

c. 350 B.C. Aristotle (384-322 B.C.), Greek philosopher, is one of the first embryologists and is considered the founder of comparative anatomy. A master of exact observation and description, he studies the day-by-day development of the fetal chick and classifies animals according to their reproductive methods. He teaches comparative anatomy by dissecting animals and using anatomical diagrams. As the greatest biologist of antiquity, he gave medicine the use of formal logic as an instrument of precision.

c. 350 B.C. Praxagoras, Greek physician, first distinguishes between veins and arteries, recognizing that there are two different kinds of

vessels. Although he is correct, he believes that these blood vessels are filled with air instead of blood. He also says that arteries are air channels that lead from the heart and taper into the nerves. He did discover the arterial pulse and recognize the connection between the brain and the spinal cord.

c. 300 B.C. Herophilus, Greek anatomist, establishes himself as the first careful anatomist and the first to perform dissections in public. Often called the father of scientific anatomy, he is the first to systematically study the anatomy of the brain and spinal cord and differentiate between nerves and blood vessels. To him we owe the clear statement that the brain is the central organ of the nervous system and the site of intelligence. He also distinguished between the brain's cerebrum and cerebellum and used a water clock to count a patient's pulse, analyzing its rate and rhythm. He was charged with human vivisection.

c. 250 B.C. Erasistratus (c.304-c.250 B.C.), Greek physician, studies the anatomy of the brain and discovers the sinuses of the dura mater. He also is the first to note cirrhosis of the liver, recognizing an association between ascites (bloated belly) and the hardening of the liver. More of a physiologist than an anatomist, he is a skilled investigator who knows the importance of the observational method and whose medicine is entirely free from any type of dogma.

c. 160 B.C. First record is made of a woman practicing medicine in China.

c. 75 Dioscorides, Greek physician, writes the first systematic pharmacopoeia. His *De Materia Medica,* in five volumes, provides accurate botanical and pharmacological information. As a physician to the Roman legions, he is able to collect plants from many regions of the ancient world. His herbal writings name and describe each plant, telling where it comes from, detailing any special features it might have, and then concluding with a discussion of the maladies or

Greek physician Dioscorides.

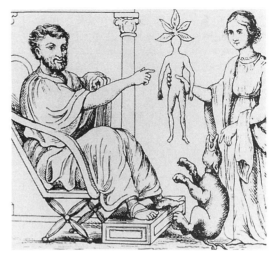

Dioscorides receiving a root of the mandrake from the Goddess of Discovery.

symptoms for which it is useful. His work is preserved by the Arabs and, when translated into Latin and printed in 1478, becomes a standard botanical reference. For a millennium and a half, he is considered the ultimate authority in botanical and herbal matters.

c. 100 Rufus of Ephesus, who lives during the reign of Trajan (98-117), first describes the hyaloid membrane of the crystalline lens of the eye. Considered one of the greatest of the Greek physicians, he is known for attributing gout to an accumulation of poisons in the body and for his belief that interrogating the patient was a fundamental stage of the examination. He used slaves to study regional anatomy but dissected only monkeys.

169 One of the first to understand the value of the pulse as a diagnostic technique is the Greek physician Claudius Galen (129-c.199). As the outstanding physician of Asia Minor, he eventually becomes physician to Roman emperor Marcus Aurelius (121-180). He founds experimental physiology and writes many medical texts that become authoritative for centuries. Much of his physiology is learned from dissecting Barbery apes. Among his many accom-

plishments is being the first to demonstrate that contraction is the only action of a muscle, and that the movement of a relaxed muscle is produced by the contraction of an opposing muscle group. He also teaches that a comprehensive physical examination should follow certain prescribed logical principles. His writings are medical gospel for over 14 centuries and his reputation is rivaled only by that of Hippocrates.

c. 178 First description of sewing an artery is made by Roman encyclopedist Aulus Cornelius Celsus. Of all his writings, only his medical works survive, and they contain the first discussion of heart attacks and insanity. His work is forgotten for centuries, but when rediscovered during the Renaissance, suddenly makes him a medical authority.

c. 375 Oribasius (325-403), Byzantine physician and pupil of Zeno, first describes the membrana tympani (of the ear) and the salivary glands. In addition to many small salivary glands in the tongue, palate, lips, and cheeks, humans have three pairs of major salivary glands surrounding the oral cavity that open into the mouth through well-developed ducts. In the human ear, the tympanic membrane, or ear-drum membrane, is located at the end of the passageway that leads from outside the head. This membrane receives sound vibrations from the outer air and transmits them to the auditory ossicles.

c. 400 Fabiola (St. Fabiola), a Christian noblewoman, founds the first nosocomium or hospital in Western Europe. After establishing the first hospital in Rome, she founds a hospice for pilgrims in Porto, Italy. About this time Christianity was becoming a force, and its practitioners placed great emphasis on helping the sick, eventually making it an ethical and religious duty upon the individual and the community. It is Fabiola who helps build the first great early hospitals.

430 Earliest recorded plague in Europe is an epidemic that breaks out in Athens, Greece.

c. 525 Aetius of Amida, Byzantine physician to Justinian I (527-565), is the first to ascribe medical cures to the use of a magnet. As royal physician to the emperor, he writes an extensive medical compilation and gives accurate accounts of several conditions and diseases. He does, however, use many charms and spells as treatment, and a magnet is only one of these.

c. 650 Paul of Aegina (625-690), Byzantine physician, is the first to practice obstetrics as a specialty. His real specialty is surgery, however, and despite the scanty anatomical knowledge of his time, his surgical achievements are indicative of the best surgeon of his time. His seven-volume *Epitome* collects the writings of ancient authors.

758 First hospital in Japan is founded.

896 Rhazes (c. 845-c.930), Persian physician and alchemist, distinguishes between the specific characteristics of measles and smallpox and gives the first accurate description of both. He is also believed to be the first to classify all substances as animal, vegetable, or mineral.

This simple but powerful classifying device was accepted, and continues today. He further subclassifies minerals into metals, spirits or volatile liquids, stones, salts, and others.

c. 900 First medical books written in Anglo-Saxon appear. *Lacnunga* and the *Leech Book* of Bald appear and have some botanical sections. These works were written well before the Norman conquest, when Anglo-Saxon medicine was characterized by the use of leeches and herbs, accompanied by some type of magic or charm. This folk medicine was a mix of East and West influences, and the plant lore derives mainly from the Roman scholar Pliny (23-79 A.D.).

1010 The School of Salerno flourishes in Italy. Begun around the start of the ninth century as the first independent medical school, it reaches its greatest influence at this time and preserves its fame to the end of the 14th century. This medical school in southern Italy grants the first medical diplomas and becomes a place to which students from Europe, Asia, and northern Africa flock. The founding of this school indicates a break with the confused medicine of medieval times in that it studies diseases in a straightforward, commonsense manner and uses rational therapy. It especially emphasizes the proper use of dietetics. It also allows Arabic medical doctrine and practice to flow to the West.

c. 1170 Roger of Salerno, Italian surgeon, also called Regear Frugardi or Roger of Palermo, writes his *Practica Chirurgiae*. As the earliest Western surgical treatise, it becomes a standard textbook. His work deals with surgical remedies for cancer, depressed fractures, and other conditions like hernia, and indicates how to treat dislocations and skin diseases. He also recommends the use of seaweed (iodine) to treat a goiter.

1250 Arsenic is first isolated by the German scholar Albertus Magnus (1193-1280). Widely distributed in nature, arsenic is usually found in association with metal ores and in many metallic sulfides. It is extremely toxic to living organisms, since it interferes with cellular metabolism. Medically, it was used with great caution in the treatment of certain diseases caused by microorganisms like syphilis. Individual susceptibility to arsenic poisoning varies greatly, and a dose that may quickly kill one person may be tolerated by another.

1296 Guido Lanfranchi, Italian surgeon also called Lanfranc, writes his *Chirurgia Magna,* in which he is the first to describe a cerebral concussion and the first to differentiate between hypertrophy and cancer of the breast. Lanfranchi lived in Paris and his work introduced many surgical innovations to the French. His chapter on the symptoms of a skull fracture is considered a classic.

1302 First formally recorded postmortem or judicial autopsy is performed in Bologna, Italy by Italian physician Bartolomeo da Varignana. This postmortem is ordered by the court to decide a case of suspected poisoning. Dissection of a human corpse was a sensitive subject at this time, and official authorization was usually required before an autopsy could be performed for teaching purposes or as in this case to discover the exact cause of a suspicious death.

1303 Simon of Genoa, Italian botanist and physician also called Simone de Cordo or Simon Januensis, dies but leaves behind his polyglot medical dictionary of drugs and simples. With the invention of mechanical printing in the next century, it becomes the first medical dictionary to be printed. It is published in Milan, Italy, in 1473 and is titled *Synonyma Medicinae sive Clavis Sanationis.* His materia medica gives information on drugs under their Greek, Latin, and Arabic names, and eventually helps to popularize some Arabic botanical terms.

JANUARY 1315 Mondino de'Luzzi (c.1275-1326), Italian anatomist, makes his first public

Woodcut of Mondino de Luzzi directing a dissection.

dissection of a female subject. He teaches anatomy at the medical school in Bologna, Italy, and writes the first book devoted entirely to anatomy. Mondino is the first outstanding anatomist worthy of that name and, as a lecturer at Bologna, he is the first to introduce the systematic teaching of anatomy into the general medical curriculum. His book *Anathomia* becomes the most used anatomical text up to the end of the 16th century. It is devoted mostly to a detailed description of how to perform a proper dissection.

1319 First criminal prosecution for body snatching is conducted. Obtaining enough cadavers for anatomical research will be a problem into the 19th century and, at this time, grave robbing proves to be the best way around laws prohibiting or restricting dissection.

1345 First apothecary shop or drugstore opens in London, England.

1348 Military medicine first begins with the treatment of wounds from newly invented firearms. Unfortunately for the wounded soldiers, the military surgeons would treat gunshot wounds by following the horrific pseudo-Hippocratic doctrine that "diseases not curable by iron are curable by fire." This often involved the use of red-hot iron and boiling oil, the latter employed directly on the wound because gunshot wounds were seen as a type of poisoned burn.

1374 As the plague spreads, the Republic of Ragusa (on the eastern shore of the Adriatic Sea) places the first known quarantines on crews of ships thought to be infected. The authorities detain a suspected ship's crew for one month before allowing them to disembark and enter the city. Called a "trentina" for its 30 days' worth of detention, it eventually becomes stretched to the more common 40 days. This leads to the now-familiar name based on "quaranta giorna" or "quarantina."

1377 First known quarantines on land are imposed at Reggio nell'Emilia, Italy, as a check against the spread of the plague. Unable to cure the afflicted or affect the course of this fast-spreading epidemic, the authorities considered prevention their only weapon. All visitors suspected of being infected by the plague are required to spend a designated amount of time at appointed places outside the city.

1388 Richard II (1367-1400), king of England, establishes the first sanitary laws in England.

1426 The lost works on medicine of Aulus

Richard II on his throne.

Cornelius Celsus, the first-century Roman encyclopedist, are rediscovered by Italian humanist and classical scholar Guarino Veronese (1374-1460). Like major scholars and encyclopedists Varro and Pliny, Celsus produced an encyclopedic work titled *De Artibus,* only one portion of which, his *De Medicina,* survived to be discovered. There is a rival claim to this discovery, and many feel it was Thomas Parentucelli (1397-1455) of Sarzana, Italy, later Pope Nicholas V, who found a manuscript copy in the papal library at St. Ambrose in Milan, Italy, in 1443.

1457 First known piece of medical printing (typographic printing) is the Mainz *Calendarium,* a purgation or bleeding calendar.

1473 First printing of Persian physician Avicenna's (980-1037) *Canon Medicinae* is made and becomes one of the most famous medical texts in the history of medicine. It is understandable that the writings of the most famous physician of the Persian world would become a standard medical text for five centuries. His *Canon* is a huge encyclopedia of medicine, and during the Middle Ages is enormously influential. With the invention of mechanical printing, it is reproduced in quantity and becomes even more well-known.

1478 First edition of *Chirurgia* by medieval surgical authority Guy de Chauliac (c.1300-1367) of France is published in Paris. Completed in 1363, it was well-known and very influential, but with the invention of typographic or mechanical printing, it passed through many editions, translations, and abridgements. As an abridged book, it becomes the "vade mecum" of surgical practice well beyond the 16th century.

1482 First Western printing of Arabian philosopher Averroës's (1126-1198) *Colliget*. This general medical encyclopedia is published in Ferrara. Although this is primarily a medical work, it is also highly philosophical. Since Averroës held that the individual is absorbed into universal nature at death and that there is no such thing as personal immortality or salvation, he was condemned in the West.

1483 Giovanni Arcolani (d. 1484), Italian surgeon also called Arculanus, publishes his book, *Practica*, in which he recommends for the first time the filling of teeth with gold. As a surgeon, he is also a pioneer of dentistry and specializes in surgery of the mouth. This work also contains descriptions of the dental instruments of his time.

1490 First printing in Latin of Greek physician Galen's (c.130-c.200) works titled *Opera*. His anatomical writings dominate until Vesalius. The most prolific of all the ancient medical writers, Galen's collected works comprise a gigantic encyclopedia of the knowledge of his time. Although this is the first printed version, the most important editions are the 1525 Aldine Greek text, the 1538 Basel edition, and the nine different editions of Latin text published in Venice by the Giunta house from 1541 to 1625.

(See also 169)

1491 First anatomical book to contain printed

illustrations is German physician Johannes de Ketham's *Fasciculus Medicinae*. As the first printed medical text with original figures, it has woodcuts illustrating writings on uroscopy, venesection, and other surgical procedures. Its illustrations are not particularly accurate or well done, but as the first illustrated medical work, it sees many editions.

1500 First successful Caesarean section is performed on a living subject. Jacob Nufer, whose job it is to neuter farm animals, performs this emergency surgical procedure on his wife, who not only survives but bears more children in her 77-year life span. At this time caesarean sections were usually done after the mother's death in an attempt to save the baby. Although Nufer is successful in this operation, it is not until the mid-17th century that surgeons are willing to attempt surgery as a sometimes necessary alternative to deliver a baby.

1502 Gabriele Zerbi (1468-1505), Italian anatomist, writes the first book considering the anatomy of the infant. In treating infantile anatomy separately as a body of medical knowledge, he acknowledges that infants are not simply very small adults, but have unique physical systems that demand special knowledge and attention.

1505 Royal College of Surgeons is first chartered in Edinburgh.

1507 Antonio Benivieni (c. 1440-1502), Italian physician, founds pathological anatomy with the posthumous publication of his book *De Abditis Nonnullis ac Mirandis Morborum et Sanitionum Causis*. In this work he explains for the first time how the dissection of cadavers can be used to study the internal causes of disease and to determine the cause of death. Although his book is exceptionally brief and discusses only twenty autopsies, it covers cases of syphilis, dysentery, gallstones, bladder cancer, gangrene, tuberculosis of the hip, and even Siamese twins.

1513 Eucharius Roesslin (c.1490-1526), Ger-

Illustration of a doctor treating a plague victim, from *Fasciculus Medicinae*.

man physician, publishes his early textbook on obstetrics called *Rosengarten*. As the first obstetrical treatise written in German after the invention of mechanical printing, it is specifically intended for midwives and, as such, is the first book to deal with midwifery independently of surgery. Roesslin writes it in the hopes of improving the deplorable state of midwifery in his time. He is eventually successful as the authorities seek to educate midwives and standardize their practices. Roesslin's book sees many editions in German and is translated into every major language.

1514 First description of injuries caused by firearms is contained in Italian physician Giovanni da Vigo's (1460-1525) *Practica Copiosa in Arte Chirurgica*. He endorses the terrible practice of using hot oil or cautery on gunshot wounds. The success of this book is far out of proportion to its value. Although it saw some 52 editions, its popularity was based on the fact that it was the only book of its time to deal with the two major surgical problems of the Renaissance—gunshot wounds and epidemic syphilis.

1516 Alessandro Achillini (1463-1512), Italian

anatomist, studies the nervous system and is the first to recognize the function of the first pair of cerebral nerves. He is also credited with the discovery of the fourth cranial nerve. His book containing this and other anatomical discoveries is published four years after his death. Achillini is an especially noteworthy writer since he was daring enough to correct certain errors in the work of Galen.

1517 Hans von Gerssdorff (1454-1517), German military surgeon, writes a field manual for treating wounds that contains the first illustration of an amputation to appear in a machine-printed book. His book, *Feldtbuch der Wundartzney*, contains some of the earliest examples of woodcut medical illustrations. The images are considered to be among the most instructive pictures of early surgical procedures.

1518 College of Physicians is first established in London.

1521 Giacomo Berengario da Carpi (1470-1530), Italian surgeon, is the first to describe the sphenoid (wedge-shaped) sinuses and the veriform (worm-shaped) appendix, as well as the first to discuss the action of cardiac valves. He is also the first to examine carefully the tympanum (middle ear cavity) and the pineal gland of the brain.

1521 First published anatomical drawings made from nature are those by Italian surgeon Giacomo Berengario da Carpi (1470-1530) in his *Commentaria*. Leonardo da Vinci's (1452-1519) earlier excellent anatomical research and drawings were not published for centuries. The reform of anatomy begins with Carpi, who is the precursor of Vesalius.

1524 First hospital in the new world is erected by Spanish explorer Hernando Cortez (1485-1547) in Mexico. The first hospital in North America is the Hôtel Dieu, built in Quebec in 1639 by the Duchesse d'Aguilon. The first hospital in what is now the United States is established on Manhattan Island in 1663.

First illustration of an amputation, from Hans von Gerssdorff's *Feldtbuch der Wundartzney*, 1517.

1525 First modern Latin translation of the work of Hippocrates (460-370 B.C.) is published in Rome.

(See also c. 400 B.C.)

1525 Gonzalo Fernandez de Oviedo y Valdes (1478-1557) of Spain publishes the first systematic description of the medicinal plants of Central America. As governor of the West Indies, he reports back to Spain on not only the plant life in the Indies, but also claims that syphilis was contracted by European sailors from their contact with natives.

1526   First Greek text of Hippocrates (460-370 B.C.) is published in Venice.

(See also c.400 B.C.)

1530   First treatise devoted exclusively to the teeth is a collection of older teachings, *Artzney Buchlein wider allerei Kranckeyten und Gebrechen der Zeen*, published anonymously in Germany.

1532   François Rabelais (c. 1494-1553), French physician and satirist, publishes the first Latin version of the aphorisms of Hippocrates. As a physician, priest, and medical lecturer, he cherished Greek ideals, and regularly ridiculed some of the more ludicrous medical treatments and superstitions of his day.

1533   Francesco Bonafede (1474-1558), Italian physician, holds the first chair of materia medica at Padua. His new teaching position is also called the chair of simples or "lectura simplicium."

(See also 1545)

1535   Mariano Santo di Barletta (1490-1550), Italian physician, gives the first account of a median lithotomy in which a surgeon makes an incision into the urinary bladder to remove a kidney stone. He becomes especially well-known for his method of removing a kidney stone via a lateral incision. Called "cutting for the stone," similar operations have been performed since ancient times, but sometimes would result in testicle damage, castration, or even death.

1535   Valerius Cordus (1515-1544), German physician and botanist, publishes his *Dispensatorium,* which is the first really adequate pharmacopoeia. Cordus participated in the Renaissance revival of the study of plants, and in his short but brilliant life described some 500 new plant species. His pharmacopoeia brought a degree of rational treatment to the use of drugs.

1536   Ambroise Paré (1510-1590), French sur-

Engraving of Valerius Cordus.

geon, makes the first exarticulation (the amputation of a limb through a joint) of an elbow joint. He is also the first to popularize the use of a truss for holding a hernia in place.

(See also 1710)

1537   Johannes Dryander (1500-1560), German anatomist and astronomer also known as Johann Eichmann, publishes his *Anatomicae . . . Corporis Humani Dissectionis.* It is one of the first to offer illustrations made directly from dissecting human cadavers. It is believed that the anatomical illustrations in this book are not simply touched-up copies of manuscript sketches done decades before, but were actually based on original observation during dissection.

1540   Thomas Raynalde of England publishes *The Byrthe of Mankind.* This obstetrical work is based on Eucharius Roesslin's *Der Swangen Frawen und Heb Amme Roszgarte,* first published in Germany in 1513. With his translation of Roesslin's landmark obstetrical textbook, Raynalde provides English midwives with at least some form of reliable birthing information.

1541   Giovanni Battista Canano (1515-1579), Italian anatomist, publishes his *Musculorum*

Humani Corporis Picturata Dissectio in Ferrara. This work is one of first anatomical books to be illustrated with copperplate engravings, which offer finer and more delicate detailing than earlier methods. The 26 copperplate illustrations of the bones and muscles of the arm and forearm surpass in realism and accuracy anything done to this point. Unfortunately, however, when Canano sees the about-to-be published woodcuts in the *Fabrica* of Andreas Vesalius (See 1543), he suppresses his own work and does not continue. He is nonetheless an important precursor to Vesalius.

1543   Andreas Vesalius (1514-1564), Flemish anatomist, publishes his *De Humani Corporis Fabrica*, the first accurate book on human anatomy. Its illustrations are of the highest level of both realism and art, and the result is a true masterpiece that ranks as one of the greatest medical works ever written. As the most commanding figure in European medicine after Galen and before Harvey, Vesalius made dissection a respectable, working science. His work was truly a revolutionary landmark since it broke completely with the past and discarded the revered Galenical tradition. Because of this, he met with resistance most of his life. His anatomical work remains the first that is both scientifically exact and artistically beautiful. His splendid woodcuts of skeletons and flayed figures standing in front of a landscape are copied for over a century.

1545   Francesco Bonafede (1474-1558), Italian physician, founds the first botanic garden in Europe in Padua, Italy. It is called an "ostensio simplicium" or demonstration of living plants. It is affiliated with the university and following this first effort, many universities begin to create gardens of their own. Botanic teaching at universities would involve outdoor excursions in the spring and fall, and apothecaries were often invited.

1556   Guillaume Rondelet (1507-1566), French naturalist and physician, builds the first anato-

my amphitheater in Montpelier, France. This facility allows the observing students to surround the dissecting instructor.

1558   Matteo Realdo Colombo (1516-1559), Italian anatomist, is the first to use living animals (dogs) in laboratory experiments (especially to study the function of the heart and lungs). With this, he achieves his major accomplishment of demonstrating that blood passes from the lung into the pulmonary vein. This is known as the lesser or pulmonary circulation. He is also the first to describe the mediastinum (the mass of tissue and organs separating the two lungs) and the first to mention the folding of the peritoneum. This large membrane of the abdominal cavity connects and supports the internal organs. It has many secreting folds and sacs which prevent friction between closely packed organs, keep them in place, and guard against infection.

1563   Bartolomeo Eustachio (1524-1574), Italian anatomist, publishes his book on teeth, *Libellus de Dentibus*, in which it is stated for the first time that the second teeth have their own dental sac and do not originate, as Vesalius believed, from the roots of the milk teeth. In this study of the finer structure of the teeth, he details both their nerve and blood supply.

1563   Garcia da Orta (c. 1500-c. 1568), Portuguese botanist, compiles the first good compendium on the materia medica of East Indians, as well as the first European manual on tropical medicine. This work is based on his extensive experience in India and includes a description, history, and evaluation of a medicinal plant's clinical effects. His work is one of the first medical treatises printed in India.

1570   Francisco Bravo, Spanish physician, writes the first medical book to be published in the New World. His *Opera Medicinalia* is published in Mexico and gives an early description of typhus.

1575   Juan Huarte de San Juan (c. 1530-1592),

Spanish physician, publishes his *Exemen de Ingenios para Las Ciencias,* which becomes the first modern attempt at systematically studying the functions of the brain and establishing a link between psychology and physiology.

1581   Francois Rousset (1535-c. 1590) publishes his *Traite Nouveau de L'Hysterotomotokie,* which is the first treatise devoted to the Caesarean section. He maintains that it is not as dangerous an operation as is commonly believed, and he details 15 successful cases.

1586   Marcello Donati (c.1538-1602) of Italy gives the first description of a gastric ulcer.

1587   Giulio Cesare Aranzio (1530-1589), Italian anatomist, gives the first description of a deformed pelvis.

1595   Andreas Libavius (1560-1616), German physician and alchemist, publishes his *Alchymia* in Frankfurt, Germany. Most now consider it to be the first chemical textbook because of its systematic treatment. Libavius is convinced totally of the significance of chemistry to medicine. He also is one of the first to suggest the transfusion of blood.

1597   Gasparo Tagliacozzi (1545-1599), Italian surgeon, performs rhinoplasty or plastic surgery of the nose. Although rhinoplasty was known to the ancient Hindus, it is Tagliacozzi who practices this surgery based on solid anatomical knowledge, and it is he who writes *De Curtorum Chirurgia per Insitionem,* the first book about it. He also performed other plastic surgical repairs of the face that were inflicted by dueling or amputation by decree.

1598   Carlo Ruini (c. 1530-1598) of Italy publishes his treatise on the anatomy and diseases of the horse, *Anatomia del Cavallo,* and founds veterinary medicine. This is also the first book dedicated entirely to the anatomical structure of a nonhuman creature.

c. 1600   Invention of the first practical obstetri-

cal forceps occurs sometime during this period and is credited to a member of the Chamberlen family. This tool, which consists of two metal blades joined together at the handle crossing, is used to deliver a child in difficult births. It is known to have been suggested by Pierre Franco of France in 1561, and it is believed that William Chamberlen (c.1540-1596) took the secret of its design with him when he fled to England from France in 1576. Both of his sons were named Peter, and it is one of them (1572-1626) who is most often credited with the invention. The family kept the forceps a closely guarded secret for more than a century.

1602   Felix Platter (1536-1614), Swiss anatomist, publishes his *Praxis Medica,* which is the first modern attempt at the classification of diseases. Based on over five decades of clinical experience, his rational classification of diseases is based on their natural history and postmortem findings.

1602   First operation to remove a foreign body from the stomach is performed by the German surgeon, Florian Matthis.

1604   Johann Kepler (1571-1630), German astronomer, publishes his book *Ad Vitellionem, Paralipomena.* He describes for the first time how the retina is essential to sight. He also details the part the lens plays in refraction, and that myopia is caused by the convergence of luminous rays before they reach the retina.

1611   Caspar Bartholin (1585-1629), Danish physician, publishes his anatomical manual in which he is the first to describe the olfactory nerve (associated with the sense of smell) as the first cranial nerve.

1624   Adriaan van den Spigelius (1578-1625), Belgian anatomist, publishes the first account of malaria.

(See also 1642)

1625   Ipecac is first mentioned as "igpecaya" by a Portuguese priest in Purchas' book *Pilgrimes.*

The dried rhizome and root of this tropical South American creeping plant is valued as an emetic (induces vomiting). By 1680, it is extensively prescribed as a secret remedy for dysentery, and in 1688 is purchased by the French government for 20,000 francs.

(See also 1648)

1628 First accurate description of the human circulatory system is offered by English physician William Harvey (1578-1657) in his landmark work, *Exercitatio Anatomica De Motu Cordis et Sanguinis Animalibus*. While a medical student in Italy, Harvey had learned that arteries and veins have one-way valves, and he puts this knowledge to use to demonstrate conclusively that blood does not oscillate back and forth as the ancients thought, but rather circulates in a one-way system. To show this, when Harvey would tie off an artery, it was always the side toward the heart that bulged with blood. When he tied off a vein, the side away from the heart bulged. Harvey also viewed the heart as a pump that was the motive force behind the blood's one-way, closed-curve circulation. Using direct observation and brilliantly understandable experiments, Harvey eventually won over even those who ridiculed his ideas.

1628 First printed description of blood transfusion is published by Giovanni Colle (1558-1630), Italian physician, in his *Methodus Facile Procurandi Tuta et Nova Medicamenta*. Although earlier writers made vague references to blood transfusions, its history really begins after William Harvey's discovery that blood circulates around the body in a closed system. Later transfusions from animal to man are seldom successful, and transfusions are eventually banned for a time by the authorities.

(See also February 1665)

1632 Marco Aurelio Severino (1580-1656), Italian surgeon also known as Severinus, publishes his *De Recondita Abscessuum Natura*,

which is the first modern manual of surgical pathology. Severino is known to have saved lives by performing tracheotomies during a diphtheria epidemic in Naples, Italy. His book details surgical remedies for all manner of lesions, neoplasms, and abscesses.

1632 Edward Hyde (1609-1674), first earl of Clarendon, describes but does not name the angina pectoris suffered by his father. This is the first recorded case of this heart disease that causes chest pain because of an oxygen deficiency.

(See also 1768)

1633 Stephen Bradwell publishes his *Helps in Suddain Accidents*, which is the first book on giving first aid to the injured.

1639 First hospital established in Quebec by the Duchesse d'Aguilon.

1641 Guerner Rolfinck (1599-1673), German physician and chemist, becomes the first professor of chemistry in Germany. When he begins to conduct two annual public dissections upon executed criminals at Jena, the practice is viewed with horror and amusement by the peasantry who jest of being "Rolfincked."

1642 First mention in the West of the disease beriberi is made in *De Medicina Indorum,* written by the Dutch physician Jacob Bontius (1598-1631). In this early work on tropical medicine, in which he also describes tropical dysentery in Java, Bontius seems aware of the beneficial effects of lemons on certain diseases.

1642 First treatise on the use of cinchona bark (quinine powder) for treating malaria is written by Spanish physician, Pedro Barba (1608-1671). This processed bark was brought to Europe from Peru in 1632 by the Jesuits, and the use of the antimalarial alkaloid had a revolutionary effect on the old-style medicine of the Galenic tradition. The treatment of malaria with quinine marks the first successful use of a chemical compound in combating an infectious disease.

(See also 1820)

1648 René Descartes (1596-1650), French philosopher and mathematician, writes *De Homine*, the first European textbook on physiology. He considers the body to be a material machine and offers his mechanist theory of life. Reducing all physiology to physics, he treats locomotion, respiration, and digestion as mechanical processes. Every organ has a mechanical counterpart: bone and muscle actions are explained by levers; teeth are similar to scissors; the chest is like a bellows; and the stomach is like a flask. Although simplistic, this mechanical explanation of body action proved to be a valuable concept in physiological research.

1648 Willem Piso (1611-1678), Dutch physician and botanist also called Le Pois, gives the first description of yaws and points out the effectiveness of ipecac against dysentery in his book *Historia Naturalis Brasiliae*. He is also among the first to become acquainted with tropical diseases, and he distinguishes between yaws and syphilis. As an infectious tropical disease caused by microorganisms, yaws results in rheumatism and tubercles, or swellings, on the skin, thus resembling syphilis.

1648 Giovanni Alfonso Borelli (1608-1679), Italian mathematician and physiologist, publishes his *Delle Cagioni delle Febbri Maligne della Sicilia negli Anni 1647 e 1648*, which contains the first full description of the iatromechanical system. This system is an elaboration of Descartes' physiological theories and it uses mechanics to explain every aspect of the body. Borelli offers mechanical explanations to the interpretation of muscular action and investigates the complex kinetics of walking, standing, and running. He also studies the mechanical action of the heart, which he says acts like a piston, and calculates its propulsive force as well as the pressure of blood in the arteries.

1651 Nathaniel Highmore (1613-1685), English surgeon, publishes his *Corporis Humani Disquisitio Anatomica*, which is the first anatomical textbook to accept Harvey's theory of the circulation of the blood.

JANUARY 24, 1656 Jacob Lumbrozo, Portuguese physician, arrives in Maryland and becomes the first Jewish physician to practice in America.

1656 Christopher Wren (1632-1723), English architect and scientist, injects opium, wine, ale, and other substances into the veins of a dog (See March 21, 1668). The significance of this experiment is not the substance injected, but the fact that Wren is able to devise a method of tapping a vein and pumping a liquid into it.

1656 Thomas Wharton (1614-1673), English physician, publishes his *Adenographia*, which is the first thorough treatment of all the then-known glands of the human body. He also tells of his discovery of the duct of the submaxillary gland that conveys saliva into the mouth. It becomes known as Wharton's duct.

1661 Nicolaus Steno (1638-1686), Danish anatomist and geologist, first describes the parotid duct (the salivary gland near the angle of the jaw). It becomes known as the duct of Steno. He announces this along with his discovery of the lachrymal (tear) duct in his *Observationes Anatomicae*.

1663 First hospital in the American colonies is established on Manhattan Island in what is now New York.

1664 Thomas Willis (1621-1675), English physician, writes his *Cerebri Anatome*, which gives the most complete and accurate account of the nervous system to date. He also first describes the eleventh, or spinal accessory, nerve as well as the communicating arteries at the base of the brain, which comes to be known as the "circle of Willis." His influential book is illustrated by the famous English architect and scientist Christopher Wren (1632-1723).

FEBRUARY 1665 First direct transfusion of blood from one animal to another is accomplished by English physician Richard Lower (1631-1691), who experiments with dogs. Two years later he transfuses the blood of a lamb into a man. These experiments are eventually banned.

(See also 1666)

1666 Jean Baptiste Denis (1643-1704), physician to French King Louis XIV (1638-1715), is the first to lose a patient following a blood transfusion. After transfusing the blood from a lamb into a human subject, Denis finds that his patient improves temporarily but then dies. Denis is later arrested and the French Chamber of Deputies soon bans such transfusions.

MARCH 21, 1668 First successful intravenous injection on a human is made by the German physician, Johann Daniel Major (1634-1693). At nearly the same time, another German doctor, Johann Sigmund Elsholtz (1623-1688), has the same success.

1668 François Mauriceau (1637-1709), French physician, publishes his important work on obstetrics. In *Traite des Maladies des Femmes Grosses*, he is the first to refer to tubal pregnancy and to correct the ancient view that the pelvic bones separate during labor. As an authoritative figure in his specialty, Mauriceau was the leading representative of the obstetric knowledge of his time, and his published works became a sort of canon for physicians.

1674 Tourniquet to stop hemorrhages is invented by the French surgeon Morel at the siege of Besançon. This early version of the word is originally applied to the handle or stick that is turned to tighten a bandage which applies pressure over a large artery to stop blood flow. Morel also calls it a "garrot" or strangler. It is used at the Hôtel Dieu hospital in Paris a

few years later, and the term tourniquet appears in English in 1695.

1679 First medical periodical published in the vernacular is the Paris *Nouvelles Decouvertes sur Toutes les Parties de la Medecine* by French physician Nicolas de Blegny (c.1646-1722). It becomes very popular and is translated into German and Latin.

1681 Gerardus Blasius (c.1626-1692), Dutch anatomist, publishes *Anatome Animalium*, the first comprehensive treatise on comparative anatomy.

1683 Joseph Guichard Duverney (1648-1730), French anatomist, conducts investigations of the inner structure of the ear and publishes the first textbook on otology (the science of the ear and its diseases).

1684 Raymond Vieussens (1641-1715), French physician and anatomist, publishes his *Neurographia Universalis,* which contains the first description of the brain's "centrum ovale" (the central white matter of the cerebrum) and makes great strides in understanding all the structures of the nervous system. His work provides the most complete description of the brain and the spinal cord to appear in the 17th century.

1686 First book in English on dentistry is written by Charles Allen of Ireland. His book, *The Operator for the Teeth*, is published in Dublin.

1696 Daniel Le Clerc (1652-1728), Swiss physician, publishes his *Histoire de la Médecine.* Called the "father of the history of medicine," Le Clerc makes the first attempt at a synthetic survey of medical history by beginning with a descriptive critique of ancient medicine and continuing to his time. His work sees many translations and remains a major contribution to the medical thought of his time.

1700 Bernardino Ramazzini (1633-1714), Italian physician, publishes the first systematic

Bernardino Ramazzini.

which is based on his microscopical observations. It is the first to distinguish between smooth muscle and striated muscle. Besides being a profound observer and enthralling lecturer, he was regarded as the master of Italian clinicians. It is said that he died so young due to overwork.

1705 Robert Henry Eliot becomes the first professor of anatomy at Edinburgh and assumes the chair at an annual salary of 15 pounds. With this appointment, the medical establishment in Great Britain acknowledges the importance of anatomical education and attempts to catch up with the continent.

1710 Henri Francois Le Dran (1685-1770) and Sauveur François Morand (1697-1773), both French surgeons, perform the first exarticulation (amputation of a limb through a joint) of the shoulder joint.

(See also 1536)

1711 First patent medicine in America is "Tuscora Rice" which is hailed as a cure for tuberculosis. Therapeutic fads and quackery are as successful in the New World as in the Old, and many traveling salesmen sell all manner of patent medicines that are at best harmless.

1718 Screw tourniquet is invented by French surgeon, Jean Louis Petit (1674-1750). Until Petit's device, there was no method during major amputations of controlling the bleeding until the surgeon could tie off all the vessels. His screw tourniquet is fixed to the lower abdomen and puts direct pressure on the main artery. This controls massive bleeding and makes amputations on the battlefield practical. They are carried out swiftly and in great numbers.

1720 Alexander Monro (1697-1767), English physician, is appointed the first professor of anatomy at the University of Edinburgh.

1723 The first modern English hospital medical school, Guy's Hospital, is founded in London by the philanthropist bookseller, Thomas

treatment on occupational diseases. His book, *De Morbis Artificum*, opens up an entirely new field of modern medicine—the diseases of trade or occupation and industrial hygiene. Ramazzini's insightful and masterful treatment of the diseases of tradesmen included ailments and hazards that plagued writers, doctors, hunters, and farmers, as well as miners, potters, painters, and stone cutters. Further, he distinguished between the hazards of their particular environment (like coal dust to miners) and the hazards of physical participation (like writer's cramp or the round shoulders of the cobbler). Ramazzini recognized the toxic effects of certain activities, noting the damage incurred to goldsmiths and physicians by their exposure to mercury. He also detailed the harmful effects on potters and painters who are regularly exposed to lead and spoke authoritatively of the diseases of tinners and colored-glass workers who handled antimony all the time. He counseled that a physician should always interrogate his patient before doing anything else, and that his first and most important question should be, "What is your work?"

1700 Giorgio Baglivi (1668-1706), Italian anatomist, publishes his *De Fibra Motrice* in Perugia,

Guy (1664-1724). This is believed to be the last general hospital to be endowed in full by one person.

1723 First comprehensive book on dentistry is written by French dentist, Pierre Fauchard (1678-1761). His book, *Le Chirurgien Dentiste, ou Traité des Dents*, discusses all aspects of dentistry and includes various prosthetic devices. He introduces crowns and root canals and the filling of hollowed-out teeth with lead or tin. Many consider him to be the founder of modern dentistry.

1726 First quantitative estimate of blood pressure is made by English botanist and chemist, Stephen Hales (1677-1761). Having studied the hydrostatics of sap movement in plants, Hales turns to the animal world and inserts a glass tube in the femoral artery of a horse. This permits him to measure directly the height of the column of blood. This is the first manometer (an instrument for measuring the pressure of a gas or liquid). Also called a tonometer, this device allows Hales to make quantitative estimates of blood pressure, the capacity of the heart, and the velocity of the circulating blood.

1734 First American colonialist to graduate from a medical school (University of Leiden) is William Bull.

1735 Boston Medical Society is first organized. This is the first of several medical societies formed in major American cities.

(See also 1766)

1735 Gaspar Casal (1679-1759), Spanish physician, gives the earliest description of the dietary deficiency disease pellagra. This nutritional disorder is caused mostly by a deficiency of niacin, a member of the vitamin B complex and characterized by what is called the classical three D's—dermatitis, diarrhea, and dementia. The reddish skin lesions are the earliest and best known characteristic, and Casal called this new disease "mal de la rosa" for "rose sickness.") (See also 1926)

1736 Claudius Amyand (c. 1680-1740), English surgeon, performs the first successful appendectomy.

(See also 1886)

1737 Albrecht von Haller (1708-1777), Swiss physiologist, first points out the function of bile in the digestion of fats. Also called gall, this golden-yellow secretion is released constantly by the cells of the liver into the common bile duct and gallbladder and is necessary for the body's proper digestion of fats. Haller is the master physiologist of his time and is one of the most imposing figures in medical history.

1739 Sauveur François Morand (1697-1773), French surgeon, performs the first excision or removal of a hip joint.

1751 Robert Whytt (1714-1766), Scottish physician, gives the first demonstration that the contraction of the eye's pupils in response to light is a reflex action (known as Whytt's reflex). His experiments also show that only a small segment of the spinal cord is necessary for reflex action in general.

1753 James Lind (1716-1794), Scottish physician, publishes his *Treatise of the Scurvy*, which contains the first proper treatment of this scourge of the sea. In this work, Lind details his experiments with 12 sailors suffering from scurvy. He puts them all on the same basic diet but then gives each one a separate, controlled item to consume. After six days, only the man who ate oranges and limes was well enough to return to work. This vitamin-deficiency disease, which killed more sailors on long voyages than did battle with the enemy, was finally eliminated in the British Navy some years after Lind's book. The Navy's practice of giving lime juice to their crews resulted in British sailors being called "limeys."

1755 The Abbe Charles-Michel de l'Epee (1712-1789) founds the first school for deaf-mutes in Paris. This pioneer of deaf-mute instruction maintains this school at his own expense and displays an intense, lifelong devotion to his pupils, even living among them.

(See also 1770)

1761 Leopold von Auenbrugger (1722-1809), Austrian physician, first describes his percussion method, which becomes a major contribution to the diagnosis and prognosis of diseases of the chest. Ever since, doctors use his thumping-on-the-chest method to diagnose the condition of a patient's chest.

1762 First medical library in the American colonies is founded at Pennsylvania Hospital.

1766 First state medical society in America, the New Jersey Medical Society, is founded. It is these professional groups that will eventually form the standards and criteria as well as help establish the laws governing the practice of medicine in the colonies and the new nation.

1768 Robert Whytt (1714-1766), Scottish physician, gives the first description of tuberculosis meningitis in children.

1768 William Heberden (1710-1801), English physician, first introduces the term "angina pectoris" to describe the painful heart condition caused by an oxygen deficiency. It is Heberden's classic account that puts the disease on a scientific basis.

1770 First medical degree in the American colonies is conferred by Kings College.

1770 The Abbe Charles-Michel de l'Epee (1712-1789) invents sign language for deaf-mutes. His work is a major contribution to developing the natural sign language of the deaf into a systematic and conventional language that can be used as a medium of instruction. He prepares a dictionary of deaf and mute manual signs, which he does not finish, but which is completed by his successor, the Abbe Cucurron Sicard (1748-1822).

(See also 1755)

1771 English surgeon, John Hunter (1728-1793), publishes his important work on dental anatomy and disease, *The Natural History of Human Teeth*, in which he first uses the terms "cuspid," "bicuspid," and "incisor." The cuspid or canine tooth comes from the Latin, "cuspis," meaning a point. Teeth with two cusps are naturally called bicuspids. The incisor tooth comes from the Latin, "incidere," to cut into. This tooth is the cutting tooth in front of the canine teeth.

1773 First insane asylum in the American colonies is founded in Williamsburg, Virginia.

1774 Thomas Percival (1740-1804), English physician, is the first English physician to recommend cod liver oil as a remedy for tuberculosis. He is best known for his 1803 publication, *Medical Ethics*, which basically establishes the rules of ethical conduct for a physician. It is this code of ethics that forms the basis for that adopted by the American Medical Association in 1847.

1775 First American textbook on military medicine is the *Plain, Concise, Practical Remarks on the Treatment of Wounds and Fractures* by the American surgeon John Jones (1729-1791). He writes it to assist the military surgeons who attend the American soldiers. Although mostly a compilation, it proves highly useful to the young military and naval surgeons of the American Revolution. Jones is a skillful lithotomist (a surgeon who surgically removes kidney stones from the bladder), and was remembered by American statesman and scientist Benjamin Franklin (1706-1790) in his will for performing a successful operation on him.

1776 Torbern Olof Bergman (1735-1784), Swedish mineralogist, and his student Karl Wilhelm Scheele (1742-1786), Swedish chem-

Thomas Percival.

ist, discover uric acid in vesical calculi. Scheele pulverizes some of these calculi, or bladder stones, and chemically studies their properties. He discovers that higher concentrations of uric acid are found in sick people than in healthy subjects.

1778   First pharmacopeia in the United States is published by Dr. William Brown of Virginia. Published in Pennsylvania, it is intended to be used by the Continental army. His 32-page pamphlet is published anonymously. Brown eventually succeeds American physician Benjamin Rush (1745-1813) as Surgeon General of the Army.

1783   First medical journal printed in America is a translation of the French *Journal de Médecine Militaire*. It ceases publication in 1790.

1789   Edward Wigglesworth, Divinity Professor at Harvard University, constructs the first American life table.

1790   James Derham becomes the first African-American to practice medicine in the United States.

1790   First specially made dental chair is intro-

duced by American dentist and artist Josiah Foster Flagg (1788-1853). He also designs an extracting forceps that is adjustable to any size tooth.

1791   Pieter Camper (1722-1789), Dutch physician and artist, produces one of the earliest works on physical anthropology. His posthumously published *Sur Les Differences que Presente Le Visage dans Les Races Humaines* establishes the bases for craniometric measurements. He also introduced what he called the "facial angle" as a criterion of race and helped establish the science of physical anthropology.

1793   Founder of psychiatry, French physician Philippe Pinel (1745-1826) is placed in charge of the Bicetre insane asylum in Paris and pioneers the humane treatment of its inmates. He frees them from being chained and initiates systematic study of their maladies. He debunks the popular belief that mental illness derives from demonic possession, and is the first to maintain well-documented case histories of mental ailments.

1794   First successful Caesarean section in the United States is performed by Jesse Bennett (1769-1842) of Virginia on his wife.

1794   John Dalton (1766-1844), English chemist, first describes color blindness. Individuals who are color blind are unable to distinguish one or more of the colors red, green, and blue. Dalton himself is color blind.

MAY 14, 1796   First vaccination against smallpox is successfully carried out by English physician Edward Jenner (1749-1823). He inoculates a boy, James Phipps, with cowpox to protect from the more severe disease smallpox. Inoculation against smallpox had been practiced for centuries in the East by introducing pus from smallpox into a cut, but this would sometimes cause a severe case instead of a mild case. Jenner knew that people who worked with cows had no fear of catching smallpox if they had contracted cowpox earlier. Jenner attempts

Edward Jenner administering the first innoculation of vaccine.

Benjamin Waterhouse.

to test this by giving the Phipps boy a case of the cowpox, which indeed proves to protect him against smallpox for the rest of his life. When the British royal family allows itself to be vaccinated, the entire population accepts it in principle and in deed. Although Jenner has no idea why vaccination works, his efforts lay the groundwork for the first defeat of a major disease.

1796 John Abernathy (1764-1831), English surgeon, first ligates the external iliac artery in a case of aneurysm. Surgeons rarely tampered with arteries at this time, given how large they are and how much pressure the blood they carry is under. In this case of an artery in the groin, Abernathy demonstrates that with knowledge and care, an artery can be stitched and repaired.

1797 First original medical journal published in the United States is the *Medical Repository*, a quarterly edited in New York. For seven years it is the country's only medical journal.

JULY 1800 Benjamin Waterhouse (1754-1846), American physician, administers the first vaccinations in the United States. Convinced that

Jenner's new vaccination methods are worthwhile, he vaccinates his four children with a virus obtained from England. Throughout the world, forward-looking physicians immediately adopt Jenner's smallpox vaccine. Waterhouse later writes that the independent-minded New Englanders' fear of the pox was so great that it caused them to endure "restrictions of liberty such as no absolute monarch could have enforced."

1800 Humphry Davy (1778-1829), English chemist, discovers the anesthetic effect of nitrous oxide or laughing gas. It becomes the first chemical anesthetic.

1801 Thomas Young (1773-1829), English physician, is the first to describe the eye condition astigmatism. Called the father of physiologic optics, he is a remarkable student of the eye and correctly determines what causes the blurring of vision that results from this corneal problem.

1802 Marie François Xavier Bichat (1771-1802), French physician, and his colleague, Karl F. Burdach, are the first to use the term "biology" to describe the science of the general properties of organisms. Despite Bichat's short life, he founds the science of histology (the study of tissues) and is the first to demonstrate that the body's organs are composed of tissue—a word he also introduces.

1805 Friedrich Wilhelm Adam Serturner (1783-1841), German chemist, first isolates morphine, the active principle of opium. In doing so, he proves the existence of organic bases containing nitrogen and lays the groundwork for alkaloid chemistry—a term he also originates.

1807 John Stearns (1770-1848), American physician, first publishes an account of the physiological effects of the drug ergot. He notes the drug's ability to stimulate and strengthen uterine contractions and to constrict small blood vessels, and he uses it to control bleeding in women who have recently given birth. Ergot is a fungal disease of grasses, especially rye, and it changes the grain into a hard, beanlike substance that constitutes the drug called ergot. Fatal poisoning can result from eating ergot in flour, and it is also the source of the powerful synthetic hallucinogenic LSD (lysergic acid diethylamide).

(See also 1938)

1807 First American textbook on obstetrics is written by Samuel Bard (1742-1821). His work, *A Compendium of the Theory and Practice of Midwifery*, is revised five times following this first edition.

1809 Ephraim McDowell (1771-1830), American surgeon, performs the first ovariotomy on a woman 47 years old. This operation to remove an ovarian cyst is successful (although conducted without anesthesia) and the patient outlives the doctor, living to be 78. His success disproves the notion that it is impossible to cut into the abdomen without death resulting, and paves the way for later surgery in the abdominal area. It is believed he performs the operation thirteen times, with eight recoveries. Later in his career, he removes a kidney stone from

James Knox Polk (1795-1849), who subsequently becomes president of the United States.

1810 Samuel Friedrich Christian Hahnemann (1755-1843), German physician, publishes his *Organon der Rationellen Heilkunst,* which offers the first complete exposition of the homeopathic method of health therapy. As the founder of homeopathy, it is Hahnemann who lays down its three major tenets: that like cures like (a disease can only be cured by a remedy that has a tendency to produce a similar disease); that medicines increase in potency with their dilution (the administration of many tiny doses is better than that of a single large one); and that a large number of nervous diseases are attributed to the condition called "Psora" (the itch). Although most of Hahnemann's doctrines do not hold up over time, his protests against medicine's more violent treatments and his advocacy of few medicines, coupled with his belief in the healing powers of nature, have beneficial effects on the practice of medicine overall.

1810 Daniel Drake (1785-1852), American physician, first describes the sometimes fatal disease called "the trembles" or the "milk sickness." The cause is eventually found to be drinking milk from cows that have fed, when grazing was scarce, on white snakeroot which contains trematol, a poisonous alcohol. Milk sickness prevails in Ohio, Indiana, and Illinois in the early 1800s. Abraham Lincoln's mother, Nancy Hanks Lincoln, is said to have died from this disease.

1810 Franz Joseph Gall (1758-1828), Austrian-French physician, publishes the first of his four volumes titled *Anatomie et Physiologie du Systeme Nerveux,* which founds a new discipline, phrenology, the study of cranial protuberances. Gall is convinced that mental functions are localized in specific regions of the brain; therefore the surface of the skull must accurately reflect the relative development of the different regions. From this conviction he founds phrenology (also called cranioscopy),

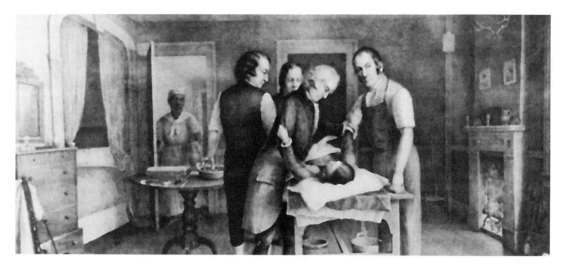

Ephraim McDowell performing the first ovariotomy.

which is the practice of determining individual intelligence and personality from an examination of skull shape. Although his concept of localization is later proved correct, his other basic premise on skull shape is proven false.

1811 Elisha North (1771-1843), American physician, publishes the first book on cerebrospinal meningitis (called "spotted fever") and recommends the use of a clinical thermometer. This disease occurred in epidemic form and was worldwide in distribution. Most often occurring in cold weather among young children, it was transmitted through droplets and produced fever, rash, and intense headaches. Sometimes temporary blindness or permanent deafness would occur; even death would sometimes result from brain swelling. North's work is considered an early American medical classic. He also establishes the first eye infirmary in the United States in New London, Connecticut. (See also 1887)

1812 First independent village medical school is founded in Herkimer, New York. This unique type of one-room medical school was formed in remote areas of the country and had two or three faculty members, usually local doctors.

1812 *New England Journal of Medicine and Surgery* is first published.

1812 Benjamin Rush (1745-1813), American physician, publishes his *Medical Inquiries and Observations Upon the Diseases of the Mind,* which is the first American treatise on psychiatry. Based on Rush's 30 years of personal observations at the Pennsylvania Hospital where he was in charge of mental patients, it sees many editions and remains for 70 years the only American text on this subject. Posterity also records that Rush was one of the most steadfast practitioners of bleeding a patient, and that even on his own deathbed, he was bled twice on his own insistence and against the better judgment of his own physicians.

FEBRUARY 28, 1818 René Théophile Hyacinthe Laënnec (1781-1826), French physician, invents the first stethoscope and develops the method of diagnosis by auscultation. He announces his discovery to the Royal Academy of Science on this date. Until Laënnec, physicians would practice auscultation by placing their ear directly on a patient's chest to listen to the heart and lungs. Laënnec found this old method to be not only ineffective but "inconvenient and indelicate," so he rolled several pieces of paper into a tube and placed one end on a patient's chest with his ear at the other. Called "mediate auscultation," this method let him hear a heartbeat with greater clarity and distinction. Further experimentation showed that a foot-long cylinder of beechwood offered best results. Although initially resisted by the medical community, Laënnec's invention eventually revolutionizes diagnostic medicine. After contributing greatly to the study of the physiology and pathology of the chest cavity, he dies fairly young of tuberculosis and overwork.

(See also 1839)

1818   Pieter de Riemer (1760-1831), Scandinavian anatomist, first uses frozen sections in anatomy. Preserving specimens by freezing enables illustrators to more accurately depict the human body in anatomical texts.

1818   Valentine Mott (1785-1865), American surgeon, is the first to ligate the innominate artery. This huge artery in the chest, only 2 in (5 cm) above the heart, provides blood to the right arm, as well as to the right side of the head and neck. Being able to tie it off during surgery in this area is necessary if the patient is not to bleed to death. Most of Mott's operations are performed in the generation before anesthesia, and this type of deep surgery required intimate anatomical knowledge, great confidence, detailed planning, and above all, speed in execution.

1818   Pierre Joseph Pelletier (1788-1842) and Joseph Bienaime Caventou (1795-1877), both French chemists, discover strychnine. This is the first alkali of vegetable origin to be discovered after morphine. Alkaloids are a class of nitrogen-containing organic compounds found chiefly in plants and with very different and powerful physiological effects. Besides morphine and strychnine, quinine, ephedrine, and nicotine are other well-known alkaloids. Used initially as a cathartic or purgative in medicine as well as a stimulant for the central nervous system, it proves very dangerous and is mostly used to kill rats. Death from strychnine poisoning is ultimately the result of suffocation or exhaustion.

1819   First public health statistics compiled in the United States are the records the Surgeon General orders kept on troop sickness and mortality. A uniform format allows collation and comparison of data.

(See also 1840)

1820   Pierre Joseph Pelletier (1788-1842) and Joseph Bienaime Caventou (1795-1877), both of France, first isolate quinine, the active principle of cinchona bark. Quinine is used chiefly in the treatment of malaria, and until World War I, it is the only effective treatment. Although it does not produce a complete cure, it works by interfering with the growth and reproduction of the malarial parasites in the infected person's blood. It is not until 1944 that total laboratory synthesis of quinine is achieved.

JUNE 6, 1822   First accurate account of the human digestion begins when a young French-Canadian named Alexis St. Martin is accidentally shot in the chest and upper abdomen. Treated by an American surgeon, William Beaumont (1785-1853), who does not believe his patient will live very long, St. Martin in fact survives a severe infection and heals in an unusual way: his stomach attaches itself to his chest wall, resulting in a fistula, or abnormal permanent opening or channel. This allows Beaumont to actually see into a working stomach and he takes advantage of this unique opportunity to study digestion first-hand. In 1833, Beaumont publishes his *Experiments and Observations on the Gastric Juice and the Physiology of Digestion*, which is based on over 200 experiments he conducts with St. Martin. By analyzing gastric juices and suspending foods into the stomach via silk threads, he concludes that the stomach acts upon food chemically and that the basic ingredient of gastric juice is hydrochloric acid.

1823   British medical journal, *Lancet*, is first founded.

1825   Pierre Fidèle Bretonneau (1778-1862), French physician, performs the first successful tracheotomy for croup. (This emergency cut made into the larynx allows a child whose windpipe has swollen shut to breathe). Bretonneau, who had previously performed this procedure successfully on animals, inserts a small, silver tube in a child's throat to save his life. Tracheotomy becomes a standard life-saving procedure for patients in severe respiratory distress and is used today.

1826   Diphtheria, a serious bacterial disease of the respiratory system and a major killer of children, is first named by French physician Pierre Fidèle Bretonneau (1778-1862). As the first to study closely its symptoms, he names it by using the Greek word for parchment to describe the yellow-gray mucous membranes that form in the throat of the infected. The disease produces toxins that are carried throughout the body, and it ravages the very young in Europe and the United States.

1826   Pierre François Olive Rayer (1793-1867), French physician, is the first to distinguish acute from chronic eczema. This dermatitis, or inflammation of the skin, is an itching, scaling, and sometimes weeping of the affected area. It is often the result of an allergy.

1827   Richard Bright (1789-1858), English physician, publishes his *Reports of Medical Cases,* which contains the first original description of the kidney disease called nephritis. He also first describes a chronic disorder of the kidney that becomes known as "Bright's disease." Although some of the symptoms of this kidney disorder were known before him, Bright is the first to connect them and make a correct observation. He is considered to be one of the greatest of modern pathologists.

1829   First American school for the blind, the Perkins School, is founded in Boston by John

Dix Fisher (1799-1850). Following this, many states establish institutions for the blind.

1829   Thomas Addison (1793-1860) and John Morgan (1797-1847), both English physicians, publish the first book in English on the physiological effects of poisons on the human body. It is titled *An Essay on the Operation of Poisonous Agents upon the Living Body.*

1831   Justus von Liebig (1803-1873), German chemist, Eugene Soubeiran (1793-1858), French pharmacologist, and Samuel Guthrie (1782-1848), American chemist, independently discover chloroform. Also called trichloromethane, it is a nonflammable, clear, colorless, and heavy liquid with an ether-like odor. It is used this same year as a potent inhalation anesthetic. It is prepared from various organic compounds, most often from alcohol or acetone. Guthrie's method of distilling alcohol with quicklime becomes the preferred, modern way of making chloroform.

1833   Philipp Lorenz Geiger (1785-1836), German pharmaceutist, and his colleague Hesse first isolate atropine from belladonna. This poisonous, crystalline substance is an alkaloid with medicinal uses. Its major drug action is a depressing or relaxing effect on the parasympathetic nervous system. Despite a large number of effects on the body, its chief use has come to be in ophthalmology, in which it is applied locally to the eye to dilate the pupil in the examination of the retina. It is found as a component of the plants of the Solanaceae family, like belladonna, henbane, and thorn apple.

DECEMBER 1834   First constituted dental society in the world, the Society of Surgeon-Dentists of the City and State of New York, is founded.

1834   Friedlieb Ferdinand Runge (1795-1867), German chemist, first prepares carbolic acid or phenol by the distillation of coal-tar oil. Despite his discovery, he does not analyze this or any other of its components. The antiseptic properties of carbolic acid make it important to medicine (although it is no longer used), but undiluted it is highly corrosive to mucous membranes and skin, and is considered a nerve poison. In the 1930s, the expanding plastics industry uses it in large amounts.

1834 Samuel George Morton (1799-1851), American physician, writes his *Illustrations of Pulmonary Consumption*, which is the first book on tuberculosis issued in the United States. His work provides a summary of knowledge about this disease.

1835 Jöns Jakob Berzelius (1779-1848), Swedish chemist, first coins the term "catalysis" to describe the agent that brings about a chemical reaction or transformation. In chemical terms, a catalyst modifies the rate of a chemical reaction, usually speeding it up, and is not consumed during the reaction. Understanding catalysis better proves valuable to medicine as it employs more and more drugs.

1837 Pierre François Olive Rayer (1793-1867), French physician, is the first to describe the disease glanders as found in humans and to prove that it is not a form of tuberculosis. Also known as farcy, glanders is an infectious disease usually found in horses and mules, but occasionally in humans. It affects the lungs and breathing, eventually forming internal lesions. Glanders in humans occurs usually through occupational contact with an infected horse. It is important to know that it is not a type of TB.

1838 Isaac Ray (1807-1881), American physician, writes the first book on the medical jurisprudence of insanity. In this work he describes a form of insanity, which he says impairs the moral sense and makes a person less responsible for his actions. It becomes known as "Ray's mania."

JUNE 1839 First dental journal, *American Journal of Dental Science*, makes its appearance in New York.

1839 First dental school in the United States and in the world, the Baltimore College of Dental Surgery, is founded by Chapin Aaron Harris (1806-1860). In 1840 it awards the first degree of Doctor of Dental Surgery.

1839 Caspar Wistar Pennock (1799-1867) of the United States invents the first flexible-tube stethoscope. This replaces the awkward wooden tube invented by French physician Rene Laennec and improves diagnoses.

(See also February 28, 1818)

1840 First national report of public health statistics (United States) is *Statistical Report on Sickness and Mortality in the Army of the United States, 1819-1839* by Samuel Forry (1811-1844).

1841 Friedrich Gustav Jakb Henle (1809-1885), German pathologist and anatomist, publishes his *Allegemeine Anatomie,* which becomes the first systematic textbook of histology (the study of minute tissue structure). His pioneering work in histology, which becomes a branch of biology dealing with the study of tissues and the differences among them, lays the groundwork for an entirely new approach to living matter, its workings, and diseases.

JANUARY 1842 First use of ether as an anesthetic during a tooth extraction is conducted by American dentist Elijah Pope under the guidance and direction of American medical student William E. Clarke. The operation, the first of its kind, is entirely successful but Clarke makes no record of it.

(See also March 30, 1842)

FEBRUARY 18, 1842 First to recognize the cause of childbed fever is English physician Thomas Watson (1792-1843), who publishes a lecture in the *London Medical Gazette* on this date. Also called puerperal sepsis, childbed fever was a disease that killed many healthy mothers after they had given birth in a maternity ward of a hospital. Watson states correctly that it is an infection that is carried from one patient to another by the person examining the women and delivering the babies. He suggests that doctors wash their hands with a chlorine solution or use gloves that they then discard. About the same time, the Hungarian physician Ignaz Philipp Semmelweis (1818-1865) reaches the same conclusion and decides to take action.

(See also 1847)

MARCH 30, 1842 First recorded use of anesthesia in surgery is accomplished by American physician, Crawford Long (1815-1878). After experimenting on himself earlier this year, Long removes a tumor from the neck of a patient who is rendered unconscious by ether. Long conducts at least eight other such operations with ether during the next few years but does not publish any account of it until 1849. By then, others had already received credit for being the first.

(See also October 16, 1846)

1842 Samuel Forry (1811-1844), Assistant United States Surgeon general, publishes *The Climate of the United States and Its Endemic Influences*, which is the first American book on medical climatology.

1842 William Bowman (1816-1892), English anatomist, first demonstrates that urine is the product of super-filtration by the membrane of the Malpighian bodies of the kidney's cortex. Bowman proves his theory that the Malpighian bodies or renal corpuscles filter the blood and separate out some of the urine. The double-walled capsule that surrounds this filtration unit is later called Bowman's capsule.

1843 Gabriel Andral (1797-1876), French physician, is the first to urge that blood be examined in cases of disease. His book, *Essais d'Hematologie Pathologique*, stresses the importance of the chemical study of the blood to find disease, and it one of the earliest books written on this new subject. His work on the laboratory

analysis of blood helps establish clinical hematology as a separate branch of pathology.

1844 Horace Wells (1815-1848), American dentist, is the first to use nitrous oxide as an anesthetic in dentistry. Wells shared a dental practice with William Thomas Green Morton (1819-1868), both of whom became familiar with the pain-killing properties of nitrous oxide, also called "laughing gas." After Morton's successful demonstration of ether anesthesia during surgery in October 1846, Wells became bitter and began to self-experiment with several other chemicals. Eventually he became unstable, was jailed for throwing acid, and eventually killed himself in jail. During this time the Paris Medical Society hails him as the discoverer of anesthetic gases.

(See also October 16, 1846)

1846 Oliver Wendall Holmes (1809-1894), American author and physician, first suggests the use of the terms "anaesthesia" and "anaesthetic" in a letter to William Thomas Green Morton (1819-1868), American dentist. Anaesthesia and anaesthetic are derived from a Greek word meaning lack of feeling.

OCTOBER 16, 1846 Ether is first used as an anesthetic in a surgical operation in Boston. William Thomas Green Morton (1819-1868), American dentist, persuades renowned American surgeon John Collins Warren (1778-1856) to give the new drug a trial procedure, and Warren successfully removes a small tumor from the neck of an anesthetized patient at Massachusetts General Hospital. The success of this painless operation makes anesthesia at once inseparable from the operating room. No longer would surgeons have to contend with a squirming, shrieking patient, nor would a patient have to undergo the torture of being cut open while awake.

(See also 1844)

1846 Arthur Hill Hassall (1817-1894) publishes his book, *The Microscopic Anatomy of the*

Human Body, which is the first textbook in English on this subject.

1846 James Marion Sims (1813-1883), American surgeon, invents the vaginal speculum. Sims uses his new tool, which resembles a bent spoon, to correct a displaced uterus of a woman who had fallen from a horse. He also uses it to operate on a vesico-vaginal fistula (an abnormal passage or canal), demonstrating how with his device he can repair this previously untreatable condition. His discovery is soon taken up in Europe and becomes indispensable for other gynecological operations.

NOVEMBER 4, 1847 James Young Simpson (1811-1870), Scottish obstetrician, introduces the first use of chloroform anesthesia in both labor and delivery. Despite this success, many patients and physicians refuse to believe that a woman should be relieved of the pains of childbirth, and Simpson wages a vigorous campaign to persuade them otherwise. The use of chloroform in delivery by Queen Victoria (1819-1901) in 1853 signaled to the world that anesthesia, and especially chloroform, was acceptable and safe during childbirth.

1847 American Medical Association is founded and holds its first convention the following year.

1847 First method to prevent childbed fever is introduced by Hungarian physician Ignaz Philipp Semmelweis (1818-1865). He is intrigued by the obvious fact that women who have their babies at home seldom contract puerperal sepsis or childbed fever, while those who deliver in a hospital attended by a physician are at very high risk and often die despite a normal birth. Semmelweis then learns from the death of a colleague who contracts blood poisoning from a minor scalpel wound he receives during an autopsy. He theorizes that doctors who go directly to their patients after examining dead bodies are responsible for transferring infectious matter from one patient to another. He

then establishes procedures in his ward requiring all patients, students, and professors of obstetrics to thoroughly wash their hands and disinfect them with a chlorinated lime solution. Although the death rate in his wards drop dramatically, Semmelweis is ridiculed and ignored by his colleagues, and it is not until Joseph Lister that he is proven correct.

1848 Hermann von Fehling (1812-1885), German chemist, invents a solution that provides medicine with its first quantitative test for sugar in urine. This becomes important in the diagnosis of diabetes mellitus, a disorder of carbohydrate metabolism resulting from an insufficiency in the production or utilization of insulin. The urine of a diabetic has an abnormally high level of sugar.

1848 First school for the mentally retarded is founded. The Massachusetts School for the Idiotic opens this year.

1848 First medical school solely for women is founded. The Women's Medical College of Pennsylvania becomes incorporated in 1850, as the Female Medical College of Pennsylvania.

1849 First woman to receive a medical degree in the United States is Elizabeth Blackwell (1821-1910). She graduates with highest honors this year from Geneva College (now a part of Syracuse University) in New York. Although she becomes the first medical doctor in America, she encounters the same prejudice after graduation that she experienced before medical school (having applied to 29 medical schools before being accepted). In May 1849, she attends La Maternite in Paris, France, but is allowed to study only as a midwife, not as a doctor. While there, she contracts ophthalmia from a sick baby and loses her left eye. After studying in London, she returns to New York but is ostracized by the male medical profession and is unable to obtain a hospital position. In 1853 she opens a dispensary in a New York slum and eventually establishes a hospital of her own.

An anatomy lecture at the Medical College for Women.

The New York Infirmary for Women and Children initially begins training nurses and eventually becomes a three-year medical college in 1864. In 1869 she returns to London and founds the London School for Women. Throughout Blackwell's ninety years, she remains a tireless crusader for the rights of women.

1849 John Snow (1813-1858), English physician, first states the theory that cholera is a water-borne disease and that it is usually contracted by drinking. He is also the first to provide experimental data in support of his water-borne theory. Cholera is often a fatal disease characterized by massive diarrhea, vomiting, and rapid and severe depletion of body fluids and salts. His concept of the spread of cholera remains valid, although its micropathogenic agents are not identified for decades. During a cholera epidemic in London in 1854, Snow breaks the handle of the Broad Street Pump, thereby shutting down what he considered to be the main public source of the epidemic.

1849 Thomas Addison (1793-1860), English physician, describes pernicious anemia and is also the first to give an accurate description of the hormone deficiency disease that results

from the deterioration of the adrenal cortex. It comes to be known as "Addison's disease" after the individual who is the first to correlate a set of its symptoms with the pathological changes in one of the endocrine glands. It is also called adrenal cortical insufficiency and is characterized by weakness, brownish pigmentation of the skin, weight loss, low blood pressure, and gastrointestinal upsets. It is treatable today with corticosteroids.

1850 Morrill Wyman (1812-1903), American physician, performs the first successful operation in the treatment of pleurisy in the United States. As an inflammation of the pleura (the lining of the thoracic cavity and lungs), pleurisy results in the generation of fluid, which causes pain and shortness of breath. Wyman treats these symptoms, although not the underlying infectious cause, by operating to evacuate the fluid.

1850 Henry Ingersoll Bowditch (1808-1892), American physician, first introduces the thoracentesis procedure. His instrument which is designed to remove fluid from the chest cavity consists of a needle, some tubing, and a suction pump. This surgical puncture of the

Elizabeth Blackwell.

chest wall can save the life of a patient with congestive heart failure.

1851 Hermann von Helmholtz (1821-1894), German physiologist and physicist, perfects the first ophthalmoscope. This new instrument, which allows the physician to see into the eye and examine the retina for disease, transforms ophthalmology into an exact science. His invention works by using a series of glass plates that serve as mirrors or reflectors to shine a beam of light into the eye, allowing the operator to peer through a magnifying lens attached to the mirror. With this, Helmholtz is able to see the retina with its blood vessels. These are the only ones in the body that are visible, and this is tantamount to being able to examine the interior of the body without any intrusive procedures. In 1847, the English mathematician Charles Babbage (1792-1871) invented the first ophthalmoscope but put it aside for testing and forgot about it.

(See also 1855)

1851 William W. Reid, American physician, first lays down the principles for the reduction of dislocations by simple manipulation. Prior to him, dislocations were reduced, or joints were

put back into place, with weights and pulleys. Reid's methods are simple and are adopted around the world almost immediately. Dislocations can only occur in the bones forming a joint, and they are usually very painful. Reid learns how to put joints back in place without using excessive force by experimenting on corpses during dissection.

JANUARY 1852 Mount Sinai Hospital, the first Jewish hospital in the United States, is founded in New York City.

1853 Charles Gabriel Pravaz (1791-1853), French physician, first publishes his description of a hypodermic syringe. This will open up an entirely new field for the administration of drugs. The key to the effectiveness of his invention is a hollow needle, which is then connected to a syringe and inserted under or into the skin.

MARCH 1854 Modern nursing practice is first founded by the work of English nurse, Florence Nightingale (1823-1910), who organizes care for the sick and wounded during the Crimean War. Through dedication and hard work, she goes on to create a female nursing service and a nursing school at St. Thomas' Hospital in 1860. Her compassion and commonsense approach to nursing sets new standards and helps create a new era in the history of caring for the sick and wounded of wartime.

1855 First woman dentist is Emmeline R. Jones.

1855 First modern laryngoscope is invented by the Spanish singing teacher Manuel Patricio Rodriguez Garcia (1805-1906), who devises this new instrument to examine the throats of his students. Garcia begins by studying his own larynx with a dental mirror, and after much research, builds his new instrument. He sends a written account of his invention to the Royal Society of London and the device becomes a permanent part of laryngology within three years. It is because of this instrument that the art of laryngoscopy is introduced into clinical practice and the medical specialty of laryngology is born.

1855 Albrecht von Graefe (1828-1870), German surgeon, first introduces the operation of iridectomy (surgical removal of part of the iris) in the treatment of iritis, iridochoroiditis, and glaucoma. As the creator of modern surgery on the eye and one of the greatest of all eye surgeons, Graefe's eye clinic becomes world famous. His pupils include nearly all the best ophthalmologists of the 19th century. As the founder of modern ophthalmology, he also develops a surgical treatment for cataract by extraction of the lens. He becomes the first to use the ophthalmoscope invented by Helmholtz. (See also 1851)

1855 First State Board of Health in the United States is founded by the State of Louisiana.

1856 Hermann von Helmholtz (1821-1894), German physiologist and physicist, publishes the first part of *Handbuch der Physiologischen Optik* (1856-1867). In this work he unites morphology, physics, and physiology into a single conceptual synthesis.

1857 Abraham Jacobi (1830-1919), American physician, occupies the first chair of children's diseases in the United States at the College of Physicians and Surgeons in New York. (See also 1860)

1857 Albrecht von Graefe (1828-1870), German surgeon, first introduces an operation for strabismus (squinting caused by one eyeball being weaker than the other). It is the result of a nonalignment of eye muscles caused by some abnormality in its nervous controls. The person with strabismus then has a "deviant" eye that may be pointed in any direction. If left untreated, it can result in monocular blindness in which the deviant or unused eye becomes functionally blind.

1857 Wilhelm Petters (1824-1889), Czech phy-

German surgeon Albrecht von Graefe, the founder of modern ophthalmology.

sician, first demonstrates the presence of acetone in the urine of diabetics. Although a normal constituent of blood and urine, it is produced in large amounts in diabetic patients. The lack of glucose in the cells of a diabetic patient activates mechanisms that increase the breakdown of fats and proteins, whose products are then burned for energy. These are partially excreted by the lungs and result in what is known as "acetone breath," an odor often mistaken for alcohol.

1858 Mary Anna Elson (1833-1884), German-American physician, is the first Jewish woman to graduate from the Women's Medical College of Philadelphia. She practices in Philadelphia and later in Indiana.

1858 Ernst Krakowizer (1821-1875), Austrian-American physician, introduces laryngoscopy in the United States and first demonstrates the vocal cords with a laryngoscope. He shows that sound is produced by the rapid opening and closing of two folds of mucous membrane that extend across the interior cavity of the larynx.

1858 The first serious, written treatment of

psychology as an experimental science is begun by German psychologist Wilhelm Max Wundt (1832-1920). His major treatise, *Beiträge zur Theorie der Sinneswahrnehmung*, is published over a period of fours years (1858-1862) and can be viewed as an outline of his life's work.

(See also 1862)

1858   First practical, powered dental drill is introduced. It is powered by a foot treadle and cannot spin very fast.

(See also 1872)

1859   Hermann Brehmer (1826-1889), German physician, establishes the first successful sanitarium for tuberculosis at Gorbersdorf in Silesia, Germany. It emphasizes exercise, fresh air, hydrotherapy, and rest. This successful establishment led to the founding of many similar institutions in mountain and seaside resorts all over the world.

1859   Albert Niemann (1806-1877) of Germany first prepares cocaine by isolating the active principle of *Erythroxylon coca,* known since ancient times as the "divine plant of the Incas." Peruvian natives had for centuries chewed on the leaves of this plant for its stimulating and exhilarating effects. After studying the white powder and noting its numbing effect on his tongue, he names it cocaine. For several decades, cocaine is used by the medical community as a pain reliever and a stimulant, and its addictive properties go unrecognized or unmentioned. Today it is used only occasionally for certain types of procedures, with the bulk of it being made and purchased illegally.

1860   Abraham Jacobi (1830-1919), German-American physician, pioneers the teaching of pediatrics in the United States and begins the first bedside medical instruction in the United States. As the first clinical professor of the diseases of infancy and childhood at the College of Physicians and Surgeons, he also helps found the *American Journal of Obstetrics and Diseases of Women and Children* and the estab-lishment of the pediatric section of the American Medical Association.

1860   Jakob von Heine (1800-1879), German orthopedist, first describes "infantile spinal paralysis," which becomes known as poliomyelitis. For many years it is known as "Heine-Medin's disease" for Heine and the Swedish pediatrician Oskar Medin (1847-1927), who describes it many years later. Polio is an acute infectious viral disease characterized by mild to severe paralysis of voluntary muscles.

(See also 1887)

1861   Guillaume Benjamin Amand Duchenne (1806-1875), French physician and neurologist, gives the first description of the progressive muscular atrophy called locomotor ataxia. As a type of muscular dystrophy, this condition is caused by a nerve disorder in the spinal column and results in muscle weakening in the legs. In his research, he explored the effects of electrical stimulation on diseased nerves and muscles and helped found both electrodiagnosis and electrotherapy. He also invented an instrument designed to remove small tissue samples from deep inside the body and can be said to have founded the diagnostic practice of biopsy.

(See also 1983)

1861   Erastus Bradley Wolcott (1804-1880), American surgeon, performs the first kidney excision as he removes a renal tumor. Although he demonstrates that a nephrectomy can be successfully performed (and that an individual can survive with one kidney), this operation is seldom performed until antisepsis becomes firmly established.

1861   Julius Homberger, American physician, founds the first journal of ophthalmology in the United States.

1861   First localization of a brain function is achieved by French surgeon and anthropologist, Pierre Paul Broca (1824-1880). He demonstrates through postmortems that damage to a certain spot on the cerebrum of the brain (the third convolution of the left frontal lobe) is associated with the loss of the ability to speak, called aphasia. This is the first clear-cut demonstration of a connection between a specific ability and a specific cerebral point of control. Broca's work stimulates brain research, and within twenty years, much of the cerebrum is mapped out and associated with parts of the body and their functions.

1861   Guillaume Benjamin Amand Duchenne (1806-1875), French physician and neurologist, is the first to recognize progressive bulbar palsy that affects the mouth and larynx (under the name labioglossolaryngeal paralysis).

1861   Ernst Brand (1827-1897), German physician, first introduces "scientific hydropathy" as a treatment for typhoid fever. Hydrotherapy is the external use of water in the medical treatment of disease or injury, and Brand specifically uses a cold-bath treatment for typhoid fever. Today, both hot and cold hydrotherapy is recognized as beneficial if used appropriately. Wet heat helps relieve pain and improves circulation. Wet cold reduces and helps to prevent swelling following an injury. Underwater massage is also used by physical therapists to heal and strengthen muscles.

JUNE 9, 1862   United States Surgeon General directs the preparation of what becomes a six-volume series titled *Medical and Surgical History of the War of the Rebellion, 1861-1865*. This is the first detailed account of the medical and surgical aspects of an army in battle. The first volume is published in 1875 and the last in 1888. This splendid collection of case histories and pathologic reports is embellished with fine plates and is a unique work in the annals of military medicine.

1862   Jules Péan (1830-1898), French surgeon, first devises the hemostatic forceps. This pincer-shaped instrument later proves crucial to surgeons who can stop hemorrhages by temporarily closing off an artery. Péan's device comes to play a major role in modern surgery.

1862   Wilhelm Max Wundt (1832-1920), German psychologist at the University of Leipzig, teaches the first university course ever given in scientific or experimental psychology. In 1879 he founds the first institute of psychology (at the University of Leipzig) and establishes the first laboratory to be devoted entirely to experimental psychology. He also founds *Philosophische Studien* in 1881, the first journal devoted to that subject. He is generally acknowledged as the founder of experimental psychology.

1862   Viktor von Bruns (1812-1883), German surgeon, performs the first laryngeal operation using a laryngoscope. A pioneer in the surgery of the larynx, he demonstrates that laryngeal tumors can be removed through the patient's mouth without making an external incision.

APRIL 9, 1864   Before a meeting at the Sorbonne in Paris, France, French chemist, Louis Pasteur (1822-1895), first announces his proof of the germ theory. Having demonstrated earlier that lactic acid fermentation is due to a living organism, Pasteur turned to the study of how microscopic life arises. Many of his famous colleagues supported the traditional view of spontaneous generation, which said that living organisms (microbes or germs) could originate from nonliving matter. In 1860, Pasteur devised an experiment that would prove his theory that dust in the air included spores of living organisms, and that it was these which multiplied when they settled on a proper medium. He boiled meat extract and left it exposed to the air, but only via a long, narrow neck that bent down and then up. This allowed unheated air into the flask, but any dust particles would settle at the bottom curve of the neck and not get in. When the meat did not decay and no organisms developed, Pasteur had proven the truth of the germ theory and disproved spontaneous generation once and for all.

Jules Péan standing beside his operating table.

1864 Joseph Janvier Woodward (1833-1884) and Edward Curtis, both of the United States, develop the process of microphotography and publish the first photomicrographs of bacteria.

AUGUST 12, 1865 First antiseptic procedure to combat infection is used by English surgeon, Joseph Lister (1827-1912). After years of study and investigation, Lister concludes that infection is caused by something outside a wound that gets in it. This is contrary to the then-held belief that the pus which formed in wounds was a necessary part of the healing process. After learning of Pasteur's discovery that microorganisms cause the death of body tissues, he decides to find a way to kill bacteria in wounds as well as to keep it out. Lister then experiments with carbolic acid (phenol), which he eventually dilutes, and on this date he uses a spray of carbolic acid during surgery on a compound fracture to prevent infection. Two years later he publishes his paper telling of his success, and lays the groundwork for antisepsis—the critical procedure that prevents the growth and multiplication of bacteria.

1865 Joseph Janvier Woodward (1833-1884) of the United States is the first to report the use

of aniline dyes for histopathological studies (the microscopic study of diseased tissue).

1866 Thomas Clifford Allbutt (1836-1925), English physician, invents the short clinical thermometer. His modern device replaces the old, foot-long instrument that requires 20 minutes to register a patient's temperature.

1867 John Staughton Bobbs (1809-1870), American surgeon, performs the first cholecystotomy—the surgical removal of stones from the gallbladder by an incision in the abdomen. Gallstones are concretions made of crystalline substances like cholesterol, bile pigments, and calcium salts. Located in the gallbladder, it may produce no symptoms, but when a stone lodges in the bile ducts, the obstruction leads to severe pain. In most instances the stones and the gallbladder are removed by surgery.

1867 Adolf Kussmaul (1822-1902), German surgeon, first introduces the intubation of the stomach. He washes out the stomach with a stomach-tube for gastric dilation and also uses it to treat gastric ulcers. Known popularly as a stomach pump, this technique is still used in cases where a person has ingested a toxic substance.

1868 Jean Antoine Villemin (1827-1892), French physician, first demonstrates that the origin of tuberculosis is not spontaneous; it is a specific infection caused by a transmitted agent. He proves that it is a contagious disease by his experiments in which he successfully transmits it from a human subject to an animal. Since his work takes place before the bacteriological era begun by Pasteur, the medical world takes little notice of his argument that some invisible agents are responsible for the transmission of this disease.

1868 Thomas Moreno y Maiz, French pharmacologist, is credited with being the first to suggest that cocaine be used as a local anesthetic.

(See also September 16, 1884)

1868 Carl Reinhold August Wunderlich (1815-1877), German physician, publishes his major work on the relation of animal heat or fever to disease. He is the first to recognize that fever is not itself a disease, but rather a symptom. He insists on accurate records of a patient's fever, and his writing forms the basis of modern clinical thermometry. Wunderlich takes advantage of the breakthroughs by physicists whose mathematical studies of heat lead him to make careful observations of temperature in relation to disease. His accurate record keeping of thermal changes in the body forms the basis for thermometry becoming a standard technique of clinical diagnosis.

1869 Mathias Eugene Oscar Liebreich (1838-1908), German physician, first recognizes the hypnotic effect of chloral hydrate. It eventually becomes used as an aid to sleep. A synthetic drug made by passing chlorine gas through alcohol, it is one of the most effective and least expensive sedative-hypnotic drugs. Liebreich assumes that it changes to chloroform once in

the body and thus acts as an anesthetic, but it does not and rather acts as depressant on the central nervous system.

1869 Friedrich von Esmarch (1823-1908), German physician, first introduces the first aid package for use on the battlefield, as well as the "Esmarch bandage," a package for use on the battlefield to control hemorrhage. As a military surgeon, he makes several contributions in the repair of gunshot wounds and helps establish the use of field hospitals. He is also the founder of military nursing in Germany.

1869 Eugenics, the study of human improvement by genetic means, is founded with the publication by English anthropologist Francis Galton (1822-1911) of his landmark work, *Hereditary Genius*. In this work dealing with the hereditary aspects of talent, he argues that superior mental abilities run in families and are therefore inherited. He argues strongly in favor of heredity over the environment. In this work, he is the first to stress the importance of applying statistical methods to biology. He is also the first to study identical twins, assuming that hereditary influences are identical and differences are attributable to their environment. Interestingly, Galton is the first cousin of Charles Darwin and is himself a child prodigy who could read at two and study Latin at four.

1870 Christian Albert Theodor Billroth (1829-1894), Austrian surgeon, conducts the first total laryngectomy or extirpation (complete removal) of the larynx on a human patient. After perfecting this operation on dogs and determining that the larynx is essential to speech but not to swallowing, he is able to remove the tumor-filled larynx of a patient and replace it with an artificial one. He also is considered the founder of modern abdominal surgery, having determined that gastric juices will not destroy the sutures nor inhibit healing.

(See also 1881)

Theodor Billroth.

1871 Carl Friedrich Otto Westphal (1833-1890), German neurologist, first describes and introduces the term "agoraphobia" as the morbid fear of open spaces. In Greek, the word "agora" was the name for an open space always found in the cities of Greece that served as a meeting ground for various activities of its citizens. No matter what it was used for, the term always meant a large, open, and public space. Westphal also introduces the term "paranoia."

1871 First American city to use a filter on its public water supply is Poughkeepsie, New York. The evidence mounts that many diseases are spread by contaminated drinking water.

1871 First physiological laboratory in the United States is established at Harvard University.

1872 William Withey Gull (1816-1890), English physician, gives the first description of arteriosclerotic atrophy of the kidney. He shows that although the kidney is affected in this case, the condition is not exclusive to the kidney since it results from the abnormal thickening and hardening of the artery walls.

1872 First electric-powered dental drill is introduced.

(See also 1957)

1873 William Withey Gull (1816-1890), English physician, first describes myxedema, a condition in which an adult suffers from a lack of the thyroid's hormone. In his definitive study, he names this state "cretinoid" because the sufferer's features are similar to those seen in cretins. He also investigates hypochondria and is one of the earliest to describe anorexia nervosa.

1873 Foundation for physiological psychology is laid by English psychologist and logician Alexander Bain (1818-1903) with the publication of his *Mind and Body: The Theory of Their Relation*. He is the first to apply to psychology the results of psychological experiments, and he pioneers physiological psychology by adopting a rigorously scientific approach to problems that others treated philosophically. He also stresses the need to better understand the brain and nervous system.

1875 First attempt in the United States at an institutional treatment of tuberculosis is a sanitarium opened in Asheville, North Carolina by a physician named Gleitsmann.

1875 A. W. Gerrard, English pharmacist, and Hardy of France first introduce the drug pilocarpine. This alkaloid obtained from the South American jaborandi shrub promotes the flow of saliva and perspiration. It comes to be used in glaucoma and conditions involving the heart.

1875 Silas Weir Mitchell (1829-1914), American neurologist, first introduces "rest cure" for the treatment of nervous diseases. Along with a prolonged stay in bed, it also calls for plentiful food and massage. During the Civil War, Mitchell works with amputees and becomes interested in the psychology when confronted with amputees who feel pain in a missing limb. He advocates psychotherapy in conjunction with his suggestion of bed rest.

1875 William Pepper (1843-1898), American physician, first describes the abnormal changes in the bone marrow that occur in pernicious anemia. In this severe, slow-developing disease caused by a vitamin B-12 deficiency, red cell production in the bone marrow is abnormal. The important red cells function abnormally and have a reduced life span. Symptoms and signs include a waxy pallor, fatigue, weakness, and progressive neurological problems.

1876 Adolph Wilhelm Hermann Kolbe (1818-1884), German chemist, accomplishes one of the first syntheses of an organic compound from inorganic substances when he converts carbon disulfide, through various steps, into acetic acid. His later work on the electrolysis of the salts of fatty acids enables him to isolate salicylic acid, which becomes a building block of aspirin. In this, he is the first to apply electrolysis to organic compounds. Electrolysis is the production of chemical changes by the passage of an electric current through an electrolyte.

1876 Louis Albert Sayre (1820-1900), American orthopedic surgeon, first introduces the use of a plaster jacket to be worn by patients with spinal deformities (Pott's disease). Also known as tuberculosis of the spine, this disease is characterized by the softening and eventual collapse of the vertebrae, often resulting in a hunchback deformity. Although Sayre's treatment does not cure the original infection, it does provide orthopedic care for the spinal column.

1876 Anthrax, a baffling disease deadly to animals and people since Biblical times, is first understood by German bacteriologist, Robert Koch (1843-1910). By comparing the blood of animals dead from anthrax with that of healthy animals, he finds distinguishing rod-shaped bodies present only in the infected blood. He

then discovers how to cultivate the anthrax bacteria outside the animal's body and consequently is able to study it at length and follow its entire life cycle. He finds that it forms spores that are resistant to outside influences, even heat or cold, and can remain suspended for a long period. Once the mystery of anthrax is solved by Koch, it becomes only a matter of time until the great French chemist, Louis Pasteur (1822-1895), discovers a vaccine to defeat it.

1876 First practical device to measure blood pressure is invented by Czech physician Samuel Siegfried Carl von Basch (1837-1905). It is also the first sphygmomanometer to determine blood pressure without cutting into an artery. The instrument consists of a small capsule with a rubber diaphragm on the bottom. The cavity of the capsule is connected to a manometer by a rubber tube. To measure blood pressure, the user presses the diaphragm on the artery until the pulse stops, and the reading on the manometer at this point indicates arterial pressure. The instrument has a very wide margin of error, however, and it is replaced eventually by an much improved and more accurate device.

(See also 1896)

1877 Patrick Manson (1844-1922), Scottish physician, first discovers filaria—a kind of parasitic worm—as the cause of the tropical disease elephantiasis. The following year he shows it is transmitted by a mosquito. With this, he is the first to discover that an insect can be host to a developing parasite that causes a human disease. He also establishes a mosquito-malaria hypothesis that facilitates the eventual discovery of malarial transmission by mosquitos. A highly experienced parasitologist, he founds the field of tropical medicine.

1877 Louis Pasteur (1822-1895), French chemist, first distinguishes between aerobic and anaerobic bacteria and introduces the concepts of aerobism and anaerobism to science. An aerobe is an organism (like certain yeasts) able to live and reproduce only in the presence of oxygen. Organisms that can live without oxygen are called anaerobic. This discovery comes from his study of the nature of fermentation, specifically of beer and lactic acid. He finds that lactic acid bacilli and butyric acid bacilli not only can live without oxygen but flourish in an atmosphere of carbon dioxide.

1878 Robert Koch (1843-1910), German bacteriologist, first publishes his landmark findings on the etiology or cause of traumatic infectious disease. It is these findings that establish the rules for properly identifying the causative agent of a disease. Koch states that the microorganism must be located in a diseased animal, and that after it is cultured or grown, it must then be capable of causing disease in a healthy animal. Finally, the newly-infected animal must yield the same bacteria as those found in the original animal. Using his rules and techniques, he goes on to isolate the specific bacteria of a number of diseases, the most famous being tuberculosis.

1878 First experimental psychology laboratory is established at the University of Leipzig by German psychologist Wilhelm Max Wundt (1832-1920). Throughout his work on human behavior, he sought ways to measure it, especially the way humans absorb sense impressions. As a pioneer in his field and the founder of experimental psychology, he taught the first course in scientific psychology in 1862 and in 1881 the first effective and regularly published journal of psychology, *Philosophische Studien*.

1879 Max Nitze (1848-1906), German urologist, invents the cystoscope, an instrument that allows visual examination of the bladder through the penis. This electrically-lighted instrument vastly improves bladder surgery and also proves useful in making urological diagnoses. He later improves his original version by introducing a system of prisms and adapts it for the use of excising bladder tumors and other operations. The modern cystoscope uses fiber optic bun-

dles that transmit light, and is a highly flexible and versatile device that is amazingly unintrusive.

(See also 1889)

1879 First hearing aid to be patented in the United States is invented by R. S. Rhodes. This fan-like device is held between the teeth and uses the principle of bone conduction. It is called the "audiophone" and is useful for certain kinds of deafness.

1879 John Shaw Billings (1838-1913), American surgeon, and Robert Fletcher (1823-1912) of England issue the first volume of *Index Medicus*. This massive bibliography of the world's medical literature is issued monthly and is initially arranged by author and subject. It continues today and is regarded as one of the primary medical bibliographies in the United States. Billings also develops the library of the surgeon-General's Office into the National Library of Medicine, the world's largest medical reference center. The example set by Billings gave great impetus to the creation of medical libraries throughout the world.

1880 Albert von Mosetig-Moorhof (1838-1907), Austrian surgeon, first introduces iodine use on surgical dressings. Iodine at high concentrations is poisonous and can cause serious damage to skin and tissue. In a dilute alcoholic solution, such as the common tincture of iodine, it has a limited use as a topical antiseptic.

1880 Karl Joseph Eberth (1835-1926), German pathologist, first isolates the typhoid bacillus. Isolating and being able to identify *Salmonella typhi*, the cause of this disease, is necessary for its proper treatment. Typhoid is a generalized and not merely intestinal infection that enters the bloodstream via food or water and reaches almost any organ of the body, causing all the symptoms of a bloodstream infection. It is not curable until the development of antibiotics.

1880 Louis Pasteur (1822-1895), French chem-

John Shaw Billings.

Clara Barton.

ist, first isolates and describes both streptococcus and staphylococcus (both in puerperal septicemia). When Pasteur turns his attention to the link between microorganisms and diseases, he begins culturing and studying these two disease-causing bacteria. He discovers that unlike bacilli and spirilla, these bacteria are round in shape. They become known as "cocci."

1881 The modern era of surgery begins as Austrian surgeon Christian Albert Theodor Billroth (1829-1894) first performs an excision of a cancerous pylorus. He successfully removes it and links the stomach directly to the duodenum at the upper end of the small intestine. Billroth achieves an operation that had been heretofore regarded as unthinkable, and intestinal surgery will soon become commonplace.

1881 Anton Woelfler (1850-1917) of Bohemia first introduces gastroenterostomy (a surgical linking between the stomach and the bowel usually necessitated by cancer). As antiseptic methods and anesthesia improve, surgeons are able to perform deep-surgery resections to remove cancerous tumors.

1881 Clara Barton (1821-1912) of the United States first organizes the American Red Cross in Washington, D. C.

1881 Steam sterilization is first introduced by the German physician Merke, who discovers the killing effect of steam on pathogenic microorganisms. His discovery is immediately taken up by the German bacteriologist Robert Koch (1843-1910) and soon becomes a standard method of assuring the cleanliness of medical instruments and proper laboratory techniques.

1882 Carlo Forlanini (1847-1918), Italian physician, first introduces the use of artificial pneumothorax (collapsing the lung) as a means of treating pulmonary tuberculosis. His new treatment of immobilizing one lung and allowing it to rest is opposed by most of the medical community, but it is eventually proven correct.

1882 First chair of neurology is established at La Saltpetiere Hospital in Paris for Jean-Marie Charcot (1825-1893), French physician. One of the founders of modern neurology, he also opens what becomes the greatest neurological clinic of its time in Europe. He is a pioneer in the study of cerebral localization. One of his students is Austrian psychiatrist Sigmund Freud

(1856-1939). It is Charcot's use of hypnosis in trying to discover an organic basis for hysteria that stimulates Freud's interest in its psychological origins.

1882 The tiny tubercle that causes tuberculosis, *Mycobacterium tuberculosis*, is first isolated by German bacteriologist, Robert Koch (1843-1910). Until this discovery, there was no certainty as to how or whether it actually spread or whether it was hereditary. Koch was able to identify the difficult-to-find bacillus by using methylene blue dye (it would not respond when stained by any standard dyes). Once identified, he was able to prove that it was indeed caused by a germ that could be carried in the air and passed from one person to another. Although Koch does not discover a successful vaccine for TB, he does develop a simple skin test to determine if a person is infected.

(See also 1924)

1882 George Miller Sternberg (1838-1915), Surgeon General of the United States Army from 1893 to 1902, makes the first microphotograph of the tubercle bacillus. As a pioneer in the study of bacteriology, he also isolates and identifies (simultaneously with Pasteur) the organism responsible for pneumonia.

1883 *Journal of the American Medical Association* is first published.

1883 Paul Gerson Unna (1850-1929), German physician, first introduces the drug ichthyol to treat skin disorders. This substance was originally prepared from a bituminous shale found in the Tyrol area, which contained the semi-fossilized remains of fish. The name comes from the Greek word for fish-oil.

1884 Sigmund Freud (1856-1939), Austrian psychiatrist, first suggests the use of cocaine as a surface anesthetic to his colleague, Bohemian ophthalmologist Carl Koller (1857-1944). Freud had studied cocaine as a treatment for

Sigmund Freud.

morphine addiction and became aware of its numbing effect on mucous membranes.

(See September 16, 1884)

SEPTEMBER 16, 1884 Carl Koller (1857-1944), Bohemian-American ophthalmologist, first uses cocaine as a local anesthetic. Following many experiments on animals, he uses it to immobilize a patient's eye during surgery. His findings are enthusiastically accepted by the medical community, and they inaugurate the era of local anesthesia.

1884 Jean Albert Pitres, French physician, gives the first classic account of agraphia, which is the pathologic loss of the ability to write.

1884 Karl Sigmund Franz Crede (1819-1892), German gynecologist, first introduces the procedure of putting a dilute solution of silver nitrate into the eyes of the newborn as a preventive measure against gonorrheal ophthalmia. He demonstrates the effectiveness of this preventive procedure by reducing the incidence of ophthalmia, which ranged from 9.2 to 13.6%, to a rate of only one or two cases a year among 1,160 infants. With the widespread introduction of silver nitrate into the eyes of the newborn immediately upon delivery, blindness among newborns from this disease is virtually eliminated.

1884 Etienne Stéphane Tarnier (1828-1897), French obstetrician, invents an incubator for the care of prematurely born infants. It is warmed by kerosene and is used at Paris.

(See also 1888)

1885 James Leonard Corning (1855-1923), American surgeon, is the first to use cocaine as a spinal anesthetic.

(See also 1884)

1885 Victor Babes (1854-1926), Austrian physician, collaborates with French bacteriologist Andre Victor Cornil (1837-1908) to publish *Les Bactéries*, the first complete treatise on bacteriology. Babes eventually has a group of bacteria named after him, *Babesia*.

1885 First use of a successful vaccine for rabies is made by French chemist, Louis Pasteur (1822-1895). Although not a common disease in the 19th century, rabies was a fatal, viral disease with no cure whose victims died a horrible death after much suffering. It was known that saliva from the bite of an infected animal could transmit the disease to humans, and Pasteur attempted to find what he believed was a causative germ. Since he was unable to find it, he believed that it was simply too small for his microscope to see. This observation was a foreshadowing of the study of viruses. Although unable to find it, he still was able to make a weakened germ by passing it through different species of animals until its virulence had lessened. Although hesitant to try his vaccine on humans, Pasteur had his mind changed by a badly mauled boy named Joseph Meister who was bitten about the face and arms. Pasteur gave his vaccine to a doctor to administer to the boy (Pasteur was not a physician and had no license to practice). The boy received 12 inoculations over several days and eventually recovered. In 1940 when the Nazis occupied Paris,

Meister was a gatekeeper at the Pasteur Institute, and when ordered by the Germans to open Pasteur's crypt, he killed himself rather than dishonor the man who saved his life.

1885 Edward Livingston Trudeau (1848-1915), American physician, founds the Adirondack Cottage Sanitarium at Saranac Lake, New York. As a place to treat incipient consumption in working men and women, it is the first of its kind in the United States.

(See also 1894)

1886 Pierre Marie (1853-1940), French physician, first gives the classic description of acromegaly, which he also names. This condition of an overactive pituitary gland results in the progressive enlargement of the patient's face, hands, and feet. Although similar to gigantism, it does not necessarily result in excessive tallness. It does, however, increasingly exaggerate facial features (like a protruding jaw or eyebrows) and is usually accompanied by osteoporosis.

1886 Marcel von Nencki (1847-1901) of Poland first introduces salol (phenyl salicylate), which is used as an antiseptic.

1886 Theodor Escherich (1857-1911), German pediatrician, publishes his treatise on the intestinal bacteria of infants, which contains the first account of *Bacillus coli* infections. His work introduces bacteriology to pediatrics. He later has one of the best known coliform bacteria, *Escherichia coli*, named after him. Although usually a harmless resident in the intestines of humans and other animals, this bacteria can cause severe inflammation of abdominal organs and membranes, and can cause shock and even death.

1886 The drug acetanilide is first introduced by Arnold Cahn and P. P. Hepp to relieve pain and fever. This white crystalline substance, obtained from the action of acetic acid upon aniline, is no longer used.

1886 Reginald Heber Fitz (1843-1913), Ameri-

Edward Trudeau.

can surgeon, first describes the pathology and clinical features of appendicitis. He details how a physician can diagnose the condition and order when an operation is required. After correlating his bedside observations with his extensive postmortem findings, he shows that the old "inflammation of the bowel" is usually peritonitis that follows a ruptured appendix. This condition had been described as early as the 16th century by the French physician Jean Fernel (1497-1558). However, it is Fitz who names the disease and who describes the normal and abnormal positions of the appendix, as well as the clinical symptoms of appendicitis, its complications, and final surgical treatment.

1886 Ernst Gustav Benjamin von Bergmann (1836-1907), Latvian surgeon, first introduces steam sterilization of instruments and dressings for surgery and establishes the modern standardized aseptic ritual.

1886 Franz von Soxhlet first suggests that milk given to infants be sterilized.

1887 Anton Weichselbaum (1845-1920), Austrian pathologist, first demonstrates that meningococcus in the spinal fluid is the cause of epidemic cerebrospinal meningitis. Meningococcus is the common name for the spherical-shaped bacterial species that enters through the nose and mouth of humans, the only natural host in which it causes disease. Meningitis results when it enters the bloodstream. (See also 1811)

1887 Augustus Desire Waller (1856-1922), French physiologist, first records the electrical activity of the human heart and lays the foundation for future electrocardiography. While using electrical methods to study emotional states, he discovers that the heart's electric currents or regular beat can be recorded by connecting electrodes placed on the surface of the body with a capillary electrometer. This is essentially the first electrocardiogram.

1887 First antitubercular dispensaries are opened in Edinburgh.

1887 Oskar Minkowski (1858-1931), Russian-German pathologist, first notes the relation between pituitary enlargement and acromegaly. Although little is known about this gland at this time, Minkowski's discovery is correct and lays the groundwork for later progress. One of the substances secreted by this gland located on the lower surface of the vertebrate brain is pituitary growth hormone, which controls the rate at which the body grows.

1887 Hermann Michael Biggs (1859-1923), American physician, writes the first published educational material on tuberculosis for the layperson. Informing the public about the nature, cause, and treatment of this common respiratory disease serves to dispel much disinformation.

1887 Oskar Medin (1847-1927), Swedish pediatrician, is the first to recognize the epidemic nature of poliomyelitis. (See also 1860)

1888 Emanuel Libman (1872-1946), American physician, presents the first proven case of an aneurysm being caused by impact. Throughout his medical career, Libman exhibits almost phenomenal powers of clinical observation. This enables him to demonstrate that an aneurysm, or bulging and thinning in the wall of a blood vessel, can be caused by a physical injury. An aneurysm is very dangerous, since once it develops it tends to grow, increasing the possibility that the vessel wall will rupture. Treatment involves surgical removal of the defect.

1888 Paul Oscar Blocq (1860-1896), French physician, first describes a condition in which a physically sound subject is unable to walk or stand because of some mental trauma or conflict suffered. It is known variously as Blocq's disease, abasia, astasia, and hysterical ataxia.

1888 First incubator for premature infants in the United States is built by William Champion Deming. Called a "hatching cradle," it is composed of two sections, one which holds the infant and the other, below, which holds 15 gal (57 l) of hot water. Similar hot-water systems are already used in Europe.

1888 Artificial drinking straws are first introduced. Public awareness of hygiene increases.

1889 Hans Buchner (1850-1902), German bacteriologist, first establishes the bactericidal effect of blood serum. He calls this discovery of albuminoids, or protective substances in the blood, "alexins." His discovery of these bloodborne natural bactericides marks the beginning of immunology theory (the study of the body's reaction to harmful agents). He is also a pioneer in gamma globulin research and the brother of Nobel Prize winning chemist, Eduard Buchner (1860-1917).

1889 The Johns Hopkins Hospital in Baltimore, Maryland first opens.

1889 Joaquin Albarran y Dominguez (1860-1912), Cuban-French urologist, invents a cystoscope that allows him to catheterize the male ureter. It is because of his work systematizing the practice of cystoscopy and improving the cystoscope—an instrument that allows visual examination of the bladder through the penis—that it becomes an accepted procedure and device.

1890 First surgical school in the United States is founded by American surgeon William Stewart Halsted (1852-1922) at Johns Hopkins University in Baltimore, Maryland.

1890 William Stewart Halsted (1852-1922), American surgeon, first introduces the use of sterile rubber gloves into the operating room. As an early convert to antiseptic ideas, he introduces antiseptic procedures into American operating rooms. His use of very thin rubber gloves that allow the surgeon to retain his delicate touch is an major innovation toward the goal of totally sterile operating conditions. During his self-experimentation with cocaine as block anesthesia, he becomes addicted and takes two years to cure himself. He then continues his career and establishes himself as a pioneer of scientific surgery, developing original operations for a number of conditions and diseases.

1890 First diphtheria antitoxin is developed by German bacteriologist, Emil Adolf von Behring (1854-1917), who uses his recent discovery of antitoxins to develop a diphtheria-fighting vaccine. Behring demonstrates that the serum of animals immunized against attenuated (weakened) diphtheria toxins can be used as a preventive inoculation in other animals and even humans. This soon becomes the accepted method of treatment.

(See also 1891 and 1901)

1891 First child is treated with the diphtheria antitoxin. The development of a vaccine that will protect the young against this usually fatal disease is a landmark event.

(See also 1890)

1892 The existence of viruses is demonstrated for the first time by Russian botanist Dmitri Iosifovich Ivanovsky (1864-1920). While studying tobacco mosaic disease, an infection that damages tobacco crops, he mashes up infected leaves and forces them through a very fine filter designed to remove all bacteria. When he discovers that the liquid that comes out of the filter is still able to infect a healthy tobacco plant, he realizes that this might indicate that there exists a pathogenic (disease-causing) agent that is smaller than bacteria. However, Ivanovsky doubts the efficiency of his filters and does not claim any discovery of what eventually come to be called viruses.

(See also 1898)

1892 Wilhem Winternitz (1835-1917) of Bohemia founds the first hydropathic establishment or hydrotherapeutic clinic at Kaltenleutgeben. The application and use of water, especially cold baths, for therapeutic purposes was very much in vogue at this time.

1892 Theobald Smith (1859-1934), American pathologist, and F. L. Kilbourne first demonstrate the transmission of bovine piroplasmosis (Texas fever) by ticks. Smith is able to prove that the severe anemia and death of Texas cattle fever is caused by an intracellular parasite transmitted by a tick. He solves this disease of cattle by infecting healthy cattle with the parasite he obtains from the host tick.

1893 First successful gastrectomy (partial or total removal of the stomach) in the United States is performed by John Montgomery Baldy (1860-1934), American surgeon. There are records of two earlier such operations in the United States—one by Conner in 1883 and the other by Augustus C. Bernays in 1887. It is not known if either was successful. The real pioneer of stomach surgery is Billroth.

(See also 1881)

1894 First major epidemic of poliomyelitis in the United States breaks out. Following the rapid spread of this infectious, viral disease that paralyzes and kills its victims, it takes little to convince everyone that this is a disease of epidemic proportions. In 1916, 27,000 Americans develop polio, nearly 7,000 of whom die.

1894 First research laboratory in the United States founded for the study of tuberculosis is established by American physician Edward Livingston Trudeau (1848-1915). It is called Saranac Laboratory.

1894 First free hospice for terminal cancer patients is established in the United States by Rose Lathrop. The words "hospice," "hospital," "hospitality," "hostel," and "hotel" all have the same origin, stemming from the Roman "hopes," meaning "host" as well as "guest." In Latin, "hospitalis" is the adjective from "hospitium" meaning a guest house.

DECEMBER 28, 1895 Wilhelm Konrad Roentgen (1845-1923), German physicist, first communicates his discovery of x rays in his paper, "Uber eine neue Art von Strahlen." This discovery not only heralds the age of modern physics but revolutionizes medicine, and one of the earliest applications of x rays is in medical diagnosis and therapy. Physicians realized almost immediately that with this new tool their diagnostic powers had taken a quantum leap. They could now detect bone fractures and see foreign objects in the body, as well as dental cavities and diseased conditions like cancer. Therapeutically, they use x rays to stop the spread of malignant growths.

1895 First introduction of lumbar puncture for diagnosis and treatment into medical practice is made by German physician Heinrich Irenaeus Quincke (1842-1922). This direct withdrawal of cerebrospinal fluid through a hollow needle inserted between the third and fourth vertebrae is performed by Quincke on 10 patients who have hydrocephalus (an accumulation of fluid on the brain). Besides relieving intracranial pressure, the procedure comes to be used for

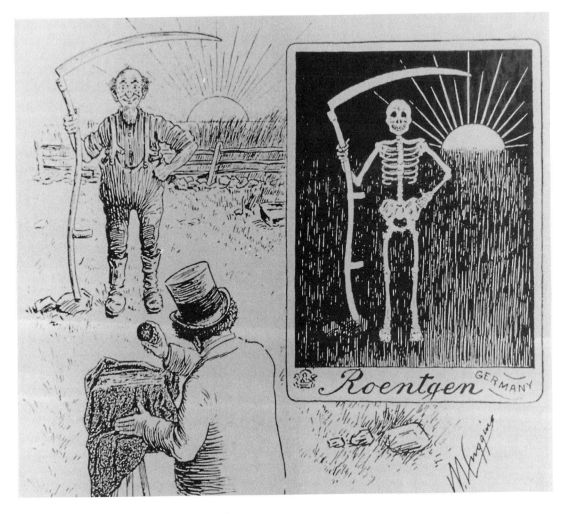

Cartoon depicting Wilhelm Roentgen's use of x rays.

obtaining a fluid sample for examination, to administer spinal anesthetics or antibiotics, and to inject air or a radioactive substance before taking an x ray.

1895 Bernard Sachs (1858-1944), American physician, publishes *A Treatise on the Nervous Diseases of Children*, which is the first book in the United States on this subject.

JANUARY 8, 1896 First radiograph (x ray) in England is made by A. A. Campbell-Swinton (1863-1930), English engineer.

FEBRUARY 3, 1896 First person in North America to use x rays as a successful adjunct to surgery is British-Canadian physicist, John Cox (1851-1923). A professor of physics at McGill University in Canada, Cox takes an x ray and locates the bullet in the leg of a young man who has been shot.

FEBRUARY 15, 1896 First radiograph (x ray) image to be published in the United States appears in the *New York Medical Record.*

FEBRUARY 28, 1896 *British Journal of Photogra-*

phy reports on the first museum artifact to be x-rayed: Egyptian mummies.

MAY 1896 First radiological journal, *Archives of Clinical Skiagraphy*, appears in London. It soon changes its name to the *Archives of the Roentgen Society*.

DECEMBER 1896 First to use x rays for a scientifically therapeutic purpose is Leopold Freund (1868-1944), Bohemian physician, who regularly x-rays a five-year old girl to treat her hirsutism (excessive body and facial hair). After twelve days, her hair begins to fall out. This case marks the first use of radiotherapy.

1896 Ludwig Rehn (1849-1930), German surgeon, is credited with being the first to suture a wound in a human heart successfully. When Rehn finally wrote a paper in 1907 detailing his surgical achievement, his former patient was still alive. There is also a claim that a surgeon named Dalton repaired a stab wound of the heart in 1891.

1896 Christiaan Eijkman (1858-1930), Dutch physician, first produces a deficiency disease experimentally (beriberi in birds). Although he begins his research assuming that a disease in chickens similar to beriberi is caused by a microorganism, he accidentally finds that it is caused by their diet of polished rice (meant for humans). When the hens are fed commercial feed, their symptoms disappear. Eijkman eventually concludes that this is a dietary-deficiency disease caused by the absence of some essential component. This becomes an important step in the discovery of trace substances called vitamins. (See also 1929)

1896 Hermann Strauss, German physician, first uses x rays to investigate a patient's gastrointestinal tract. Being able to actually see such internal problems as tumors or blockages in the small or large intestine enables physicians to make a diagnosis and prescribe treatment with much higher confidence.

1896 First practical use of agglutination is made by Max von Gruber (1835-1927), German bacteriologist, and Herbert Edward Durham (1866-1945), English bacteriologist. Agglutination is the clumping action by the blood serum of an immunized subject. This phenomenon of blood clumping together occurs when the antibodies it contains react to something introduced, like bacteria. Gruber is the first to introduce the term, and the diagnostic test he and his colleague develop is later called the "Gruber-Durham reaction."

1896 Bartolomeo Gosio of Italy first identifies and isolates mycophenolic acid—an antibiotic and a toxic product that comes to be studied in the 1970s for its antitumor properties.

1896 The mercury sphygmomanometer, which is the precursor of the modern blood-pressure instrument, is invented by Scipione Riva-Rocci (1863-1903) of Italy. His device uses an arm band, which is inflated until the blood flow through the arteries is no longer detected. Air is then released from the band, and when the pulse reappears, the pressure is measured on a mercury manometer. This device proves accurate but limited, since it only measures systolic pressure (pressure within the artery when the heart is contracting). The later use of a stethoscope to listen to both the sound of maximum (systolic) and minimum (diastolic) pressure is suggested by Russian physician, Nikolai Korotkoff.

JULY 1897 First whole-body radiograph (x ray) of a living person taken in a single exposure is made by American neurologist William James Morton (1845-1920).

1897 Emil Hermann Fischer (1852-1919), German chemist, first synthesizes caffeine, theobromine, xanthin, guanine, and adenine. These are members of a group of substances or compounds called purines, and almost all knowledge of this group is attributable to Fischer. Purines prove to be an important part of nucleic acids, which later prove to be the key molecules of living tissues.

1897 Raymond Jacques Adrien Sabouraud (1864-1938), French dermatologist, first cultivates the acne bacillus. He disproves the common notion that perspiration is responsible for the growth of acne (an inflammatory disease of the sebaceous or oil glands of the skin). His research marks the beginning of being able to treat this condition, which usually occurs in teenagers.

1897 Carl Zeiss (1816-1888), German instrument manufacturer, first is credited with developing the stereomicroscope for use in microsurgery. In 1933, he also introduces the standard L-stand microscope construction.

1898 Theobald Smith (1859-1934), American pathologist, makes the first clear distinction between human and bovine tubercle bacilli. Smith corrects the work of German bacteriologist, Robert Koch (1843-1910), who reported there was little difference in the tubercule infecting humans and animals. Smith demonstrates that there are basic differences in the biology and morphology (the form and structure of an organism without regard to function) of the two types.

1898 Edward Parker Davis and H. Varnier are the first to use x rays for obstetrical diagnoses. Although this new diagnostic tool often provides much-needed information about the situation in utero, it is learned much later that overuse can result in severe and permanent damage to the fetus.

1898 Heroin is first introduced by Hermann Dreser. While searching for a substitute for morphine, which sometimes killed patients, he heats morphine with acetyl chloride and discovers diacetylmorphine, a morphine derivative he considers an ideal substitute for relieving pain. It is introduced into medical practice under the trade name Heroin and is advertised, among other ways, as "the sedative for coughs." It becomes an extremely popular drug very quickly, but within four years it is discovered that while its pain-relieving power is four to eight times that of morphine, it is one of the world's most addictive substances. Although the United States and most other nations forbid the manufacture and importation of heroin, it makes up a large portion of the total illicit traffic in narcotics.

1898 First tuberculosis clinic in the United States opens in New York.

1898 First state-run sanitarium for tuberculosis in the United States opens in Massachusetts.

1898 Viruses are first discovered by Dutch botanist Martinus Willem Beijerinck (1851-1931). He performs the same filtering experiment on tobacco mosaic disease as did Ivanovsky in 1892, but he believes in his experiments and publishes his bold conclusions—that the disease is caused by an agent that is not bacterial but is nonetheless alive. He calls this new agent a "virus" which is Latin for "poison." His theory eventually proves Pasteur correct in his speculation that there exist disease-causing agents that are simply too tiny to see with a microscope.

1898 Ben-Gay ointment for sore muscles is first discovered by French pharmacist, Jules Bengue, and becomes a commercial success.

JUNE 1899 First proven cure of a cancer patient by x-ray treatment is made in Sweden by Tage Anton Ultimus Sjögren (1859-1939), Swedish surgeon. The patient's squamous cell carcinoma (skin cancer) of the cheek is cured by x rays combined with a minor operation. This is an early form of radiotherapy or radiation therapy that is still used today to destroy certain types of cancer cells.

1899 Aspirin, or acetylsalicylic acid, is first introduced on the market by the German pharmaceutical firm Farbenfabriken Bayer. This mild analgesic relieves headache and muscle and joint aches. It also reduces fever and reduces swelling associated with arthritis. Its blood-thinning properties make it useful for many heart patients. It is available without a prescription and is probably the most widely used of drugs.

1899 George Henry Falkiner Nuttall (1862-1937), American biologist, first summarizes the role of insects, arachnids, myriapods, and especially ticks as transmitters of bacterial and parasitic diseases. Ticks are among the most significant parasites of wild and domestic animals and are carriers of several serious diseases, from relapsing fever to the better known Rocky Mountain spotted fever.

1899 First International Dental Congress is held in Paris.

1900 The different types of human blood are first discovered by Austrian-American physician Karl Landsteiner (1868-1943). He finds that there are different types of blood groups that differ in the capacity of their serum to agglutinate or clump together. By 1902, Landsteiner has clearly divided human blood into four main groups, which he names A, B, AB, and O. Once this is done, it becomes a simple task to show that in certain combinations transfusions would work, while in others, the red cells would clot and possibly have fatal results. With the ability to blood-type both patient and donor, blood transfusions become safe.

1900 Ernst Wertheim (1864-1920), German surgeon, first introduces the radical hysterectomy, a radical operation for cancer of the cervix. Although some hysterectomies had been performed decades earlier, this operation removes the complete uterus (including the cervix), and after Wertheim, becomes a recognized surgical procedure.

1900 First specialized course on orthodontics (dentistry that deals with teeth irregularities and their correction with mechanical aids) is offered by American orthodontist, Edward Hartley Angle (1855-1930). Considered to be the founder of the modern practice of orthodontics, he devises the first simple and logical classification system for malocclusions and pioneers the movement against the indiscriminate extraction of permanent teeth. He founds the first school and college of orthodontia, organizes the American Society of Orthodontia in 1901, and founds the first orthodontic journal in 1907.

1901 Joseph Everett Dutton (1874-1905), English physician, and his colleague John Lancelot Todd (1876-1949) discover the parasite *Trypanosoma Gambians* that is responsible for the African sleeping sickness disease. They are the first to recognize the existence of this worm-like parasite in human blood. It is later shown that the tsetse fly is the host for this parasite. Dutton dies young while studying African relapsing fever.

1901 Emil Adolf von Behring (1854-1917), German bacteriologist, becomes the first recipient of the Nobel Prize for Physiology or Medicine. He receives this new award for his work on serum therapy, especially its application against diphtheria.

1901 Yellow fever, the sometimes fatal infectious disease of tropical and subtropical areas, is first proven by American military surgeon Walter Reed (1851-1902) to be transmitted by a mosquito. Reed had carefully studied the disease, which had killed more American soldiers in Cuba during the Spanish-American War in 1898 than had Spanish guns, and believed that it was not transmitted by bodily contact or by clothing or bedding. Instead, he theorized that the germ of yellow fever was transmitted by a mosquito. To prove this, he runs a controlled experiment using volunteers who live under the same conditions, except that one group is pro-

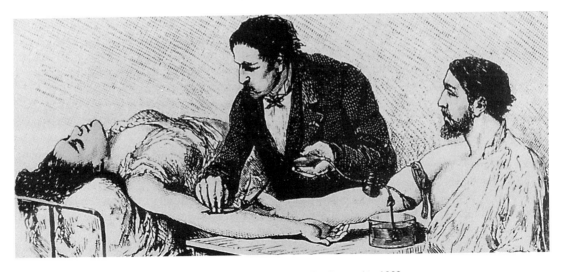

Illustration of the first direct transfusion of blood, performed by Dr. Roussel in 1882.

tected from mosquitos. The unprotected group contract the disease while the protected group remained healthy. Reed demonstrates that yellow fever (which destroys liver cells and causes jaundice, thus accounting for its name) is an entirely preventable disease. When a campaign to locate and destroy the breeding sites of the Aedes mosquito is conducted, the disease all but vanishes from Havana. Reed has little time to enjoy his success, however, as he dies of acute appendicitis in 1902.

1902 Harvey Cushing (1869-1939), American surgeon, performs the first nerve suture. As one of the founders of modern neurosurgery, Cushing originates scores of innovative surgical techniques that serve to advance the practice of operating on the brain.

1902 Alexis Carrel (1873-1944), French surgeon, first introduces his method of delicately sewing blood vessels together end-to-end. Carrel demonstrates how a perfect end-to-end linking of blood vessels can be achieved by using a new triple-threaded stitch that converts a round opening into an equatorial triangle, thus permitting, upon tightening, an excellent seal without any leakage or danger of clotting.

Carrel is a highly innovative practitioner who revolutionizes surgery on the vascular system.

1903 Svante August Arrhenius (1859-1927), Swedish chemist, first uses the term "immunochemistry." His work on the quantitative laws of biological chemistry lead him to study the chemical relationship of toxins and antitoxins, shedding new light on an emerging field.

1903 Willem Einthoven (1860-1927), Dutch physiologist, invents the string galvanometer from which the electrocardiograph is derived. Theorizing that the healthy heart works with a regular rhythm that is reflected electrically, he guesses that perhaps a departure from that normal electric rhythm might indicate when something is wrong. He then builds the first string galvanometer as a diagnostic tool. It consists of a delicate conducting thread stretched across a magnetic field. A current flowing through the thread causes it to move at right angles to the magnetic lines of force, with the distance it moves proportional to the strength of the current. His device is sufficiently sensitive and could record graphically the varying electrical potential of the heart's contracting muscles. He coins the term electrocardiogram to describe this process. His device remains an important tool in the diagnosis of heart disease.

1904 Paul Ehrlich (1854-1915), German bacteriologist, discovers a microbial dye, called trypan red, that helps destroy the trypanosomes that cause such diseases as sleeping sickness. This is the first such active agent against trypanosomes (parasitic protozoa). Ehrlich reasons correctly that if the dyes he developed achieved their staining effect by somehow binding to certain molecules of the cells, then this same mechanism might be employed to destroy certain of those molecules without damaging normal cells. This notion becomes the basis for the development of modern chemotherapy.

1904 Ernest François Auguste Fourneau (1872-1949), French chemist and pharmacologist, first synthesizes amylocaine. A cocaine derivative, it is used eventually as a local anesthetic known as stovaine.

1904 Ernst Ferdinand Sauerbruch (1875-1951), Prussian surgeon, advances the possibilities for chest surgery with his invention of a pneumatic chamber at reduced atmospheric (negative) pressure, which prevents the lung from collapsing. He also invents a positive pressure cabinet in which the patient breathes compressed air while the pleural cavity (the lining membrane of the chest wall) is opened at ordinary atmospheric pressure. Using this allows surgery of the chest and esophagus to advance.

1904 First radical operation for prostatic cancer is performed by the American urologist Hugh Hampton Young (1870-1945). In aggressively removing as much of the prostate gland as possible, Young shows some success in prolonging the lives of many of his older, male patients.

1904 Novocaine is first developed as a safe substitute in local anesthesia. Also known as procaine hydrochloride or simply procaine, it is a synthetic organic chemical generally used in a

1-10 % saline solution and administered by injection. Unlike cocaine, it is not toxic, addictive, or irritating. It is used today primarily as a dental anesthesia.

1905 Jules Jean Baptiste Vincent Bordet (1870-1961), Belgian bacteriologist, and his colleague, Octave Gengou, discover the bacillus of whooping cough. Because of this, the bacterium responsible for whooping cough becomes known as *Bordetella pertussis*. Bordet goes on to discover a method of immunization against this dreaded childhood disease.

1905 Vicks VapoRub is first invented by American pharmacist, Lunsford Richardson. His distinctive-smelling external cold remedy becomes a marketing success when he changes the name from its original Richardson's Croup and Pneumonia Cure Salve. He renames his product after his brother-in-law, Dr. Joshua Vick.

1905 First use is made of organized group therapy as a means of treating patients with psychological problems. This method is not widely used until after World War II, when the much larger number of patients almost requires group therapy.

1905 First intelligence tests are developed by French psychologist Alfred Binet (1857-1911). Using pictures, inkblots, and other visual devices, he paves the way for projective testing, by which conclusions are drawn from a subject's response to visual material. He also develops his highly influential scales for measuring schoolchildrens' intelligence primarily by noting the differences between normal and subnormal individuals.

1906 First diagnostic blood test for syphilis is introduced by German bacteriologist August von Wasserman (1866-1925). He develops a test for the antibody produced by persons infected with the protozoan *Spirochaeta pallida* which is the causative agent of syphilis. With this discovery of a universal blood-serum test for syphilis, the basic tenets of immunology are now extended to diagnosis. The "Wasserman test" soon becomes a standard test.

1906 Frederick Gowland Hopkins (1861-1947), English biochemist, first argues that certain "accessory substances" in food are necessary to sustain life. This theory of trace substances becomes the starting point of further work on vitamin requirements, which leads to the discovery of what come to be called vitamins. Years earlier, Hopkins had isolated the amino acid tryptophan from protein and showed that it and certain other amino acids could not be manufactured by animals from other nutrients and had to obtained from their diet.

1906 Stanton Abeles Friedberg (1875-1920), American surgeon, performs the first operation for the removal of a foreign body from the respiratory tract with the aid of a bronchoscope and an esophagoscope. An esophagoscope is an instrument for inspecting the interior of the esophagus (which connects the pharynx to the stomach). A bronchoscope is a thin, tubular instrument to inspect the interior of the trachea and bronchi (of the lungs).

1907 First cutaneous or skin reaction test for the diagnosis of tuberculosis is introduced by Clemens Peter von Pirquet (1874-1929), Austrian physician. His diagnostic test for TB consists of scratching a small area of skin, introducing a drop of tuberculin, and watching for a reddened reaction. The main problem with this early test is that the physician has little control over the amount introduced. Pirquet also is the first to suggest the word "allergy" to denote hypersensitivity or a capacity for reaction.

1907 Charles Franklin Craig (1872-1950), American physician, and Percy Moreau Ashburn (1872-1940), American surgeon, work in the Phillipines and are the first to prove that dengue fever (also called "breakbone fever" and "dandy fever") is caused by a virus. This acute, infectious, mosquito-borne disease can be com-

pletely incapacitating but is usually never fatal. It gets its popular name from the odd gait its victims have, since they feel extreme pain and stiffness in their joints.

1908 First skin test to determine a patient's susceptibility to diphtheria is developed by Hungarian pediatrician, Bela Schick (1877-1967). He discovers that by injecting a small amount of diphtheria toxin into the skin, he can measure the level of immunity (by the amount of skin reaction obtained). Redness at the site of injection indicates a positive reaction (and the absence of an antibodies for diphtheria).

1908 Reuben Ottenberg, American physician, reports the first blood transfusion in which tests for compatibility are done. Transfusion could not become a useful or safe procedure until the blood group antigens and antibodies were discovered, permitting routine typing of donors and recipients. Undesirable transfusion reactions or even death can occur if a patient receives the incorrect blood type.

1908 Psychological school of behaviorism is founded by American psychologist John Broadus Watson (1878-1958). Presiding over a laboratory for psychological studies at Johns Hopkins University, he forcefully asserts that psychology is the science of human behavior which, like animal behavior, should be studied under exacting laboratory conditions. He eventually promotes conditioned responses as the ideal experimental tool and views conditioning as the basis of thinking and learning.

1909 Willis D. Gatch, American surgeon, invents a surgical bed with an adjustable steel frame for supporting the patient in a sitting or a semisitting posture for postoperative care.

1910 Edward Bright Vedder (1878-1952), American physician, first discovers the ability of emetin to kill the amoeba that cause amebic dysentery. Emetin is a crystalline alkaloid extracted in the form of a bitter, white powder from the ipecac root. Although it was used for nearly century as a remedy for dysentery, it is not until Vedder that its amebicidal properties are known.

1910   First model of blood coagulation is offered by American physiologist William Henry Howell (1860-1945). He studies the role of antithrombin (a clot-preventing substance) and thromboplastin (a protein that initiates the clotting of blood) and greatly contributes to an understanding of the factors involved when blood coagulates. This proves especially significant when transfusions become more commonplace during World War I.

1910   James Bryan Herrick (1861-1954), American physician, first describes sickle-cell anemia, a disease caused by a congenital anomaly in the blood that appears mostly in those of African descent. Specifically, this disease is the result of an abnormal hemoglobin that does not react normally to oxygen deficiency. The red cells become distorted and elongated, and assume a rigid, sickle-like shape. It is usually associated with chronic anemia and bone and kidney changes. The sickle-cell trait refers to the inheritance of an abnormal gene from only one parent (which does not cause any illness). The anemia results from the inheritance of the abnormal gene from both parents.

1910   Emanuel Libman (1872-1946), American physician, publishes the first and classic description of bacterial endocarditis or inflammation of the heart's lining. He is also the first to employ the phrase "subacute bacterial endocarditis." The inflammation may be associated with a noninfectious disease or may be caused by infection, and is usually located in the valves of the heart. The acute form mostly results from infection with a disease-causing organism (like Staph or gonorrhea), and the subacute is the more common, chronic form.

1910   Isador Henry Coriat (1875-1943), American psychiatrist and neurologist, publishes his *Abnormal Psychology*, the first systematic treat-

ment in English of this subject. As agreement over what constitutes "abnormal" behavior eventually breaks down among psychological professionals, this term goes out of use, although the particular phenomena are still studied.

1911   First pure, in vitro culture of the syphilis-causing *Treponema pallidum* organism is achieved by Japanese bacteriologist Hideyo Noguchi (1876-1928). In this same year he first introduces the luetin or skin test for syphilis by drawing a substance he calls luetin from the *Treponema pallidum*. Using this, his diagnostic skin test can reveal latent or congenital syphilis. Noguchi dies of yellow fever contracted while researching the disease in Ghana.

1911   Russell A. Hibbs (1869-1932), American surgeon, first introduces the spinal fusion operation. His technique of removing the cartilage and periosteum (connective tissue) of adjacent vertebral parts allows for the fusion of a diseased or fractured spine. Until this surgical technique, there was no reliable way to support a diseased or fractured spine.

1911   R. A. Lambert and F. M. Hanes perform the first successful in vitro cultivation of tumoral cells. They achieve this using sarcomas (malignant tumors of connective tissue) from mice and rats. This technique proves very useful in the advancement of oncological research.

1912   Elmer Verner McCollum (1879-1967) and Marguerite Davis, both American chemists, discover Vitamin A in butter and egg yolk. It is discovered at the same time by American biochemist Thomas Burr Osborne (1859-1929) and Lafayette Benedict Mendel (1872-1935), American physical chemist, but McCollum and Davis publish first. This fat-soluble vitamin is necessary for normal bone development and the health of certain epithelial tissues, especially the retina in the eye.

1912   George Charles de Hevesy (1885-1966), Hungarian-Danish-Swedish chemist, and Friedrich Adolf Paneth (1887-1958), German-British chemist, first use radioactive indicators experimentally in living systems. It is Hevesy who develops the idea of using radioactive isotopes as tracers, since compounds containing an isotope retains its chemical and physiological properties, but can be traced in mixtures or molecules throughout the body. Hevesy realizes the enormous potential this holds for studying the network of metabolic reactions within a living organism.

(See also 1923)

1912   Alexis Carrel (1873-1944), French-American surgeon, succeeds in producing the first true cell culture (chick embryo fibroblasts). Since he is able to keep tissue alive, his research allows for the development of cells to be followed.

1912   James Bryan Herrick (1861-1954), American physician, first describes a myocardial infarction, commonly known as a heart attack.

1912   Gestalt psychology is founded by German psychologist and philosopher Max Wertheimer (1880-1943). Convinced that the segmented approaches of academic psychology are inadequate to properly study human behavior, he first originates the concept of Gestalt psychology, which attempts to examine the total, structured forms of mental experience. Gestalt psychology is characterized by its emphasis on the dynamics and interrelatedness of the elements of the whole.

(See also 1923)

1913   Johannes Andreas Grib Fibiger (1867-1928), Danish pathologist, produces the first nonviral tumor in a laboratory experiment on rats. This shows that cancers can arise from an external (e.g., chemical) source.

1914   Robert Barany (1876-1936), Austrian physician, is awarded the Nobel Prize for Physiology or Medicine for first devising a number of tests that can diagnose vestibular (inner ear) problems like vertigo.

1914   Henry Hallett Dale (1875-1968), English biologist, first isolates acetylcholine—a substance that will prove crucial to the discovery of the chemical transmission of nerve impulses.

1916   Jay McLean (1890-1957), American physiologist, first isolates the anticoagulant called heparin. As a substance that inhibits the coagulation or clotting of blood, it proves useful for keeping blood for transfusions from clotting, for treating heart conditions that may cause fatal clots, and during some surgical operations. It is administered by injection.

1917   Walter Edward Dandy (1886-1946), American neurosurgeon, first introduces ventriculography with air. This diagnostic procedure is an x-ray process that images the brain's cerebral ventricles by injecting air into them, which displaces their cerebrospinal fluid.

1918   First blood and serum banks are established in Europe during World War I. Landsteiner's 1900 discovery of the human blood groups made blood transfusions safe and practical, and the massive numbers of casualties caused by this long war makes them a necessity. Many lives are saved because of these blood banks.

1919   Mercurochrome is first introduced by Hugh Hampton Young (1870-1945), American urologist, and colleagues E. C. White and E. O. Schwarz. This red organic mercury-containing dye is used as an antiseptic and germicide and becomes very popular for treating minor cuts.

1919   Kurt Huldschinsky, German physician, first demonstrates the curative effect of sunlight (ultraviolet rays) in the treatment of rickets. A vitamin D deficiency disease of childhood, rickets causes deformation of bones due to a defective deposit of calcium salts in their growing ends.

1919   Francis Gilman Blake (1887-1952) and James Dowling Trask (1890-1942), both

American physicians, first demonstrate that measles is caused by a virus.

1919 Ernest Sachs (1879-1958), American neurosurgeon, receives the professorship of neurosurgery at the Washington University School of Medicine (St. Louis, Missouri). This is the first such professorship awarded anywhere in the world.

1920 Louis Rehn (1849-1930), German surgeon, performs the first successful pericardiectomy. This heart operation on the thickened pericardium (the sac around the heart) has since saved many lives.

1920 Hearing aids that are electrically amplified bone conductor types are first introduced in England by S. G. Brown.

1921 Thomas Hunt Morgan (1866-1945), American geneticist, offers the first estimate for the size of genes. His studies on the fruit fly (*Drosophilia*) move genetic research to the forefront of science. His work establishes such major concepts as the linear array of genes on chromosomes, the exchange of parts between chromosomes, and the interaction of genes in determining traits, including sexual differences.

1921 Rorschach test for diagnosing mental disorders is first introduced by Swiss psychiatrist Hermann Rorschach (1884-1922). While a student at the University of Zurich, he initially attempts to understand the reactions of individuals to inkblots. He then continues his work on hundreds of mental patients and develops the Rorschach inkblot test, in which a person is asked to describe what he or she sees in ten inkblots. Responses are scored as to the location of what is seen in the blot. Interpretation is now considered overly subjective and the test is eventually attacked as unreliable for diagnosis and prognosis.

1921 George Tully Vaughan, American physician, performs the first successful ligation (tying) of the abdominal aorta. Taken from the

Latin word "ligare," meaning to bind or tie, ligation of the major vessels proves to be the most difficult to execute correctly.

1922 Insulin is administered for the first time to a patient with diabetes mellitus. This becomes possible after the Canadian physiologists Frederick Grant Banting (1891-1941) and Charles Best (1899-1978), develop a method in 1921 of extracting insulin from the human pancreas and then injecting it into the blood of diabetics to lower their blood sugar.

1923 Max Wertheimer (1880-1943), German psychologist and philosopher, publishes *Untersuchungen zur Lehre der Gestalt*, which first popularizes the concept of Gestalt psychology. This school of psychological thought attempts to examine the total, structured forms of mental experience.

(See also 1912)

1923 George Charles de Hevesy (1885-1966), Hungarian-Danish-Swedish chemist, first uses radioactive tracers to follow the path of a substance in an organism. He follows the course of radium D (the natural radioactive isotope of lead) through plant organisms. He continues his research during the next decade and, using a radioactive isotope of phosphorus, discovers much about the dynamic state of the body's physiological processes.

1923 The cause of scarlet fever and the means of its prevention are discovered by George Frederick Dick, American physician and bacteriologist, and his wife, Gladys R. Henry Dick. They find it is a streptococcic disease caused by the hemolytic streptococcus, and they prepare a toxin for immunization. They also develop a diagnostic skin test (the Dick test) to determine susceptibility to scarlet fever. Also called scarlatina, it is an infectious disease that marks the skin with red spots and whose complications are frequent and serious.

1923 American surgeon Evarts Ambrose Graham (1883-1957) and Warren Henry Cole, American physician, introduce the first successful cholecystography. They inject an opaque dye and are able to x-ray the gallbladder, seeing the outline of this organ and the presence of any gallstones. Later, the dye is given by mouth.

1924 Ko Kuei Chen, Chinese-American pharmacologist, and Carl Frederick Schmidt, American physician, first introduce ephedrine, an alkaloid obtained from the leaves of the common horsetail plant. Similar to epinephrine but less powerful, it has the advantage of being taken orally. It dilates the bronchi in the lungs and is used as a decongestant and for asthma. This marks the introduction to Western medicine of a drug called "ma hung" that had been used in China for 5000 years.

1924 Albert Léon Charles Calmette (1863-1933), French bacteriologist, and Camille Guerin (1872-1961), French biologist, invent an antituberculosis vaccine using attenuated live bacilli and successfully vaccinate children. For fifteen years, the two men cultured tubercle bacilli from cattle and found that over many generations they had developed a strain that was no longer virulent. They tested an even weaker strain (called Bacillus Calmette-Guerin or BCG) on humans and found it made them immune to TB. Although successful, the use of a live virus makes the United States and Great Britain wary, and the vaccine is used only for those at risk.

1925 Edward Bright Vedder (1878-1952), American physician, publishes *The Medical Aspects of Chemical Warfare*. It is the first American text to consider the modern medical problems of gas warfare. During World War I, both the Germans and the Allied forces introduced a succession of poison gases. Chief among these chemical weapons were chlorine, phosgene, and mustard gas. These and other gases caused numerous casualties.

1925 Florence Rena Sabin (1871-1953), Ameri-

can anatomist and biologist, becomes the first woman member of the National Academy of Sciences in the United States. During her long and exceptional career, she was also the first female admitted to Johns Hopkins Medical School (1896) and later became the first female professor there.

1925 American Army Colonel Calvin H. Goddard begins his pioneering work in forensic ballistics. Over several decades he develops the first accurate method for comparing a fired bullet with one experimentally fired from a suspected weapon. He also studies rifling characteristics and the effects of projectiles on the human body.

1925 First sympathectomy as a remedy for hypertension is conducted by Leonard George Rowntree, American physician, and Alfred Washington Adson (1887-1951), American surgeon. This extensive operation which severs the pathways of the sympathetic nervous system by removing a portion of nerve, offered only temporary results and was eventually replaced by drug therapy. This later chemical interruption of the nerve pathway is as effective as the old, surgical way and less invasive.

1926 Joseph Goldberger (1874-1929), American pathologist, and colleagues discover the cause and cure of the disease pellagra. This disease leads to a severe skin condition, diarrhea, erratic behavior, and even coma and death. He discovers that it is caused by a diet totally deficient in niacin, a member of the Vitamin B complex. Like scurvy, pellagra joins the list of diseases that are not due to contagion or microbes, and can be avoided simply by an adequate diet.

(See also 1735)

1926 Johannes Andreas Grib Fibiger (1867-1928), Danish bacteriologist, is awarded the Nobel Prize for Physiology or Medicine for his discovery of the Spiropter carcinoma. He is the first scientist to induce cancer in laboratory animals. By showing that the cancerous tumors he is able to induce in the stomachs of mice and rats undergo metastasis (spread from one tissue or organ to another), he adds significant evidence to the argument that some cancers are caused by tissue irritation. His work is of profound importance to cancer research and leads directly to the study of chemical carcinogens (cancer-causing agents).

1926 Louis Hopewell Bauer (1888-1964) of the United States Army School of Aviation publishes *Aviation Medicine*, the first American textbook concerned with the prevention and treatment of an aviator's medical problems.

1926 Peter Muhlens (1874-1943) of Germany and colleagues introduce the first synthetic antimalarial drug called Plasmoquin or Plasmochin. This search for a synthetic drug is spurred by the German government's worries during World War I that their quinine supplies from the Dutch East Indies would be cut off. Ample supplies of quinine are essential for the conduct of any military campaign in tropical countries.

1926 George Richards Minot (1885-1950) and William Parry Murphy, both American physicians, first introduce their raw liver diet in the treatment of pernicious anemia. This often fatal condition is characterized by a reduction of red blood cells and/or a lack of sufficient hemoglobin (the oxygen carrier). Its most obvious signs are pallor of the skin and fingernails. Minot discovers that ingestion of a half pound of raw liver a day dramatically reverses this disease, and he eventually succeeds in preparing effective liver extracts that can be taken orally. It is only in 1948 that medicine discovers that the necessary factor in liver is Vitamin B-12.

(See also 1948)

1926 Dutch biochemists Barend C. P. Jansen and Willem F. Donath first isolate vitamin B-1 (thiamine). They conduct a search to identify the unknown substance in certain foods that prevents the deficiency disease beriberi and eventually are able to isolate a crystalline material that cures polyneuritis (the inflammation of the arms and legs that occurs with beriberi). They call this substance "aneurine," but it eventually becomes known as thiamine or Vitamin B-1.

1926 Otto Heinrich Warburg (1883-1970), German biochemist, is the first to determine that cancer cells derive energy from lactic acid fermentation and can therefore be damaged by radiation. His studies of the respiratory mechanisms of cancerous tissues shows that their oxygen uptake is distinctly less than in normal cells and that these abnormal cells get their energy by a process known as glycolysis. Despite his discovery of the distinctive metabolic pattern of cancer cells and their susceptibility to radiation, neither Warburg nor others since have been able to pinpoint the cause of cancer.

1927 Julius Wagner-Jauregg (1857-1940), Austrian psychiatrist and neurologist, is awarded the Nobel Prize for Physiology or Medicine for his discovery of the first successful therapy for syphilitic paresis (partial paralysis or weakness) of the insane by infecting patients with malaria. After noting that persons suffering from certain nervous disorders show marked improvement after contracting an infectious fever, he deliberately induces such a condition in the insane, choosing malaria because it is controllable with quinine. This radical treatment of giving mental patients an infectious disease and a fever is the first example of shock therapy. Although Wagner-Jauregg uses malaria, his treatment later evolves into modern shock therapy which uses drugs or electric shock.

1927 Ruth Tunnicliff (1876-1946), American scientist, first introduces a serum to be used against measles. Also called rubeola, it is an infectious disease caused by a virus that primarily infects children. A red rash on the face, neck, trunk, and extremities accompanies a high fever, headache, and cough. Despite

Tunnicliff's efforts, an effective live-virus vaccine is not developed until after the 1954 isolation of the measles virus.

(See also 1954)

SEPTEMBER 1928 First of the wonder drugs, penicillin, is discovered by Scottish bacteriologist Alexander Fleming (1881-1955). While searching for a safe antibacterial substance, Fleming grows staphylococcus cultures in petri dishes, which he accidentally leaves uncovered overnight. When he returns the next day, he notices that a spore of mold had entered the culture and grown on it. His scientist's eye, however, also notices that in a ring around the mold are dead and dying staphylococcus microbes. He then isolates the mold and identifies it as one called *Penicillium notatum*, closely related to what grows on stale bread. He also finds that the mold produced some type of substance, as it grew, that destroyed bacteria. He calls this substance penicillin. Fleming meets with no success, however, when he tries to isolate and identify this substance, and since he also is unable to obtain much help in his search, his discovery of what is the first antibiotic goes unexploited until his colleague, Howard Florey, at Oxford's School of Pathology, takes up his cause and isolates it ten years later.

(See also 1938)

1928 George Nicholas Papanicolaou (1883-1962), Greek-American physician, invents the Pap test, a simple and painless smear procedure for the early detection of cervical and uterine cancer. After making a microscopic study of vaginal discharge cells in pigs, he conducts human studies and observes cell abnormalities in a woman with cervical cancer. This leads him to develop a method of detecting cancer cells through cytology or microscopic cell examination. By 1943 his method wins wide acceptance because it allows for detection of cancer in its presymptomatic stage, when the disease can best be treated. Cancer of the cervix in its earliest stages is almost 100% curable, and 80% of uterine cancer cases detected by a Pap test can be cured.

1928 Albert Szent-Györgi (1893-1986), Hungarian-American biochemist, first isolates and describes ascorbic acid (Vitamin C). In searching for the substance in some foods that makes them effective against scurvy, he extracts from certain fruits and vegetables, an agent he calls hexuronic acid. In 1931 he shows this to be identical to Vitamin C (ascorbic acid). Because the body stores very little of this necessary vitamin, a daily dietary source is required. Following Szent-Györgi's discovery, the vitamin is synthesized and made easily available.

1929 Christiaan Eijkman (1858-1930), Dutch physician, is awarded the Nobel Prize for Physiology or Medicine for his discovery that the disease beriberi is the result of a nutritional deficiency. He is also the first to experimentally establish a deficiency disease. Frederick Gowland Hopkins (1861-1947), English biochemist, is also awarded the Nobel Prize for Physiology or Medicine for his discovery of growth-stimulating vitamins.

1929 Werner Forssmann (1904-1979), German surgeon, invents the first practical system for cardiac catheterization. This insertion of a tube with radio-opaque dye into the heart makes for more accurate heart diagnoses without surgery. As a pioneer in heart research, Forssmann first tests this technique on himself.

1929 Philip Drinker (1893-1977), American industrial hygienist, invents the iron lung. This machine performs the function of the muscles that control breathing and is one of the first inventions that keeps people alive who are unable to breathe. Initially called the "Drinker tank respirator," this device saves the lives of those polio patients whose breathing muscles are paralyzed. The patient is enclosed from the neck down in an airtight metal box, the inside of which is reduced in air pressure. This simulates the negative pressure in the chest cavity that occurs during inhalation. This decreased air pressure forces air through the patient's nose and mouth and into the lungs.

1931 Francis C. Wood and Charles C. Woolferth, both Americans, first demonstrate the use of precordial leads (electrodes directly over the heart) for electrocardiograms. This is one of many refinements in instrumentation and technique that makes electrocardiography one of the most useful diagnostic tools in medicine. It becomes highly accurate, easy to interpret, and relatively inexpensive. Today, it is even portable.

1931 The first, albeit imperfect, electron microscopes begin to be developed. An early, working model is developed by German engineers, Ernst Ruska and Max Knoll. Knowing that conventional microscopes rely on visible light (and are thus restricted from detecting an image smaller than one wavelength of light), they design a microscope that uses a beam of electrons instead of a beam of light. Their major technical breakthrough occurs with the use of a magnetic coil to act as a lens and focus the beam. Although this early model can only magnify a few hundred power, it demonstrates that electrons can be used in microscopy, offering medicine a valuable and highly versatile research tool.

(See also 1939)

1931 Kenneth Merrill Lynch, American pathologist, reports what is believed to be the first fatal case of asbestosis in the United States. Asbestos has been used since Roman times, and by the 1930s, it could be found in everything from clothing to child's clay. Asbestos fibers are small and light and easily inhaled or ingested. Since the human body is not able to rid itself of them, they remain and build up in the lungs over time, puncturing the air sacs and reducing elasticity. This irreversible condition of decreasing ability to exchange oxygen is known as asbestosis. The first recorded case of this disease in asbestos workers is reported in the 1890s in England.

1931 Alka-Seltzer is first marketed in the United States as a remedy for headache and upset stomach. Introduced by Miles Laboratories, it is a large tablet composed of an antacid, aspirin, and a bubbling agent. It becomes very successful and is popularly believed to be a cure for hangovers.

1931 First controlled clinical studies on the use of marijuana (*Cannabis sativa*) are conducted on American soldiers in Panama. Marijuana is a green herb from the flower of the hemp plant and is a mild, natural hallucinogen usually smoked in cigarettes or pipes. The study concludes that the drug is not addictive, produces no persistent neurological or mental changes, and is used mainly by men who are incompetent soldiers. By 1937, however, the United States places restrictions on the natural drug, which are extended to its synthetic counterpart in 1968. By that time, most countries considered its possession and sale illegal. While the drug does have some minor medical uses, it is mostly used as a reality-distorting agent and as such, can produce psychological dependence. Researchers now say that the major argument against its use is that it serves as a "gateway" drug to other, more addictive and dangerous substances.

1933 First blood bank in the United States is founded by American bacteriologist Earl W. Flosdorff (1904-1958) and American microbiologist, Stuart Mudd. Before blood banks, the physician determined the patient's blood type and arranged for appropriate friends and relatives to donate blood. Flosdorf and Mudd show the practicality of storing fresh blood for future needs.

(See also 1937)

1933 The first total pneumonectomy (complete removal of the lung) for cancer is conducted by American surgeon Evarts Ambrose Graham (1883-1957) and American physician Jacob Jesse Singer (1885-1954). The successful execution of such a complicated and radical surgical procedure indicates how far surgery has progressed. Graham's success also shows that a patient can survive with only one lung.

1934 Philipp Ellinger (1888-1952), German biochemist, and Walter Koschara discover Vitamin B-2 (riboflavin) and establish its chemical formula. Known since the late 1800s as a yellowish substance found in milk, this vitamin is known to function as part of the enzyme systems concerned with the oxidization of carbohydrates and amino acids. Lack of B-2 leads to lesions and cracks of the mouth and tongue as well as eye disturbances. Dairy products and leafy vegetables are a good natural source. With this discovery of the chemical formula, the vitamin can be synthesized.

1935 First hospital for drug-addicted patients is founded in Lexington, Kentucky.

1935 John Heysham Gibbon, Jr., American surgeon, demonstrates for the first time that life can be maintained by an external pump acting as an artificial heart. During his research on a heart-lung machine, he finds that roller pumps are gentle enough to minimize both clotting and damage to blood cells. He also learns that centrifugal force can be used to spread the blood in a layer thin enough to absorb the required amounts of oxygen. He accomplishes this during surgery on a dog, laying the groundwork for future work on humans.

(See also July 3, 1952)

1935 Antonio Caetano de Abreu Freire Egas Moniz (1874-1955), Portuguese surgeon, performs the first lobotomy. Also called a leukotomy, this operation cuts into the skull and severs the nerve fibers connecting the patient's thalamus with the frontal lobes of the brain. This procedure opens an entirely new field called psychosurgery, in which mental disturbances are treated by means of brain surgery. This surgery is applied to patients showing chronic and severe mental conditions that result in severe distress, depression, or aggressiveness, as

well as to relieve pain caused by an incurable illness. It is usually employed as a last resort and eventually is done away with once tranquilizers and other mind-affecting drugs are discovered.

1935 Gerhard Domagk (1895-1964), German chemist and pathologist, produces the first effective sulfanilamide derivatives, which prove valuable in fighting bacterial infections. While systematically surveying for both new dyes and new drugs, he finds that one particular dye, Prontosil red, possesses antibacterial properties against streptococcal infection in mice. He determines that the dye is converted in the body into sulfanilamide, a member of a group of chemical substances known by the same name. Although drugs in this group are only mildly disinfectant in vitro (outside the body), they are strongly protective in vivo (within the living body).

(See also 1936)

1936 Perrin Hamilton Long (1899-1965), American physician, and Eleanor Albert Bliss, American bacteriologist, first introduce sulfa drugs in the United States. They conduct studies of Prontosil in a series of wide-scale applications. Sulfa drugs work by suppressing the growth of folic acid, which is what invading bacteria live on in human cells. In these early years of use, sulfa drugs are applied directly to open wounds and given orally. Although they can also be toxic, they offer an important weapon in the fight against infection-causing bacteria.

1936 Cortisone is first isolated by Tadeus Reichstein, Polish-Swiss chemist. During his work on hormones of the adrenal cortex, he describes and names "substance F." This proves identical with a substance found by American biochemist Edward Calvin Kendall (1886-1972) who at the same time isolates nine related steroid hormones from adrenal cortical extracts, one of which (Compound E) is later named cortisone. An organic compound found in small amounts in the body, cortisone is found to have anti-inflammatory properties, and by 1948 is used in the treatment of rheumatoid arthritis. The necessary dosages are much larger than the amount normally present in the body, however, and it can lead to a swelled appearance, increased gastric activity, and metabolism upsets. Continued research has limited some of its side effects, and it becomes an important drug to organ transplantation by assisting the prevention of rejection.

1936 First centrifuge to study acceleration effects on the human body is built by Harry G. Armstrong and John W. Heim of the United States. As aeronautics makes several major technical advances between the wars, the study of kinetic stresses on pilots creates a new field of research called aerospace medicine. Aircraft pilots can encounter severe acceleration and deceleration forces during banking turns, nosing over and pulling out of a dive, ejection, and parachute landings. Postwar experimental research in astronautics (space flight) places even greater stresses on the body during launch and reentry. Accelerative forces are stated in terms of multiples of the force imposed by the earth's gravity or "G."

1936 Frank Macfarlane Burnet (1899-1985), Australian immunologist, isolates the first mutant bacteriophages, which are tiny, bacteriolytic virus that will play a major part in future genetics theory. He also discovers a method for identifying bacteria by the viruses or bacteriophages that attack them and develops a now-standard technique of culturing viruses in living chick embryos. His work permits the large-scale production of vaccines.

1937 First successful allogenic (same species) corneal transplant is conducted by the Russian V. P. Filatov. The cornea is the transparent front of the eyeball and serves as the major refracting medium. Filatov's successful surgical demonstration makes the transplantation of the cornea from a dead donor to a live recipient one of the oldest and most common transplantation procedures. Transplantations are especially useful for hereditary corneal diseases and for burns of the cornea.

1937 George Porter Robb, American physician, and Israel Steinberg first introduce angiocardiography. They devise a method of seeing the chambers of the heart and other large vessels after injecting a contrast medium. X-raying radiopaque material as it flows through the heart is used to determine its condition as well as its major vessels. Among the conditions detectable are such internal situations as arteriosclerosis (thickening and loss of elasticity of the arteries) or damage to the heart from a myocardial infarction (heart attack).

1937 First major blood bank is established by Bernard Fantus at the Cook County Hospital in Chicago. The discovery of the human blood groups in 1900 made it possible to proceed safely and effectively with blood transfusions. This, in turn, creates a demand for an inventory of ready-to-use blood.

1937 First antihistamines are discovered by Daniele Bovet, Swiss-French-Italian pharmacologist. Histamines in the body are the cause of allergic reactions, and Bovet searches for a natural counteragent to them. Finding that none exists, he investigates structurally similar compounds and their neutralizing agents. This research pays off, and he synthesizes the first antihistamine, thymoxydiethylamine. This proves too toxic for humans, and in 1944 he succeeds with an antihistamine (pyrilamine) that effectively neutralizes allergic reactions. This work lays the basis for the safe, effective synthesis of antihistamines. In later years, he also first synthesizes curare and develops a method to use it during surgery as a muscle relaxer. For his work on both, he wins the Nobel Prize in 1957.

1937 Lung cancer in cigarette smokers is first described by American surgeons, Alton Ochsner and Michael Ellis DeBakey, who suggest that smoking causes cancer. At this point in time, smoking is already a worldwide phenomenon,

Daniele Bovet.

and by the end of World War II it begins to rapidly become almost universal. Despite increasingly strong social, religious, and medical arguments, the use of tobacco creates a major American industry. By the 1980s, however, cigarette smoking is associated with not only lung cancer but coronary artery disease, chronic bronchitis, lip cancer, and emphysema. By the 1990s, the connection to lung cancer is almost absolute and it is demonstrated that nicotine and related alkaloid substances furnish the narcotic effects of smoking and make it addictive. During these years, the United States Federal Government bans smoking from its buildings and facilities, and all life insurance companies have separate and much higher rates for individuals who jeopardize their health by smoking.

1938 Howard W. F. Florey (1898-1968), Austrian-English biochemist and Ernst Boris Chain (1906-1979), German-English biochemist, first isolate and purify penicillin, originally discovered by Scottish bacteriologist Alexander Fleming (1881-1955) in 1928. While searching for antibacterial substances, they obtain a culture of Fleming's original mold and are able to extract and purify the penicillin. Tests on animals reveal that it is nontoxic as well as an

effective antibiotic. It also does not harm living cells or interfere with the activity of white blood cells. Subsequent trials with humans are so successful that the United States sponsors an efficient method of mass-producing penicillin to treat wounded soldiers during World War II. Penicillin proves a true miracle drug, and is the most powerful of the antibiotics, being used to treat syphilis, meningitis, and pneumonia.

(See also 1943)

1938 Albert Hoffman of Switzerland first synthesizes lysergic acid diethylamide, known as LSD-25. By 1943, Hoffman discovers its potent ability to influence mental processes. A chemically synthesized hallucinogenic drug, it can be derived naturally from the ergot alkaloids. Given its ability to produce marked deviations in human behavior, it is initially used in medicine to induce certain mental states that resemble actual psychotic diseases. A serious side effect, however, is the transient reappearance of the induced psychotic reaction. Until it becomes a highly controlled drug, it is used experimentally in the treatment of neuroses, alcoholism, narcotic addiction, and with autistic children. It becomes a trendy, mind-altering (and illegal) drug in the 1960s and 1970s. Studies have suggested a link between use of the drug and chromosomal and genetic damage.

1939 First attempts at ultramicrotomy (superthin sections for microscopic analysis) are made. As the electron microscope proves an increasingly important tool to medical researchers, the preparation of these ultra-thin specimens becomes imperative.

(See also 1931)

1939 Paul Hermann Müller (1899-1965), Swiss chemist, discovers the insect-repelling properties of DDT (dichlorodiphenyltrichloroethane)—which was first synthesized in 1873. He spends four years searching for the "ideal" insecticide (one that kills a large number of insects rapidly but does little damage to plants or animals and which is chemically stable and long-lasting) before he tests DDT. During World War II, it is used successfully against fleas, lice, and mosquitos and later proves extremely effective for agricultural purposes. It is eventually determined to be a powerfully harmful environmental pollutant and is subjected to severe restrictions. By the 1960s, science learns that not only do some species of insects develop resistance to DDT, but that its high stability leads to its accumulation in insects and consequently in the animals, mainly fish and birds, that consume them. DDT begins to be replaced by less toxic and shorter lasting agents and is eventually banned by many countries.

(See also January 1944)

1940   Karl Landsteiner (1868-1943), Austrian-American physician, and Alexander Solomon Wiener, American immunohematologist, discover the Rhesus-factor in blood. Named for the species of monkey in which it is discovered, this discovery of the Rh blood groups proves to have a connection with a newborn disease called erythroblastosis fetalis. The Rh factor is the basis of a series of events that can occur in the mother's blood and the child she carries and which can result in this infant disease.

FEBRUARY 1, 1941   Penicillin is first given to a human to treat a condition. An American policeman with an advanced staphylococcus infection complicated by streptococcus is given penicillin for five days. Despite his desperate condition, abscesses, deep bone damage, eye destruction, and lung damage, the infection is temporarily arrested and his condition improves. He soon dies, however, when his doctors run out of the new drug and cannot make more fast enough. It becomes clear that given in massive doses, penicillin can deal with the most difficult infection, but also that it must be administered until the disease is eliminated.

(See also 1943)

1941   The anticoagulant drug dicumarol is first identified and synthesized by American biochemists Mark Arnold Stahmann, Karl Paul Link, and by C. F. Huebner. During the 1920s, a cattle disorder was discovered in which livestock that had eaten hay from spoiled sweet clover bled to death. During the 1930s, it was found that a substance in the clover reduced the activity of prothrombin, a clotting factor in the blood. Once this clotting factor was isolated, this American team was able to synthesize it as dicumarol. The availability of this and other anticoagulants allows them to be used to keep blood transfusions from clotting, to treat conditions involving dangerous blood clots, like cerebral thrombosis and coronary heart disease, and during surgical operations on heart valves.

1942   C. Auerbach in England discovers the first chemical mutagen called yperite. The identification of this and other substances that can cause structural changes in genes offers science a new tool for investigating the structure of genes via induced mutations.

1943   First test-tube synthesis of glycogen is achieved by Carl Ferdinand Cori (1896-1984) and Gerty Theresa Radnitz Cori (1896-1957), Czech-American husband and wife biochemists. This leads to the 1947 Nobel Prize for Physiology or Medicine for their discovery of the role sugar plays in the metabolism of animals. Being able to synthesize glycogen means that they understand carbohydrate metabolism in the body. They find a conversion cycle in which liver glycogen is converted to blood glucose that is reconverted to glycogen in muscle, where its breakdown to lactic acid provides the energy used in muscle contraction. This in turn eventually leads to an understanding of the influence of hormones on the interconversion of sugars and starches in animals.

1943   Penicillin is first used on a large scale by the United States Army in the North African campaigns. Data obtained from these studies show that early expectations for the new drug are correct, and the groundwork is laid for the massive introduction of penicillin into civilian medical practice after the war.

JANUARY 1944 DDT (dichlorodiphenyltrichloro-ethane) is given its first major field test by the United States Army in Naples, Italy, where it stops a typhus epidemic. DDT effectively kills the typhus-carrying lice, marking the first time that a winter typhus epidemic has been stopped.

(See also 1939)

NOVEMBER 29, 1944 First successful operation to remedy the "blue baby" syndrome is conducted by American surgeon Alfred Blalock (1899-1964). He uses a technique based on the research of the American pediatric cardiologist Helen Brooke Taussig (1898-1986). A founder of pediatric cardiology, she becomes a pioneer in the surgical repair of congenital heart disease. "Blue babies" are infants born with insufficient oxygenation of their bloodstream, and Taussig uses fluoroscopy and x rays to develop her theory that this condition is caused by a nonfunctioning artery. She develops with Blalock what becomes known as the "Blalock-Taussig Shunt," a surgical procedure that is a breakthrough in cardiac surgery and saves thousands of babies. Taussig later contributes to the health of thousands of babies by going to Germany and investigating reports of birth defects caused by the anti-nausea drug thalidomide. She is instrumental in the American banning of that birth defect-causing drug.

1944 First compilation of worldwide epidemiological, sanitary, and public health data is published by the United States Army. Titled *Global Epidemiology: a Geography of Disease and Sanitation*, this work begins the modern concept of global medicine—an increasingly relevant and important subject as international travel becomes almost commonplace. Disease has never respected borders of any kind, and today's epidemiologists often must be able to track the course of a disease across continents.

1944 Archer J. P. Martin and Richard L. M. Synge, both English chemists, first develop the technique of paper chromatography for which they share a Nobel Prize in 1952. With this method of being able to separate very closely related compounds, others are able to determine the number of particular amino acids in protein molecules and the scheme of photosynthesis is eventually worked out.

1944 First eye bank is established. As a repository for eye tissue available to ophthalmic surgeons for the purpose of restoring sight by corneal transplants, eye banks are usually non-profit organizations supported entirely by contributions and public funds. They obtain most of their eye tissue from hospitals and personal pledges and also support ongoing research into the most common causes and possible prevention of sight impairment.

1945 First kidney dialysis machine is invented by Dutch-American Willem Johan Kolff. His artificial kidney machine keeps patients with kidney failure alive by filtering out urea from their blood. The first hemodialysis on a human also is successfully conducted this year. It works on the dialysis principle in which smaller molecules are separated from larger ones by diffusion through a semipermeable membrane. Kolff's machine diverts blood from an artery, usually the wrist, and brings it into contact with one side of a membrane. As this happens, dissolved substances in the blood like urea and inorganic salts pass through into a sterile solution on the other side. The blood's red and white cells, as well as its platelets and proteins, cannot penetrate the membrane because they are too large. Dialysis is soon improved using ultrafiltration and becomes a common procedure for patients with temporary or chronic kidney failure. It also buys the necessary time for those awaiting a transplant organ.

1945 Guy Henry Faget (1891-1947), American physician, first introduces the drug promin which proves effective against leprosy. Until this time, there were no truly satisfactory drugs available to treat leprosy—a chronic infectious disease caused by a bacillus which results in lesions and gross deformity. Promin, a sulfone

DDT was first sprayed over lakes to kill mosquito larvae and other pests during the late 1940s.

drug that is a synthetic compound, is initially used to treat tuberculosis. Later tried by Faget in the treatment of leprosy, it proves successful, having both a social and a medical impact on this disease which for centuries had made its victims absolute social outcasts.

1945 Edwin J. Pulaski of the United States Army establishes the Surgical Research Unit which becomes the first American center for the study of burn patients.

1946 First drug developed specifically for the treatment of cancer is prepared by American pathologist and physician Cornelius Packard Rhoads (1898-1959). He develops an alkylating compound which is derived from the nitrogen mustard used as a chemical warfare agent. This drug works against cancer cells by binding with chemical groups in the cell's DNA and preventing it from reproducing. Alkylating compounds prove important to cancer chemotherapy.

1947 Theodore E. Woodward, American physician, reports the first specific cure of typhoid fever with chloramphenicol while in Malaya. This is one of the earliest antibiotics to be synthesized and proves relatively simple to make

after its isolation from a mold this same year. It also proves valuable in the treatment of typhus which, like typhoid fever, does not react to penicillin.

1948 First antimetabolite in cancer chemotherapy is used by American physician Seymour Morgan Farber. He uses methotrexate (aminopterin) to treat leukemia. This form of the B-vitamin, folic acid, is found to be effective in disrupting the metabolism of cancer cells and also proves useful in combating breast cancer and other forms of cancer.

1948 American zoologist and student of sexual behavior, Alfred Charles Kinsey (1894-1956) first publishes his *Sexual Behavior in the Human Male*. This landmark study of sexual behavior is followed by his 1953 book, *Sexual Behavior in the Human Female*. His work is based on 18,500 personal interviews and describes a wide range of behavior. Despite some problems with his sampling and statistical techniques, these studies are widely recognized as the most comprehensive and significant survey of the norms, extent, and variability of American sexual behavior. In 1947 Kinsey founds the Institute for Sex Research, a nonprofit corporation affiliated with Indiana University which continues his pioneering scientific study of human sexual behavior in the United States.

1948 Team of American chemists, Edward Lawrence Rickes, N. G. Brink, F. R. Koniusky, T. R. Wood, and Karl Folkers, first crystallizes Vitamin B-12. Researchers had been searching since the 1930s for the active principle in liver that allows consumption of it to cure pernicious anemia. Despite their purification of the vitamin into tiny red crystals, its molecular size and structure is so large and complicated that B-12 is not synthesized until 1971. It becomes the last of the vitamins to be discovered.

(See also 1926)

1949 First gastric suction tube is invented by I. J. Wood. Although this invention is used solely to remove gastric juice from the stomach, modern medicine puts the simple electric-powered suction to a variety of uses. It is used to clear air passageways of mucous obstructions, to keep an area free of blood during surgery, and to gently evacuate the ear of excess cerumen (earwax).

1949 Prostaglandins are discovered by American pharmacologist Charles Christian Lieb (1880-1956) and R. Kurzrock. These series of fatty acids are found in high amounts in semen but are also in other organs and tissues. Considered as hormones, they have a number of physiological roles, like lowering blood pressure, and are still being studied. It is also thought that they regulate excitability of the central nervous system.

1949 Lithium is first used in the treatment of psychiatric disease. Australian psychiatrist John Cade studies manic-depressive patients and theorizes that they may be metabolizing an overabundance of uric acid. He injects uric acid into guinea pigs in the form of lithium salts and lithium carbonate and confirms his theory. Shortly thereafter, Danish physician Mogens Abelin Schou treats his patients suffering from manic symptoms with lithium and reports good

results. Lithium is adopted by many American psychiatrists in the 1970s.

1950 A. C. Finlay and Gladys Lounsbury Hobby of the United States discover the antibiotic drug oxytetracycline, later given the trade name Terramycin. Unlike penicillin, an antibiotic developed from a mold, Terramycin (which is part of the family that includes erythromycin and aureomycin) is developed from bacteria. The tetracycline group of antibiotics are broad-spectrum drugs that can have side effects and are used more judiciously than the penicillin or cephalosporin groups.

1950 Kenneth D. Orr of the United States Army leads a Cold Injury Research Team in Korea that obtains the first epidemiological data on frostbite. As a result of these studies, both military and civilian cold weather living is made safer and more functional.

1950 First human to survive a kidney transplantation is Ruth Tucker, a 49-year-old American woman dying from chronic uremia. The American surgeon, Richard Lawler of Chicago, transplants a kidney from a cadaver into his patient who survives for a short time. Lawler finds, upon autopsy, that the new kidney has not only stopped functioning but has turned into a shriveled mass of dead tissue. This is the result of immunological rejection in which the body's white cells attack and destroy a graft of foreign tissue.

(See also December 23, 1954)

1951 First United States Army helicopter detachments whose primary mission is casualty evacuation become operational in Korea. In this conflict between North and South Korea (June 1950-July 1953), American soldiers are the principal non-Korean combatants. In two years they will evacuate over 17,000 wounded by helicopter. Prior to this conflict, ground casualties are moved overland to field hospitals. This marks the beginning of modern military evacuation techniques, made possible by improvements in air transport. The helicopter revolutionizes military medicine and saves countless lives.

1951 Joseph E. Smadel, Kenneth Goodner, Fred R. McCrumb, Jr., and Theodore E. Woodward, American physicians, are the first to demonstrate that broad spectrum antibiotics can cure septicemic and pneumonic types of human plague. These broad spectrum drugs called tetracyclines are developed from bacteria and prove effective in combating this potentially lethal infectious disease. These antibiotics do have side effects, however.

1951 First successful oral contraceptive drug is introduced. Gregory Pincus (1903-1967), American biologist, collaborates with American biologist Min-Chueh Chang and American physician John Rock (1890-1984), and employs a synthetic hormone that renders a woman infertile without altering her capacity for sexual pleasure. They use a synthetic progesterone called progestin and find that it suppresses ovulation. Clinical trials begin in 1954, and in 1960 the United States Food and Drug Administration approves it as the first contraceptive pill. It soon is marketed in pill form and effects a social revolution with its ability to separate the sex act from the consequences of impregnation.

(See also 1954)

1952 Paul M. Zoll, American cardiologist, introduces the first practical cardiac pacemaker. Zoll theorizes that he can use the heart's natural responsiveness to electrical stimulation to treat cases of heart block (when the sinus node responsible for sending impulses to the heart muscles suddenly stops). His initial attempt at passing an electrode down the esophagus is not successful and he soon develops an external pacemaker that sends an electric shock to the heart through electrodes placed on the patient's chest. Although correct in principle, his early pacemaker requires that both the patient and the device be close to an electrical outlet. The

shocks are also painful to the patient's chest muscles. Pacemakers become really practical with the invention of the transistor which allows them to be so reduced in size as to be implantable. Today's pacemakers adjust themselves to the patient's exercise level and have microprocessors that are even reprogrammable using radio-frequency signals.

1952 Paul M. Zoll, American cardiologist, first develops the technique of externally stimulating a heart with electric shock in cases of cardiac arrest. After taking advantage of the heart's natural responsiveness to electrical stimulation to invent the electric pacemaker this year, he employs the same principle to try to save lives in an emergency cardiac situation. During some heart attacks, a condition occurs in which various muscle groups in the heart beat independently and without any rhythm, meaning they pump no blood despite their rapid quivering. This is called fibrillation, a usually fatal condition. In 1956, Zoll introduces the external defibrillator, a device that provides an external electrical impulse to the chest and shocks the muscles back into uniform, rhythmic beating.

JULY 3, 1952 First mechanical heart pump is used successfully on a human being undergoing heart surgery. Designed by American surgeon Forest D. Dodrill, with help from General Motors engineers, the portable electric pump performs the heart's functions while it is undergoing surgical repair. Called the Dodrill-GMR Mechanical Heart, this portable electric pump moves blood throughout a patient's body and allows the heart chamber to be emptied of blood while surgery is performed. It takes two years of work to design and build a pump that is durable and dependable yet gentle, since blood is very fragile and easily destroyed. This breakthrough marks the beginning of today's increasingly sophisticated cardiac bypasses, valve repairs, and even transplants.

SEPTEMBER 1952 Charles Anthony Hufnagel, American surgeon, inserts the first artificial heart valve. Heart valves are tissue flaps inside the heart that open only one way and allow blood to flow from one chamber to another and then close to prevent any blood from leaking back. Valve disease or failure can cause stress on the heart and result in eventual cardiac failure. The only effective treatment in these cases is valve replacement. Hufnagel's breakthrough invention to prevent backflow is a tube-and-float device inserted in a patient's descending aorta. Hufnagel shows the feasibility of what becomes an increasingly prevalent surgical procedure.

1952 Nobel Prize for Physiology or Medicine is awarded to Selman Abraham Waksman (1888-1973), Russian-American microbiologist, for his discovery of streptomycin, the first antibiotic effective against tuberculosis. Until Waksman discovers this important antibiotic, the only treatment for tuberculosis was prolonged bed rest and nutritious food. The drug proves so safe and effective that it nearly eliminates the disease in the United States. The fact that TB is making a comeback both in America and the rest of the world in the 1990s has more to do with the lack of medical treatment than with the efficacy of the drug. In fact, streptomycin is found to be active against 70 different bacteria which did not respond to penicillin, including infections of the abdomen, urinary tract, pelvis, and meninges.

FEBRUARY 1952 Douglas Bevis, English physician, publishes an article describing his use of amniocentesis in Rh-factor cases. This process uses a needle to obtain samples of amniotic fluid (in which the fetus floats) from a mother's womb. This is the first use of amniocentesis for a fetal diagnosis. Bevis chemically analyzes the iron and urobilinogen content of the fluid to determine the possibility of hemolytic (blood) disease in the unborn child. After Bevis shows the feasibility of diagnosing by amniocentesis, others pursue it and not only determine fetal sex with the procedure but are able to determine whether a fetus is affected by Down's

Microbiologist Selman A. Waksman.

syndrome. Amniocentesis is usually conducted during the 16th to 18th weeks of gestation when there is a sufficient amount of amniotic fluid for sampling. The discovery of unwanted conditions in the fetus does not always mean an abortion is recommended by the physician, and sometimes the diagnosis of a fetal disorder allows it to be treated in the uterus before the baby is born. Hundreds of hereditary diseases can now be diagnosed through amniocentesis. (See also 1968)

MAY 6, 1953 First successful open-heart operation using a heart-lung machine is conducted by American surgeon John Heysham Gibbon, Jr. (1903-1974). Gibbon spends years developing his pump-oxygenator that shunts blood from the veins through a catheter to a machine that supplies the blood with oxygen and then pumps it back into the arteries. He uses this machine to keep a patient alive while he operates directly on the heart, closing an opening between the atria. The era of open heart surgery begins with this operation.

1953 Michael Ellis DeBakey, American surgeon, performs the first successful carotid endarterectomy. This surgical removal of a plaque buildup from within the lining of an artery is a delicate and risky operation. As a pioneer in the development of surgical procedures that correct defects and diseases of the circulatory system, he also develops an efficient method of correcting aortic aneurysms by grafting frozen blood vessels to replace diseased ones. He eventually progresses from grafts to plastic tubing.

DECEMBER 23, 1954 Joseph Edward Murray, American surgeon, conducts the first successful organ transplant as he transfers the kidney of one twin to another. Murray makes a bold application of the principle, known since the 1930s, that identical twins share a common genetic identity, and that an organ from one is not considered to be foreign tissue and thus is not rejected.

(See also April 5, 1962)

1954 First vaccine for measles is developed by John Franklin Enders (1897-1985), American micrologist, and Thomas Peebles, American pediatrician. This infectious disease of childhood is caused by a virus. Along with symptoms of high fever, headache, cough, and a red rash, complications can arise if there is a secondary bacterial infection. After Enders isolates the measles virus he is able to find an attenuated (weakened) strain that is suitable for a live-virus vaccine. Although an immunization program using this is started, a truly practical and successful vaccine is not achieved until 1963.

1954 First safe and effective anti-poliomyelitis vaccine which prevents paralytic polio is developed by Jonas Edward Salk, American virologist. Working on polio since 1949, Salk confirms that there are three virus types responsible for the disease and then experiments with ways to kill the viruses yet retain the ability to produce an immune response. By 1952, he produces a dead virus vaccine that works against all three. He then tests the vaccine on monkeys, then on children who have had the disease, and finally on his own children, none of whom ever had

polio. His vaccine is tested nationally this year and proves between 60 and 90% effective. It is soon followed by the Polish-American virologist Albert Bruce Sabin's (1906-1993) development of a live virus vaccine that is also the first oral vaccine. This proves more effective than Salk's dead virus vaccine and has longer-lasting immunity. Sabin's oral vaccine is still used today.

1954 Frank Milan Berger, Czech-American physician and pharmacologist, first uses meprobamate to treat anxiety. This white, crystalline solid is prepared for commercial use from the related alcohol propanediol. It becomes popular in the treatment of neurotic conditions and acts as a central nervous system depressant, selectively targeting the spinal cord and the higher centers of the brain. It is marketed as Miltown after the name of the village in New Jersey where it is developed. High doses taken for long periods usually results in addiction.

1954 First clinical trials of oral contraceptives are conducted.

(See also 1951)

c. 1955 First physician to advocate the wide use of x rays to screen women for breast cancer is American physician and radiologist Jacob Gershon-Cohen. He begins a five-year study of more than 1,300 women and finds that women diagnosed early through mammography (x-ray imaging of the breast) have a better recovery rate than those whose disease is discovered at a later date.

(See also 1967)

1955 Chlorpromazine is first used to treat psychiatric disorders. Also called Thorazine, this is a powerful, antipsychotic drug that comes to be used primarily to treat schizophrenia. Developed by a French pharmaceutical company, it is found to profoundly alter a patient's mental awareness, giving previously agitated individuals an almost detached sense of calmness. Along with lithium and other powerful antipsychotic

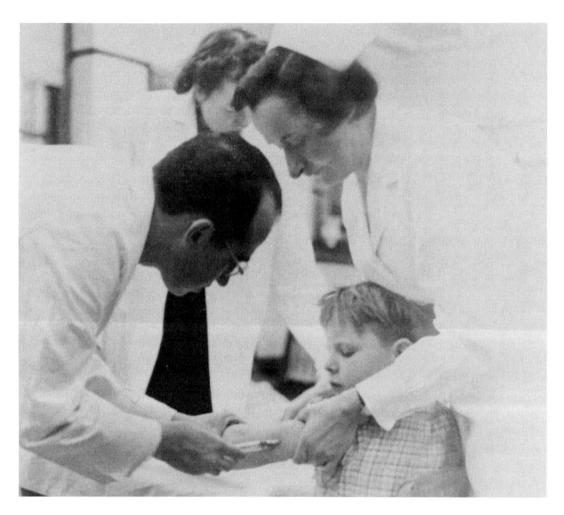

Jonas Salk innoculating a six-year-old during the 1954 polio vaccine field trials.

drugs, chlorpromazine initiates the age of tranquilizers. These powerful drugs become widely used for treatment of institutionalized patients and offer physicians a way of controlling unacceptable behavior. These drugs are difficult to overdose on and are not addictive, although patients build up a tolerance over time and need higher doses. They are not preferred by drug abusers, since they make non-psychotics feel uncomfortable rather than calm or pleasurable. The development of chlorpromazine marks the first time a drug is discovered that specifically targets the central nervous system without profoundly affecting other behavioral or motorized functions.

1957 Alick Isaacs (1921-1967), Scottish virologist, demonstrates that antibodies act only against bacteria. This important discovery means that antibodies are not one of the body's natural forms of defense against viruses. This knowledge leads eventually to the search for how the body attempts to defend itself against a virus and to the discovery of interferon this same year by Isaacs and his colleague, Jean Lindenmann of Switzerland. They find that in most living things, the generation of a small amount of protein is the first line of defense against a virus. Lindenmann and Isaacs describe this protein as an "interfering protein" and give it the name interferon. Unlike antibodies that take several days to form, interferon is produced within hours of a viral invasion. Although interferon holds great promise in the treatment of viral diseases, it is found to be species-specific (only human interferon will work on people) and is produced by the body in minute quantities. The goal of interferon research is a large-scale, genetically engineered product that can be targeted to fight cancer and specific viral diseases.

1957 First clinically successful high-speed dental drill is introduced. Spinning at a rate of 300,000 rpm, this drill greatly reduces patient discomfort and shortens the amount of time a dentist must work to excavate a tooth.

1958 Ultrasonic waves are first recorded as being used to make in utero observations of a fetus. Ultrasound works because sound waves of very high frequencies can easily and harmlessly penetrate human flesh. As they pass into the human body, they encounter different materials (bone, internal organs) which cause the waves to reflect back to the source with different signatures according to what they encounter. Using high-frequency sound waves, physicians can harmlessly observe the condition of an unborn child.

1958 First use is made of closed-chest cardiac massage combined with mouth-to-mouth respiration for cardiac resuscitation. Now popularly known as CPR (cardiopulmonary resuscitation), this technique provides a means of keeping a stricken individual's heart pumping oxygenated blood to the brain. It becomes standard procedure used by rescue and emergency crews in cases of heart attack or life-threatening situations like drowning or electric shock in which a person's heart stops. These crews carry special ventilation bags or airway tubes to avoid contracting disease. Proper CPR techniques are now taught to increasing numbers of people.

1958 First international textbook on surgery, *Emergency War Surgery*, is published by NATO (North Atlantic Treaty Organization).

1958 Ultrasound is first used in obstetrics to examine an unborn fetus. In use after World War II primarily to test machine parts by using beams of high frequency sound waves to detect cracks, ultrasound is believed by English physician Ian Donald to have a future in obstetrics as a replacement for x rays. By 1957 he builds and tests an ultrasonic device for medical use which he tests successfully by using sound waves to correctly diagnose a patient's heart condition. Ultrasonics works with sound waves of varying frequencies (from one to ten megahertz) which penetrate tissue harmlessly. Because different types of tissue reflect sound waves back to the source differently, physicians can compile a picture of what exists inside. Improvements in ultrasound techniques have made it the most common method for examining a fetus and have helped obstetricians gain valuable information in treating individual pregnancies.

1958 Denis Burkitt, Irish physician, publishes his first account of what becomes known as "Burkitt's lymphoma." The lymphatic system is a network of vessels within the body that collects fluid that has been filtered by capillaries from the spaces between tissue cells. It removes bacteria from the tissues, produces white blood cells needed to fight infection, and returns the fluid it collects, including escaped blood proteins, to the bloodstream. Lymphoma is cancer of the lymphatic system. Burkitt discovers a childhood form of lymphoma in Uganda and ascribes its cause to an insect-borne virus. This virus-caused disease is also prevalent in Southeast Asia.

1961 Irving S. Cooper, American neurosurgeon, first uses a freezing technique known as cryosurgery to freeze and destroy damaged tissue within the brain of patients with Parkinson's disease. Using liquid nitrogen, extremely low temperatures of about -200°F (-130°C) are achieved which quickly destroy any tissue it touches. As an alternative to traditional surgery, cryosurgery can usually be used without general anesthesia and can be applied to difficult to reach areas. It is also fairly safe. This entirely new way of performing surgery leads to cryosurgery being used in delicate eye surgery to remove cataracts from the lens of the eye as well as to repair detached retinas.

1961 First physicist to win the Nobel Prize for Physiology or Medicine is Georg von Bekesy (1899-1972), Hungarian-American physicist. He receives the award for his discovery of the physical means by which sound is analyzed and communicated in the cochlea (a portion of the inner ear). Working on telephone and long-distance communication problems, he becomes interested in the mechanics of human hearing and brings the techniques of physics to the study of physiological acoustics. He builds mechanical models of the cochlea and discovers that vibrations transmitted to the fluid in the cochlea set up traveling waves in the basilar membrane (which is made up of 24,000 fibers that stretch across the cochlea). He finds that it is the shape of these sound waves that goes to the brain and is interpreted as sound. Bekesy views the cochlea as a kind of frequency analyzer, an electronic device that measures and interprets the frequency of waves.

APRIL 5, 1962 Joseph Edward Murray, American surgeon, transplants a cadaver kidney into a human subject. This is the first successful renal allograft between unrelated subjects. His patient lives for 17 months. This is also the first time that chemical immunosuppression using the drug azathioprine is used.

1962 First eye surgery with a laser is performed. Lasers are used to operate on the eye's retina to repair tears in it or to remove unwanted, dis-

eased tissue. When a laser incision in the cornea is compared to that made by a surgical instrument, the edge of the laser wound is smooth and regular, showing little disruption of tissue. The cut made by a razor-sharp surgical blade, however, is ragged and irregular. A new concept to appear in the past few years is the prospect of using lasers to reshape the cornea, the transparent covering over the pupil and iris. Because both the cornea and the lens focus light on the retina, changing the cornea's optical characteristics by reshaping has the potential for correcting refractive errors of farsightedness, nearsightedness, and astigmatism.

1963 Daniel Carleton Gajdusek, American pediatrician, describes the first slow-virus infection to be identified in humans. He studies the disease kuru, a degenerative, often fatal, disease of the central nervous system that occurs only among the Fore peoples of New Guinea. It begins with an unsteady gait and progresses to tremors, writhing, and involuntary movements. Theorizing that the disease has a viral cause, he realizes that the tribe's practice of honoring their dead by eating their brains is an ideal mechanism for viral transmission. He conducts kuru studies with apes and discovers that the disease is caused by a slow-moving virus that takes at least two years to appear. His work on slow viruses has led others to postulate that they are really a new type of infectious particle called a prion (a protein that has the ability to cause an infectious disease). The prion hypothesis is still being tested and debated.

(See also 1966)

1963 First human liver transplant is performed by Thomas Earl Starzl, American surgeon. The recipient survives for 23 days. Such operations are experimental at this time, since the problem of suppressing the body's immune system in order to prevent organ rejection is not yet solved.

(See also July 1976)

1963 Valium, the world's most widely used tranquilizer, is first developed. As a barbiturate, it is a mood-altering drug that manipulates the central nervous system. It is a minor tranquilizer, often prescribed to calm people and help them sleep while they are trying to cope with a traumatic event. Excessive use, however, may hinder a person's coping skills, and the mixing of Valium and alcohol can prove fatal.

1966 Daniel Carleton Gajdusek, American pediatrician, transfers for the first time a viral disease of the central nervous system from humans to another species. The viral disease kuru is found in New Guinea and is spread by the ritual eating of the deceased's brain. In order to prove that the disease is transmitted by a slow virus, Gajdusek implants pieces taken from the brains of kuru victims into apes and finds that the disease begins to appear in the apes more than two years later.

(See also 1963)

DECEMBER 3, 1967 First successful human heart transplant is performed by South African surgeon Christiaan Neethling Barnard in South Africa. The recipient, a 53-year-old grocer named Louis Washkansky, lives for 18 days. He receives the heart of a 25-year-old woman who had been killed by an automobile. Barnard performs four more heart transplants over the next few years, and although one patient lives 593 days, none is considered really successful since each dies of infection after receiving massive doses of immunosuppressants. Although Barnard masters the surgical techniques of implanting a new heart to replace a diseased one, he stops performing the operation until the problem of a safe immunosuppressant drug is resolved. His landmark surgery underscores the many legal, medical, and ethical questions surrounding the issue of organ transplantation.

(See also July 1976)

1967 Mammography, a form of x-ray detection for breast cancer, is first introduced as a widely used diagnostic tool. It becomes a standard screening method advocated by the National Cancer Institute in the United States. The advantage of mammography is that it can detect tumors while they are still very small and are most easily treated. Some critics claim that the procedure does more harm than good, exposing women to unnecessary levels of radiation. Guidelines set by the Institute recommend the procedure for older women and for those under forty who are at special risk to develop breast cancer.

(See also c.1955)

1967 First coronary artery bypass surgery utilizing a vein is performed by Rene G. Favaloro, American surgeon. Plaque blockage of the coronary arteries that supply the heart with blood can cause severe chest pain and in some cases heart attack. Favaloro devises a method of grafting a vein from the patient's leg around a blocked portion of a coronary artery, bypassing it and creating an alternative blood-supply pathway. This technique takes advantage of new developments in arteriography (images of the heart) and microsurgery. This procedure becomes very popular in the 1970s.

1968 Carlo Valenti, American physician, makes the first diagnosis as to whether a fetus is affected by Down's syndrome by analyzing fetal cells taken from a mother's amniotic fluid. Also known as Mongolism, this genetic disorder results in mental retardation, heart defects, a shortened lifespan, and a characteristically broad, flat face with up-slanted eyes. Individuals with this disorder have chromosome 21 represented triply rather doubly, so that there are 47 rather than the normal 46 chromosomes in most cells. Valenti employs amniocentesis to culture fetal cells and analyze their chromosomes. The telltale extra chromosome 21 indicates that the fetus has this genetic disease.

(See also February 1952)

APRIL 3, 1969 First license for a vaccine to prevent rubella is granted. Also called German measles, this fairly mild viral infection generally affects children and young adults. Although not usually dangerous to them, it can have tragic results if contracted by a woman during the first three months of her pregnancy. As early as 1941, it was known that it can cause serious birth defects, from mental retardation to heart disease and cerebral palsy. The rubella virus was first propagated in a laboratory culture by American physician Thomas Huckle Weller, in the 1960s. After this success, researchers from Harvard University and the Walter Reed Army Medical Center develop an attenuated (weakened) virus suitable for a vaccine. Although this early vaccine causes some side effects in adults, it generally provides long-lasting immunity to the disease and eventually is improved to the point where it is included in the standard MMR (measles, mumps, rubella) vaccinations required of all school-age children in the United States.

APRIL 1969   First artificial heart is implanted in a human being. Denton A. Cooley, American surgeon, implants a mechanical heart made of silicon. This temporary plastic device keeps the patient alive for 65 hours until a human heart is implanted to replace it. The patient dies 38 hours after this second operation due to pneumonia and kidney failure. Since the heart and its functions can be viewed primarily as a pump that circulates blood throughout the body, many believe at this time that an artificial heart will be superior to organ transplantation. However, the results of repeated implantation of several different types of artificial hearts reveal what appear to be insurmountable complications: infections, patient immobility, and stroke-causing blood clots.

c. 1969   First computerized axial tomography device (CAT scan) is constructed by Allan MacLeod Cormack of the United States. His initial model uses a thin beam of x rays aimed at one internal section of the body but repeated from many different angles. This method allows him to combine many x-ray pictures into one complete, three-dimensional view. However, he lacks a system to process the large

amount of data produced for a single CAT scan, and his findings and theories receive little attention. The computer eventually solves Cormack's data-processing problems.

(See also 1971)

1970   Implantation of the first cardiac pacemaker is accomplished. This battery-powered device is inserted surgically into the chest of a patient.

1971   The first diamond-bladed scalpel is introduced to surgery in Great Britain. The diamond bit on these new surgical tools are synthetic diamonds which, although too small and impure to be used as gems, prove excellent for precision cutting.

1971   First mass screening laboratory for heroin use is set up in Vietnam by the United States Army. The program, which samples urine for the drug, is later expanded to include amphetamine and barbiturate screening, and serves as a model for civilian programs.

1971   First computerized axial tomography (CAT) machine is installed in a hospital in Nimbledon, England. It incorporates the new design by Godfrey Newbold Hounsfield, English electrical engineer, who uses computers to collate the x-ray data and create a tomographic image which offers a detailed, sharp map of a particular cross-section of the body. Like the American Cormack, who developed the first tomographic device (See also c.1969), Hounsfield had been working on his own CAT scanner and solved Cormack's data problems by using computers. Assembling a complete picture involves a huge amount of information, since CAT machines use up to 300 x-ray scanners taking 300 snapshots each, resulting in almost 90,000 x-ray "slices." CAT scans are nearly one hundred times more efficient than x-ray machines and significantly reduce the need for dangerous, exploratory surgery, particularly of the brain.

1972   First satisfactory artificial hip is designed

by English orthopedic surgeon John Charnley (1911-1982). He spends years searching for a successful method of replacing hip joints made painful from rheumatoid arthritis. Early attempts at hip replacements were far from satisfactory, and Charnley tests new materials, achieving initial success in the 1960s with low-friction Teflon. He continues to improve his artificial hip and in 1972 perfects his technique by using high-density polyethylene to build the hip socket. His studies on the control of infection after hip surgery contribute to the success of his replacement procedure for severe hip degeneration. This operation is performed on a regular basis today, with new materials like titanium contributing to its effectiveness. In the early 1990s, a professional athlete, American football and baseball player Bo Jackson, undergoes a hip replacement and returns to actively compete in professional baseball.

1974 Heimlich maneuver is introduced as first aid for choking. Developed by American chest surgeon Henry J. Heimlich, this method of dislodging an obstruction in a person's throat is a recognized life-saving technique and is endorsed by the American Medical Association's Commission on Emergency Medical Services. When a person is choking and able to stand, the rescuer stands behind the victim, encircles the victim's waist with his arms, and places the thumb side of his fist above the choker's navel and below his rib cage. The rescuer then grasps his fist with his free hand and thrusts upward into the victim's abdomen. The procedure can also be performed on people who are seated or lying on their backs. Dislodging an obstruction quickly is crucial, since within four minutes, brain anoxia (insufficient oxygen reaching the brain) can result in permanent brain damage or death.

JULY 1976 Jean François Borel, a microbiologist at the Swiss pharmaceutical firm Sandoz, publishes the structure and properties of a new immunosuppressant called cyclosporin a. This leads to clinical trials of a drug that proves to be an immunosuppressant whose action specifically inhibits graft rejection while allowing the bulk of the immune system to function normally and fight general infection. While searching for new antibiotics in 1969, he analyzes soil samples containing a fungus from southern Norway where he is vacationing and discovers cyclosporin. Tests reveal that cyclosporin acts to inhibit the immune system's T-lymphocyte activity that detects and attacks foreign invaders. That is, it selectively interferes with only one subpopulation of the immune system, allowing the rest to do their work. In 1983, the United States Food and Drug Administration approves the drug for use in all transplant operations, and successful transplantation becomes commonplace. The drug must be taken indefinitely and can, in high doses, produce lymphomas (tumors) and have a toxic effect on the kidneys, but it has proven to be a life-saver to thousands.

SEPTEMBER 16, 1977 First balloon coronary angioplasty is performed by Swiss surgeon Andreas Gruentzig. He modifies an older heart catheter method by using a small balloon that is inflated after being placed into the blocked portion of the artery. Inflation of the balloon opens up the blockage by compressing it and widens the artery. Although not always a permanent solution, this technique saves lives and avoids elaborate open-heart surgery.

1977 Earliest documented AIDS (Acquired Immune Deficiency Syndrome) victims in the United States are two homosexual men in New York who are diagnosed as suffering from Karposi's sarcoma. In addition to an increase in this rare type of cancer, physicians in New York and California report a variety of infections including pneumonia among previously healthy young homosexual men. The unusual character of all these diseases, typically associated with a failure of the immune system, leads epidemiologists to search for a common link.

(See also 1981)

1978 First test-tube baby (conceived by in vitro fertilization) is born in England. This is the first human being conceived outside the human body to be born. English physicians Patrick Steptoe and Robert G. Edwards develop a method of stimulating ovulation with hormone treatment, then retrieving the nearly mature ova (eggs) and placing them in culture to mature. They then add capacitated male sperm to the petri dish containing the eggs, and fertilization takes place. After undergoing division, the eight-celled embryo is implanted in the mother's uterus and develops normally. The first human being produced by this technique is a baby named Louise Brown. Since this initial success, in vitro fertilization (IVF) has become a widely used method of treating certain cases of infertility, and more than 3,000 such babies have been conceived in this manner. Several improvements and variations on IVF have also been developed.

1978 First Magnetic Resonance Image (MRI) of the human head is taken in England. Invented by American physicist Raymond Damadian in 1971, this new imaging technology is called "nuclear magnetic resonance spectroscopy" and is first used to study the molecular structures of chemicals. Later experiments on parts of the body revealed that, linked with a computer, the device produced a vivid, three-dimensional cross section showing layers of skin, muscle, and bone and was apparently harmless to patients. It is found to be especially useful for imaging the brain since an MRI is made possible by the water content of the body, and it can thus distinguish gray matter from white, from blood, muscle, etc. Since bone contains very little water, it does not appear on the MRI monitor and does not interfere with this internal image. MRI technology has since revolutionized medical diagnosis. It is painless, noninvasive, and provides accurate information.

FEBRUARY 7, 1980 Shock waves are first used successfully on a human patient to break up kidney stones. Called "extracorporeal shock wave lithotripsy," it is developed and performed by a team in West Germany. After the patient is placed in a tub of water and the location of the stone identified with x ray, a series of shock wave pulses strike the stone and are reflected, causing mechanical stress and fracturing. They are disintegrated into small enough pieces to be passed through the urinary tract. This new technique gains worldwide acceptance by urologists and patients who do not have to undergo surgery. The first lithotripter in the United States is installed in 1984.

1980 Charles Weissmann of Switzerland produces the first genetically engineered human interferon, making large-scale production possible. Interest in interferon is strongly revived following the discovery that it slows or stops tumor growth in mice, and the demand for it increases significantly. This technique indicates one way of making it in large enough quantities.

(See also 1957)

1981 AIDS (Acquired Immune Deficiency Syndrome) is officially recognized by the United States Centers for Disease Control, and the first clinical description of this disease is made. It soon becomes recognized that AIDS is an infectious disease characterized by a failure of the body's immunologic system and caused by a virus that is spread almost exclusively by infected blood or body fluids. The virus in question comes to be known as the Human Immune Deficiency Virus (HIV) for which there is still no cure or vaccine. Although exact epidemiological data is not available, health officials throughout the world believe AIDS to be approaching a pandemic. Researchers expect the burden of AIDS, in both human suffering and resources, to continue to grow.

1981 First successful combined heart-lung transplant is achieved by American surgeons Norman E. Shumway and Bruce A. Reitz. In three previous attempts at a heart-lung transplant, the longest survival was 23 days. The surgeons use the new immunosuppressive drug cyclosporin on their 45-year-old patient who survives and returns to a normal life.

DECEMBER 2, 1982 First Jarvik 7 artificial heart is implanted by William DeVries, American surgeon. The gravely ill patient, Barney Clark, lives for 112 days. This procedure is the first total heart replacement intended for permanent use. The replacement heart is designed by American physician Robert Jarvik and is a plastic and titanium pump powered by compressed air delivered through two tubes inserted in the patient's abdomen. After Clark dies, four other Jarvik 7 implants are carried out. Each patient dies, including William Schroeder, who lives for 620 days. The Jarvik 7 and other artificial hearts soon fall out of use as surgeons concede that transplants are far superior solutions.

1982 First commercial product of genetic engineering is approved by the United States Food and Drug Administration. Eli Lilly & Company is permitted to market human insulin that is produced by bacteria. Called "Humulin" and produced jointly by Lilly and the Genentech Corporation, the product is identical with human insulin, which is a hormone produced in the pancreas that helps convert carbohydrates to simple sugar glucose and which regulates its level in the blood. Since the 1950s, it is known that insulin is made up of two chains (A chain and B chain). In this new process, the A and B chains are produced separately in different strains of *E. coli* bacteria. Each chain is separated from the bacteria and purified, after which the two are combined chemically and repurified. Genetically engineered insulin does not trigger allergic reactions. This new process of making insulin also assures a steady supply.

1982 First brain tissue transplants are used for Parkinson's disease. They are not successful in alleviating the condition.

(See also April 1987)

1983 Researchers at Harvard Medical School

discover a gene marker for Duchenne muscular dystrophy, a sex-linked recessive disorder. Muscular dystrophy describes a condition in which healthy muscle cells die and are replaced by fat and connective tissue, resulting in a progressive weakening and wasting and finally death. By this time, it is known that MD is caused by a genetic abnormality that causes a deficiency of a particular protein in the membrane of the muscle fibers. In 1986, researchers find that the defective gene responsible for Duchenne muscular dystrophy is located on the short arm of the X chromosome. This discovery opens the possibility of implanting new muscle cells that contain the correct form of the gene.

(See also May 25, 1988)

1983 Nancy Wexler and James Gusella discover a gene marker for Huntington's chorea. This relatively rare and invariably fatal disease is hereditary and results in the degeneration of brain tissue. It is characterized by spasmodic movements of the arms, legs, and face and by the gradual loss of mental faculties. In most cases these symptoms initially appear in middle life. A genetic disorder like this disease is the result of the loss, mistaken insertion, or change of a single nitrogen base in a DNA molecule. Being able to identify the gene marker for this or any other hereditary disease enables genetic counselors to inform at-risk individuals of their options. It also suggests the potential for gene therapy to correct a genetic disorder.

(See also 1993)

1983 Researchers first reveal their findings that a bacterium causes ulcers of the stomach and small intestine. This leads to a major reevaluation of such factors as stress and diet in the treatment of ulcers. Ten years later, the medical community believes that the common peptic ulcer is most likely caused by a bacterium, *Heliobacter pylori*, which can be treated with antibiotics.

1983 Blood is drawn through the umbilical cord from an unborn fetus and is used for the first time to diagnose a disease. Until this technique is developed, fetal blood sampling for diagnostic purposes was done with a fetoscope. Since it was inserted surgically through the uterine wall, this highly invasive procedure had many risks. This new method retrieves samples of fetal blood from the umbilical cord, using a needle guided by real-time ultrasound imaging. Called cordocentesis or percutaneous umbilical blood sampling (PUBS), this procedure is done when blood is needed to diagnose a hereditary blood disorder.

1984   William H. Clewall, American surgeon, performs the first successful surgery on a fetus in utero. With the ability to diagnose prenatal problems has come the ability to correct some of them in utero. Some of the correctable anatomic malformations are urinary tract obstruction, diaphragmatic hernia, and hydrocephalus.

1984   First successful gene therapy is accomplished in a mammal. This successful introduction of a normal gene into an animal to compensate for a defective one suggests the feasibility of using genetic engineering in a therapeutic manner on human patients.

(See also 1991)

1985   Lasers are used for the first time to unblock clogged arteries. A laser is a device that produces an extremely intense monochromatic, nondivergent beam of light capable of generating intense heat when focused at close range. Research on the use of lasers in medicine began around 1962, and lasers became useful for delicate eye surgery, removing birthmarks from the skin, and sealing blood vessels during operations to prevent bleeding. Now they are used to reopen blocked arteries by vaporizing fatty deposits.

1986   First genetically engineered (or recombinantly-produced) vaccine approved for human use is the hepatitis B vaccine. Compared to hepatitis A, also known as infectious hepatitis

which is spread by the oral-intestinal route, hepatitis B or serum hepatitis is mainly transmitted by blood transfusion or body fluid exchange. There are also two other types of hepatitis, but all are caused by viruses. To create dependable, sufficient supplies of a preventive vaccine, researchers make this genetically engineered product by inserting part of the B virus gene into baker's yeast cells which then produce large amounts of viral protein. This protein resembles the infectious hepatitis virus but lacked the part that would cause disease in humans. The United States Food and Drug Administration gives its approval.

APRIL 1987   First successful neural (brain) tissue transplants for the treatment of Parkinson's disease are achieved. This operation consists of taking a portion of tissue from the patient's adrenal gland and placing it into the caudate nucleus, an area deep inside the brain that is believed to control body movement. Although there is an initial flurry of enthusiasm and optimism for this procedure, this is tempered when the procedure is repeated in a number of patients with much less dramatic results. Subsequent studies on laboratory animals indicate that the future of this type of tissue transplantation may rest with fetal tissue taken from the part of the brain that manufactures dopamine, since the cardinal feature of Parkinson's disease is dopamine depletion. Another promising avenue for investigation is the use of cultured cells. Tissue transplantation to the brain has unlimited potential that is only now beginning to be explored.

APRIL 1988   First patent for a genetically engineered animal is issued for a "transgenic nonhuman mammal"—a genetically altered mouse that is predisposed to cancer. Although early bioengineering research was geared to producing cows that gave more milk or pigs that provide leaner bacon, this research involves the manipulation of animal genes to benefit human health. The mouse is called "transgenic" because it received a human gene (in this case, one that predisposed it to a disease). In 1992, scientists patent a strain of mice that carry a human gene making them resistant to viral infections, and in March 1993, geneticists breed the first two pigs with genetically modified hearts developed from human genes.

MAY 25, 1988 Louis Kunkel, Eric P. Hoffman, and colleagues announce that sufferers from Duchenne muscular dystrophy lack the protein dystrophin. After discovering the defective gene responsible for the disease, they find that the protein dystrophin, produced by the normal gene, is absent from the cells of Duchenne muscular dystrophy sufferers. They believe it may now be possible to diagnose the disease in its earliest stages. They also believe that experiments on mice with a defective gene, who were treated with new cells containing correct copies of the gene and began producing dystrophin, may be extended to humans.

1988 A scanning tunneling microscope (STM) produces the first clear image of molecular structure. The development of this powerful device makes the electron microscope obsolete. Using a tiny needle made of diamond, tungsten, or silicon, it relies on the subject's ability to conduct electricity through its needle. Although there are other powerful optical tools available, such as the field ion microscope, only the STM is flexible enough to peer into the structure of biological molecules and to be used on living cells.

SEPTEMBER 1990 First patient to receive human gene therapy to treat a genetic defect is a four-year-old American girl suffering from ADA deficiency. She is the first federally approved recipient of this new therapy. Because of this disease, adenosine deaminase (ADA) deficiency, her immune system is unable to fight infection. She receives a series of intravenous infusions of her own immune cells that have been "gene-corrected" by the introduction of normal genes encoding functional human ADA. The treatment has to be repeated

once a month because, like all body cells, lymphocytes (infection-fighting cells) die and must be replaced. The altered cells succeed in raising the girl's immune function to near normal levels.

(See also October 1995)

1992 First birth of a normal, healthy baby following preimplantation diagnosis. This new technique screens embryos that are fertilized in the laboratory prior to implantation in the mother's uterus. In this case, both parents are known to be carrying a genetic defect that is associated with cystic fibrosis. Physicians remove cells from the embryo three days after fertilization and screen them for genetic abnormalities. Following reimplantation, a baby girl is born free of both cystic fibrosis and the aberrant gene.

1992 First transplantation of a baboon's liver into a man whose liver was destroyed by hepatitis B is achieved. The patient dies of an infection 71 days later. When a second patient, a 62-year-old man, dies in February, 1993, 26 days after receiving a baboon's liver, there are many demands for such experiments to stop. Both are completed under the direction of American surgeon Thomas Earl Starzl. Starzl is a pioneer in organ transplantation and argues that the future of transplants lies in using animal parts. This may mark the beginning of animal-to-human transplants.

1993 Researchers in France announce that they have succeeded in constructing the first rough map of all the human chromosomes. As part of the Human Genome Project, the international effort to identify, map, and sequence every human gene, success here means that the entire project is on track for its target completion date of 2006.

1993 Teams of researchers in the United States and Europe discover the gene for the disease Huntington's chorea. They find a genetic "mistake" on chromosome 4 that gives affected individuals this hereditary neurological affliction. Although investigators had known for some time that the gene is located near the tip of the short arm of chromosome 4, the gene itself eluded detection for almost a decade. This effort is also noteworthy for its understanding of RFLP (restriction-fragment length polymorphism) which is a distinctive stretch of DNA that differs in length among individuals. These have enabled scientists to trace a disease through generations of an affected family and ultimately to the actual DNA responsible for the defect.

(See also 1983)

OCTOBER 20, 1995 First success of gene therapy is reported in an article published in *Science*.

Written by American researcher R. Michael Blaese of the National Center for Human Genome Research, the article details how, five years after two young girls were given immune system genes they lacked, they are healthy and thriving. Many scientists say that these results offer the best evidence yet of the potential of gene therapy, a technique in which doctors give patients healthy genes to replace the defective ones inherited from their parents or to enhance the action of genes they already have. This early enthusiasm should be tempered somewhat since it remains unclear how much of the girls' improvement can be attributed to their new genes and how much is due to a new drug, PEG-ADA, that they also have been taking.

(See also September 1990)

Physics

c. 2650 B.C. First great stone tomb, the Step Pyramid, is built by Egyptian scholar and architect Imhotep. As the first of the pyramids and the oldest known monument of hewn stone, it is 200 ft (61 m) high with a base 388 ft (118 m) by 411 ft (125 m). Its successful completion on the ancient site of Memphis evidences the ability of ancient engineers and their familiarity with the basics of mechanics.

2637 B.C. Earliest recorded reference to the use of magnetic influence is made in China during the reign of emperor Huang Ti. Reference is made to the use of a directional needle during wartime. Also called the "Yellow Emperor," he is one of China's three great legendary heroes and symbolizes a particular stage in the development of its technology. Among the many developments attributed to him is the first use of a compass.

c. 600 B.C. First to systematically study magnetism is Greek philosopher Thales of Miletus (624-546 B.C.). He is also the first to notice the electrification of amber by friction. He finds that when amber is rubbed, it becomes capable of picking up light objects. This process becomes known as triboelectrification. The Greek name for amber is *elektron* and it gives rise to many words we use today in connection with

electricity. Thales also writes one of the earliest treatises on physics and proposes water as the basic substance of the universe.

c. 575 B.C. Term "first principle" is first introduced by Greek philosopher Anaximander (610-c.546 B.C.). He states that the first principle is one, invisible thing he calls the unlimited or the infinite, and argues it is not one of the elements but rather something of an entirely different nature that comes out of the heavens.

c. 550 B.C. First to discover the mathematical ratios of the musical intervals is Greek philosopher Pythagoras (c.582-497 B.C.). Finding that shorter strings produce higher notes, he finds that the relationship of pitch can be simply correlated with length. If one string is twice as long, it emits a sound an octave lower. If the length ratio is three to two, the musical interval is called a fifth, and if it is four to three, the interval produced is called a fourth. He also posits that numbers and the proportions they contain (which he calls harmonies) are the first principle of nature.

c. 525 B.C. First to propose that air is the single, fundamental element of the universe is Greek philosopher Anaximenes (c.570-c.500 B.C.). He says that when compressed air takes the form of water and earth, and when rarified, it becomes fire. He also states that eternal motion causes the changes we see regularly.

c. 500 B.C. First to state that fire is the fundamental element of the universe is Greek philosopher Heraclitus (c.540-475 B.C.). He also asserts that all things are in constant motion and that nothing is ever lost. These ideas on the role of fire in nature and on the continuous state of change of matter influence the physical concepts of the Stoics.

c. 475 B.C. First to suggest that matter can be neither created nor destroyed is Greek philosopher Parmenides. This is an anticipation of the modern conservation of energy law. He holds that the multiplicity of existing things, as well as their changing forms and motion, are but an appearance of a single eternal reality. This gives rise to the Parmenidean principle saying "All is one."

c. 450 B.C. Greek philosopher Zeno (c.490-c.425 B.C.) first offers his four paradoxes, all of which seem to disprove the possibility of motion as our senses know it. These famous and controversial paradoxes go by names taken from the account of Greek philosopher Aristotle (384-322 B.C.). He names them the "Achilles," the "dichotomy," the "arrow," and the "stadium." Although the conclusions of each seem absurd, they are all based on Zeno's assumption that space and time are infinitely divisible, and this serves to stimulate further scientific thought and discussion.

c. 450 B.C. First to state the rule of causality, that every natural event has a natural cause, is Greek philosopher Leucippus. He is also credited with being the creator of atomism since he is the teacher of Greek philosopher Democritus (c.470-c.380 B.C.). He believes that every substance is a gathering of atoms or countless, small bodies of varying size and form. These are eternal, unchanging, and indivisible.

(See also 425 B.C.)

c. 450 B.C. Anaxagoras (c.500-c.428 B.C.), Greek philosopher, offers one of the first atomic theories, saying that all matter consists of atoms or "seeds of life." Although he states that all things can be infinitely divided, he also says, at the beginning of creation, the portions of what became the universe were so small that they could not be seen.

c. 450 B.C. First to state that all matter is composed of four essential ingredients (fire, air, water, and earth) is Greek philosopher Empedocles. He explains that two forces, Love and Strife, interact to bring together and to separate the four substances. His theory of the physical world is offered in the attempt to overcome the opposing school which emphasizes the unity of all things. In stating that nothing comes into being or is destroyed but that things are merely transformed according to the ratio of the four substances to one another, he anticipates the modern law of conservation of energy.

c. 425 B.C. First fully stated atomic theory, that all matter consists of infinitesimally tiny particles that are indivisible, is offered by Greek philosopher Democritus (c.470-c.380 B.C.). He states that these atoms are eternal, uncaused, and unchangeable, although they can differ in their properties. They can also recombine to form new patterns. His intuitive ideas contain much that is found in modern theories of the structure of matter and foreshadows theories of the indestructibility of matter and of the conservation of energy.

(See also 1473)

c. 325 B.C. Theophrastus (c.372-c.287 B.C.), Greek botanist, first observes the attractive properties of tourmaline, a precious crystal that the Swedish botanist Carolus Linnaeus (1707-1778) will later name "Lapis electricis." These attractive properties are the result of the crystal possessing what comes to be known as piezoelectricity. Piezoelectricity occurs when a crystal is subjected to mechanical pressure or heat and a positive electrical charge appears on one side and a negative charge on the opposite.

(See also 1756)

c. 300 B.C. First to argue that a body accelerates as it falls is Strato (c.340-c.270 B.C.), Greek physicist also called Straton of Lampsacus. He says a falling body moves more quickly with each successive unit of time. He agrees with Greek philosopher Aristotle (384-322 B.C.) that heavier bodies fall faster than light ones. Although he also agrees with Aristotle that a vacuum does not exist in nature, he describes methods for creating one. He becomes famous for his doctrine of the void, saying all substances contain void and differences in the weight of substances are caused by differences in the extension of the void.

c. 270 B.C. First to state that all matter is composed of five basic elements is Zou Yan of China. He argues that the five basic elements are earth, fire, water, metal, and wood.

c. 220 B.C. Principle of buoyancy is discovered by Greek mathematician and engineer Archimedes (c.287-c.212 B.C.). In his *Treatise on Floating Bodies*, he tells of his discovery of this principle with which he is able to determine if a crown is pure gold or not. He measures the amount of water it displaces and compares it to the quantity displaced by an equal amount of actual gold. He also works out mathematically the principle of the lever, showing that weights and distances are in inverse proportion, and develops the notion of a center of gravity. With all of this, he is regarded as the founder of the science called "statics."

c. 10 First to consider atmospheric refraction is Greek astronomer Cleomedes. He discusses the optical properties of water and says a ring on the bottom of an empty vessel just hidden by the edge, becomes visible when the vessel is filled with water. He then suggests that in the same way the sun may be visible when it has actually gone a bit below the horizon.

c. 50 Principle of the motive power of steam is first established by Greek engineer Hero, who builds many steam-powered devices. His most famous device is a hollow sphere to which two bent tubes are attached. When water is boiled in the sphere, steam escapes through the tubes and the sphere whirls rapidly around. This is a demonstration of what we now recognize as the law of action and reaction. He also writes on the various simple machines, describing the lever, pulley, wheel, inclined plane, wedge, and screw. Writing on the nature of air, he shows it is a compressible substance that takes up space. It will be another 1,500 years until these ideas are known again.

c. 850 First great Arab author in physics is physicist Al-Kindi (c.801-866). He writes a treatise on optics and the reflection of light and studies meteorology, the tides, and specific weights. Known also as "the philosopher of the Arabs," he concerned himself both with philosophical questions and scientific matters (applied as well). He is known to have written over 270 works.

c. 1025 First to maintain the correct theory of vision or human sight is Arab physicist Alhazen (c.965-1038). He rejects the prevailing idea that people see because their eyes send out a light which reflects back from the object, and argues that light comes from the sun or another source and reflects from the object into the eye. He works with lenses and attributes their magnifying effect to the curvature of their surfaces and not to any inherent property of the substances of which they are composed. He also studies all aspects of light, especially reflection and refraction, and discusses the rainbow. Altogether, his work represents the beginning of the scientific study of optics.

c. 1150 First known description of a machine for perpetual motion is offered by Bhaskara (1114-c.1178), Indian astronomer and mathematician. He discusses this concept in his

Siddhantasiromani, which is one of the great mathematical masterworks of India.

(See also c. 1235)

1180 First European to make reference to the directional ability of magnetism is English scholar Alexander Neckam (1157-1217). He does not invent the compass but mentions its use in Europe in one of his books.

c. 1200 First Western reference to the magnetic compass is made by German epic poet Hartmann von Aue, who mentions this device in one of his epic stories of the Royal court.

c. 1235 First to suggest a perpetual motion device using an overbalanced wheel is French architect and engineer Villard de Honnecourt. He leaves behind an album of drawings containing sketches of this and other machines— some practical, others theoretical. The idea of an overbalanced wheel as a source of perpetual motion is based on the notion that more energy can be extracted from falling weights on one side of the wheel than is required to raise those weights on the other. This false belief is, in turn, based on the incorrect notion that such weights deliver more effort if they are farther removed from the center of rotation than if they are closer.

c. 1250 First treatise on the study of motion from a purely kinematic point of view is written by Flemish physicist Gerard of Brussels. His *Liber de Motu* takes up and elaborates the kinematic ideas of Greek mathematician Euclid and Greek mathematician and engineer Archimedes (c.287-c.212 B.C.). Kinematics is the study of motion considered in itself, apart from its cause.

1269 First suggestion that magnetism might be converted to kinetic energy is made by Petrus Peregrinus, French scholar also called Pierre de Maricourt. He speculates that a motor that uses magnetic force to keep a planetarium moving might be constructed. This year he writes a small treatise on the compass titled *Epistola de Magnete* and discusses the nature and fundamental properties of magnetism. This is believed to be the first full, Western account of a compass and its workings. He also explains a compass would work better if the magnetic sliver or needle were put on a pivot instead of floating on a piece of cork.

c. 1325 Ockham's razor, also called the law of economy, is first popularized by English scholar William of Ockham (c.1280-1349). This principle states that when two theories equally fit all observed facts, the one requiring the fewest or simplest assumptions is to be accepted as the more valid. Although stated earlier by others, he mentions the principle so frequently it takes on his name. He uses it to argue strongly for the importance of empiricism and to uphold the notion that the simplest explanation is often the best. Also called the law of parsimony, it is used later by Italian astronomer and physicist Galileo Galilei (1564-1642), French mathematician Pierre Louis Moreau de Maupertuis (1698-1759), and Austrian physicist Ernst Mach (1838-1916), among others.

c. 1350 First major revision of Greek philosopher Aristotle's (384-322 B.C.) theory of motion is made by Jean Buridan (c.1300-c.1385), French philosopher. He refutes the Aristotelian notion that an object in motion requires a continuous force, and maintains that only an initial impetus is required. His impetus theory states that the mover imparts to the moved a power, proportional to the speed and mass, which keeps it moving. He also correctly theorizes that air resistance progressively reduces the impetus, and that weight can add or detract from speed. He anticipates Newton's first law of motion by saying that the celestial bodies stay in motion in this manner.

1440 First useful hygrometer is invented by German mathematician and scholar Nicholas of Cusa (1401-1464). In writing on his studies of hydraulics, he mentions not only a hygrometer which measures moisture but a bathometer for measuring depths in water.

1473   First Latin translation is made of *De Rerum Natura,* written by the Roman philosopher and poet Lucretius (c.95-c.55 B.C.). It is through this translation that the atomism of Greek philosopher Democritus (c.470-c.380 B.C.) becomes known in the West. Lucretius holds that all things are composed of atoms, including the mind, the soul, and even the gods. He argues mankind lives in an evolutionary universe of which the gods play no real role. In fact, his views are essentially antireligious.

(See also c. 425 B.C.)

c. 1500   First to describe capillary action (liquid crawling up the side of a tube) is Italian artist Leonardo da Vinci (1452-1519). Also called capillarity, it is the result of surface forces. The rise of water in a thin tube inserted in water is caused by forces of attraction between the molecules of water and the glass walls and among the molecules of water themselves. These attractive forces just balance the force of gravity of the column of water that has risen. He also experiments with hydrostatics and diffraction and offers a version of the principle of inertia (which will not come until the time of Galileo). He displays an ingenious insight and develops ideas for five simple machines.

(See also 1709)

1537   First book written on the theory of projectiles or ballistics is *La Nouva Scientia* by Italian mathematician Niccolo Tartaglia (1499-1557). A book on artillery science, it contains Tartaglia's theorem which states that the trajectory of a projectile is a curved line and that the maximum range at any speed of its projection is obtained with a firing elevation of 45°. However, he also believes incorrectly that a cannon ball falls straight downward after being propelled forward and losing its thrust.

1543   First Latin translation of several previous-

ly little-known works by Greek mathematician and engineer Archimedes (c.287-c.212 B.C.) are made and published by Italian mathematician Niccolo Tartaglia (1499-1557). This work contains his two basic works on statics and hydrostatics and details the propositions of Archimedes on the equilibrium of planes and on floating bodies. Both are demonstrated using an abstract mathematical approach. These works represent the application of geometrical analysis to statical theorems of mechanics and are responsible for much of the progress that follows in those fields.

1558   First published reference to communicating at a distance using a magnetic telegraph is made by Italian natural philosopher Giambattista della Porta (1535-1615). In his *Magia Naturalis,* he offers practical, theoretical, and experimental sections on optics and magnetism. He specifically states that, "to a friend that is at a far distance from us and shut in prison, we may relate our minds; which I doubt not may be done by two mariner's compasses, having the alphabet writ about them ...."

1581   First to note that steel, as is traditionally believed, does not change its weight when magnetized is English navigator and instrument maker Robert Norman. In his work on the lodestone called *The Newe Attractive,* he discusses the known properties of the magnet and describes his discovery of the "magnetic dip," suspending a compass needle to allow vertical movement and noting that it points down toward the earth. This is the first European publication of well-authenticated descriptions of the magnetic variation or declination made from actual observation. It is later used by English physician and physicist William Gilbert (1544-1603).

(See also 1600)

1581   Principle of isochronism, or the regularity of the pendulum, is discovered by Italian astronomer and physicist Galileo Galilei (1564-

1642). He observes a hanging lamp that is swinging back and forth and notes that the amount of time it takes the lamp to complete an oscillation remains constant, even as the arc of the swing steadily decreases. He then suggests that the principle of the pendulum may have an application in regulating clocks.

(See also 1673)

1586 Law of inclined planes is first stated by Belgian-Dutch mathematician Simon Stevin (1548-1620). The law states that less weight on a steep slope can balance more weight on a gentler slope. He also publishes a report of an experiment in which he refutes the Aristotelian doctrine that heavy bodies fall faster than light ones. His work on displacement is the first since antiquity, and he founds hydrostatics by demonstrating that the pressure on a liquid varies according to how high above the surface it is and not upon the shape of the container that holds it. This becomes a fundamental principle of hydraulics. He also offers demonstrations that eliminate many standard arguments in favor of the possibility of perpetual motion.

1587 Law of falling bodies is first stated by Galileo Galilei (1564-1642), Italian astronomer and physicist, who shows that a body's rate of fall is independent of its weight. He uses a gently sloping inclined plane to demonstrate this once and for all and later states correctly that all objects will fall at the same rate in a vacuum. He also demonstrates that a body can move under the influence of two forces at one time. He proves these and many other of his claims using the geometric methods of the Greeks and contributes much to the eventual downfall of Aristotelian physics.

1591 First Westerner to note that snowflakes are hexagonal (six-sided) is English mathematician Thomas Harriot (1560-1621). He does not publish his findings, but may have used some sort of magnifying device since he de-

scribes showing off such wonders to the Native Americans he encountered during his 1585 trip to America. He mentions "a perspective glass whereby was showed many strange sights." The Chinese, however, have known this fact about snowflakes from the second century B.C.

(See also 1611)

1600 First to use the terms electric attraction, electric force, and magnetic pole is William Gilbert (1544-1603), English physician and physicist. This year he publishes his *De Magnete*, a full account of his extensive investigations on magnetic bodies and electrical attraction, and which is considered the first great work on physical science produced in England. He suggests the earth itself is a great, round magnet, and he actually builds a model or "terrella," which is a lodestone ground into a spherical shape. This "little earth" functions as a model from which he is able to transfer his findings directly to the earth itself. Using this, he also shows how the "magnetic dip" phenomenon works. Gilbert is regarded by most as the father of electrical studies.

1611 Johann Kepler (1571-1630), German astronomer, publishes the first description of the hexagonal nature of snowflakes. In a pamphlet whose translated title is "On the Six-Cornered Snowflake," Kepler proves highly intrigued by the fact that all elements of snow appear to be hexagonal. The hexagonal form of these atmospheric ice crystals is an outward manifestation of an internal arrangement in which the oxygen atoms form an open lattice or network with a hexagonally symmetrical structure.

1611 Laws of refraction are first stated by German astronomer Johann Kepler (1571-1630) in his *Dioptrice*. With this work, Kepler founds the science of modern optics. He studies the newly invented telescope and applies geometric optics to the study of lenses and combinations of lenses, making a true theory of the telescope possible.

1621 Mathematics of the refraction of light are discovered by Dutch mathematician Willebrord van Roijen Snell (1580-1626). He finds when a ray of light passes obliquely from a rarer into a denser medium (such as from air into water), it is bent toward the vertical. Although this was known over 1,500 years ago by the Greeks, Snell offers a general mathematical relationship to express this refraction of light by relating the degree of the bending of light to the properties of the refractive material. This is a key discovery in optics but it goes unpublished.

1629 First to describe electrical repulsion is Italian physicist Niccolo Cabeo (1585-1650). He publishes his *Philosophia Magnetica* in which he continues English physician and physicist William Gilbert's experimental work on magnetism (1544-1603).

1635 First indication that the earth's magnetic field slowly changes over time is made by English astronomer and mathematician Henry Gellibrand (1597-1636) in his book, *A Discourse Mathematical of the Variation of the Magneticall Needle*. His readings made in London over many years demonstrate this variation, and to date, there is no satisfactory explanation for this occurrence.

(See also 1722)

1638 Foundations of modern mechanics are first laid by Galileo Galilei (1564-1642), Italian astronomer and physicist, in his *Discorsi e Dimostrazioni Mathematiche Intorno a Due Nuove Scienze*. In this work he formulates what comes to be known as the first law of motion (or the law of inertia), as well as the laws of cohesion and strength of materials, and of the pendulum. He is the first to show that if a structure continues to increase in all its dimensions equally it will grow weaker. Using what is known as the square-cube law, he shows that if a deer grows to the size of an elephant and keeps its exact proportions, it would collapse, for its legs would have to be thickened far out of proportion to

support its great body. He also provides a definition of momentum, and details the steps or stages of what comes to be known as the experimental method. In this, Galileo is credited with establishing the modern experimental method. This work also introduces the practice of proving or disproving a scientific theory by conducting tests and observing the results and eventually spells the death of Aristotelian physics.

1639 First work on impact (the action of two bodies in collision) is *De Proportione Motus* written by Johannes Marcus Marci von Kronland (1595-1667), Bohemian physician. Apart from the properties of the materials of the two objects, two factors affect the result of impact: the force and the time during which the objects are in contact.

1643 First to create a sustained vacuum is Italian physicist Evangelista Torricelli (1608-1647). He creates this first artificial vacuum in the process of inventing the barometer. He fills a 4-ft (1 m) glass tube with mercury and inverts it onto a dish of mercury. He observes that not all the mercury flows out and that over time, the level remaining in the tube varies. He concludes correctly that these changes are caused by atmospheric pressure. He further theorizes correctly that a vacuum is created above the mercury in the tube and that the mercury is held at a given level not by the vacuum, but by the pressure of air pushing down on the mercury in the dish. With this one experiment, he demonstrates the existence of a vacuum, explains why pumps then in use can move liquids vertically only to a certain height, and creates an instrument capable of measuring air pressure.

1644 Cartesian physics is first fully stated as French philosopher and mathematician René Descartes (1596-1650) publishes his *Principia Philosophiae*. At the core of its principles (among which are the nonexistence of the vacuum, the constancy of the quantity of motion, and the infinite speed of light) is an essentially mechanistic view of the world. He attacks and rejects

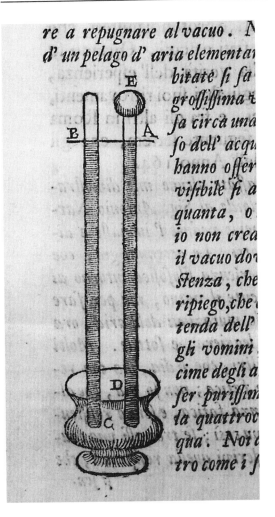

re a repugnare al vacuo. N
d' un pelago d' aria elementa
bitatē fi fa
groffiffima t
fa circa una
fo dell' acqu
hanno offer
vifibilē fi a
quanta, o
io non crea
il vacuo do
ſtenza, che
ripiego, che
tenda dell'
gli vomini
cime degli a
ſer puriffim
da quattrod
qua. Noi
tro come i f

Illustration of Torricelli's barometer.

ern scientific thought by restricting its modes of explanation to purely causal considerations and mechanical analogies.

1648 Variability of atmospheric pressure is first demonstrated by French mathematician and physicist Blaise Pascal (1623-1662). Although he is not able to participate in this experiment, which requires an ascent up the Puy de Dom mountain, Pascal conceives and directs this test which demonstrates that air pressure decreases as altitude increases. He sends two Torricelian barometers up the mountain, and their mercury columns drop as he predicts. This means if the mercury column is actually held up by air pressure (if one goes higher), then there should be less air above to push down on the mercury outside the vacuum, and the column should therefore drop. Pascal's experimenters show the mercury does indeed drop 3 in (8 cm) in a 1 mi (2 km) ascent. This is an experimental verification of the barometric principles discovered in 1643 by Italian physicist Evangelista Torricelli (1608-1647).

1648 Johannes Marcus Marci von Kronland (1595-1667), Bohemian physician, publishes *Thaumantias Liber De Arcu Coelesti Deque Colorum Apparentium Natura* in which he is the first to discuss the diffraction of light. Although he is its discoverer, it does not become a recognized fact for nearly 20 years.

(See also 1665)

most Aristotelian physical concepts and sees the entire universe acting as a great machine.

1647 First book written explicitly to spread and defend the notions of Cartesian physics is the *Fundamenta Physices* written by Henri Duroy (1598-1679), Dutch physician also called Deroy. Based on the writings of French philosopher and mathematician René Descartes (1596-1650), Cartesian physics is essentially a mechanistic approach to nature and in particular to physical phenomena. This mechanistic interpretation of nature profoundly influences mod-

1650 Otto von Guericke (1602-1686), German physicist, constructs the first air pump with which he is able to create a vacuum. He conducts several experiments using evacuated spheres that show a vacuum's properties. He demonstrates that animals cannot live in a vacuum nor will a candle burn or a bell be heard. The pump he devises is hand operated and like a water pump, but with its parts well-fitted to be reasonably airtight. His experiments are sometimes conducted with great fanfare and in public, and his achievements become well-known and widespread.

1653 Pascal's principle is first formulated by French mathematician and physicist Blaise Pascal (1623-1662). While studying fluid pressure by exerting pressure on water and noting how that pressure is transmitted against the walls of a closed vessel, he concludes the pressure is transmitted undiminished throughout the fluid and it pushes at right angles to all surfaces it touches. This notion that the pressure at any point in a liquid is the same in all directions forms the basis of the hydraulic press (which he also describes in theory). What comes to be known as Pascal's principle is not published until a year after his death.

1657 Accademia del Cimento is founded in Florence, Italy. Supported by the Medici family and the Grand Duke of Tuscany nearly a decade before its formal creation, this early scientific society is organized by two pupils of Italian astronomer and physicist Galileo Galilei (1564-1642), Italian mathematician Vincenzo Viviani (1622-1703), and Italian physicist Evangelista Torricelli (1608-1647). They invite many of the leading men of science to come together and share in experimental physics. Together they work out new research methods, invent new instruments, and devise better standards of measurement. They also publish in 1666 a collection of their experiments titled *Saggio di Naturali Esperienze*. When finally translated into Latin in 1731, this work of experimental physics becomes the laboratory manual of the 18th century.

1660 First machine to generate an electrical charge is invented by German physicist Otto von Guericke (1602-1686). He invents a frictional electrical device which generates static electricity. His hand-rotated globe of sulfur mechanizes the act of rubbing and accumulates static electricity. Since it can be discharged and recharged indefinitely, he is able to conduct several electrical experiments with it, generating sizeable electric sparks and most notably establishing the principle of electrical repulsion. Perhaps his largest contribution is simply the enthusiasm for electrical experimentation his work creates.

1662 Robert Boyle (1627-1691), English physicist and chemist, discovers the law which states air is not only compressible, but this compressibility varies with pressure according to a simple inverse relationship. He conducts his experiments by trapping air in the short, closed end of a J-shaped, 17-ft (5 m) long glass tube into which he pours mercury to close off the bottom. He discovers that by adding more mercury, the additional weight of the fluid squeezes the trapped air more closely together and that its volume decreases. He then finds the volume varies inversely with the pressure, so that if he doubles the mercury weight, the volume shrinks to one-half. The implications of Boyle's law are that air and other gases have atoms that are widely spread apart.

1665 Robert Hooke (1635-1703), English physicist, publishes his *Micrographia* in which he compares light with waves in water. This is the first serious opposing theory to the Pythagorean concept of light as a stream of particles. It also contradicts the theory of English scientist and mathematician Isaac Newton (1642-1727), who states that since light moves in straight lines and casts shadows, it consists of a stream of particles moving from the luminous object to the eye.

1665 Discovery of the diffraction of light is revealed in the posthumous publication of *Physico-Mathesis de Lumine, Coloribus, et Iride*. Written by Italian physicist Francesco Maria Grimaldi (1618-1663), this work describes an experiment in which he lets a beam of light pass through two narrow apertures, one behind the other, and then onto a blank surface. When he notes the band of light on the surface is slightly wider than it should be, he concludes that the light has been bent slightly outward. He names this phenomenon diffraction. Although the question of whether light is composed of waves or particles is beginning to be raised about now, the controversy continues for over a century without any reference to Grimaldi's discoveries.

1666 Isaac Newton (1642-1727), English scientist and mathematician, first entertains thoughts on gravitation while on leave from school due to the London plague. Newton himself states that it is while watching an apple fall to the ground that he begins to wonder if the same force that pulls the fruit downward also holds the moon in its grip. Unable to fully calculate a satisfactory proof of his idea at this time, he puts this problem aside for nearly 20 years.

(See also 1684)

1669 Optical phenomenon called double refraction is discovered by Danish physician Erasmus Bartholin (1625-1698). He notes that images seen through Icelandic feldspar (calcite) are not only doubled, but when the crystal is rotated, one image remains still while the other whirls with the crystal. He calls the stationary light the "ordinary beam" and the moving image the "extraordinary beam." This discovery is not explained by Bartholin or anyone else until polarized light becomes better understood in the early 19th century.

1672 Isaac Newton (1642-1727), English scientist and mathematician, publishes his letter on light in the Royal Society's *Philosophical Transactions*. This letter, which is his first scientific publication, details his prism experiments of 1666 and offers findings that reveal for the first time the true nature of light. He recounts how he let a ray of sunlight enter a darkened room through a small hole and then passed the ray through a prism onto a screen. The ray was refracted and a band of consecutive colors in rainbow order appeared. He then passed each separate color through another prism and noted that although the light was refracted, the color did not change. From this, he deduces that sunlight (or white light) consists of a combination of these colors. Later he elaborates further on this ground-breaking experiment in his 1704 book, *Opticks*.

1673 Principle of isochronicty is first applied to regulate a clock by Dutch physicist and astronomer Christiaan Huygens (1629-1695). This year he publishes his *Horologium Oscillatorium* in which he details his invention of the pendulum or "grandfather" clock. He employs this principle stated by Italian astronomer and physicist Galileo Galilei (1564-1642) nearly a century before and ingeniously adapts it to the inner workings of a clock. His device begins the era of accurate timekeeping so important to the advancement of physics. His highly original work not only demonstrates great mechanical ability but superior mathematical theorizing as well.

(See also 1581)

1675 First recorded observation of barometric light is made by French astronomer Jean Picard (1620-1682). This is a light or electric glow that appears in the vacuum above the mercury in a barometer when it is moved about. This luminous glow is the result of an electrical charge that takes place with a variety of rarified gases that are trapped in the tube. Shaking is essential since electrification is caused both by the splashing of the mercury and by its movement over the glass surface.

1676 Edme Mariotte (1620-1684), French physicist, independently discovers Boyle's law and adds an important qualification to it. Like English physicist and chemist Robert Boyle (1627-1691), he notes that air expands with a rising temperature and contracts with a falling temperature, but adds that the inverse relationship between temperature and pressure only holds if the temperature is kept constant. Because of this, Boyle's law is called Mariotte's law in France.

(See also 1662)

1678 Francesco Redi (1626-1697), Italian physician, is the first to communicate the fact that the electric shock of the torpedo fish may be transmitted to the fisherman through the line and rod. These cartilaginous fish have a pair of

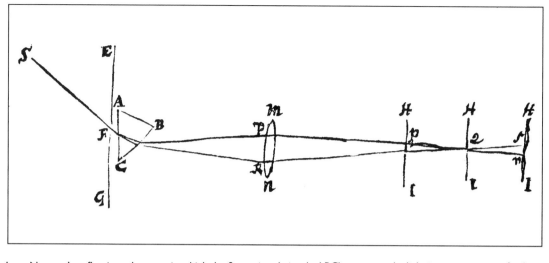

Isaac Newton's reflecting telescope, in which the first prism (triangle ABC) separates the light into a spectrum of colors and the second prism refracts a single color beam.

electric organs near their head capable of shocking and stunning their prey. Bioelectricity, or electric activity in living tissue, is a cellular phenomenon dependent on the cell membrane. The membrane acts like a capacitator, storing energy as electrically charged ions on opposite sides of the membrane. This stored energy is then available for rapid use.

1678 Law of elasticity is first stated by English physicist Robert Hooke (1635-1703) in his *Lectures...Explaining the Power of Springing Bodies*. Now known as Hooke's law, this says the force tending to restore a spring to the equilibrium position is proportional to the distance by which it is displaced from that equilibrium position. While experimenting with systems of springs, he finds that the amount of weight added is proportional to the distance the spring stretches, so that a 4 lb (2 kg) weight will stretch a spring twice as far as a 2 lb (1 kg) weight. Stress is therefore proportional to the strain.

1684 Isaac Newton (1642-1727), English scientist and mathematician, provides the first summary exposition of his theory of gravitation in a memoir entitled *De motu corporum*. When

asked by his friend, English astronomer Edmond Halley (1656-1742), how the planets would move if there was a force of attraction between bodies that weakened as the square of the distance, Newton tells him they would move in ellipses, since he calculated this in 1666. Halley urges Newton to resume work on this problem, and he soon expands this memoir after 18 months of work into his 1687 book, *Philosophiae Naturalis Principia Mathematica*.

1686 Edmond Halley (1656-1742), English astronomer, develops a fairly reliable formula linking the altitudes of various localities with the atmospheric pressure measured there. His altimetric formula is one of the first practical applications of the new barometric discoveries. It is on the basis of Halley's formula that instruments known as altimeters are eventually built.

1687 First systematic work on theoretical physics is English scientist and mathematician Isaac Newton's (1642-1727) *Philosophiae Naturalis Principia Mathematica* in which he states his three laws of motion. The first enunciates the principle of inertia: a body at rest remains at

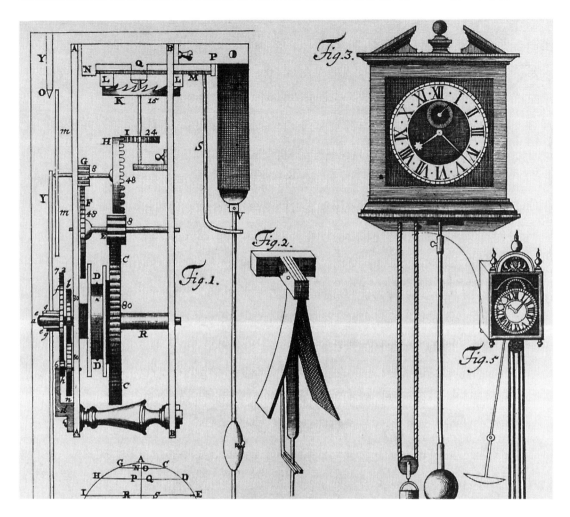

Diagram of Christiaan Huygen's pendulum clocks.

rest and a body in motion remains in motion at a constant velocity as long as outside forces are not involved. The second defines force as the product of mass and acceleration. This gives the first clear distinction between the mass of a body and its weight. The third law of motion states that for every action there is an equal and opposite reaction. From these laws, Newton is able to deduce his law of universal gravitation. He shows it is directly proportional to the product of the masses of two bodies (earth and moon) and inversely proportional to the square

of the distance between their centers. This landmark work opens the era of classical mechanics.

1690 Wave theory of light is first stated by Danish physicist and astronomer Christiaan Huygens (1629-1695). He argues in his *Traité de la Lumière* the unpopular theory that sees light as a longitudinal wave that undulates in the direction of its motion much as a sound wave does. He is opposed by English scientist and mathematician Isaac Newton (1642-1727) who argues that light is a particle (because there is a vacuum between the earth and the sun, and he cannot see how a wave can travel where there is nothing to wave). Huygens counters by saying that there must be some sort of very subtle fluid in space.

1694 Jacques Bernoulli (1654-1705), Swiss mathematician also called Jakob, discovers the equation of the elastic curve of a beam. This is the curve of the neutral surface of a structural member subjected to loads which cause bending. Three years earlier, he studies the catenary (the curve formed by a cord suspended between its two extremities), and it is soon applied in the building of suspension bridges. In 1695, he also applies calculus to the design of bridges.

1701 Term "acoustics" for the study of sound is first introduced by French mathematician Joseph Sauveur (1653-1716). He examines the relations of musical scale tones, discovers overtones in strings, and gives a correct explanation of beats. He also performs experiments by which he determines rates of vibration with considerable precision.

1702 David Gregory (1661-1708), Scottish mathematician and astronomer, publishes his *Astronomiae Physicae et Geometriae Elementa*, the first astronomy text based on English scientist and mathematician Isaac Newton's (1642-1727) gravitational principles.

1709 First to study capillary action experimentally is Francis Hauksbee (c.1666-1713), English physicist. In his *Physico-Mechanical Experiments on Various Subjects* he makes the first accurate observations of the effects involving the attractive forces between a liquid and a solid. It is this action that causes water to rise within thin tubes and to spread over a flat surface.

(See also c.1500)

1714 First really accurate thermometer is invented by Gabriel Daniel Fahrenheit (1686-

1736), German-Dutch physicist. His use of mercury instead of alcohol means temperatures far above the boiling point of water and well below its freezing point can be measured (since alcohol has a low boiling point). He also invents the Fahrenheit temperature scale at which the freezing point of water is 32° and the boiling point is 212°. He arrives at these by adding salt to water to find the lowest freezing point which he calls zero. Physicists and other scientists would now possess a precision instrument for measuring temperature.

1722 Daily variation of magnetic declination is discovered by English optician and instrument maker George Graham (1673-1751). Because the earth's geomagnetic poles do not coincide with the geographic poles, there is a degree of departure between the two called magnetic declination. It is also sometimes called magnetic variation.

(See also 1635)

1729 Founder of photometry, the measurement of light intensities, is Pierre Bouguer (1698-1758), French physicist. This year he publishes his *Essai d'Optique sur la Gradation de la Lumière* in which he makes some of the earliest measurements in astronomical photometry. He also investigates the absorption of light in the atmosphere and formulates what comes to be known as Bouguer's law. This concerns the attenuation of a light beam upon passage through a transparent medium. He also is the first to attempt to measure the horizontal gravitational pull of mountains.

1729 First to use the term "physics" in place of "natural philosophy" is Dutch physicist Pieter van Musschenbroek (1692-1761). He also builds the first crude testing machine using a steelyard. This is a form of balance in which the object is suspended from the shorter arm of a lever, and its weight found by moving a counterpoise along the arm to produce equilibrium.

1729 First to divide material into "electrics"

(conductors) and "non-electrics" is Stephen Gray (c.1696-1736) of England. His research demonstrates that the difference in electrical conductivity depends on what the object is made of and not, as thought, such irrelevant things as color or shape. He then shows that metal conducts and silk thread does not. The next year he demonstrates that the human body is a conductor. He also discovers conductors can be prevented from doing so if they are placed on non-conducting material like resin.

(See also 1742)

1733 First to suggest the use of conductors as a means of protection against lightning is Johann Heinrich Winckler (1703-1770), German physicist. He also produces an improved frictional electrical machine that substitutes a piece of leather pressed against the glass by a spring instead of using a dry human palm. This proves that the human body does not play an essential part in the conduction of electricity as believed.

1733 Charles Francois de Cisternay Du Fay (1698-1739), French physicist, discovers that two electrified objects sometimes attract and sometimes repel each other. He notes that their means of being charged seems to account for the difference, and he states therefore that there are two types of electricity—"resinous" and "vitreous." He then formulates the basic electrical law that "like charges repel and unlike charges attract." Later, American statesman and scientist Benjamin Franklin (1706-1790) calls these two types of electricity "positive" and "negative."

(See also 1751)

1738 Bernoulli's principle is first stated by Swiss mathematician Daniel Bernoulli (1700-1782). In his comprehensive work on fluid flow, *Hydrodynamica*, he demonstrates the principle that as the velocity of fluid flow increases, its pressure decreases. Bernoulli also becomes the first to attempt to explain the behavior of gases with changing pressure and temperature. While

several others had earlier observed and described this phenomenon, none had yet to attempt to explain it scientifically. Bernoulli treats the problem mathematically, and although he does not solve it completely, he makes a good start.

1742 First to use the word "conductor" to describe those substances that allow electricity to flow is John Theophile Desaguliers (1683-1744), French-English physicist. In his *A Dissertation Concerning Electricity*, he also names nonconducting materials "insulators."

1742 Andrew Gordon (1712-1751), Scottish physicist, abandons the use of glass globes to produce electricity and is the first to use a glass cylinder which he is able to rotate rapidly. He achieves 680 revolutions a minute with his new device and can thus generate enough electricity to badly shock a person.

1742 New temperature scale is first described by Swedish astronomer Anders Celsius (1701-1744). He applies a new scale to his thermometer by dividing the temperature difference between the boiling and freezing points of water into an even 100° (with 0 at the freezing point and 100 at the boiling point). Compared to the Fahrenheit scale whose freezing and boiling values are the oddly chosen 32 and 212, the simplicity of a system in which a positive reading means water and a negative value means ice is obvious and attractive to all. His system is adopted initially as the "Centigrade" scale (from the Latin for "hundred steps"), but is eventually converted to the "Celsius" scale by international agreement in 1948. Oddly, the United States is the lone exception, as the entire world adopts Celsius and it alone retains Fahrenheit.

1744 Principle of least action is first stated by French mathematician Pierre Louis Moreau de Maupertuis (1698-1759). This year he publishes a paper titled "Accord de differentes lois de la nature qui avaient jusqu'ici paru incompatibles," in which he states that physical laws include a rule of economy in which action is a minimum. Maupertuis argues that this principle shows that nature chooses the most economical path for moving bodies, rays of light, and other things.

1744   First demonstration of the ignition of inflammable substances by an electric spark is made by Christian Friedrich Ludolf of Germany.

1745   First use of electricity to relieve muscle sprains is made by German physician Christian Gottfried Kratzenstein (1723-1795).

1745   First "Leyden jar" is discovered by Dutch physicist Pieter van Musschenbroek (1692-1761). This device for storing electricity is also discovered independently this same year by German physicist Ewald Georg von Kleist (1700-1748). Musschenbroek makes his accidental discovery when he places water in a glass bottle suspended by insulating silk cords. He then leads a brass wire connected to an electrical machine through a cork into the water. One of Musschenbroek's students named Cunaeus (or Cuneus) picks up the container while touching the brass wire and nearly dies from the electric shock he receives. This startling demonstration shows just how much of an electric charge can be stored in this bottle or jar. Kleist has a similar shocking experience and retires from any further such work. Because Musschenbroek continues to experiment and further popularize his discovery, and because he does his work at the University of Leyden, this electricity-storing device comes to be called a "Leyden jar." It becomes immediately adopted by other experimenters who can discharge its electricity at will, and is of future importance as a prototype of capacitators, which are widely used in radios, television, and other electrical and electronic equipment.

1746   First in France to experiment with the new Leyden jar is French physicist Jean Antoine Nollet (1700-1770). He performs electrical

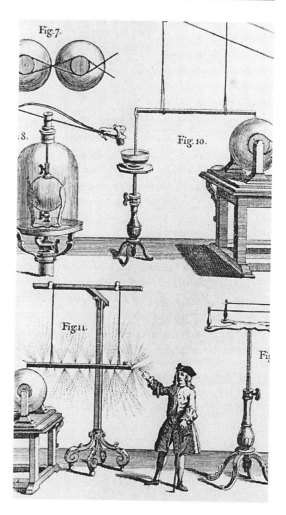

Several apparatus used to perform early experiments in electricity.

demonstrations before the king and is responsible for much of the popularization of electricity and electrical phenomena.

1746   First to make powerful bar magnets is Gowin Knight (1713-1772), English natural philosopher and physician. Concerned with compasses and navigation, he rubs a bundle of magnets on an iron bar to create super magnets which he then uses to magnetize compass needles for longer periods of time. These are needed as the range of England's ships keeps increasing.

Knight's super magnets are used by the British Navy for the next 80 years.

(See also June 10, 1766)

1746 First to recognize that the air or atmosphere is almost constantly charged with electricity even in clear weather is Louis Guillaume Le Monnier (1717-1799), French naturalist.

1746 First to strengthen the charge of a Leyden jar by applying a coating of tinfoil or sheet-lead is John Bevis (1693-1771), English physician and astronomer. He is also the first to observe that the charge increases as larger jars are used.

1746 First to investigate the passage of a current through rarified gas and to discover that conductivity increases is William Watson (1715-1787), English physician and botanist. He publishes his *Experiments on the Nature of Electricity* this year in which he announces his discoveries.

1750 John Michell (1724-1793), English geologist, publishes *A Treatise on Artificial Magnets* in which he anticipates French physicist Charles Augustin de Coulomb (1736-1806) in his first explanation of the inverse-square law for repulsive forces of magnetism. It is Coulomb, however, who fully explains the law. Michell also later anticipates Coulomb in his invention of the torsion balance—a device to measure extremely small forces. Unfortunately, Michell dies before he can use his device to measure the strength of the gravitational constant, which English physicist and chemist Henry Cavendish (1731-1810) achieves in 1798.

(See also 1777 and 1785)

1751 First major theoretical work on electricity is written by American statesman and scientist Benjamin Franklin (1706-1790). He publishes a two-volume work, *Experiments and Observations on Electricity Made at Philadelphia in America*, in which he formulates a theory of general electrical "action" also called his single fluid theory of electricity. His theory is adopted as

are several of the terms and concepts he introduces, like positive and negative charge.

(See June 1752)

JUNE 1752 First to prove that electricity generated in the atmosphere and artificially on earth are identical is American statesman and scientist Benjamin Franklin (1706-1790). He flies a kite carrying a pointed wire in a thunderstorm and attempts to test his theory that atmospheric lightning is an electrical phenomenon similar to the spark produced by an electrical frictional machine. He attaches a silk thread to the kite with a metal key at the end, and as lightning flashes, he puts his hand near the key which sparks just as a Leyden jar would. He proves his point in this extremely dangerous experiment. Franklin is wise enough to connect a ground wire to his key. Two other scientists attempting to duplicate the experiment neglect this ground wire and are killed when struck by lightning.

1752 First lightning rod is invented by American statesman and scientist Benjamin Franklin (1706-1790). After performing his famous kite experiment in June of this year and proving that lightning is identical to electricity generated on earth, he puts his discovery to good use. Deciding that lightning strikes a building because the structure gathers charge during a storm, and realizing that a Leyden jar discharges very easily if a sharp needle is attached to it, he theorizes that a sharp metal rod on top of a building that is grounded properly would silently dissipate any charge into the ground. He is proven correct and lightning rods appear first in America and then spread to Europe.

(See also 1778)

1756 First observation of what comes to be known as magnetic storms is made by John Canton (1718-1772), English physicist. He begins three years of careful weather observations and finds that on days when the aurora borealis is very noticeable, a compass needle becomes irregular.

1756 First scientific examination of the electric properties of a tourmaline crystal is made by Franz Maria Ulrich Theodor Hoch Aepinus (1724-1802), German mathematician and physicist. He shows that heat, in the form of a temperature between 99°F and 212°F (37°C and 100°C), is necessary for its attractive powers.

1759 First systematic attempt to apply mathematics to the theory of electricity and magnetism is made by Franz Maria Ulrich Theodor Hoch Aepinus (1724-1802), German mathematician and physicist. His book *Tentamen Theoriae Electricitatis et Magnetismi* is one of the most original and important books in the history of electricity. It is the first reasoned exposition of electrical phenomena based on action-at-a-distance.

1760 Johann Tobias Mayer (1723-1762), German astronomer, is the first to state that magnetic action follows the law of inverse squares. This states that the force acting between two magnetic poles separated from each other is proportional to the magnitude of each of the two poles, divided by the square of the distance between them.

1760 First to devise methods for measuring light intensities accurately is Johann Heinrich Lambert (1728-1777), German mathematician. He coins the word "photometry" in his book, *Photometria*, which lays the foundation for this new discipline of measuring visible light.

JUNE 10, 1766 First English patent in the field of electricity and magnetism is obtained by Gowin Knight (1713-1772), English natural philosopher and physician. His patent is for the construction of compasses "so as to prevent them from being affected by the motion of the ship." He finds the British navy's compass needles are made of wires that are crudely magnetized and inaccurate. He develops a needle with this patent incorporating a type of steel

that increases magnetic movement and improves suspension. His new compass becomes a standard for the Royal Navy for nearly the next century.

(See also 1746)

1766 First device used to measure electric potential is invented by Horace Benedict de Saussure (1740-1799), Swiss physicist. His electrometer is able to measure electrical potential difference by means of electrostatic forces.

1766 First chart containing information on magnetic inclination is made by Swedish physicist Johann Carl Wilcke (1732-1796). He later conducts experiments that help introduce the concept of specific heat.

(See also 1781)

1769 Electrical concept of potential is first introduced by Alessandro Giuseppe Antonio Anastasio Volta (1745-1827), Italian physicist. In what is his first work, the young Volta publishes his *De Vi Attractiva Ignis Electrici ac Phaenomenis Indepentibus*.

1775 Hair hygrometer is first developed by Swiss physicist Horace Benedict de Saussure (1740-1799). He uses a human hair that moves and indicates a change in the humidity. In 1783, he publishes his *Essais sur L'Hygrometrie* in which he details this invention and discusses the general principles of hygrometry.

1776 First quantitative measurements of the earth's field of magnetic force are made by Jean Charles de Borda (1733-1799), French mathematician. He is also the first to accurately establish a knowledge of a third element of terrestrial magnetism, i.e., its intensity. Comparative intensities are obtained by measuring the vibration periods of a vertical needle placed in the magnetic meridian. This completes the introduction of the three elements of the terrestrial field—variation, dip, and intensity.

1777 First practical torsion balance is invented

by French physicist Charles Augustin de Coulomb (1736-1806). This device is used to measure an extremely small force by quantifying the amount of twist it produces in a thin, stiff fiber. He finds that since the amount of torsion (twisting) is proportional to the amount of force (and weight is a force), his device can be used to measure tiny weights with considerable delicacy. He then puts his new instrument to work on his electrical experiments.

(See also 1750 and 1785)

1778 Giuseppe Toaldo (1719-1798), Italian meteorologist, first introduces the lightning rod in the Venetian States.

(See also 1752)

1780 Earliest known observations of the magnetic dip in the United States are made by Samuel Williams, American professor in Massachusetts.

(See also 1581)

1781 Concept of specific heat is independently introduced by Scottish chemist Joseph Black (1728-1799), and German physicist Johann Carl Wilcke (1732-1796). This theory states that different bodies in equal masses require different amounts of heat to raise them to the same temperature. Specific heat is the ratio of the quantity of heat required to raise the temperature of a body one degree to that required to raise the temperature of an equal mass of water one degree. In the early 19th century, it is discovered that measurements of specific heats of substances will allow calculation of their atomic weights.

1783 First proof that kinetic energy is lost in the collision of bodies is offered by Lazare Nicolas Marguerite Carnot (1753-1823), French military engineer and mathematician, in his *Essai sur Les Machines en Général*. In this systematic exposition of mechanics and the concept of force, Carnot emphasizes the notion that force works in imperceptible degrees and is not apt to produce sudden variations.

1785 First of seven papers in which the basic laws of electrostatics and magnetism are established are published by French physicist Charles Augustin de Coulomb (1736-1806). Using his newly invented torsion balance, he determines that English scientist and mathematician Isaac Newton's (1642-1727) law of inverse squares also applies to electrical and magnetic attraction and repulsion. He states that the degree of attraction or repulsion depends on the amount of the electric charge or the magnetic pole strength (Coulomb's law). This means that electrical forces obey a rule similar to that of gravitational forces.

(See also 1750 and 1777)

1787 First to observe the contact electrification of metals is English physicist and inventor Abraham Bennet (1750-1799). He also invents the gold-leaf electroscope which is considered to be the most sensitive and most significant of the early instruments used for detecting the presence of electricity. It consists of a glass cylinder covered with a brass cap that has an opening for wires. Inside the cap is a tube with two strips of gold leaf which, when an electric charge is brought near, stand apart at an angle (the charge causes them to repel each other).

1787 First to make an accurate estimate of the degree of expansion of a gas is French physicist Jacques Alexandre Charles (1746-1823). He repeats the gas-expansion work of his countryman, French physicist Guillaume Amontons (1663-1705), and is the first to offer the rule that the volume of a given quantity of gas is proportional to the absolute temperature if the pressure is held constant. This rule is called both Charles' law and Gay-Lussac's law. Since Charles does not publish his discovery, French chemist Joseph Louis Gay-Lussac's (1778-1850) demonstration in 1802 that different gases expand by equal amounts with a rise in temperature, induces many to give him the honor.

1787 First to discover the quantitative relationships that rule the transmission of sound is German physicist Ernst Florens Friedrich Chladni (1756-1827). His discovery is publicized in his *Theorie des Klanges,* published this year. He also creates "Chladni's figures" by spreading sand on thin plates and vibrating them, producing complex patterns from which much is learned about vibrations. In later work, he discovers the longitudinal vibrations in a string or rod, as well as their application to the determination of sound velocity in solids. He determines the velocity of sound in gas other than air by filling organ pipes with the gas and determining the resulting pitch. As the first to investigate torsional vibrations in rods and to determine the absolute rate of vibration of bodies, he is considered the father of acoustics.

1790 First to report on the negative electricity of waterfalls is German-Swiss physicist Johann Georg Tralles (1763-1822). While in the Alps, he notices that his electrometer reacts when near a fine spray. He continues to experiment to understand this phenomenon, and eventually attributes it to the evaporation of the spray.

1790 Friedrich Albrecht Carl Gren (1760-1798), German chemist and physicist, founds the *Journal der Physik* as a periodical for the "mathematical and chemical branches of natural science." It is succeeded by the *Neues Journal der Physik* and is known today as the *Annalen der Physik.*

1791 "Animal electricity" is first proposed by Italian anatomist Luigi Galvani (1737-1798). Observing that the muscles of a dissected frog would twitch when touched by an electric spark from a Leyden jar, he concludes after years of experimentation, that animal tissue contains a heretofore unknown innate, vital force. He calls this force "animal electricity," and says that it activates nerve and muscle when touched by metal probes. He argues that this is a different form of electricity than that produced during a lightning storm or by a Leyden jar or torpedo fish. He builds an entire anatomical theory

around animal electricity. Although he is proved partly wrong by Italian physicist Alessandro Giuseppe Antonio Anastasio Volta (1745-1827), he is a pioneer in the field of electrophysiology and his work leads to the invention of the voltaic pile, the first battery. His name also becomes a household word.

(See also March 20, 1800)

1791 Theory of exchanges is first offered by Swiss physicist Pierre Prévost (1751-1839). He correctly states that all bodies of all temperatures radiate heat, noting that the cold does not flow from the snow to the hand, but rather it is the loss of heat and not the gain of cold that makes the hand feel cold. This is an important and fertile distinction to the later science of thermodynamics.

(See also 1824)

1798 First to calculate the earth's mass is Henry Cavendish (1731-1810), English chemist and physicist. He does this by obtaining what English scientists and mathematician Isaac Newton (1642-1727) had not provided—a value for the gravitational constant. He builds a model with light-weight balls and large, heavy ones and uses a sensitive wire to calculate the strength of the attraction between the two. Once he obtains this constant value, he calculates the earth's mass at nearly what we know it to be today. This classic calculation comes to be known as the Cavendish experiment.

1798 Heat as a form of motion is first stated by English-American physicist Benjamin Thompson, Count Rumford (1753-1814). Observing the intense heat generated when a cannon is bored with a dull drill, he is led to question the prevailing "caloric" notion of heat (that heat is a fluid or liquid form of matter). In order to disprove the existence of caloric, he weighs an amount of water in both liquid and solid form and reasons that if caloric exists, the water should weigh more at a higher temperature (since it should have more caloric). Finding no difference in weight, he states that caloric is either weightless or nonexistent. His work eventually overturns the old caloric theory and establishes the beginnings of the modern theory that heat is a form of motion.

1799   First known suggestion of the chemical origins of the electrical discoveries of Italian anatomist Luigi Galvani (1737-1798) is made by Italian chemist Giovanni Valentino Mattia Fabbroni (1752-1822). He does not propose an alternative theory to galvanism, however.

MARCH 20, 1800   Voltaic pile or first real battery is invented by Italian physicist Alessandro Giuseppe Antonio Anastasio Volta (1745-1827). His work duplicating the "animal electricity" experiments of Italian anatomist Luigi Galvani (1737-1798) leads him to discover that it is the contact of dissimilar metals that causes the electricity. Galvani argues that the electricity originates in the muscles themselves. Volta then uses bowls of salt solution connected by arcs of metal dipping from one bowl to the next, with one end of the arc being copper and the other tin or zinc. This produces a steady flow of electrical current. He later streamlines this by arranging suitable pairs of metallic plates in a certain order and separates them by pieces of leather soaked in brine, this creating a "pile" or battery that produces a continuous and controllable electric current. As an electrical pioneer, he is later honored by having the electromotive force that moves the electric current named the "volt" after him.

1801   First to demonstrate that the human eye sees only three colors is English physicist and physician Thomas Young (1773-1829). He shows that to the eye, all other colors are combinations of red, green, and blue. This three-color theory is later refined by German physiologist and physicist Hermann Ludwig Ferdinand von Helmholtz (1821-1894) and is referred to as the Young-Helmholtz three-color theory. Today's color photography and color television make use of this theory.

1801   Ultraviolet region of the spectrum is discovered by German physicist Johann Wilhelm Ritter (1776-1810). Knowing silver chloride breaks down and turns black in the presence of light, and that the blue end of the spectrum brings this about more efficiently than the red, he then explores the region beyond the blue or the violet (where nothing is apparent to the eye), and finds that an invisible light also blackens silver chloride. He concludes that radiation invisible to the eye must exist. This section of the spectrum adjacent to the violet is now called "ultraviolet" for beyond the violet. In this same year, English chemist and physicist William Hyde Wollaston (1766-1828) makes a similar discovery, but does not pursue its implications as rigorously and as far as Ritter.

1802   First demonstration of the wave nature of light is provided by English physicist and physician Thomas Young (1773-1829). He performs his classic experiment on interference in which sunlight is made to pass through two pinholes in an opaque screen. He then allows the two separate beams of light that emerge from the holes to overlap and finds that when they spread and overlap, the overlapping region forms a striped pattern of alternating light and dark. This phenomenon can only happen, he argues, if light, like sound, is a wave and not a particle. Streams of particles could simply not produce interference effects. With a wave interpretation of his observations, he is also able to make the first quantitative values of the length of light waves.

1805   Electroplating is first developed by German physicist Johann Wilhelm Ritter (1776-1810). Coating one metal with another, also called gilding, is usually done to cover a strong metal, like iron, with an attractive one, like gold. This had been achieved by either pounding one onto another or by bonding when both were in a molten state. After the invention of the electric battery, Ritter finds if a current is passed through copper sulfate, it separates the molecules of copper, making the metallic cop-

Alessandro Volta was able to produce and sustain electric current by a "crown of cups" (top) and a series of metal piles. These voltaic piles were the first batteries.

per able to be plated to other metals. Electroplating is not used on a sizeable scale for another 35 years, and becomes used mostly for plating gold and silver.

1807 Thomas Young (1773-1829), English physicist and physician, first uses the word "energy" in its modern, scientific sense meaning the property of a system that makes it capable of doing work and as proportional to the product of the mass of a body and the square of its velocity.

1807 First to treat shear as an elastic strain is Thomas Young (1773-1829), English physicist and physician. His *Lectures on Natural Philosophy* marks the beginnings of the concept of practical elasticity in the science of materials. Shear is an action or stress that results from applied forces and which causes two contiguous parts of a body to slide away from each other.

1808 First to observe the polarization of light is French physicist Etienne Louis Malus (1775-1812). He accidentally discovers that light he sends through a piece of Iceland spar crystal (a doubly refracting crystal) is split into two beams. Trying to explain this phenomenon, he says each beam is aligned with a mystical "pole of light" which he describes as analogous to the poles of a magnet. He therefore describes the two beams as being "polarized." Despite believing that light is made up of particles and not waves, he establishes polarization as a general property of light. Polarization eventually proves useful not only in the study of light but in today's use of polarized lenses that block and reflect direct sunlight.

(See also 1812)

1810   First artificial ice is created by John Leslie (1766-1832), Scottish physicist and mathematician, who freezes water using an air pump.

1811   First known report of the daily periodicity of the electricity intensity in the atmosphere is made by Gustav Schubler (1787-1834), German meteorologist.

1812   Rotary dispersion of polarized light is discovered by French physicist Jean Baptiste Biot (1774-1862). While experimenting with double refraction, or what is called polarized light, he finds the degree of rotation of the plane of polarization depends on the color of the light. This work leads Biot to another discovery.

(See also 1815)

1815   Jean Baptiste Biot (1774-1862), French physicist, discovers that polarized light, when passing through an organic substance in liquid or solution, rotates either clockwise or counterclockwise depending upon the optical axis of the substance. He suggests that this is due to an asymmetry of the molecules that make up the substance. He continues his work on polarization and makes further discoveries.

(See also 1835)

1815   Brewster's law is first stated by Scottish physicist David Brewster (1781-1868). Experimenting with polarized light, he finds that

a beam of light can be split into a reflected portion and a refracted portion that turn at right angles to each other, each of which is then completely polarized. Known as Brewster's law, it contributes strongly to the argument that light consists of transverse waves, since neither the particle theory nor the longitudinal theory can explain it.

JULY 1820   Hans Christian Øersted (1777-1851), Danish physicist, experiments with a compass and electricity and demonstrates that a current of electricity creates a magnetic field. After bringing a compass needle near a wire through which a current is passing, he observes the needle slightly deflect. By moving his compass needle around the wire, he discovers that the magnetic field is circular and that the needle always points at a right angle to the field. He announces his discovery in a short article. This is the first time a real connection can be shown between electricity and magnetism, and it founds the new field of electromagnetism.

AUGUST 1820   First to apply advanced mathematics to electrical and magnetic phenomena is French mathematician and physicist André Marie Ampère (1775-1836). Within a week of hearing about the discovery made by Danish physicist Hans Christian Øersted (1777-1851), he conducts experiments that allow him to extend Oersted's work and formulate one of the basic laws of electromagnetism. He discovers that two parallel wires each carrying a current attract each other if the currents are in the same direction, but repel each other if in the opposite direction. He then concludes that magnetism is the result of electricity in motion. His mathematical theory is able to explain a magnet's properties and behavior, and essentially founds the science of electrodynamics (now called electromagnetism). Ampère's law relates the electromotive force that is created by the currents in two parallel conductors to the product of their currents and the distance between the conductors. He also builds a galvanometer (to measure the flow of electricity) and is the first to distinguish between the rate at which current is passed and the force (volts) that pushes it. In his honor, the quantity of electric current passing a given point in a given time is measured in amperes.

SEPTEMBER 1820 First rudimentary galvanometer is devised by German physicist Johann Salomo Christoph Schweigger (1779-1857). After learning of the connection between magnetism and electricity demonstrated by Danish physicist Hans Christian Ørsted (1777-1851), he realizes that compass needle deflection can be used to measure current strength. He then builds his "multiplier," a compass-box with a coil of wire wrapped around it in the direction of the needle. This device contains the germ of the modern galvanometer which is used to measure extremely small electrical currents.

MAY 14, 1821 First to develop a theory of elasticity is French engineer Claude Louis Marie Henri Navier (1785-1836). This year he publishes his paper, "Memoire Sur Les Lois de L'Equilibre et du Mouvement des Corps Solides Elastiques" which contains a set of general equations for the equilibrium and vibration of an elastic solid. This work lays the foundation for a modern theory of elasticity.

1821 First to determine the wavelength of light is Joseph von Fraunhofer (1787-1826), German physicist and optician. This year he publishes the first of two papers in which he describes his method of using closely spaced thin wires or gratings to refract light. He details how wavelength can be measured via a mathematical relation between the wavelength of diffracted light and the distance between the wires. Fraunhofer is the first person to take an analytical approach toward the construction of these diffraction gratings, and he is credited with making them into a precision instrument for obtaining the spectrum of light (splitting it into a band of its component colors). He goes on to measure the positions of the more prominent lines he observes and proves that they always fall in the same portion of the spectrum whether the light source is direct or reflected sunlight. These lines come to be called Fraunhofer lines.

1821 First to observe the conversion of heat into electricity (thermoelectricity) is Russian-German physicist Thomas Johann Seebeck (1770-1831). He notes that an electric current will flow continuously in a circuit made up of two different metals if they are kept at different temperatures. Since he does not fully understand this phenomenon, he does not pursue its implications. However, what eventually is known as the "Seebeck effect" becomes the foundation of all future work on thermoelectricity. This effect is caused by heat flowing from hot metal to cold, thus generating an electric current. Until the early 20th century, the only major application of this effect was the thermocouple (a type of regulating thermometer), but after World War II, Seebeck circuits are used to turn waste heat into power. The most advanced designs use heat from decaying radioactive isotopes to operate satellites.

1822 Mathematical theory of stress is first offered by French mathematician Augustin Louis Cauchy (1789-1857). Stress in physics is rationalized as the quantitative expression of a condition within an elastic material due to deformation, or strain, brought about by external forces. Cauchy studies the mechanics of materials and contributes several major aspects of the overall concept, such as the introduction of the stress tensor and the differential equations representing the balance of force.

1822 Jean Baptiste Joseph Fourier (1768-1830), French mathematician, publishes his *Théorie Analytique de la Chaleur,* which is an influential work of mathematical physics. Studying how heat flows from one point to another through a particular object, he offers what becomes known as Fourier's theorem. This states that any periodic oscillation can be broken down into a series of simple, regular wave motions, the sum of which is the original, complex oscillation. In other words, it can be expressed as a mathematical series made up of trigonometric functions. His theorem proves highly useful in the study of sound, light, and wave phenomena, with their mathematical treatment being known as harmonic analysis. His book also contains the first statement that a scientific equation must involve a consistent set of units, and thus begins dimensional analysis (in mathematics).

1824   First to consider quantitatively the manner in which heat and work are interconverted is French physicist Nicolas Léonard Sadi Carnot (1796-1832). His highly original work, *Réflexions sur la Puissance Motrice du Feu*, introduces the important concept of cyclic operations and the principle of reversibility. Attempting to understand the behavior of heat, so as to be able to maximize a steam engine's efficiency, he proposes an equation stating that the maximum efficiency of any heat engine depends only on the difference between the hottest and coolest temperature within the engine. Carnot's theory is recognized as the first true theory of thermodynamics (the study of heat movement), and although this work founds the science of thermodynamics, it is neglected for ten years. In this work, Carnot also makes a major contribution to physics by defining work as "weight lifted through a height." The modern restatement of this important concept defines work as any force applied through a distance against resistance.

(See also 1834)

1825   First electromagnet is invented by English physicist William Sturgeon (1783-1850). Beginning with the work of French mathematician and physicist André Marie Ampère (1775-1836), he wraps a wire many times around a horseshoe-shaped iron core. When he sends a current through the wire, each coil reinforces the other since they form parallel lines with the current running in the same direction. The resulting strong magnetic force can lift nine pounds or twenty times its weight and only

works while the current is running. Later Sturgeon invents an improved galvanometer and founds the first English journal devoted entirely to electricity.

1826   First systematic textbook on the strength and mechanics of materials and on structural analysis is written by French engineer Claude Louis Marie Henri Navier (1785-1836). His book, *Resume... sur L'Application de la Mecanique a L'Etablissment des Constructions et des Machines*, also contains the flexure formula.

1827   Ohm's law is first proposed by German physicist Georg Simon Ohm (1787-1854). Experimenting with various conductors of electricity, he uses wires of different length and diameter and discovers that a long, thick wire passes less current than a short, thin wire. He continues to experiment and finally states what becomes his law: that the amount of current passing through a wire is inversely proportional to the length and directly proportional to the thickness. Once accepted by the scientific community, his law makes it possible for scientists to calculate the amount of current, voltage, and resistance in circuitry, thus eventually establishing the science of electrical engineering.

1829   Thermopile or thermomultiplier is invented by Italian physicist Leopoldo Nobili (1784-1835). This improved device is a sensitive instrument for measuring radiant heat and is based on the Seebeck effect. In 1831, Nobili's colleague, Italian physicist Macedonio Melloni (1798-1854), improves this thermopile and finds there are different kinds of radiant heat and heat rays, just as there is variety among the different rays of light.

1829   First to give the modern scientific definitions of the terms "work" and "kinetic energy" is French physicist Gustave Gaspard de Coriolis (1792-1843). In his *Du Calcul de L'Effet des Machines* he defines the kinetic energy of an object as half its mass times the square of its velocity. Work done upon an object is equal to the force upon it multiplied by the distance it is moved against resistance. His later work concerning motion on a spinning surface results in his observations of the circular motions of winds and ocean currents. The effect of the earth's rotation on the circulation of air and water at or near the earth's surface comes to be known as the Coriolis force.

AUGUST 1831 Electromagnetic induction is discovered by English physicist and chemist Michael Faraday (1791-1867). After laboring for ten years to achieve the opposite of what Danish physicist Hans Christian Øersted (1777-1851) had done—to convert magnetism into electricity—he finally produces for the first time an induction current in a metal object by using a magnet. Further work leads to the generation of a continuous current by rotating a round copper disk between two poles of a horseshoe magnet. This is the forerunner of both the dynamo (electric generator) and the transformer. With such devices, mechanical energy can be converted into electrical energy and will lead to powering and interconnecting the modern world. In August 1830, American physicist Joseph Henry (1797-1878) actually discovered the principle of electromagnetic induction, but putting off further confirmations until the following August, Henry is shocked to learn that year that Faraday had already announced his own discovery.

1832 Faraday's laws of electrolysis are first stated by English physicist and chemist Michael Faraday (1791-1867). He is able to reduce electrolysis (the process by which an electric current is passed through a substance to effect a chemical change) to quantitative terms by stating that the mass of the substance liberated at an electrode during electrolysis is proportional to the amount of electricity going through the solution; and that the mass liberated by a given amount of electricity is proportional to the atomic weight of the element liberated and inversely proportional to the "combining pow-

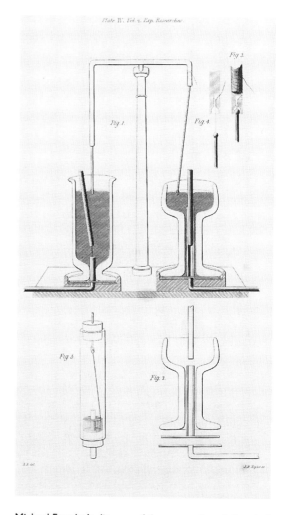

Michael Faraday's diagram of the conversion of electrical energy into mechanical rotation using a bar magnet.

er" of the element liberated. These two laws establish the very close connection between electricity and chemistry.

1833 Johann Karl Friedrich Gauss (1777-1855), German mathematician, introduces a measurement of terrestrial magnetism that is expressed in absolute units. This is the first time such units—length, mass, and time—are used outside mechanics. Gauss works with German physicist Wilhelm Eduard Weber (1804-1891) and they organize a worldwide system of stations for the systematic observation of terrestrial magnetism.

1834 First to employ Carnot's theory of thermodynamics is French engineer Benoit Pierre Clapeyron (1799-1864). Making use of the principles laid down by French physicist Nicolas Léonard Sadi Carnot (1796-1832) ten years before, he publishes his own mathematical treatment of Carnot's non-mathematical work. This effort offers a new perspective in that it discovers the important relationship between the heat of vaporization of a fluid, its temperature, and the increase in volume involved in its vaporization. His work contains what eventually becomes the second law of thermodynamics.

1834 First general description of self-induction is offered by Russian physicist Heinrich Friedrich Emil Lenz (1804-1865). He investigates electrical induction and discovers that a current induced by electromagnetic forces always produces effects that oppose those forces. American physicist Joseph Henry (1797-1878) also independently discovers that a coil carrying electric current not only induces a flow in another coil but can induce a current in itself, but it becomes known as Lenz's law because he elaborates it further. Self-inductance eventually becomes an important part in the design of electric circuits.

1834 Peltier effect is first stated by French physicist Jean Charles Athanase Peltier (1785-1845). He discovers that at the junction of two dissimilar metals, an electric current will produce heat or cold, depending on the direction of current flow. This comes to be known as the Peltier effect and is used in devices for measuring temperature. With the later discovery of new conducting materials, it is eventually put to use in refrigeration.

1835 Polarimetry is founded by Jean Baptiste Biot (1774-1862). After extensive experimentation with polarized light, he shows how the hydrolysis (chemical decomposition in water) of sucrose can be traced and followed by noting the changes in its optical rotation. This breakthrough leads to the establishment of polarimetry in 1846 by French chemist Louis Pasteur (1822-1895), who shows his fellow chemists that the ability of a substance to affect polarized light is closely related to its chemical structure.

1836 First reliable source of electric current is invented by John Frederic Daniell (1790-1845), English chemist and meteorologist. Unlike the recently-invented zinc-copper voltaic battery, the Daniell cell does not decline rapidly but supplies an even current with continuous operation. Daniell introduces a barrier between the zinc and copper and is therefore able to stop the formation of hydrogen which impairs battery function.

1840 Photovoltaic effect is first demonstrated by French physicist Alexandre Edmond Becquerel (1820-1891). While investigating the solar spectrum and electricity and magnetism, he discovers that when light induces certain chemical reactions it can produce an electric current. This leads him to invent a device that can measure light intensity by determining the strength of the current produced between two metal plates. Today's solar cells employ the photovoltaic effect.

1842 First attempt at obtaining a fairly accurate mechanical equivalent of heat is made by German physicist Julius Robert Mayer (1814-1878). His experiment involves a horse-powered mechanism that stirs paper pulp in a caldron. He compares the work done by the horse with the temperature rise in the pulp, and suggests something similar to what is now recognized as the notion of conservation of energy. His experiments are not carefully conducted nor his accounts very detailed, so he gets no credit for being one year ahead of English physicist James Prescott Joule (1818-1889) on this and four years ahead of German physiologist and physicist Hermann Ludwig Ferdinand von Helmholtz (1821-1894) on the conservation of energy. Circumstances dictate that he gets little credit for either until his later years.

(See also 1843)

1842   Doppler effect is first stated by Austrian physicist Christian Johann Doppler (1803-1853). He studies a common but unexplained phenomenon—that the pitch of a sound varies as the source moves toward or away from the listener. Knowing that sound waves exist and pitch depends on how far apart those waves are, he theorizes that if a sound source were moving toward a listener, the waves in front of the source would be squeezed together, creating a higher frequency. Similarly, the waves behind would be lengthened, creating a lower frequency. He is then able to work out the mathematical formula governing this shift, and once demonstrated, it becomes known as the Doppler effect. Later, both sonar and radar employ this principle, as do astronomers who prove light can undergo a Doppler shift.

1843   First accurate value for the mechanical equivalent of heat is given by English physicist James Prescott Joule (1818-1889). This year he publishes his value for the amount of work required to produce a unit of heat (which is known as the mechanical equivalent of heat). He uses four increasingly accurate methods of determining this value, and by using different materials as well, he also establishes that heat is a form of energy regardless of the substance that is heated. The value of the mechanical equivalent of heat is generally represented by the letter J, and a standard unit of work is called the joule. His argument that the various forms of energy—mechanical, electrical, and heat— are basically the same and can be changed, one into another, forms the basis of the law of conservation of energy, the first law of thermodynamics.

(See also 1847)

1843   First electrical "bridge" is built by English physicist Charles Wheatstone (1802-1875). His very sensitive device provides the solution for determining an unknown resistance in a circuit. His electrical "bridge" circuit includes three resistors of known value along with the unknown resistance, and uses a calibrated variable resistor to adjust the components in the circuit. Because the components have specific relations to each other, he is able to use one component to measure another. Knowledge of resistance is important since resistance impedes the flow of electricity. Being able to measure it enables one to know how much current is needed to perform a given task.

1847   First detailed and clear explanation of the concept of conservation of energy is offered by German physiologist and physicist Hermann Ludwig Ferdinand von Helmholtz (1821-1894). In his famous paper "Die Erhaltung der Kraft," he considers the sun the source of all energy and demonstrates that energy cannot be created spontaneously nor can it simply vanish, but that it is either used or released as heat. In stating that the total amount of energy in the universe is constant, meaning that it can be neither gained nor lost (although it can be converted from one form to another), he offers science one of the most basic laws of nature. The law of conservation of energy is also known as the first law of thermodynamics.

1848   Concept of absolute zero is first introduced by Scottish mathematician and physicist William Thomson, Lord Kelvin (1824-1907). He confronts the dilemma posed by Charles' law, which states that when a gas is cooled from 32°F (0°C), its volume decreases by 1/273 for every degree drop. The dilemma is if the temperature were reduced to -459°F (-273°C), then the volume of the gas would be reduced to zero, and no one can explain how matter can take up no volume. Thomson explains this by stating it is not the volume that reaches zero but the motion of the gas's molecules that stops at zero, and it effectively takes up no space. Thomson then names -459°F (-273°C) as absolute zero since no further reduction in temperature could occur. He also introduces a new temperature scale with this ultimate low point as its zero. On the Kelvin scale, the freezing point of water is 273 K and the boiling point is 373 K.

1850 Second law of thermodynamics is first offered by German physicist Rudolf Julius Emmanuel Clausius (1822-1888). This law states that heat can never pass of its own accord from a colder to a hotter body. As the French physicist Nicolas Leonard Sadi Carnot (1796-1832) earlier showed, some of the energy generated by a steam engine is lost as heat and is not turned into useful work. Clausius finds, mathematically and not experimentally, that this is true of any energy conversion. He says that some energy is always lost as heat and that heat can never be converted completely to any other form of energy. Establishing the second law of thermodynamics implies theoretically that at some point in a far distant future the universe may simply have no useful energy left at all. He later calls this "entropy."

(See also 1865)

MARCH 1851 William Thomson, Lord Kelvin (1824-1907), Scottish mathematician and physicist, publishes *On the Dynamical Theory of Heat* in which he explores the work of French physicist Nicolas Léonard Sadi Carnot (1796-1832) and deduces that all energy tends to run down and dissipate itself as heat. This is another form of the second law of thermodynamics and is advanced further by German physicist Rudolf Julius Emmanuel Clausius (1822-1888) about the same time. Kelvin's work is considered the first 19th-century treatise on thermodynamics.

(See also 1850)

1851 First experimental proof of the earth's rotation is given by French physicist Jean Bernard Leon Foucault (1819-1868). After realizing that a pendulum has a tendency to stay swinging in one plane, he conceives a brilliant experiment to prove a moving earth. He conducts a spectacular demonstration before a huge crowd at the Pantheon in Paris in which he suspends a large iron ball from a wire that is more than 200 ft (61 m) long. After the pendulum is set swinging, a spike on the bottom of the ball scratches a line in the sand on the floor. As times passes, the line shifts and the crowd realizes that since the pendulum could not arbitrarily change direction once it began to move, then the shift had to be the result of the earth rotating. The spectators gasp at the realization that they are watching the earth rotate under the pendulum.

1852 Gyroscope is invented by French physicist Jean Bernard Leon Foucault (1819-1868). Learning from his pendulum experiment of the year before, he finds that just as a pendulum swings in an unchanging plane, so a large sphere in rotation has a similar tendency to maintain the direction of its axis. He then sets a wheel within a heavy rim in rotation and finds that it not only holds its axial direction, but when tipped, it is set right again by the force of gravity. This suggests that a gyroscope can be used like a compass as an indicator of true north.

1852 Term "fluorescence" is first coined by George Gabriel Stokes (1819-1903), British mathematician and physicist, to describe the visible glow produced by materials exposed to ultraviolet rays (an invisible form of light). He works with fluorescence and phosphorescence to study ultraviolet light since both are caused when a chemical's atoms become "excited" by ultraviolet radiation. Fluorescent chemicals lose their glow almost immediately but phosphorescent materials can glow for several hours after exposure. He is the first to show that quartz is transparent to ultraviolet light, whereas ordinary glass is not. He also suggests his law of viscosity which describes the motion of a solid sphere in a fluid as well as Stokes' theorem on vector analysis.

1852 James Prescott Joule (1818-1889), English physicist, and Scottish mathematician and physicist William Thomson, Lord Kelvin (1824-1907), jointly discover that when a gas is allowed to expand freely, it will drop in temperature. This becomes known as the Joule-Thomson effect and is explained by the fact that gas molecules have a slight attraction for other gas molecules. It is in overcoming this attraction during expansion that individual molecules lose energy (temperature). This effect eventually proves very important in achieving extremely low temperatures and forms the basis of the modern science of cryogenics.

1853   William Holms Chambers Bartlett (1804-1893), American astronomer, publishes his *Elements of Analytical Mechanics* which is the first mechanics text for technical students written by an American author.

1853   Concept of potential energy is first introduced into physics by Scottish engineer William John Macquorn Rankine (1820-1872). Also called the energy of position, it is stored energy that depends upon the relative position of various parts of a system. Potential energy is a property of a system and not of an individual body or particle, thus a spring has more potential energy when it is compressed or stretched.

1855   First Geissler tubes become possible as German inventor Heinrich Geissler (1814-1879) devises a mercury air pump with no moving parts that can evacuate a chamber more thoroughly than anything to date. By moving a column of mercury up and down, the vacuum above the column is used to very slowly suck out the air in an enclosed vessel until the vacuum within the vessel approaches that above the mercury. With this new pump he is able to produce highly improved vacuum tubes. Called Geissler tubes, they make possible a more advanced study of electricity and eventually of the atom. They lead to the discovery of the cathode ray tube and the electron.

1858   First successful experiments with Geissler tubes are conducted by German mathematician

and physicist Julius Plucker (1801-1868). Using the new improved vacuum tubes, he forces an electric current through a vacuum and observes fluorescent effects. This bright, stream-like glow between the electrodes is brighter than anything produced before. He also finds that the glow responds to a magnetic field and actually shifts position. This discovery proves extremely important because it suggests that the stream crossing the vacuum is composed of particles and not rays. This is the first suggestion of an awareness of subatomic particles.

1859   First electric battery that can be charged after discharge is built by French physicist Gaston Plante (1834-1889). This rechargeable "storage battery" has lead plates immersed in sulfuric acid and is essentially similar to the batteries used by today's automobiles. Prior to his invention, batteries had to be entirely replaced when they died.

1859   Kirchhoff's law is first stated by German physicist Gustav Robert Kirchhoff (1824-1887). After using the spectroscope to discover the relation between emission and absorption spectra, he conducts further experiments with light and discovers that when light passes through a gas, those wavelengths are absorbed which that gas would emit when incandescent. This becomes known as Kirchhoff's law, and it is eventually used to analyze light from any source, including the sun and the stars, and identify its constituent elements.

1860   "Blackbody" concept is first introduced by German physicist Gustav Robert Kirchhoff (1824-1887). A blackbody is a purely theoretical object that is both a perfect absorber and a perfect emitter of radiation. Kirchhoff points out that such a body that absorbs all radiation falling on it, of whatever wavelength would, if heated to incandescence, emit all wavelengths. This hypothetical concept is later adopted by German physicist Max Karl Ernst Ludwig Planck (1858-1947), who uses it with enormous success as the basis of his quantum theory.

1861 "Heat death" of the universe is first suggested by Scottish mathematician and physicist William Thomson, Lord Kelvin (1824-1907) in his *Physical Considerations Regarding the Possible Age of the Sun's Heat*. He offers this theme as the inevitable consequence of the concept of entropy or the principle of the dissipation of energy. Eventually, he says, all of the energy in the universe will eventually be converted to unusable heat. The finite amount of heat in the universe will be used up. This notion is not widely held, as many feel that different parts of the universe may be governed by different laws of physics.

1863 John Tyndall (1820-1893), Irish physicist, publishes his *Heat as a Mode of Motion*, the first popularization of the dynamic theory of heat (heat as molecular vibration). His book goes through several editions and also makes the law of conservation of energy understandable. In 1872 and 1873 he travels the United States giving lectures that further the popular understanding of science.

1865 Term "entropy" is first introduced by Rudolf Julius Emmanuel Clausius (1822-1888) who further refines his second law of thermodynamics. Defining entropy as the ratio of heat content in a closed system to its absolute temperature, he states that the heat given off would not be reconverted into usable energy and that all of the energy would eventually be used up, filling the system with waste heat. Modern theorists have explored this concept with varying results. Some say it implies the inevitable heat-death of the universe, while others say that we have no way of knowing if the laws of physics we understand even apply to the rest of the universe. Clausius selects the word entropy from the Greek word for transformation.

(See also 1850)

1866 "Kundt's tube" is first developed by German physicist August Adolph Eduard Eberhard Kundt (1839-1894) to study the velocity of sound in different gases. He finds he can make

accurate measurements of the speed of sound in the air by dusting the inside of his tube with fine powder which is then disturbed by traveling sound waves. After experimenting, he determines that he can calculate the velocity of sound in the material making up the tube (a solid) or in the gas contained in the tube, by studying the pattern of disturbance.

1869 Tyndall effect is first stated by Irish physicist John Tyndall (1820-1893). Knowing that a beam of light passing through pure water or a transparent crystal is not dispersed or interfered with, he studies light by passing it through different solutions. He finds that when a beam of light passes through a colloidal solution (one containing tiny particles), the beam is dispersed. Called the Tyndall effect, it is caused by the particles interfering with the light and inducing some to bounce away in all directions. He continues this research and examines the effect colloid solutions have on different wavelengths of light. With this, he later explains the blue of the daytime sky and the colors of sunrise and sunset.

(See also 1871)

1871 The Cavendish Laboratory is founded at Cambridge University in England. Its first director is the Scottish mathematician and physicist James Clerk Maxwell (1831-1979). A great deal of the development of modern physics takes place at this laboratory which is endowed by the family of English chemist and physicist Henry Cavendish (1731-1810).

1871 Relation of the variation of light-scattering with wavelength is first stated mathematically by English physicist John William Strutt, Lord Rayleigh (1842-1919). Working with the recently discovered Tyndall effect, he develops a mathematical equation that accounts for the way light waves of different lengths are scattered by minute particles, showing that this scattering is inversely proportional to the fourth power of the wavelength. His work confirms earlier opinions that dust in the atmosphere scatters the sun's light in a certain way, making the sky appear to have different colors at different times.

(See also 1869)

1873 Electromagnetic theory is first stated by Scottish mathematician and physicist James Clerk Maxwell (1831-1879). After linking magnetism and electricity and then proving they are distinctly related, he calculates mathematically the transmission speed of both electromagnetic and electrostatic waves. When his calculations indicate that both are around 186,300 mi (299,757 km) per second, he realizes instantly that this is coincidental with the speed of light. Maxwell then makes the theoretical leap and identifies light as a form of electromagnetic radiation that travels in waves. Although unable to prove his theory at this time, he nonetheless uses it to make unheard of predictions. His idea that there are wavelengths below infrared will be proven by radio waves and radar, and his reference to those above ultraviolet will be demonstrated once x rays and gamma rays are discovered. This year he publishes his unification of the three main fields of physics—electricity, magnetism, and light—in his landmark, two-volume work, *A Treatise on Electricity and Magnetism*.

1873 Van der Waals equation is first stated by Dutch physicist Johannes Van der Waals (1837-1923) in his doctoral thesis. Although little is known at this time about the structure of molecules, Van der Waals' dissertation accurately describes the molecular forces that dictate gas behavior. Realizing that the existing gas laws were rather crude predictors of gas behavior and that some gases, like hydrogen and helium, do not always follow these laws, he goes to the molecular level and introduces an equation involving two new constants or numbers for each gas that take into consideration molecular structure and attraction. He then goes beyond this and, working with the temperature, pressure, and volume of a gas at its critical point (where the gas and liquid forms become equal

in density and cannot be told apart), he develops another equation in which new constants are not needed and which will hold for any gas. With his new equation he is able to validate the work of others. It also leads to the modern science of cryogenics.

1873 Capillary electrometer is invented by French physicist Gabriel Lippmann (1845-1921). This instrument for measuring extremely small voltages (as small as one thousandth of a volt) later becomes useful in some of the earliest electrocardiograms.

1874 Crystal rectifiers are discovered by German physicist Karl Ferdinand Braun (1850-1918). While studying mineral metal sulfide, he notes that certain crystals transmit electricity in one direction more easily than in the other. He recognizes that these crystals (called rectifiers) are useful in converting alternating current (which travels in two directions and forever doubles on its own tracks) into direct current (which travels in one direction). They are first used in crystal-set radios whose range they also improve. Generations later, they return in an improved form in solid-state circuitry systems.

1876 Henry Augustus Rowland (1848-1901), American physicist, establishes for the first time that a moving electric charge or current is accompanied by electrically charged matter in motion (and produces a magnetic field). He demonstrates this by attaching pieces of tin foil to a glass disk, placing an electric charge on the tin, and allowing the disk to rotate rapidly. When the disk deflects a magnet in the proper manner, he proves his point.

1876 First to use the name "cathode rays" is German physicist Eugen Goldstein (1850-1930). He applies the name to the luminescence produced at the cathode in a vacuum tube. He is also one of the first to observe (in 1886) the positive rays produced when an electrical current is passed through a gas in a glass ("cathode") tube. He names these rays

"Kanalstrahlen", and they are called "channel rays" or "canal rays" in English. The study of these rays leads to the eventual discovery of the proton.

1876 "Phase rule" is first stated by American physicist Josiah Willard Gibbs (1839-1903). This year he begins a series of papers in which he deals with the principles of thermodynamics and applies them to the complex processes involved in chemical reactions. He also offers an elaborate equation that becomes known as the "phase rule" and which governs the several phases of matter. Gibbs considers each state of matter a phase and each substance a component. Thus, a system of ice and water would be one component in two phases, while solid and dissolved sugar in water would be two components in two phases. He then takes all of the variables involved in a chemical reaction—temperature, pressure, energy, volume, and entropy—and includes them in one equation. His work is difficult to understand, complicated, multidisciplinary, and is not really appreciated until the 1890's.

1877 First to successfully liquefy gases is French physicist Louis Paul Cailletet (1832-1913). By combining low temperatures with high pressure, he produces small amounts of liquid oxygen, nitrogen, and carbon monoxide. He begins by compressing a gas and then cooling it as much as possible. He allows it to expand and in expanding, it cools drastically. He then uses what is called the "cascade" process which reduces temperature step by step. With this method, one liquefied gas is used to cool a second gas that has a lower critical temperature, then the second gas, when liquefied, is used to cool another gas with an even lower critical temperature, and so on. Many claim that Swiss chemist Raoul Pierre Pictet (1846-1929) has priority over Cailletet, but it is known that Pictet liquefies gases at about the same time and uses a more elaborate process which results in greater quantities being produced.

1879 First quantitative rate for the cooling of hot bodies is offered by Austrian physicist Josef Stefan (1835-1893). Particularly concerned with the nature of blackbody radiation (a black body is a purely theoretical object that is both a perfect absorber and a perfect emitter of radiation), he finds that the energy radiated from a black body is proportional to the fourth power of its temperature. Called the Stefan-Boltzmann law, it states that the amount of heat radiated by a surface is proportional to the fourth power of the absolute temperature. Austrian physicist Ludwig Boltzmann (1844-1906) gets his name included because he shows mathematically in 1884 that this law can be deduced from thermodynamic principles.

1879 Hall effect is first announced by American physicist Edwin Herbert Hall (1855-1938). Working with many different types of conductors, he finds that an electric current flowing through a gold conductor in a magnetic field produces an electric potential perpendicular to both the current and the field. Thus the strength of the potential is directly related to the strength of the magnetic field and creates a transverse current in the conductor. Hall's discovery stimulates others to begin their own studies and eventually leads to the discovery of three other transverse effects.

1880 Electrons are first seen (unknowingly) by English physicist William Crookes (1832-1919). While experimenting with his own much improved Geissler tubes (which he invents in 1875 and which become known as Crookes tubes), he observes a green glow that appears when an electric current is introduced and the tube evacuated. He also notes that when he places a pivoting vane in the tube, it turn slightly, as if it were in the current of an invisible stream. Although he explains this effect as being caused by negatively charged particles, he cannot explain what these particles are. In 1897 English physicist Joseph John Thomson (1856-1940) proves these to be electrons.

1880 Piezoelectricity is discovered by French chemist Pierre Curie (1859-1906), who collaborates with his brother Jacques. Working with crystals, they note that certain crystals (like quartz), when compressed or stretched, create an electric charge proportionate to the amount they are stressed. They then discover when a current is applied to certain faces of a quartz crystal it expands. It also vibrates when an alternating current is applied. They name this generation of electricity by stress "piezoelectricity" from the Greek word "to press." Crystals with piezoelectric properties are soon used as an important part of sound-electronic devices like microphones and record players.

APRIL 12, 1883 Zygmunt Florenty von Wróblewski (1845-1888), Polish physicist, and Karol Stanislaw Olszewski (1846-1915), Polish chemist and physicist, announce their development of a method of making liquid oxygen in usable quantities. They also liquefy nitrogen and carbon monoxide and are the first to obtain at least a fine mist of liquid hydrogen.

1883 George Francis Fitzgerald (1851-1901), Irish physicist, first suggests a method of producing radio waves. From his studies of radiation, he concludes that a rapidly oscillating (i.e., alternating) current should result in the radiation of electromagnetic waves. This prediction is later verified experimentally by Heinrich Rudolf Hertz (1857-1894) in 1888 and used in the development of wireless telegraphy.

1883 Edison effect is first described by American inventor Thomas Alva Edison (1847-1931). After inserting a small metal plate near the filament of a light bulb, he finds that the plate draws a current when he connects it to the positive terminal of the light bulb circuit, even though the plate is not touching the filament. This passage of electricity from a filament to a plate of metal inside an incandescent light bulb eventually comes to be called thermionic emission. Although Edison patents this idea in 1884 and describes it in the technical literature,

he can find no immediate practical use, so he does not pursue it. This turns out to be Edison's one purely scientific discovery and it becomes a major factor in the invention of the vacuum tube as well as an important phenomenon in the study of the structure of matter during the next decade.

(See also 1911)

1885 Balmer series is first proposed by Swiss mathematician and physicist Johann Jakob Balmer (1825-1898). Studying the solar spectra and working entirely from experimental data, he discovers a simple numerical formula with which the prominent series of lines in the visible region of the hydrogen spectrum can be correlated to wavelength (which is the inverse of frequencies). Not a theoretician, he does not attempt to explain the reasons for his discovery. A generation later, Danish physicist Niels Henrik David Bohr (1885-1962) offers a model that provides the theoretical basis needed to explain Balmer's equations and uses it in his theory of the internal structure of the atom.

1886 Concept of molecular surface tension is first introduced by Hungarian physicist Roland von Eötvös (1848-1919). His study of the earth's gravitational field, which leads him to invent a precise torsion balance, results in his proof that inertial mass and gravitational mass are equivalent. Inertial mass is the amount of acceleration produced in a body through the application of a force of a given magnitude. Gravitational mass depends upon the intensity of an object's gravitational field at a given distance. Eötvös reasons that if these two are identical, then objects in a given gravitational field will always drop in a vacuum at the same rate, regardless of mass. His extremely delicate measurements made with his torsion balance find this to be so. This proves to be a major principle in German-Swiss-American physicist Albert Einstein's (1879-1955) general theory of relativity.

1887 First to note the sudden change in the

movement of the airflow over a moving object that occurs as it approaches the speed of sound is Austrian physicist Ernst Mach (1838-1916). He conducts extensive aerodynamic experiments studying the conditions that occur when a solid object and air are in rapid motion relative to each other. The speed of sound is the natural rate at which the air's molecules move, and when an object is moving through the air at speeds higher than that natural rate, the molecules literally get shoved aside faster than they want to go. This creates a bunching up of sound waves that then expand to produce a sudden clap. Because of his discovery, the speed of sound in air (under certain temperature conditions) is called Mach 1. Mach 2 is twice that speed and so on.

1887   Non-existence of "ether" in space is first demonstrated by the failure of the Michelson-Morley experiment. German-American chemist Albert Abraham Michelson (1852-1931) collaborates with American chemist Edward Williams Morley (1838-1923) to test the age-old hypothesis that the earth moves through a "luminiferous" ether which is a supposedly invisible, motionless, light-carrying element that exists outside or above the earth's atmosphere. They use Michelson's interferometer—which splits a beam of light in two, sends each on a different path, and then reunites them—to test out this idea. If the ether exists, Michelson states, then one of the beams traveling against the ether would be slowed. When they conduct their beam-splitting experiment, they can detect no difference in the velocity of light in any direction under any circumstances. Since Michelson strongly believes in the existence of the ether, he considers the experiment a failure. What it is, however, is the evidence needed to overturn the old ether theory once and for all. It also forces others to recognize that an explanation for the invariance of the speed of light is needed.

1888   Radio waves are generated for the first time by German physicist Heinrich Rudolf

Hertz (1857-1894). Devising a simple wire loop detector that can detect and measure an electromagnetic wave, he is able to prove experimentally the 1873 prediction of Scottish mathematician and physicist James Clerk Maxwell (1831-1879) that electricity is propagated in wave formations and to verify his hypothesis that light is an electromagnetic phenomenon. The work of Hertz not only discovers radio waves and leads eventually to radio communication but brings together the three main fields of physics—electricity, magnetism, and light.

1889   First practical photoelectric cells that can measure the intensity of light are developed by German physicists Johann Philipp Ludwig Julius Elster (1854-1920) and Hans Friedrich Geitel (1855-1923). They study the photoelectric effect (when an electric current is created upon the exposure of certain metals to light) and invent the modern photoelectric cell by modifying a cathode-ray tube. Their affordable cell makes it possible for many industries to develop photoelectrical technology and leads directly to the development of television.

1891   James Alfred Ewing (1855-1935), English physicist, discovers and names the phenomenon of hysteresis, the property of magnetic materials to resist becoming magnetized or demagnetized. After observing that in electromagnets using alternating current, the magnetization of the metal lags behind the changing of the current flow, he theorizes that all molecules are like tiny magnets and thus explains hysteresis as a resistance of the molecules to rearranging themselves in alignment with the new direction of magnetic force.

1891   Electron is first suggested by Irish physicist George Johnstone Stoney (1826-1911) for the what he calls "units of electricity." Since the laws of electrochemistry are explained by supposing that electricity is not a continuous fluid but consists of particles of fixed minimum charge, Stoney suggests that this minimum electric charge be called an electron. In 1897, when it is discovered that cathode ray particles carry a similar charge, the name "electron" is applied to them.

1892   First thermos bottle is built by Scottish chemist and physicist James Dewar (1842-1923). Studying the liquefaction of gases, he faces the problem of keeping the gases cold long enough to analyze them. Since liquid oxygen kept in a flask absorbs heat from the air and turns back to a gas, he puts the flask with liquid gas in it inside a larger flask and creates a vacuum between them. He finds that the vacuum prevents the transfer of energy that occurs through conduction or convection, and that heat will not penetrate nor will cold escape. He continues to improve his device and it comes to be known as the Dewar flask. Since his flasks keep hot liquids hot and cold liquids cold, they become adapted to everyday use and are known as the thermos bottle.

APRIL 25, 1894   Guglielmo Marconi (1874-1937), Italian electrical engineer, uses the method of producing radio waves invented by German physicist Heinrich Rudolf Hertz (1857-1894) and builds a receiver to detect them. After experimenting with Hertz's spark-gap generator and building an improved detector, he succeeds in sending his first radio waves thirty feet to ring a bell. The next year, his improved system can send a signal 1.5 mi (2.4 km).

(See also 1896)

NOVEMBER 5, 1895   X rays are discovered accidentally by German physicist Wilhelm Konrad Roentgen (1845-1923), initiating the modern age of physics and revolutionizing medicine. While working on cathode ray tubes and experimenting with luminescence, he notices that a nearby sheet of paper coated with a luminescent substance glows whenever the cathode ray tube is turned on. He becomes especially intrigued when he sees that the paper is glowing despite being blocked by a piece of cardboard. Some experimentation reveals to him that some

sort of radiation is emerging from the cathode ray tube that is highly penetrating but invisible. Having no idea what the nature of these rays is, he calls them x rays for unknown. Finally, he works feverishly for seven weeks, continuing to experiment, and near the end of the year feels prepared to report publicly the basic properties of these unknown rays.

(See also December 28, 1895)

DECEMBER 28, 1895   Discovery of x rays is first announced by German physicist Wilhelm Konrad Roentgen (1845-1923), who submits a paper documenting his discovery. He tells how this unknown ray or radiation can affect photographic plates, and that wood, paper, and aluminum are transparent to it. It can also ionize gases and does not respond to electric or magnetic fields nor exhibit any properties of light. This discovery leads to such a stream of groundbreaking discoveries in physics that it has been called the beginning of the second Scientific Revolution.

(See also 1901)

1895   Curie point is first discovered by French chemist Pierre Curie (1859-1906). Studying the effects of temperature on magnetism, he finds there is a certain critical temperature above which magnetic properties disappear. He then formulates a correlation between these properties that becomes known as the Curie's law.

MARCH 1, 1896   First observation of natural radioactivity is made by Antoine Henri Becquerel (1852-1908), French physicist. While studying fluorescent materials to see if they might emit the newly discovered x rays, he discovers instead that uranium produces a natural ray. He finds this out when he uses potassium uranyl sulfate (containing uranium atoms) as the fluorescent material and discovers that it fogs photographic film on a sunless day. He realizes that whatever rays the material is giving off have nothing to do with either sunlight or fluorescence. In 1898, Polish-French chemist Marie

Sklodowska Curie (1867-1934) studies these rays and names the phenomenon "radioactivity." (See also 1901)

JUNE 2, 1896   First patent in the history of radio is obtained in England by Italian electrical engineer Guglielmo Marconi (1874-1937). After considerably improving his new device and being turned down when he shows it to the Italian government, he travels to London hoping to interest the British Navy in the potential of wireless communication. By this time, he has founded corporations in England and the United States and has sent and received a radio signal over 9 mi (14 km).

(See also December 12, 1901)

1896   Johann Philipp Ludwig Julius Elster (1854-1920) and Hans Friedrich Geitel (1855-1923), both German physicists, study the newly discovered radioactivity and demonstrate that external effects do not influence the intensity of radiation. They are also the first to characterize radioactivity as being caused by changes that occur within the atom.

1896   Charles Edouard Guillaume (1861-1938), Swiss-French physicist, discovers a new alloy he names invar. While searching for a durable, inexpensive metal out of which to construct standards of length and mass, he discovers an alloy of iron and nickel in the ratio of 9 to 5 that changes its volume only slightly as the temperature changes. This proves to be the unchanging metal he is searching for since its properties make it an ideal material for precision instruments. He chooses the name invar for its "invariable" properties, and it comes to be used in the manufacture of balance wheels and hair springs.

1896   First cloud chamber is developed by Scottish physicist Charles Thomson Rees Wilson (1869-1959). Trying to duplicate cloud effects, he devises a way of allowing moist air to expand in a closed container. This expansion cools the air so that it becomes supersaturated, and when he uses dust-free air, the air remains supersaturated. Following the discovery of x rays, he introduces radiation and later discovers that charged particles leave useful tracks that can be studied. These tracks reveal whether the charge is positive or negative and how massive the particle is. They also indicate collisions of particles with molecules or other particles. The Wilson cloud chamber proves to be indispensable in the study of nuclear physics and eventually leads to the development of the bubble chamber.

1897   Oscilloscope is invented by German physicist Karl Ferdinand Braun (1850-1918). He modifies a cathode ray tube so that the streaming electrons are affected by the electromagnetic field of a varying current. This results in the green fluorescent spot formed by the stream of speeding particles shifting in accordance with the electromagnetic field set up by a varying current. It results in his invention of the oscillograph or oscilloscope, whose ability to finely vary an electric current makes it the first step toward television and radar. The device is so named because the spot can follow and reveal the oscillations of the field. Braun's invention is the ancestor of the modern television screen.

1897   First to use a radio antenna is Russian physicist Alexander Stepanovich Popov (1859-1906). He experiments with the newly discovered Hertzian (radio) waves and by attaching an antenna to a coherer, becomes the first to send long distance radio signals—from ship to shore over a distance of 3 mi (4.8 km). Since he concentrates primarily on the meteorological aspects of his invention (detecting lightning strikes), he does not exploit the full potential of radio waves for communication as does Italian electrical engineer Guglielmo Marconi (1874-1937). However, he does work on methods of receiving radio signals over long distances a year before Marconi does.

1897   Electron is discovered by English physicist Joseph John Thomson (1856-1940). Although many scientists believe that cathode rays emit a stream of negatively charged particles, they are unable to deflect the rays with an electrical field (which should occur if they are correct). Thomson produces rays in a much-improved vacuum tube he builds and is then able to produce the necessary deflection. He conducts further cathode ray experiments and concludes that the rays consist of negatively charged "electrons" that are smaller in mass than hydrogen atoms, the smallest then known. He goes on to propose the first theory of atomic structure stating that these sub-atomic electrons are the constituents of all matter. Although an inadequate theory, it proves a good beginning.

1897 Alternating current is first made understandable to the average engineer and electrician by German-American electrical engineer Charles Proteus Steinmetz (1865-1923). This year he completes his major theoretical work, *Theory and Calculation of Alternating Current Phenomena*, which works out in complete mathematical detail the intricacies of alternating current and its advantages over direct current. Unlike direct current which flows in one direction, alternating current oscillates forward and backward at a specific frequency. The work of Steinmetz makes a very complex situation understandable and eventually leads to the final victory of alternating current over direct current, serving to found the modern electrical industry.

1899 Alpha and beta rays are first named by British physicist Ernest Rutherford (1871-1937). By using powerful magnetic fields, he finds that nuclear radiation can be separated into at least two parts. When forced to pass between two strong magnetic poles, the radiation separates into two distinct streams, one of which is attracted to the south magnetic pole and the other to the north. He then suggests calling these two forms "alpha rays" and "beta rays."

OCTOBER 15, 1900 First structure designed according to the principles of acoustics, Boston Symphony Hall, opens. These principles are discovered by American physicist Wallace Clement Ware Sabine (1868-1919) in 1895. When asked by Harvard University to study the acoustics of a new lecture room in which no one could be heard, Sabine finds a new way of measuring the absorptivity of sound in a particular room. He then discovers that the duration of reverberation multiplied by the total absorptivity of the room is a constant that varies in proportion to the room's volume. This forms Sabine's law which, if followed, assures that a room has enough reverberation to give it strength but not so much as to interfere with actual hearing.

1900 Quantum theory is first stated by German physicist Max Karl Ernst Ludwig Planck (1858-1947) in his classic and revolutionary paper on quantum physics. In this paper he tells of his discovery that light or energy is not found in nature as a continuous wave or flow but is emitted and absorbed discontinuously in little packets or "quanta." Further, each quantum or packet of energy is indivisible. Planck's entirely new notion of the quantum seems to contradict the mechanics of Newton and the electromagnetics of Maxwell, and to replace them with new rules. In fact, his quantum physics are the new rules of a new game of physics—the game of the very fast and the very small. His theory is soon applied by Einstein and incorporated by Bohr and becomes the watershed mark between all physics that comes before it (classical physics) and all that is after it (modern physics).

1900 Radon is discovered by German physicist Friedrich Ernst Dorn (1848-1916). He demonstrates that the newly discovered element radium gives off a radioactive gas he calls radium emanation, as well as producing radioactive radiation. This proves to be the first demonstrable evidence that in the radioactive process, one element is actually transmuted into another.

1900 Gamma rays are first described by French physicist Paul Ulrich Villard (1860-1934). While studying the radiation given off by uranium, he notes besides the already-discovered alpha and beta rays, that a third type of radiation exists. Further analysis shows it is unaffected by a magnetic field and is more energetic than beta rays. He also finds that where alpha rays are stopped by a few centimeters of air, and beta rays by a few centimeters of aluminum, gamma rays are unusually penetrating and can only be stopped by a relatively thick piece of lead. It is some time before it is realized that since the wavelengths of gamma rays are like very short x rays, they are a form of electromagnetic radiation that is even more energetic than x rays. It is not known how they get their name.

DECEMBER 12, 1901 First transatlantic radio transmission is achieved by Italian electrical engineer Guglielmo Marconi (1874-1937), who successfully sends a radio signal from England to Newfoundland. Although the system is still only useful for sending Morse code, its achievement generally is considered by most as the day radio is invented. Marconi proves his many critics wrong in their belief that since radio waves seem to travel in only straight lines, they would travel no further than the horizon. He sends his signals over 2,000 mi (3,218 km) and demonstrates that radio is viable for vast distances and will become an important form of communication.

1901 First Nobel Prize in Physics is awarded to German physicist Wilhelm Konrad Roentgen (1845-1923) for his discovery of X rays.

(See also November 5, 1895)

1901 National Bureau of Standards is first established in the United States. This Department of Commerce agency develops measurement standards and techniques for American science and industry and for other government agencies. It also works to maintain, improve, and apply fundamental systems of measurement, including those of length, time, and

mass. Although it has no regulatory authority, it helps private organizations promote the establishment of voluntary standards for commerce and industry. In 1988 its name is changed to the National Institute of Standards and Technology.

1901 Pyotr Nicolaievich Lebedev (1866-1912), Russian physicist, uses extremely light mirrors in a vacuum and is able for the first time to measure the radiation pressure exerted by light. This is also an experimental confirmation of Scottish mathematician and physicist James Clerk Maxwell's (1831-1879) theory of electromagnetism. Lebedev also uses this to explain why comet tail bend away from the sun, explaining that the light pressure exceeds gravitation.

1901 First scientist from India to become internationally known is physicist and plant physiologist Jagadis Chandra Bose (1858-1937). He devises extremely sensitive instruments that are able to demonstrate the minute movements of plants to external stimuli and to measure their rate of growth. He pioneers what becomes biophysics and makes improvements on the coherer, an early form of radio detector.

1901 First clear understanding of the atom as something more than a featureless sphere is achieved by French physicist Antoine Henri Becquerel (1852-1908). He studies the rays emitted by the natural substance uranium and concludes that the only place they could be coming from was within the atoms of uranium. This understanding also implies that atoms are a dynamic reality that might contain electrons. Becquerel's discovery of radioactivity and his focus on the uranium atom make him the father of modern atomic and nuclear physics.

1902 First prediction of the existence of the ionosphere is made independently and almost simultaneously by English physicist and electrical engineer Oliver Heaviside (1850-1925), and British-American electrical engineer Arthur Edwin Kennelly (1861-1939). The ionosphere is an electrically conductive layer in the upper atmosphere that reflects radio waves. They theorize correctly that wireless telegraphy works over long distances because a conducting layer of charged particles in the atmosphere exists that reflects radio waves and allows them to follow the earth's curvature instead of traveling off into space.

(See also 1924)

1903 William Ramsay (1852-1916), Scottish chemist, and Frederick Soddy (1877-1956), English chemist, discover that helium is continually produced by naturally radioactive products. They demonstrate that the breakdown of both uranium and radium will produce helium atoms.

1904 Lorentz transformations are first suggested by Dutch physicist Hendrik Antoon Lorentz (1853-1928). Considering the negative results of the 1887 Michelson-Morley experiment, he attempts to explain its results by assuming that matter, consisting of electrons, decreases in length as it moves. The faster it moves, the more the matter is shortened. While this is not noticeable at everyday velocities, speeds approaching that of light make this a significant phenomenon. One consequence of this theory is that nothing can exceed the speed of light since no object can have a negative length (it can have a zero length which it has at exactly the speed of light). These mathematical formula that he uses to describe the increase of mass, shortening of length, and dilation of time are called Lorentz transformations. They eventually serve to form the basis of German-Swiss-American physicist Albert Einstein's (1879-1955) special theory of relativity.

1904 Bertram Borden Boltwood (1870-1927), American chemist and physicist, discovers that radium is a decay product of, or is descended from, uranium. This means that radioactive elements are not independent but that one is descended from another, forming a radioactive series.

1904 X rays are first shown to be transverse waves by English physicist Charles Glover Barkla (1877-1944). He spends two years proving that x rays can be polarized and are therefore similar to transverse light waves and not the longitudinal waves of sound. This leads him in 1906 to discover that each particular element produces its own kind of x rays and that different gases affect the intensity of x rays differently. He further finds that the intensity of this difference is proportional to the gas's position on the periodic table and is thus proportional to its atomic weight. This breakthrough leads eventually to the completion of the notion of atomic number and becomes the cornerstone of even greater understanding of atomic structure.

1904 Boundary layer concept is first offered by German physicist Ludwig Prandtl (1875-1953). He discovers that as liquid flows in a tube, a film layer adjacent to the wall forms and does not flow as fast as the rest of the liquid. Called the viscous boundary layer, this concept also applies to the surface of a body moving in air or water. This, in turn, eventually leads to an understanding of skin friction drag and of the way in which streamlining reduces the drag of airplane wings and other moving bodies. His work becomes the basic material of aerodynamics.

MARCH 17, 1905 Dual nature of light is first stated by German-Swiss-American physicist Albert Einstein (1879-1955). Assuming that light travels in quanta—following the new theory of German physicist Max Karl Ernst Ludwig Planck (1858-1947)—Einstein says it therefore has particle-like properties and is not solely a wave that requires some material (like the old notion of the ether) to do the waving. Thus, prior to stating his special theory of relativity, he states that light has a dual, wave-particle, quality.

JUNE 30, 1905 Special theory of relativity is first offered by German-Swiss-American physicist Albert Einstein (1879-1955), who submits a paper on the subject titled "Zur Elektrodynamik bewegter Korpen." It states that the speed of light is constant for all conditions and that time is relative or passes at different rates for objects in constant relative motion. Making a startling mental turnaround from traditional wisdom, Einstein does not seek to explain why the velocity of light remains unchanged but rather postulates it as a universal constant, independent of the state of motion of the observer or that of the light source. This is a fundamentally new and revolutionary way to look at the universe in that it posits no fixed point in it. This new world picture soon replaces the old Newtonian system.

(See also September 27, 1905)

SEPTEMBER 27, 1905 Equation $E=mc^2$ is first posited by German-Swiss-American physicist Albert Einstein (1879-1955). In this his second paper on relativity, he includes this famous equation stating the relationship between mass and energy. In the equation, E is energy, m is mass, and c is the velocity of light. This equation transforms the traditional concept of mass as a fixed, inert thing into something that can be transformed into energy, since mass and energy are but different aspects of the same phenomenon. Coupled with his earlier special theory of relativity, Einstein abolishes concepts of absolute space and time and revolutionizes physics to its core, altering its most primary conceptions.

(See also June 30, 1905)

1905 Third law of thermodynamics is first stated by German physical chemist Hermann Walther Nernst (1864-1941). Also described as the determination of chemical equilibrium, it proves to be a powerful tool for determining the feasibility of many chemical reactions. After working for eleven years on the fact that chemical reactions behave very differently near absolute zero, Nernst succeeds in approaching absolute zero to within one degree and finds that free energy and heat content within the system approach each other as the temperature drops.

Theoretically they should meet at absolute zero, but thermodynamic principles clearly show that equal heat and equal energy are an impossibility. Nernst holds this to be a proof of the impossibility of attaining absolute zero and hails this as the third law of thermodynamics. His law has wide applications and supports the emerging theories of quantum physics.

1905 High pressure physics is first opened for investigation by American physicist Percy Williams Bridgman (1882-1961). He builds useable high-pressure apparatus by designing seals that squeeze together tighter as the pressure increases. Limited therefore only by the strength of the material making up the chamber itself, he soon achieves a pressure of 20,000 atmospheres (128 tons to the square inch). His continued improvements allow him to attain even higher pressures with which he is able to study new forms of solids.

DECEMBER 24, 1906 Music and voice are successfully transmitted via radio waves for the first time by Canadian-American physicist Reginald Aubrey Fessenden (1866-1932). His transmissions are picked up by wireless receivers. This becomes possible by his invention of the modulation of radio waves. This system sends out a continuous signal rather than a pulsed one, with the amplitude of the waves varied or modulated. These modulations are then sorted out, amplified, and reconverted into sound at the receiving end. On this Christmas Eve, Fessenden broadcasts his own voice, violin, and recorded music, marking the first radio broadcast in history. Later in 1907, he makes the first trans-Atlantic voice communication and the first trans-Atlantic two-way connection between the United States and Scotland.

1906 First woman to teach at the Sorbonne in France is Polish-French chemist Marie Sklodowska Curie (1867-1934). She assumes the professorship of her husband, French physicist Pierre Curie (1859-1906), after he is killed in a traffic accident. She continues his lectures at the exact point at which they were interrupted.

1907 Ferromagnetism is first explained by French physicist Pierre Weiss (1865-1940). He states that although all atoms are made up of charged particles and magnetic properties always accompany electric charge, only iron and a few related metals show strong magnetic properties. Non-iron metals that have such properties are called ferromagnetic. He explains this phenomenon by stating that iron and these other ferromagnetic materials have an atomic structure that form small "domains" of a certain polarity all pointing in the same direction. Iron contains such "domains" that point in various directions, but when some external magnetic field forces them to be aligned, they become a single, strong magnetic force. This explanation still applies.

1907 First American to win a Nobel Prize in one of the sciences is German-American physicist Albert Abraham Michelson (1852-1931). Although remembered for his painstaking measurements of the velocity of light made in 1887, he receives the award "for his optical precision instruments."

1908 Jean Baptiste Perrin (1870-1942), French physicist, uses German-Swiss-American physicist Albert Einstein's (1879-1955) formula for Brownian motion and, by actually counting the distribution of suspended particles of gum resin in a liquid, finds that he can for the first time calculate the approximate size of atoms and molecules by observation. This is the final proof that atoms are real entities and not just convenient fictions.

1908 First direct process of color photography is devised by French physicist Gabriel Lippmann (1845-1921). This wholly original method is based on the principle that light rays can overlap and create a mutlihued "interference." This phenomenon accounts for the color produced on soap bubbles and pools of oil. Lippmann

employs this phenomenon in photography by applying a light-sensitive emulsion to a glass plate on the side opposite the camera lens and allowing the mirror effect to create the interference. He produces weak color photographs but his process has many practical drawbacks and cannot be transformed into a commercial product.

1908 First Geiger counter is developed by German physicist Johannes Hans Wilhelm Geiger (1882-1945) and British physicist Ernest Rutherford (1871-1937). Rutherford's research on energetic particles emitted by radioactive substances is made much easier as his assistant, Geiger, invents the first successful device to detect and later record these particles. He builds a cylinder containing a gas under high electric potential. When a high-energy subatomic particle enters, it sets off a mass of ionization that produces a momentary electric discharge (so arranged to give a click sound). This first successful counter of individual alpha rays is improved by Geiger in 1913 (so as to count beta particles as well), and again in 1928 in collaboration with Walther Müller, to produce a more sensitive instrument with a longer lifetime that is an all-particle detecting device.

1910 First indication that ordinary elements may also exist as isotopes is given by English physicist Joseph John Thomson (1856-1940). While measuring deflections of positive rays in a cathode tube and calculating the masses of various atoms, he obtains results suggesting that the gas neon has ions of two different types. Isotopes are different elements (equal in nuclear charge but different in mass) produced in radioactive transformations that occupy the same place in the periodic table. This is the first confirmation that isotopes are possible, as predicted by English chemist Frederick Soddy (1877-1956).

(See also February 18, 1913)

MAY 7, 1911 Theory of the nuclear atom is first announced by British physicist Ernest Rutherford (1871-1937). He states that the atom contains a very small, positively charged nucleus at its center and that it also contains all the protons of the atom and therefore virtually all of its mass. This is surrounded by negatively charged electrons which are very light and which pose no barrier to the passage of alpha particles. This concept replaces the old idea of the atom as a featureless, indivisible sphere and lays the foundation for the development of nuclear physics.

1911 Superconductivity phenomenon is discovered by Heike Kamerlingh-Onnes (1853-1926), Dutch physicist, as he studies the properties of certain metals subjected to the low temperatures of liquid helium. Aware that metals tend to lessen their resistance to an electric current as the temperature drops, he finds that some metals, like mercury and lead, undergo a total loss of electrical resistance at temperatures close to absolute zero. He also discovers that a form of liquid helium called "helium II" is produced that has properties unlike any other substances.

(See also c.1927)

1911 First congress of physicists is sponsored by Belgian chemist Ernest Solvay (1838-1922). It is held in Brussels, Belgium, and its subjects are radiation theory and quanta.

1911 Basic principle of electron emission used in vacuum tubes is first stated by English physicist Owen Willans Richardson (1879-1959). He proves that electrons are emitted from hot metal and not from the surrounding air as many had supposed. This phenomenon is also called the Edison effect and it soon makes possible the vacuum tube. This same year, Richardson also proposes a mathematical equation that relates the rate of electron emission to the absolute temperature of the metal. Called Richardson's law, it later comes to be an important aid in electron tube research and technology.

(See also 1883)

JUNE 1912 First experimental observations of

the diffusion of x rays by crystalline structures are made by German physicists Max Theodor Felix von Laue (1879-196), Walter Friedrich (1883-1968), and Paul Knipping. Since previous attempts with diffraction gratings had failed to indicate any bending of x rays, they use a crystal whose regularly spaced atoms act as a natural grating, and obtain spectral results that enable them to measure the length of x rays. This shows that x rays are light rays of very short length and demonstrates their electromagnetic nature. This also marks the beginning of studies on the physics of solids as an analysis of the periodic and regular disposition of atoms in a crystal.

FEBRUARY 18, 1913 Isotopes are first named by English chemist Frederick Soddy (1877-1956). He observes that an atom that loses an alpha particle or a beta particle changes in a very predictable way into another kind of atom. According to what it loses, its charge changes and it occupies a different but predictable place on the periodic table. Thus an atom in position 90 on the table could have been produced by an alpha decay of element 92 (since an alpha has a 2 charge) or a beta decay of element 89 (charge of 1). This means that atoms with the same number can still have different atomic weights. He calls this phenomenon "isotopes" meaning roughly, "the same place." Today, isotopes are considered forms of an element that have the same atomic number (the same number of protons) but different atomic weights (different number of neutrons).

1913 Atomic numbers of the elements is first proposed by English physicist Henry Gwyn-Jeffreys Moseley (1887-1915). After comparing the wavelengths of over 50 elements with their respective atomic weights, he discovers that the characteristic wavelength x rays all decrease regularly and in an orderly pattern as their atomic weight increases. He then discovers that this means that a particular element has the same number of electrons as its nuclear charge (which he calls "atomic number"), and that the position of an element in the periodic table might be indicated by its nuclear charge as well as by its atomic weight. He also realizes that arranging elements by atomic number would give scientists, for the first time, foreknowledge of how many elements remain to be discovered and where in the periodic table they would fall.

1913 Presence of ozone in the upper atmosphere is first demonstrated by French physicist Charles Fabry (1867-1945). Using an interferometer of his own design, he also demonstrates that solar ultraviolet radiation is filtered out by an ozone layer in the upper atmosphere. Ozone is now known to be important as a screen against most of the sun's harmful ultraviolet radiation. Seventy-five years after this discovery, a "hole" in the protective ozone layer over Antarctica is discovered via satellite.

1913 First application of quantum theory to the atom is made by Danish physicist Niels Henrik David Bohr (1885-1962). His idea offers an entirely different theoretical model for the hydrogen atom. Searching to understand how an atom radiates light, he incorporates the quantum postulate of German physicist Max Karl Ernst Ludwig Planck (1858-1947) and states that electrons orbit the nucleus in fixed orbits, giving off quanta (separate packets of energy) when jumping from one orbit to another. When an electron loses energy it would drop to a lower orbit nearer the nucleus, and when it absorbs energy, it would rise to a higher orbit. This is the first reasonably successful attempt to explain the internal structure of the atom via spectroscopy (as well as explain spectroscopy by that structure).

1913 Value of the electric charge of a single electron is first determined by American physicist Robert Andrew Millikan (1868-1953). In his "oil drop" experiment, he allows a tiny drop of oil to be charged electrically while dropping under the influence of gravity and against the pull of a charged plate above it. From time to time, a droplet would pick up an electron or an

ion produced by x rays he introduces. This additional charge causes the droplet to move up or down slightly, and by adjusting the charge on the electric plate, he keeps it in suspension. It is this amount by which the charge has to be adjusted that is the equivalent of the charge on the single ion (electron). This becomes the final proof of the particulate nature of electricity.

1914 Wavelength of x rays is determined for the first time by English physicist William Henry Bragg (1862-1942) and his son, Australian-English physicist William Lawrence Bragg (1890-1971). Together they invent the science of x-ray crystallography with which they are able to measure the wavelength of x rays. They also develop a mathematical system for analyzing the information they obtain. For this, the Braggs become the first and only father-son combination to share a Nobel Prize (for physics in 1915).

1916 Powder crystallography is invented by Dutch-American physical chemist Peter Joseph Wilhelm Debye (1884-1966). He shows that solid substances can be used in their powdered form for the x-ray study of their crystal structure. This breakthrough eliminates the difficult step of trying to prepare good crystals by other methods.

1918 Mass spectrograph is first developed by English chemist and physicist Francis William Aston (1877-1945). His new device uses electrical and magnetic fields to separate a beam of positive cathode rays into distinct lines and can separate atoms of different mass so as to measure their weight with remarkable accuracy. With his new instrument, he discovers that most stable elements are mixtures of isotopes, differing in mass but not in their chemical properties. This means that different forms of the same element can exist. Now called isotopes, these differ in mass but not in chemical properties.

(See also 1920)

1919 First man-made nuclear reaction is created by British physicist Ernest Rutherford (1871-1937). When he bombards nitrogen gas with alpha particles, the nitrogen atoms struck by alpha particles are converted to oxygen atoms and protons (hydrogen nuclei), thus achieving the first form of artificial fission. He succeeds, through atomic bombardment with subatomic particles, in altering an atomic nuclei and transforming one element into another. For investigating radioactivity, discovering the alpha particle, and developing the nuclear theory of atomic structure, Rutherford becomes known as the father of nuclear physics.

(See also April 1925)

1919 Barkhausen effect is first stated by German physicist Heinrich Barkhausen (1881-1956). He discovers that subjecting iron to a continuously increasing magnetic field produces actual noise that can be heard as a series of clicks if magnified. This is eventually explained by the fact that magnetization of iron increases in little jumps rather than smoothly. Further, since iron is also shown to consist of minute "domains" that line up to become a magnet, one of these "domains" rubs against another and the vibrations make the actual noise.

(See also 1907)

1920 Whole-number rule is first formulated by English chemist and physicist Francis William Aston (1877-1945). Using his mass spectrograph, which enables him to separate atoms of different mass and accurately measure their weight, he is able to state the mass of oxygen defined as exactly 16, all other isotopes have masses that are very nearly whole numbers. This notion becomes fundamental to the development of nuclear theory.

(See also 1918)

1923 Particle-wave hypothesis for electrons is first proposed by French physicist Louis Victor Pierre Raymond de Broglie (1892-1987). Using and combining the mathematical formulas laid down by German-Swiss-American physi-

cist Albert Einstein (1879-1955) and German physicist Max Karl Ernst Ludwig Planck (1858-1947), he posits that an electron or any other subatomic particle can behave either as a particle or as a wave. His theory is proven in 1927.

1923 Compton effect is first stated by American physicist Arthur Holly Compton (1892-1962). While accurately measuring the wavelengths of scattered x rays, he discovers some of the rays have lengthened their wavelength in the process of scattering. He accounts for this by stating that a "photon" of light strikes an electron, which recoils and subtracts some energy from the photon, therefore increasing its length. It is Compton who coins the term photon to describe light as a particle. This particulate or corpuscular nature of light seems to contradict the new wave theory, but Compton shows how, through quantum theory, light can be seen as both a particle and a wave, depending upon the type of reaction.

1924 First to refract x rays with a prism is Swedish physicist Karl Manne Georg Siegbahn (1886-1978). He develops improved spectroscopic techniques that allow him to accurately determine the wavelengths of x rays. These improvements in x-ray spectroscopy are such that he is able to produce an actual x-ray spectra for each element. He finds that when x rays are diffracted by a crystal into a series of ordered wavelengths, certain lines or groups of lines appear that are characteristic of certain elements. The fact that he is able to diffract x rays with a prism means that this form of radiation is similar to light.

1924 Effect of ionosphere on radio waves is first stated by English physicist Edward Victor Appleton (1892-1965). Working on the problem of fading radio signals at night, he demonstrates that this interference is caused by the radio signal reaching a certain spot via two different routes and thus out of phase. While one radio beam goes directly to a certain spot, the other bounces off (Kennelly-Heaviside) layers of charged particles. His experiments allow him to calculate that layer's height at about 60 mi (97 km) high. Later he discovers that there are charged layers at an even greater altitude, and these found at about 150 mi (241 km) up come to be called the Appleton layers.

(See also 1902)

JANUARY 1925 Exclusion principle is first stated by Austrian-American physicist Wolfgang Pauli (1900-1958). Further developing the notion of electron spin, he suggests that no two electrons can have the same set of four quantum numbers. This means two electrons with the same quantum numbers cannot occupy the same atom. This principle develops into one of the most powerful basic descriptive tools in physics and comes to explain the chemical properties of elements.

APRIL 1925 First photographs of a nuclear reaction in progress are taken by English physicist Patrick Maynard Stuart Blackett (1897-1974). To achieve this he uses a Wilson cloud chamber and takes more than 20,000 photographs of over 400,000 alpha particle tracks and eventually observes eight actual collisions of an alpha particle and a nitrogen molecule. His sophisticated experiment shows conclusively exactly what the crude detection device employed by Ernest Rutherford indicted in 1919—that nitrogen has been transmuted into oxygen.

(See also 1919)

1925 Cosmic rays are first named by American physicist Robert Andrew Millikan (1868-1953). This year he and his associate George H. Cameron (1902-1977) attempt to detect the background radiation that constantly bombards the earth at extremely high speeds. They lower an electroscope into a lake and are able to detect radiation that is more powerful than any known emission. Measurements taken from different locations assure them that the source is from outer space, and Millikan thus chooses to call this radiation "cosmic rays." He is not, however, able to explain what they are.

JANUARY 1926 Schrodinger wave equation is first postulated in a paper by Austrian physicist Erwin Schrödinger (1887-1961). Seeking to improve the Bohr model of the atom, he develops an equation that describes the existence of standing electron waves in an atom. A standing wave consists of an exact number of wavelengths. His model states that a standing electron wave is stable and electrons traveling in such a wave do not emit energy. In contrast, an electron traveling in a wave consisting of a fractional number of wavelengths is in an unstable condition. He finds his model of "wave trains" predicts the same orbital positions as does the Bohr model, but also provides it with a solid theoretical basis. His mathematical development of wave mechanics becomes so satisfactory it finally places the quantum theory of Max Karl Ernst Ludwig Planck (1858-1947) on a firm mathematical basis.

FEBRUARY 1926 Fermi-Dirac statistics are first introduced by Italian-American physicist Enrico Fermi (1901-1954), and independently by English physicist Paul Adrien Maurice Dirac (1902-1984). This theory describes, in quantum mechanics, the statistical behavior of a system of indistinguishable particles in which each of the available discrete states can be occupied by only one particle. This theory is thus able to explain the behavior of "clouds" of electrons and also shows that gas particles obey the exclusion principle of Austrian-American physicist Wolfgang Pauli (1900-1958). Although discovered independently, the theory is named in honor of both scientists.

JUNE 1926 Nondeterministic or probabilistic view of quantum mechanics is first offered by German-British physicist Max Born (1882-1970). He publishes a paper arguing that one can never state specifically where an electron will be found at any one point in time, but one can predict the probability of finding the electron at any given movement. In working out the mathematical basis of quantum mechanics, Born gives electron waves a probabilistic interpretation that proves to be a pioneering way of looking at subatomic particles.

OCTOBER 1926 Term "photon" is first introduced by Gilbert Newton Lewis (1875-1946), American chemist, to describe the minute, discrete energy packet of electromagnetic radiation that is essential to quantum theory. The concept comes into general use after American physicist Arthur Holly Compton (1892-1962) demonstrates in 1923 the corpuscular nature of x rays. Photons are subatomic particles that travel at the speed of light, have no electric charge, and are carriers of the electromagnetic field. The term comes from the Greek "phos, photos" for light.

SEPTEMBER 1927 Idea of complementarity is first introduced by Danish physicist Niels Henrik David Bohr (1885-1962). In general, this principle states that certain atomic phenomena can be considered in each of two mutually exclusive ways, with each way staying valid in its own terms. In physics, he applies it specifically to the simultaneous wave and particle behavior of both light and electrons, and says that although it is impossible to observe both wave and particle aspects simultaneously, together they present a more complete description than either of the two taken alone. He contrasts this situation on the atomic and subatomic level with the large-scale level of the ordinary world we know (in which such wave and particle characteristics are incompatible and not complementary). His work serves to reconcile some of the apparent conflicts between classical physics and quantum physics. It also comes to be used by other disciplines besides atomic physics.

1927 Clinton Joseph Davisson (1881-1958), American physicist, has a laboratory accident which eventually leads him to discover that electrons or subatomic particles behave like waves. While bombarding a nickel target with electrons, an explosion occurs that leads him eventually to heat the nickel. Doing this, he

later discovers that heating changes the crystalline structure of the nickel which, in its new form, not only reflects but diffracts electrons as well. Since diffraction is a property of a wave, this finding confirms the theory of electron waves put forth by French physicist Louis Victor Pierre Raymond de Broglie (1892-1987) in 1923. Later this year, English physicist George Paget Thomson (1892-1975) independently confirms de Broglie's hypothesis.

c. 1927 First to solidify helium is Willem Hendrik Keesom (1876-1956), Dutch physicist, as he applies external pressure in combination with temperatures below 3K. He also describes the unusual properties of the form of liquid helium called helium II. Its heat capacity changes abruptly and all internal friction disappears. Normal helium is a unique gas in that it remains a gas at colder temperatures than any other element, and is the only substance known that refuses to freeze solid even at absolute zero. Keesom adds high pressure to the super-cold temperatures and finally achieves what was thought impossible.

1927 Uncertainty principle is first postulated by German physicist Werner Karl Heisenberg (1901-1976). Also known as the principle of indeterminacy, this specifically states the impossibility of accurately and simultaneously determining two variables of an electron (such as its position and momentum). More generally, it states that when working at the level of atom-sized particles, the very act of measuring such particles significantly affects the results obtained. Philosophically, this is a troubling notion, for it calls into question much of our traditional belief in straight-forward cause and effect. Scientifically, however, it contributes to a better understanding of how the universe actually works, and it serves to explain many aspects of physics and the universe that would otherwise be meaningless or incomprehensible.

1928 First application of quantum mechanics to the study of nuclear structure is made by Russian-American physicist George Gamow (1904-1968). Studying the problem of why uranium nuclei cannot be penetrated by alpha particles, he develops the quantum theory of radioactivity. This is the first theory to successfully explain the behavior of radioactive elements, some of which decay in seconds and others decay only after thousands of years.

1928 Raman effect is first stated by Indian physicist Chandrasekhara Venkata Raman (1888-1970). After the Compton effect shows that x rays tend to lengthen their waves when diffracted, Raman sets out to show that the same occurs with any electromagnetic radiation, including visible light. While investigating light scattering, he finds that a change in the wavelength of light occurs when a beam of light is deflected by molecules. He also finds that the exact wavelengths produced in the scattering depend upon the nature of the molecules doing the scattering. Because of this, "Raman spectra" prove useful in determining what the molecules are as well as some of the finer details of their molecular structure.

1929 First successful particle accelerator is developed by English physicist John Douglas Cockcroft (1897-1967) and Irish physicist Ernest Thomas Sinton Walton. They devise an arrangement of condensers that build up very high potentials and can produce a beam of protons accelerated at over 500,000 electron volts. Until their invention, the only means of changing one element into another was by bombarding it with alpha particles from a natural source such as a radioactive substance.

(See also April 1932)

1929 Coincidence counter is first devised by German physicist Walther Wilhelm Georg Franz Bothe (1891-1957). This new method of studying cosmic rays uses two Geiger counters to detect cosmic rays. With one of them placed above the other, an event is recorded only if both counters sense it virtually simultaneously. Being vertically aligned eliminates de-tection of anything but cosmic rays (which streak down from above). The coincidence counter allows the measurement of extremely short time intervals, and Bothe uses it to demonstrate that the laws of conservation and momentum are also valid for subatomic particles.

1930 First cyclotron is built by American physicist Ernest Orlando Lawrence (1901-1958). He invents a device that accelerates protons, electrons, and other subatomic particles by having them spin in spirals, giving them an additional boost each time around. It is the first accelerating device that uses a magnetic field to force particles into a spiral orbit of increasing radius and energy. This first device is small, but Lawrence builds successively larger ones and gives physicists a major tool with which many advances are made.

1931 Existence of neutrinos is first posited by Austrian-American physicist Wolfgang Pauli (1900-1958). He seeks to explain the dilemma in which an energy loss occurs when a beta particle is fired out of a nucleus (beta particle emission). Pauli then suggests the existence of an uncharged particle (neutrino) with little or no charge that is carried away from an atomic nucleus when a beta particle is emitted. He also says that the energy is divided between the electron and the other particle (neutrino) in a random manner. He further explains that since the electron has all the electric charge available, the other particle (neutrino) has to be neutral; and that since all the kinetic energy of an electron can be converted into only a tiny quantity of mass, the other particle (neutrino) has little or no mass. His prediction is proved correct in 1955 when a neutrino is finally detected by Reines and Cowan. The other, uncharged particle is later named neutrino (Italian for "little neutral one") by Italian-American physicist Enrico Fermi (1901-1954).

(See also 1955)

MAY 27, 1931 First to penetrate the stratosphere are Swiss physicist Auguste Piccard

(1884-1962) and Dr. Paul Kipfer. They ascend in a pressurized balloon and reach an altitude of 55,563 ft (16,940 m). Piccard's balloon has a revolutionary design, with its airtight cabin equipped with pressurized air. During this flight that reaches nearly ten miles high to where the atmospheric pressure is about one tenth of that at sea level, they conduct their physics experiments.

MAY 1931 First antiparticles are postulated by English physicist Paul Adrien Maurice Dirac (1902-1984). After his mathematical calculations indicate that an electron can have two different types of energy states—one positive and one negative—he infers that since the electron is negatively charged, there should exist a similar, positively charged particle. He suggests that there must exist a positive twin to the electron that has the positive charge of a proton but with a mass equal to that of an electron. Conversely, he says there should also be a particle with a negative charge like an electron but with the mass of a proton. Dirac's theory is confirmed with the 1932 discovery of the positron, and leads to the notion of antimatter.

FEBRUARY 27, 1932 Neutron is discovered by English physicist James Chadwick (1891-1974) who announces the news this day. Although it is known that bombarding light elements like beryllium and lithium with alpha particles results in an intense form of radiation, no one can identify it. Chadwick directs a beam of this unknown radiation at a piece of paraffin and observes the ejection of protons. He then concludes that the radiation must consist of particles with no charge and a mass equal to that of the proton. He calls this neutral (no charge) particle a neutron. This discovery is of major importance to nuclear physicists since, unlike the proton, electron, and other subatomic particles, the neutron is not repelled by either the nucleus or the orbital electrons in an atom. It thus proves to be a much more efficient "bullet" for atomic scientists who seek to initiate nuclear reactions.

APRIL 1932 First artificial transmutation is conducted by English physicist John Douglas Cockcroft (1897-1967) and Irish physicist Ernest Thomas Walton, who use their new particle accelerator to bombard lithium and produce two alpha particles (having combined lithium and hydrogen to produce helium). This is the first nuclear reaction that has been brought about through the use of artificially accelerated particles and without the use of any form of natural radioactivity. This proves highly significant to the creation of an atomic bomb.

(See also 1929)

1932 Positron is discovered by American physicist Carl David Anderson (1905-1991). This is the name for the positively charged electron that English physicist Paul Adrien Maurice Dirac (1902-1984) had predicted in 1930. Anderson devises a cloud chamber with a lead plate dividing it that slows down cosmic rays just enough to make them easier to interpret. He then finds the track of a particle whose mass is the same as a negatively charged electron, but whose charge is positive. He suggests calling this first form of antimatter to be discovered a positron.

1932 Magnetism is first considered in light of the new quantum theory by American physicist John Hasbrouck Van Vleck (1899-1980). In his book, *The Theory of Electrical and Magnetic Susceptibility*, he uses modern quantum mechanical theory to explain the nature of paramagnetic and ferromagnetic materials. He is regarded as the father of modern magnetism.

1934 First example of "artificial radioactivity" is produced by husband-and-wife team and French physicists Frédéric Joliot-Curie (1900-1958) and Irene Joliot-Curie (1897-1956). They bombard aluminum with alpha particles and discover that it continues to emit radiation even after the bombardment ceases. They then find that the reaction results in a transformation, and that they have produced an isotope of phosphorus (a radioactive form of phosphorus).

They soon learn that radioactivity is not only confined to heavy elements like uranium, but that any element can become radioactive if the proper isotope is prepared. For producing the first artificial radioactive element they win the Nobel Prize in chemistry the next year.

1934   Nuclear chain reaction idea is first conceived by Hungarian-American physicist Leo Szilard (1898-1964). This notion of a nuclear chain reaction is one in which a neutron induces an atomic breakdown, which releases two neutrons which break down two more atoms, and so on). Although his method uses beryllium rather than uranium and would be impractical, it is correct in principle. He keeps it a secret, foreseeing its importance in making nuclear bombs, but it is soon discovered by Hahn and Strassman.

(See also January 1939)

1934   First nuclear fission reaction is produced by Italian-American physicist Enrico Fermi (1901-1954). He bombards uranium, the heaviest known element, with neutrons and obtains not only a new element, but also a number of other products he is unable to identify. What he eventually discovers is that he has not only created the first synthetic element, number 93, but that he has also produced the first nuclear fission reaction. During this experiment, he also discovers that neutrons that pass through paraffin or water are more effective at initiating nuclear reactions than those that do not because they are slowed down and spend more time in the vicinity of the target nuclei.

(See also December 2, 1942)

1934   Cherenkov effect is first stated by Russian physicist Pavel Aleksetyevich Cherenkov. Also called the Cerenkov effect, it describes the blue glow of the water surrounding the core of a nuclear reactor. This phenomenon is discovered after Cherenkov's ingenious experiments to prove that it is not a form of luminescence. He concludes that the light emitted when high-energy radiation passes through a transparent

material, like water, is caused by the particles moving at a speed greater than the speed of light. When this occurs, it throws back a "wake" of light. The effect is similar to that of the sonic boom caused by an aircraft when it moves through the air faster than does the wave of sound it is producing. Despite his experimental accomplishments, he is unable to provide a theoretical explanation for the effect named after him.

1935   First demonstrable case of the conversion of energy into matter is offered by English physicist Patrick Maynard Stuart Blackett (1897-1974). He demonstrates how when gamma rays pass through lead, they sometimes disappear and give rise to a positron and an electron. This is a dramatic and very understandable example of the transformation of matter into energy and as such, is a confirmation of German-Swiss-American physicist Albert Einstein's (1879-1955) famous mass-energy equivalence equation, $E=mc^2$.

1935   Uranium-235 is discovered by Canadian-American physicist Arthur Jeffrey Dempster (1886-1950). Using his improved mass spectrometer (or spectrograph), he discovers a rare isotope that has been overlooked by others. It turns out to be the most significant of all, since it is this isotope that will undergo fission and be used by the United States to build the atomic bomb. When bombarded with neutrons, the atomic nucleus of uranium-235 splits into roughly two equal parts and releases large amounts of energy and additional neutrons (which continue the reaction).

1936   Meson is discovered by American physicist Carl David Anderson (1905-1991). While studying cosmic radiation on Pike's Peak in Colorado, he observes the track of a particle whose lifetime is only a few millionths of a second but which is 130 times more massive than an electron, yet only a quarter as massive as a proton. Because of these properties, Anderson suggests calling this new particle a "mesotron."

The name is soon shortened to "meson" and then later it becomes known as a "mu meson" or "muon" to distinguish it from another type of meson later discovered. Mesons are eventually understood to be a very heavy particle related to the electron.

JANUARY 1939 Otto Hahn (1870-1968), German physical chemist, and Fritz Strassman, German chemist, describe their uranium-bombardment experiment but carefully avoid suggesting that the uranium nucleus actually breaks in two. Both are chemists and know that such an assertion contradicts traditional chemical beliefs. Despite this lack of assertion, they are among the very first to witness and understand what comes to be called the "fission" process.

(See also February 11, 1939)

FEBRUARY 11, 1939 Term "fission" is first used by Austrian-Swedish physicist Lise Meitner (1878-1968) and her nephew, Austrian-English physicist Otto Robert Frisch to describe the splitting apart of the uranium nucleus and the release of energy that accompanies it. Although the discovery of this phenomenon is made by German physical chemist Otto Hahn (1870-1968) and German chemist Fritz Strassman, and published the month before, they do not use the word "fission" nor actually describe the breaking apart of the uranium nucleus. Meitner and her nephew, however, develop a theory as well as a name for Hahn's discovery, and their explanation of fission proves to be the critical science needed to drive the new technology. This will prove an essential link in the development of the atomic bomb.

1939 First to experimentally confirm Austrian-Swedish physicist Lise Meitner's (1878-1968) theory on uranium fission is American physicist John Ray Dunning (1907-1975). In early 1940, he also separates small quantities of two uranium isotopes (uranium-235 and uranium-238) and shows that the more rare type, uranium-235, is the element that undergoes fission.

(See February 11, 1939)

1939 Liquid drop model of the atomic nucleus is first proposed by Danish physicist Niels Henrik David Bohr (1885-1962). This model is essential to his theory of the mechanism of fission as he suggests that the protons and neutrons that make up the atomic nucleus can be viewed as similar to the molecules that make up a tiny drop of water. He says that the fission of the nucleus behaves somewhat like the breaking apart of a liquid droplet. His prediction that it is the uranium-235 isotope that undergoes fission is soon proven correct when the United States begins serious work on an atomic bomb.

1940 Donald William Kerst, American physicist, builds the first betatron. This particle accelerator spins electrons (a cyclotron spins protons) in a circle to study high-energy particles. Since electrons are so much lighter than protons, they must be whirled at much greater velocities, and the betatron uses a doughnut-shaped hollow ring and two sets of magnets rather than a spiral-shaped ring.

(See also 1949)

1941 Superfluidity is first described by Russian physicist Peter Leonidovich Kapitza (1894-1984). While studying the properties of helium-II or liquid helium that exists at temperatures below 2.2K, he finds that it conducts heat more than 800 times more efficiently than copper. He coins the word "superfluidity" to describe this incredible conductivity. Superfluidity is the ability of a substance to flow even more easily than do gases, climbing up the walls of a container and through a sealed lid, and it is not identical with superconductivity. In 1946, Kapitza refuses to work on the Soviet Union's nuclear weapons program and is put under house arrest for seven years. Years later, he willingly works on their space program.

(See also 1911)

DECEMBER 2, 1942 First self-sustaining chain reaction is produced by Italian-American physicist Enrico Fermi (1901-1954), who heads a Manhattan Project team at the University of

Chicago. A controlled chain reaction is produced in an "atomic pile" made up of uranium and uranium oxide with graphite blocks. This pile is actually the first nuclear reactor and consists of the fuel (uranium), a moderator (graphite), and control rods (made of cadmium). Control rods absorb neutrons released during fission, and as the rods are slowly withdrawn from the reactor, more and more neutrons become available to start the reaction. Once it is actually demonstrated that a nuclear reaction is doable, they are able to stop it from going out of control by reinserting the rods. With this, the atomic age begins.

1943   First operational nuclear reactor is activated at the Oak Ridge National Laboratory in Oak Ridge, Tennessee. This facility is created as part of the United States effort to build an atomic bomb, and its specific task is to explore the best methods of separating the fissionable uranium-235 from uranium-238.

1944   Theoretical basis for quantum electrodynamics (QED) is first worked out by three physicists, all working independently of one another. Richard Feynman (1918-1988) and Julian Seymour Schwinger of the United States and Sin-Itiro Tomonaga (1906-1979) of Japan each perform theoretical work in which the behavior of electrons is worked out mathematically with far greater precision than ever before. In its modern form, QED is a successful theory used with a high degree of precision to predict such phenomena as the magnetic momentum of the electron and other particles, the detailed structure of spectroscopic lines, the properties of cosmic ray showers, superconductivity in metals, and superfluidity of helium.

JULY 16, 1945   First atomic bomb is detonated by the United States near Almagordo, New Mexico. The successful experimental bomb generates an explosive power equivalent to between 15 and 20 thousand tons of TNT. Within a month, the United States is ready to deliver this weapon and use it to try to end the war in the Pacific.

(See also August 6, 1945)

AUGUST 6, 1945   First use of nuclear power as a weapon occurs as the United States destroys the Japanese city of Hiroshima during World War II with a nuclear fission bomb based on uranium-235. Three days later a plutonium-based bomb destroys the city of Nagasaki. Japan surrenders on August 14 and World War II ends.

DECEMBER 24, 1946   First self-sustaining nuclear reactor in the Soviet Union goes into successful operation. It is directed by Igor Vasilevich Kurchativ (1903-1960), Russian physicist.

1946   George Dixon Rochester and Clifford Charles Butler, English physicists, discover "V particles" in a cloud chamber. These strange, new particles are photographed in a cloud chamber as making V-shaped traces. The role of these new elementary particles is not understood until American physicist Murray Gell-Mann's concept of "strangeness" is introduced.

(See also 1953)

1946   Leptons are first named by Dutch physicist Abraham Pais, and Danish physicist Christian Moller. As physicists begin to classify particles, Pais and Moller conduct research on high energy particles and suggest the name "lepton" for lightweight subatomic particles that are governed by the weak interaction. The weak interaction is one of the basic forces of nature and is the force involved in radioactive decay. Electrons, muons, and neutrinos are all subject to the weak interaction and are called leptons. The name comes from the Greek word for "weak."

1947   Holography is first conceived by Dennis Gabor (1900-1979), Hungarian-British physicist and electrical engineer. He works out the idea of holography (as lenseless, three-dimensional photography) and begins to develop its basic techniques. It involves splitting a beam of light resulting in a three-dimensional image that conveys far more information than an ordinary photograph. Because available light sources are too weak or too diffuse and the technology of optics has not advanced enough, he must wait until the 1960 invention of the laser to make his idea commercially feasible.

1947 Pion or pi-meson is discovered by English physicist Cecil Frank Powell (1903-1969). As a particle with a mass somewhere between a proton and an electron, the pion interacts readily with protons and neutrons and decays very quickly, although carrying a strong force.

(See also February 1948)

1947 Lamb shift is first found by American physicist Willis Eugene Lamb, Jr. Although the lines in the spectrum of hydrogen appear to be single dark lines, they are actually many fine lines close together. No one actually sees this hyperfine structure until Lamb applies his elegant new methods to measure the spectrum lines and finds their positions to be slightly different from what had been predicted. He shows there is actually a splitting of energy levels, later called the Lamb shift, that necessitates a revision of theory to fit these newly-discovered facts. His discovery spurs further refinements in the quantum theories of electromagnetic phenomena.

FEBRUARY 1948 Cesare Mansueto Giulio Lattes, Brazilian physicist, and Eugene Gardner, American physicist, confirm the existence of heavy and light mesons and detect the first artificially produced pion.

(See also 1947)

1949 First electron synchrotron goes into full operation at the University of California at Berkeley to study high-energy particles. Like the betatron, it consists of a doughnut-shaped hollow ring into which particles are injected after being accelerated by a linear accelerator.

Then they are given additional energy by magnets inside the ring. This device produces electron beams of 320 MeV energy.

(See also 1940)

1949 First atomic clock is built by American physicist Harold Lyons. He builds a device that is able to harness and utilize the natural vibration frequencies within certain atoms and molecules. By linking a clock's oscillator with the unfaltering resonance frequencies of atoms and molecules, atomic clocks achieve unparalleled accuracy. Recent clocks are driven by the atomic frequency of the electrons found in the element cesium and achieve such a level of precision that scientists can time events to a millionth of a trillionth of a second.

1950 Optical pumping is first developed by German-French physicist Alfred Kastler. In this system, atoms are excited by light or radio waves which can then be used to better study atomic structure. Kastler finds that by illuminating atoms with frequencies of light they are capable of absorbing, they momentarily attain a high energy state and then return to emitting light. He also discovers that from the manner of emission, facts concerning atomic structure could be easily deduced. This technique leads directly to the development of masers and lasers.

1950 Leo James Rainwater, American physicist, first suggests the correct idea that the nucleus of an atom is not spherical but rather spheroidal. This is an important breakthrough in understanding the internal structure of the atomic nucleus.

NOVEMBER 1, 1952 First thermonuclear device is exploded successfully by the United States at the Bikini Atoll in the South Pacific. This hydrogen-fusion bomb (H bomb) is the first such bomb to work by nuclear fusion and is enormously more powerful than the atomic bomb exploded over Hiroshima on August 6, 1945.

1952 First bubble chamber is constructed by American physicist Donald Arthur Glaser. While studying new "strange" particles, he finds he needs a medium of higher density than the vapor cloud available in a cloud chamber. He builds a device that is the reverse of a cloud chamber and uses a liquid ready to boil rather than a vapor ready to condense. His glass container is filled with superheated diethyl ether in which ions form when high energy radiation passes through. This new method proves far more sensitive than cloud chambers and when improved, proves to be a powerful detection device for particle physicists.

DECEMBER 1953 First working model of a maser (Microwave Amplification by Stimulated Emission of Radiation) is designed by American physicist Charles Hard Townes. This new device produces an intense beam of radiation of a precise wavelength in the microwave part of the spectrum, and finds use in atomic clocks and receivers in radio telescopes. It also leads to the development of the laser, called the optical maser since it amplifies light rather than microwave radiation.

1953 Theory of "strangeness" is first posited by American physicist Murray Gell-Mann. As more and more new particles are discovered, with some behaving oddly and having unexpected properties, he studies them and arrives at a way to organize and make sense of this "particle zoo." He demonstrates that each group has a special property he calls "strangeness" that depends on the properties of the average electric charge. He assigns each a strangeness number that can explain some of their behavior. A similar concept is developed independently by Japanese physicist Kazuhiko Nishijima.

(See also February 1964)

1955 Antiprotons are first detected by Italian-American physicist Emilio Segre and American physicist Owen Chamberlain. Using the new bevatron accelerator at the University of California, they bombard copper atoms with protons and produce and detect the complimentary particle of a proton called an antiproton.

1955 Neutrinos are detected and observed for the first time by American physicists Frederick Reines and Clyde Lorrain Cowan, Jr. As extremely tiny particles with no electrical charge and probably no mass, neutrinos rarely interact with matter and are nearly impossible to detect. They are detected when Reines and Cowan focus on one particular neutrino reaction and ignore the many other reactions simultaneously occurring.

1956 First continuous maser is designed by Dutch-American physicist Nicolas Bloembergen. As a cousin to the laser, the maser (Microwave Amplification by Stimulated Emission of Radiation) is a device used to generate and amplify radio and light waves. Until this invention, masers could not be operated continuously and had to pause to raise their energy level each time they worked. This version is a solid-state maser that can steadily produce energy.

JANUARY 15, 1957 Principle of parity is first announced to be flawed by Chinese-American physicist Chien-Shiung Wu. Basing her studies on the theoretical work of two Chinese-American physicists, Chen Ning Yang and Tsung-Dao Lee, she and her colleagues declare that the parity principle, which had been universally accepted for more than 30 years, was incorrect. This principle states that, on a nuclear level, an object and its mirror image will behave the same way (rather than the expected mirror-image reversal that one gets in classical physics). Wu performs the crucial electron-spinning experiments for Yang and Lee and finds that the parity principle does not hold. Disproving this principle eventually allows new and more correct views about neutrinos to emerge.

1957 Tunnel effect or tunneling is discovered by Japanese-American physicist Leo Esaki. Tunneling is a quantum phenomenon in which particles also have wave properties which allow them to penetrate an energy barrier that, in classical physics, should prevent their passage. Exploring the new field of solid-state physics, Esaki works with tiny crystal rectifiers (semiconductor diodes) and finds that at times the resistance decreases with current intensity instead of increasing as expected. He experiments and finds that this effect can be made more pronounced by making the diode junction as small as possible and by doping (coating with impurities) the semiconductor very heavily. The barrier-crossing electrons he is able to produce have a number of practical applications such as a very small, high-speed switch in solid-state electrical circuits.

1958 Mossbauer effect is discovered by German physicist Rudolf Ludwig Mossbauer. Studying the emission of gamma rays, he finds how to make gamma rays with an extremely narrow wavelength that can then be used to measure any wavelength very accurately. This effect is later used for the measurement of the magnetic field of atomic nuclei and the study of other properties of solid materials. It also finds applications in verifying the theory of relativity posited by German-Swiss-American physicist Albert Einstein (1879-1955).

MAY 1960 First working laser is constructed by American physicist Theodore Harold Maiman. He uses a ruby cylinder that emits a coherent and monochromatic light (a single wavelength all in a single direction). He finds that it can travel thousands of miles as a beam without dispersing, and that it can be concentrated into a small, superhot spot. Laser is an acronym for Light Amplification by Stimulated Emission of Radiation which actually describes not the beam of light but the machine used to create the beam. Modern applications of the laser include delicate surgery, especially upon the eye, communications, and even weaponry.

1960 First proton synchrotron opens at the Brookhaven National Laboratory, Long Is-

land, New York. It is called the 3-GeV Cosmotron.

1961 Nuclear structure of protons and neutrons is first determined by American physicist Richard Hofstadter. Using a linear accelerator, he is able to "see" within individual protons and neutrons and finds that these particles are not simple points of matter but have complicated structures like that of the atom itself. He announces that they are made up of a central core of positively charged matter, around which are two shells of mesonic material. In the proton, these meson shells are both positively charged, and in the neutron, one is negatively charged in such a way that the overall charge is zero. Hofstadter's work implies the existence of new types of subatomic particles and marks an important step in the progression toward even more fundamental discoveries.

FEBRUARY 1964 Murray Gell-Mann, American physicist, first introduces the concept of "quarks" as he postulates the existence of unusual particles that carry fractional electric charges. He proposes that protons are not themselves fundamental particles but that they are composed of quarks. Gell-Mann's models based on quarks prove useful for their predictive value, and thirty years later, quarks are still being investigated as to whether they are ultimate particles or not. Gell-Mann chooses the unusual name from a line in *Finnegan's Wake*, written in 1939 by Irish novelist James Augustine Joyce (1882-1941): "Three quarks for Master Mark."

1974 First grand unified theory is developed by American physicists Howard M. Georgi and Sheldon Lee Glashow. Their unified theory ties the strong and electroweak theories together and attempts to show how both forces can be thought of as manifestations of a single basic force. Their theory accounts for the strong, weak, and electromagnetic forces as parts of a single force that broke apart when the universe cooled down after the Big Bang.

OCTOBER 1986   Teams of physicists in the United States and Germany observe independently for the first time individual quantum jumps in individual atoms.

1986   Superconductivity in ceramic materials is first demonstrated by Swiss physicist Karl Alexander Müller and German-Swiss physicist Johannes Georg Bednorz. They discover an oxide combination that, at around 30K (or 30° above absolute zero), is the highest known temperature for superconductivity reached top date. Prior to this, superconductivity was attainable only at impractical temperatures near absolute zero (0° K). Their breakthrough opens an entire new field of ceramic investigation.

1991   New material called Nitromag is discovered by a team of scientists in Ireland. Belonging to the generation of permanent magnetic alloys, its composition is such that magnets made of this material remain useful at much higher temperatures, allowing them to be used in more severe environments.

1994   Tokamak fusion reactor generates over 10 million watts of power. Although it still consumes more power than it generates, for the first time the hydrogen isotopes it uses as fuel are partially heating themselves with energy released by a fusion reaction. This self-heating is but a fraction of what is needed for a real self-sustaining reaction, but this first sign marks a significant advance toward the dream of fusion power.

1995   New state of matter, a Bose-Einstein condensate, is first produced by American physicists Carl Wieman and Eric Cornell. By producing conditions extremely close to absolute zero, they produce this new state of matter that appears as only a brief blob on a TV screen. This single, fuzzy megaparticle is what happens to atoms when put in temperatures of 20 nanokelvins, which is 36 billionths of a degree Fahrenheit above absolute zero. This is the coldest temperature ever produced and at this state, atoms slow down and smear. This means that while their speed is able to be determined, their location is indefinite since they become wavelike.

Transportation

c. **7500** B.C.   Reed boats are first developed in Mesopotamia and Egypt, while in northwestern Europe, dugout canoes are used.

c. **5000** B.C.   Earliest known ship illustration is found at Hierakonpolis in Egypt. It shows a reed boat with a steering oar that works from the port (left) side.

c. **4500** B.C.   Oldest long-distance highway begins to be regularly used. Later known as the Royal Road by the Persians, it takes travelers on a 1,755-mi (2,826-km) journey beginning in Susa (now in Iran) to the Turkish ports of Ephesus and Smyrna.

c. **4000** B.C.   Yoke is used possibly for the first time in the Near East. This wooden bar or frame rests on the shoulders or withers of draft animals and is tied to their neck or horns to assure that they pull together.

c. **3500** B.C.   Wheel is invented at the dawn of the Bronze Age. The first wheeled vehicles are believed to have appeared in Sumer (now Iraq) at this time.

c. **3500** B.C.   Earliest illustration of a sail dates from this period. It is painted on the outside of a funeral vase found near Luxor, Egypt. The sail is fixed to a single mast and there is a shelter aft or toward the rear.

c. **3500** B.C.   Imperial road system first begins to be developed in China. Totaling about 2,000 mi (3,220 km) of road, it is well-built and often surfaced with stone, and distinctive in its crookedness.

c. **3100** B.C.   Earliest bridge on record is believed to have been built on the Nile River by Menes, the first king of unified Egypt.

c. **3000** B.C.   Slide-car, a type of improved and more permanent travois, is first used to carry objects. It has two poles or shafts that are attached to an animal's side, with its other ends dragging on the ground. The poles are kept apart by crossbars and the cargo is carried on them.

c. **2650** B.C.   Oldest find of a complete ship, known as the Cheops burial ship, is discovered in the modern era and is dated to this time. It is a wood-planked ship over 141 ft (43 m) long, and its refinement indicates substantial experience with timber construction.

c. **2300** B.C.   Bridles (strap-like headgear an animal wears so it can be controlled) and bits (the hard part of a bridle that is held in the animal's mouth) made of horn or bone are first introduced in Mesopotamia. Oxen are guided by a rope or ring through their nose.

c. 2250 B.C.   Principle of a railway, as a track that guides vehicles along it, are first known and used in Babylonian times. They build grooved stone wagonways about 5 ft (152 cm) apart.

c. 2180 B.C.   First known tunnel of any significance that is used to transport men and goods is built under the Euphrates River by the Babylonians in Mesopotamia. After diverting the river from its usual bed to a temporary channel, they dig a 3,000-ft (914 m) long tunnel under the river bed and line it with bricks and asphalt.

c. 2000 B.C.   Horses are first tamed by the nomads in the steppes of what is now Iran. Although not as strong as an ox and difficult to harness, they prove well-suited to pulling a light, wheeled vehicle that becomes the chariot. Horses are eventually trained to allow men to ride on their backs.

c. 1900 B.C.   Earliest roads in Europe, called the "Amber Routes," begin to be made and used by Etruscan and Greek traders to transport amber and tin from north of Europe to the Mediterranean and Adriatic. Four separate routes have been identified.

1352 B.C.   Egyptian King Tutankhamen dies this year and is buried with his chariots. When the intact tomb is discovered in 1922 by English archaeologist Howard Carter (1873-1939), working parts of dismantled wooden chariots are found. The chariot was first introduced to Egypt around 1600 B.C. by Asiatic invaders, the Hyksos.

c. 1000 B.C.   Kite is invented in China. As the true ancestor of the airplane, it can be considered the first type of heavier-than-air device to fly. Since the propulsion is supplied by the kite being tilted or inclined to the wind, it is really a tethered glider.

681 B.C.   First recorded illustration of a bireme (a ship with two banks of rowers) is found in the ruins of Sennacherib's palace in Assyria.

621 B.C.   Earliest Roman bridge about which a record exists is the "Pons Sublicus" or Bridge of Piles. It figures in the legendary feats of the Roman hero Horatius Cocles. This wooden bridge is generally considered to be the first of eight bridges across the Tiber River at Rome that were built by the Romans.

c. 600 B.C.   Greek galley is first introduced as the major fighting ship of this time. It is a long, low, sleek vessel that is lightly built and propelled by oars. The later biremes and triremes are larger versions of this ship.

c. 500 B.C.   First set of laws governing conduct at sea, the Lex Rhodia is adopted by most of the maritime nations. Named after Rhodes, the capital and island in the Mediterranean, it becomes the basis for all subsequent maritime jurisprudence.

c. 400 B.C.   "Cercurus" type of ship first appears in Greek waters. This light, fast, round-bottomed boat becomes the standard cargo carrier of the time. Its mast carries a large, square sail and it is aided by a single bank of 20 oars.

c. 325 B.C.   Alexander III the Great (356-323 B.C.), king of Macedonia, is first reported to have used "boots" or "sandals" on his horses' feet to protect them in rough terrain. These are an early form of horseshoes. About 50 B.C., the Roman poet Catullus (c.84-c.54 B.C.) mentions U-shaped metal plates that protect a horse's hooves. Around 900 A.D., real horseshoes made of iron begin to appear. These increase a horse's speed and extend the distance they can cover.

312 B.C.   Construction first begins on the "Via Appia" or Appian Way under the direction of the Roman censor and dictator Appius Claudius or Caecus (the Blind). One of 29 great military roads that radiate from Rome, it divides in two at Beneventum and eventually extends 410 miles (660 km) down the Adriatic coast. Probably the most famous road of all time, it embodies the heights reached by the Romans who are the first scientific road builders.

280 B.C.   First recorded lighthouse, the Pharos of Alexandria, is built. Ptolemy II Philadelphus of Egypt (308-246 B.C.) erects a 400-ft (122 m) marble tower on top of which a fire is kept burning for ships.

c. 50 B.C.   Earliest known attempt to determine either the speed of a ship or its distance covered (known in English as a "log") is made by Roman architect Marcus Vitruvius Pollio (c.75-c.25 B.C.). He uses a waterwheel fixed to the hull of a ship which carries a drum filled with pebbles. Each time the wheel revolves, one pebble falls into a box to be counted and from this, an estimate of the distance traveled can be made.

c. 200   Traditional Arab trading vessel called a "dhow" first appears. It is equipped with fore-and-aft sails of the triangular, lateen type that allow it to sail across the direction of the wind as well as with it. Supposedly modeled on the shape of a whale, the dhow is an efficient and speedy vessel.

c. 450   Stirrups first appear as used by the Mongolian armies of Atilla the Hun (king of the Huns from 434 to 453). The great advantage provided by this hanging sling or ring into which a horserider can put his feet is that it provides him with a more comfortable, secure ride. In terms of combat, however, it revolutionizes infantry tactics and is the technological basis for shock combat. With the lateral support provided by the stirrup, the horse and rider become one. The rider can use his hands to do battle and can deliver a fearsome blow that has the entire weight of the horse and rider behind it. With his feet securely held, the rider will not himself be knocked off by such a blow.

500   Modern harness or horse collar is first de-

picted in frescoes in Kansu, China. This device goes over an animal's head and neck and sits on its shoulders rather than pressing on its windpipe, allowing the animal to throw all its weight into pulling.

c. 500   Clinker-built ships first appear in Scandinavia. Clinker ships are made to weather high seas and are typically northern boats. Unlike carvel-built ships that are southern and have planks that are placed edge-to-edge, these have external planks that are lap-jointed.

c. 700   Lateen, or triangular, sail first becomes established in the Mediterranean. Its evolution is pioneered by the Arabs. As a "loose-footed" sail or one with great flexibility of movement, it can be adjusted quickly to a large variety of angles to better catch the wind. In a sea with no trade winds like the Mediterranean, such a sail proves ideal.

984   First canal or water lock for which there is clear evidence is built by Chinese engineer Ch'iao Wei-Yo on the Grand Canal. A lock is a chamber enclosed by movable gates and placed between two stretches of water which vary in height from each other. Water is then admitted through the sluices to raise the boat, or water is drained through sluices to lower the boat to the proper new canal level. Although the lock is invented in China, it is used rarely there and its later development occurs mainly in the West.

c. 1100   Road system of the Inca empire in South America first begins as the Incas establish themselves at Cuzco (Peru) and start two parallel roadways extending from what is now Quito, Ecuador, to points south. As the Incas do not have the wheel, their road traffic consists entirely of people on foot and pack animals (llamas). They have a swift courier system and also build suspension bridges with wool or fiber cables.

1295   Junk, a classic Chinese sailing vessel and probably the first ship to have a central stern rudder (to steer), is described by Italian adventurer Marco Polo (1254-1324) upon his return to Venice from China. As a flat-bottom ship with a high stern (end) and forward-thrusting bow, the junk is considered the most aerodynamically efficient of all the large ocean-going sailing ships.

1388 Sidesaddle is invented by Queen Anne of England (1366-1394), the wife of King Richard II (1367-1400). Since it is considered scandalous for a woman to straddle a horse as a man would, the queen's new saddle allows a woman to sit with both legs on the same side of the horse. Prior to this, whenever women rode horses, they usually rode behind a man, sitting on "pillions," which were little side-seats placed directly behind the man's saddle. The inventive queen basically adapts stirrups to the pillion and eliminates the need for a man and his saddle.

1545 Universal joint is invented by Italian mathematician and physician Girolamo Cardano (1501-1576). Later called the cardan, this joint allows for the relative angular movement of two shafts whose geometric axes converge at a single point. It later becomes essential for automobiles, which use it to link two turning shafts whose positions may vary in relation to each other.

1600 First land vehicle that does not use muscle power of any sort as the source of its propulsion is built by Belgian-Dutch mathematician Simon Stevin (1548-1620). His wind-propelled "zeylwagen," or sail wagon, is built of wood and canvas and is a kind of two-masted ship on wheels. It is steerable and is said to cover 50 mi (80 km) in only two hours.

1610 Hackney coach, a four-wheeled coach drawn by two horses and seating six people, first appears in England.

1612 Great Sauk Trail or the Potawatomi Trail is discovered in America by the French. A Native American overland route connecting Canada and the Mississippi River Valley, it is named for the Sac Indians who, with the Fox Indians, use this path in their travels. It branches at the "place of the strait" (in French, "place du detroit"—now Detroit) and runs west between the Checagou (Chicago) and Illinois Rivers.

1620 Cornelius Jacobszoon Drebbel (1572-1633), a Dutch inventor, invents the first navigable submarine. His "diving boat" is made of wood and is propelled in the Thames River in England by oars. Air is supplied by two tubes that use floats to keep one end above water.

c. 1650 Continuous sound of the post-horn is heard for the first time. This horn is blown, playing the same six-note tune, by stagecoaches throughout Europe, warning every other vehicle that they have the right of way. Stagecoaches get their name by going "in stages" to their destination, using fresh horses at each stage.

MARCH 18, 1662 First "bus" begins operations in France. This scheduled service operates horse-drawn vehicles that seat eight and are known as "carrosses a cinq solz." It is owned by a company formed by French scientist and philosopher Blaise Pascal (1623-1662). It leaves regularly and on time, whether full or empty.

1667 Light, two-wheeled carriage called a cabriolet makes its first appearance. This doorless, hooded, one-horse carriage is first used in France and becomes very popular in the 18th century. It is often used as a cab for hire.

MAY 1681 Languedoc Canal in France, also known as the Canal du Midi, first opens and links the city of Toulouse with the Mediterranean. This 150-mi (241 km) waterway is considered to be the greatest feat of civil engineering between Roman times and the 19th century. It has 100 locks, three long aqueducts, and one tunnel. It becomes the pioneer and model of the modern European canal.

NOVEMBER 14, 1698 First Eddystone Lighthouse is built and becomes the first to be exposed to the full force of the sea. Designed by English engineer Henry Winstanley (1644-1703), it marks the dangerous submerged reef off Plymouth, England. Built of timber, it is destroyed in a violent storm on November 26, 1703. The second is built in 1708 by English engineer John Rudyard. Anchored firmly to the reef with iron rods driven deep into rock, it stands firm until destroyed by fire in 1755. The third is built by English engineer John Smeaton (1724-1792) in 1759. Made out of interlocking stones, its new design revolutionizes lighthouse and tower design, and it stands firm until erosion forces its replacement in 1882. That year, a fourth is completed by James N. Douglass of England who generally follows Smeaton's design, but he makes the lighthouse taller. He also uses a cofferdam which allows his crews to work on the sea's bottom and build a very solid foundation. It still stands today.

SEPTEMBER 14, 1716   First lighthouse in America, Boston Lighthouse, is built on Little Brewster Island in Boston Harbor, Massachusetts.

c. 1750   Conestoga wagons become the first preferred vehicle for hauling freight in America. They travel only by day and do not change their six heavy horses. Their large hoops, strong hemp cover, and dory shape foster a legend that they were originally designed by a carpenter. The wagons originate in Lancaster, Pennsylvania, and have a very high body that makes fording streams easier.

1755   Streetcar is invented by Englishman John Outram. Built as a vehicle for public transport, it runs on cast iron rails and is pulled by two horses.

1757   First oscillating lighthouse is built by Swedish inventor Jonas Norberg (1711-1783). In order to distinguish the Swedish lighthouse at Korso from a nearby light, he builds a mechanism that turns the light horizontally, giving the effect of a flashing light.

JUNE 11, 1764   Oldest original lighthouse still

standing and in use in the United States, the Sandy Hook Lighthouse in New Jersey, begins operations.

1769   First self-propelled vehicle is designed by French inventor Nicholas Joseph Cugnot (1725-1804). He builds a steam-fueled car or truck for transporting the French army's cannons. Commissioned by the government, this huge, three-wheeled vehicle is able to go about 2.5 mph (4 kph)but has to stop every 15 minutes for its boiler to be refilled. It is the first such vehicle built for actual, practical use and not for experimentation. It proves nearly impossible to steer.

SEPTEMBER 7, 1776   First operational submarine is operated by American Ezra Lee under New York Harbor. Designed by American inventor David Bushnell (1742-1824), this unique turtle-shaped vessel is intended to be propelled underwater by an operator who hand-turns its propeller. Made of oak and covered with tar, it submerges by taking on water, and its air supply lasts about 30 minutes. Lee tries to screw a mine into the hull of a British ship to blow it up in the harbor, but is unable to do so.

1777   First cast iron bridge opens in Coalbrookdale, England. Built by two English ironmasters, Abraham Darby III (175?-1791) and John Wilkinson (1728-1808), this bridge crossing the Severn River consists of a single 140-ft (43 m) semicircular arch made up of five arched ribs. Iron will soon replace stone in bridge-building.

1782   Lighthouse illumination is revolutionized as Swiss inventor Aime Argand (1755-1803) builds an oil lamp whose circular wick and glass chimney give a brilliant flame. Combined with the parabolic reflectors coming into use—hundreds of mirror fragments set in a curved plaster of Paris mould—lighthouses can for the first time produce a beam of very high intensity from a relatively small light source.

JUNE 4, 1783   Joseph Michel Montgolfier (1740-

Illustration of a manned hydrogen balloon.

1810) and his brother Jacques Etienne (1745-1799) give the first public demonstration of their hot-air balloon by sending up a large model made of linen lined with paper. The balloon is 30 ft (9 m) in diameter when inflated and open at the bottom so it can be filled with hot air from a ground fire. It rises about 6,000 ft (1,829 m) and stays up for 10 minutes.

NOVEMBER 21, 1783 First free or untethered human flight takes place when Jean Francois Pilatre de Rozier (1756-1785) and the Marquis François Laurent d'Arlandes (1742-1809) fly as high as 500 ft (152 m) and travel 5 mi (8 km) in a Montgolfier balloon, floating for about 25 minutes across Paris.

DECEMBER 1, 1783 Jacques Alexandre Cesar Charles (1746-1823), a French physicist, and his colleague named Robert make the first manned trip in a hydrogen balloon, going 27 mi (43 km) from Paris. Modern ballooning derives many of its features from Charles, who fits his balloon with a valve, a suspended basket or car, ballast or weight for stability and control, and a barometer which acts as an altimeter.

JANUARY 7, 1785 First crossing by air of the

English Channel is accomplished by Jean Pierre Blanchard (1753-1809) of France and John Jeffries (1744-1819) of England. They fly from Dover to Calais in a nightmarish, two-hour crossing during which their balloon is at times hit by waves.

JUNE 15, 1785 First balloon fatalities occur as Jean Francois Pilatre de Rozier (1756-1785) and Pierre Ange de Romain, both of France, are killed near Boulogne while attempting a France-to-England crossing of the Channel. Twenty-five minutes after taking off, the balloon catches fire and crashes. As Rozier was the first man to fly less than two years before, he is now the first to die.

JULY 1786 First successful American steamboat makes a short trip on the Delaware River. Built by American inventor John Fitch (1743-1798), this small vessel is a barge with a steam engine that operates, using linking beams, six vertical oars or paddles on each side.

1790 First regular steamboat service begins during the summer as American inventor John Fitch (1743-1798) operates a one-boat line between Pennsylvania and New Jersey. During its brief summer run, his steamboat totals between 2,000-3,000 mi (3,218-4,827 km) and reaches speeds as high as 7-8 mph (11-13 kph). The service does not make a profit and Fitch never has a chance to try again.

1795 First engineered and planned road built in the United States, the privately built toll turnpike from Philadelphia to Lancaster, Pennsylvania, is completed. Sixty-two miles (100 km) long, it is surfaced with broken stone and gravel and becomes the best built and most important American road of its time.

OCTOBER 22, 1797 Modern parachute is born as André Jacques Garnerin (1770-1823) of France makes the first successful human parachute descent from the air, jumping from a hydrogen balloon 2,300 ft (701 m) over Paris. The parachute resembles a ribbed parasol with a diameter of nearly 40 ft (12 m) when fully opened.

1801   First passenger-carrying road vehicle is the four-wheel steam car built in England by Cornish engineer Richard Trevithick (1771-1833). It carries several people, with a combined weight of over one ton, and the driver sits in front and steers two small front wheels.

1802   First mechanical ship's log is patented by Englishman Edward Massey. It consists of a small rectangular box containing a rotating wheel. The mechanism is towed astern (at the rear), hauled in and read, and reset at each change of course. A log indicates a ship's speed and distance covered.

(See also 1878)

1803   First steam locomotive is built by English engineer Richard Trevithick (1771-1833). It has a single, horizontal cylinder mounted inside the boiler and has flat wheels designed to run on rails. There is no evidence that it actually runs at this time.

1804   Oliver Evans (1755-1819), American inventor, builds an amphibious steam-powered dredge he names the *Orukter Amphibolos*. He drives it 1.5 mi (2.4 km) through the streets of Philadelphia and then into the Schuylkill River. More than a self-propelled vehicle, it looks like a flat-bottomed boat with wheels and serves as a working dredge, equipped with chain buckets, digging devices, and paddles to move it through the water. Since Evans drives this device on land, it is the first powered road vehicle to operate in the United States.

1806   Beaufort Scale is first proposed by English Commander Francis Beaufort (1774-1857). Used to measure wind intensity on a scale from 0-12, it makes possible the standardization of sailing records. It is adopted by all navies in 1854 and by the international meteorological community in 1874.

1806   Isaac de Rivaz (1752-1828) of Switzer-

land builds the first vehicle powered by an internal combustion engine. His four-wheel wagon has a single, vertical cylinder that is powered by hydrogen carried in a tank over the rear wheels. He reasons correctly that a controlled explosion inside an open-top cylinder will power the vehicle's axles. His car runs slowly but stalls frequently. He builds a similar, improved version in 1813.

AUGUST 1807   First practical, reliable steamboat begins its trials on the Hudson River. Built by American inventor Robert Fulton (1765-1815), the 150-ft (46 m) *Clermont* steams from New York to Albany in 32 hours. This is compared to the four days required by a sailing sloop. The steam engine has a single vertical cylinder, which, through cranks and gears, drives two 15-ft (4.5 m) paddlewheels, one on each side of the hull. Commercial trips start in September, and they prove so successful that Fulton builds another steamboat.

JULY 25, 1814   George Stephenson (1781-1848), English engineer, builds his first locomotive and makes a successful run. He improves existing designs and goes into business building locomotives.

(See also September 27, 1825)

MARCH 17, 1816   First steamboat to cross the English Channel is the 63-ft (19 m) British steamer *Majory*. When it actually makes the trip, its name is *Elise*, since it is bought by a French firm. The trip takes an uncomfortable 17 hours as it fights a stiff gale.

1817   Forerunner of the bicycle is first patented by Carl von Drais de Sauerbraun (1785-1851) of Germany. Called a "Laufmaschine" or a "running machine," it is little more than a curved beam set above two wooden wheels. The front wheel is steerable and the vehicle scoots along as the rider pushes off the ground with his feet. The vehicle becomes all the vogue for a while in Paris and London and is called a

The first steam engine wagon, built by Oliver Evans.

"Draisienne" in French and a hobby horse in English.

1822   Santa Fe Trail, the first of the pioneer roads connecting the American frontier with the far west, begins to be regularly used.

1822   A strong pencil-beam of light that is ideal for lighthouses is first produced by the French physicist Augustin Fresnel (1788-1827), whose work on the nature of light leads to new lenses. His prism system, which replaces the older method of mirrors, results in all the light emitted from a source being refracted into a horizontal beam. This revolutionizes the effectiveness of lighthouses.

1823   First macadam road in the United States is laid over the Boonesborough Turnpike Road between Hagerstown and Boonsboro, Maryland. Construction follows the details given by Scottish inventor John Loudon Macadam (1756-1836) in his book. The road is compacted by a 2-3-ton (1.8-2.7-metric ton) cast-iron roller.

1824   Joseph Aspdin (1779-1855), English bricklayer, first patents what he calls Portland cement, a material produced from a synthetic mixture of limestone and clay. He calls it Portland because of its resemblance when set to Portland stone. Cement is used to make mortar and concrete, and becomes a widely used and essential material for roads and bridges.

FEBRUARY 1825   First steam locomotive to run on rails in the United States is built by American inventor John Stevens (1749-1838). He builds a half-mile circular track in his backyard in Hoboken, New Jersey, and experiments with his new machine. Stevens is one of the first to be convinced that railroads and not canals are the overland transport system of the future for America.

SEPTEMBER 27, 1825   First steam-hauled, public passenger train ceremoniously opens the Stockton & Darlington Railway in England. English engineer George Stephenson (1781-1848) uses one of his improved locomotives to steam 21 mi (34 km), pulling 12 wagons of goods and 21 wagons carrying nearly 600 excited passengers. For the first time in history, land transportation is possible at a rate faster than any horse can run. His line turns a profit and Stephenson's locomotives begin a transportation revolution throughout the world.

OCTOBER 26, 1825 Erie Canal, stretching more than 350 mi (563 km) from Buffalo on Lake Erie to the Hudson River at Albany, New York, officially opens. This first great American engineering work is organized by New York mayor DeWitt Clinton (1769-1828), and built by American engineer James Geddes (1763-1838). It has 82 locks, takes eight years to build, and costs $7 million. When finished, it connects by boat the Atlantic Ocean and the Great Lakes.

MAY 5, 1826 First modern railway system (in that it is operated entirely by steam locomotives) is incorporated. It is with this line—the Liverpool & Manchester Railway—that the railroad becomes established as a means of regular public transport that rivals and finally passes the coach and canal companies. The engineer is George Stephenson (1781-1848) of England. The entire railway opens on September 15, 1830. It has double tracks and operates passenger trains on a timetable. It still runs today.

1826 First American "buggy" is made. This is the most typical American horse-drawn carriage and the best known. Modeled after a German wagon design, it owes its extreme lightness to its frame and wheels, which are made of hickory wood. It has four small wheels and elliptical springs, front and rear, mounted at right angles to the front.

1828 First successful use of planet gearing is made in Paris. This device for transmitting power from one rotating shaft to one or more others through small gears that mesh with the outer teeth of the larger gear will become an important factor in the development of automobiles. Only with such a device can the turning power of an engine be transmitted to the wheels.

1831 First railroad tunnel in the Western Hemisphere, the Staple Bend Tunnel, is begun near Jamestown, Pennsylvania. It is part of a rail-road/canal transportation route that runs from Philadelphia to Pittsburgh. It still stands today but has not been used since 1852.

1832 First urban streetcar in the United States is built by American wagon builder John Stephenson and goes into operation in New York. This 30-passenger, stagecoach-on-rails is pulled by horses. Similar horsecar services are begun in Boston (1856) and Philadelphia (1858).

1833 First clipper-ship style vessel, the American *Ann McKim*, is launched at Baltimore, Maryland. "Clipper" ships do not have one particular design but are faster, bigger sailing ships and are built to "clip" days off a voyage and thus compete with steamships. Their major design development is the bow change from a blunt "U" to a sharp "V".

APRIL 20, 1836 First narrow-gauge public railroad, the Festiniog Railway in Wales opens. Narrow gauge is any track less than standard gauge (4 ft. 8.5 in [144 cm]). It is horse-drawn.

1836 English "hansom" carriage is invented by Joseph A. Hansom (1803-1882). This carriage is a light, open two-seater that is driven from a highly elevated box placed behind the passengers. They replace the hackney coaches.

APRIL 4, 1838 First Atlantic crossing entirely under steam begins as the British steamship *Sirius* leaves Cork Harbor for New York. It is made possible by the 1834 invention by Samuel Hall (1781-1863) of England of surface condensers that use fresh water instead of sea water. Although the ship must carry a huge supply of fresh water, this system does not have to be cleaned out every four days (as the sea water system requires).

1838 First brougham coach is made in England by Robinson and Cook for Lord Brougham, then Lord Chancellor. This ancestor of many 19th century coaches has several innovations, one of which is its extremely low-slung center body which is sunk deep enough to permit easy entrance from ground level.

1839 First step in the development of the bicycle from the wooden hobby horse to a mechanically operated, two-wheeled vehicle is made by Scottish blacksmith Kirkpatrick Macmillan. He appears to be the first to discover that two wheels placed in line could be balanced and could be propelled by treadles and cranks fitted to one of the axles.

1840 First railroad disc-and-crossbar signals are introduced on the Great Western Railway in England. They are at right angles to each other, so when the crossbar faces the train, it means "stop." When turned 90 degrees, the disc means "line clear."

MARCH 28, 1843 William Samuel Henson (1805-1888) of England publishes the design and receives the patent for his "Aerial Steam Carriage." This is the first reasoned, formulated, and detailed design for a propeller-driven aircraft. Because this monoplane design receives world-wide publicity, it can be said to have greatly influenced the basic configuration of the modern airplane.

JUNE 15, 1844 Charles Goodyear (1800-1860), American inventor, first patents "vulcanizing" of rubber. While experimenting on how to prevent rubber from becoming stiff in the cold and soft and sticky in hot weather, he accidentally drops some India rubber mixed with sulfur on a hot stove and discovers that it becomes dry and flexible in either temperature. His name and discovery are now synonymous with automobile tires.

JULY 26, 1845 First propeller-driven steam liner to cross the Atlantic, the *Great Britain* leaves Liverpool for New York. Built by English engineer Isambard Kingdom Brunel (1806-1859), it is also the first large, ocean-going ship to be made of iron. It is the largest ship in the world when completed.

c. 1850 First and only hinged ship, the *Connector*, is built in England. This unusual ship is built in three sections with articulated or hinged joints that allow it to be split up into three parts to load or discharge cargo. Essentially a variation of the barge train used on inland waterways, it is designed to ride easier in rough seas because the ship's sections will undulate with the waves. Unfortunately, sideways rolling will easily tear the sections apart, and the ship is never tried at sea.

SEPTEMBER 24, 1852 First powered, manned airship is flown by its builder, French engineer Henri Giffard (1825-1888). Powered by a steam propeller, the airship flies at about 5 mph (8 kph) and covers 17 mi (27 km) from Paris to near Trappes, France. This craft marks the beginning of the practical airship.

1852 Emile Loubat, French engineer, invents the system of placing horse-drawn streetcar rails within the street pavement itself, making them flush with the street and posing much less of an obstacle to coach and wagon traffic. By 1860, horse-drawn railcars are rapidly replacing the omnibuses or coaches, since the former is a more efficient use of horsepower. A team of horses can pull vehicles of 2 tons (1.8 metric tons) or more with up to 50 passengers on rails. This is double the weight and number of a horse-drawn omnibus.

1852 First "safety" elevator is invented by American inventor Elisha Graves Otis (1811-1861). Otis designs a safety brake that will keep an elevator from falling even if the cable holding it is completely cut. His first safety guard consists of a used wagon spring on top of the hoist platform and a ratchet bar attached to the guide rails. If the cable snaps, the tension is released from the spring and each end immediately catches the ratchet, locking the platform in place and preventing a fall.

1853 George Cayley (1773-1857), English engineer, builds a full-size airplane and tests it in

Yorkshire. He persuades his coachman, John Appleby, to make a glide across a valley. This can be considered the first, unpowered manned flight in a heavier-than-air vehicle.

MARCH 6, 1855 First train to cross a span suspended by wire cables goes over the Niagara Bridge. This is the first successful railway suspension bridge in the world. Designed and built by American engineer John Augustus Roebling (1806-1869), its two decks are suspended from four cables which rest atop two masonry towers.

1856 Henry Bessemer (1813-1898), English metallurgist, announces his discovery of what becomes known as the "Bessemer process," a quick, inexpensive way of making steel. Up to now, steel was made by an extremely costly method of first making cast iron, converting it into wrought iron, and then adding carbon to make steel (which has the hardness of cast iron and the toughness of wrought iron). Using his new converter which held molten pig iron, Bessemer finds that forcing a blast of air into it made mass production of affordable steel possible. The availability of cheap steel begins the era of large-scale building and transportation ventures.

SEPTEMBER 1, 1859 First Pullman railroad sleeping car begins service on an overnight trip between Minnesota and Illinois. It is invented by George M. Pullman (1831-1897), an American cabinet maker. In 1864, his newly designed sleeping car becomes the forerunner of the modern railroad sleepers.

APRIL 3, 1860 Pony Express, the first rapid overland mail service to the Pacific Coast, officially opens. Regularly changing horses and riders, the pony riders cover 250 mi (402 km) in a 24-hour period. Although famous and colorful, the system is a financial disaster and is put out of business by the completion of the transcontinental telegraph system.

c. 1860 American "covered wagon" called the

The Bessemer process of making steel provided a less expensive wrought iron.

"prairie schooner" first takes settlers to the far west. Similar to a Conestoga wagon, they have a flat, box body instead of the tipped-up, boat design of the Conestoga. These are the main means of westward migration for many years.

1860   First practical gas engine or internal combustion engine is built and patented by Belgian inventor Jean Joseph Etienne Lenoir (1822-1900). His new engine is a converted double-acting steam engine with slide valves to admit the explosion mixture and to discharge exhaust products. This two-stroke cycle engine uses a mixture of coal gas and air and proves to be smooth-running and durable. His success spurs other inventors to examine the possibilities of this new engine over steam.

(See also 1862)

1861   First to suggest driving the front wheel of a velocipede or bicycle directly by cranks or pedals on the axle is Pierre Michaux, his son, Ernest, and their employee, Pierre Lallement. They set up shop in Paris and in 1866 introduce a new model made of elegant wrought iron. Its front wheel is much larger than the rear, so that each revolution of the front pedals will carry the rider farther. These bicycles become very popular.

1862   Nikolaus August Otto (1832-1891), German inventor, first experiments with a four-cycle system on his four-cylinder engine. The four cycles or movements of the piston (induction, compression, ignition, and exhaust) will become the basis of the internal combustion engine. When a modern engine is running, every cylinder goes through this same sequence of events or cycle hundreds of times a minute.

JANUARY 10, 1863   First underground city passenger railroad or subway opens to traffic in London, England.

1864   Siegfried Markus (1831-1898), Austrian inventor, produces a wooden cart powered by a two-cycle gas engine. Some claim this to be the first automobile. By 1875, he builds a greatly improved version with a four-cycle gasoline engine that drives the rear axle. It has a wheel for steering and goes only one speed since it has no gears.

1867   First elevated railroad in the United States begins operations in New York City. This method of actually raising the tracks above street level is the cheapest way of providing a right-of-way for inner suburban trains in large city centers.

JANUARY 23, 1869   George Westinghouse (1846-1914), American engineer, applies for an air-brake patent. His new power braking system uses compressed air as the operating medium and far surpasses any other method. It is first used on a passenger train, and its final improvement in 1872 makes Westinghouse a rich man as he adds an important element of safety to train travel.

MAY 10, 1869   First transcontinental United States railroad is completed with the Golden Spike ceremony. The railroad joining the Atlantic and Pacific Coasts are linked at Promontory Point, Utah, north of the Great Salt Lake.

NOVEMBER 17, 1869   Suez Canal, a sea-level waterway across the Isthmus of Suez in Egypt, first opens. Built by French diplomat Ferdinand de Lesseps (1805-1894), it takes five years to organize and plan and ten years to construct. It extends 105 mi (168 km), using several lakes, and is an open cut without locks. Following improvements and widening, it becomes one of the busiest canals, handling a significant portion of the world's sea traffic.

1871   Andrew Smith Hallidie (1836-1900), American engineer, invents the underground continuous moving cable and mechanical gripper for the underside of streetcars. His new electric cable car runs two to three times faster than a horse-drawn rail car and can go up the steepest grades. By 1873, his system is running cable cars in San Francisco and other American cities. In 1964, this system is still running in San

Francisco and becomes the first moving National Historic Landmark.

1871   First wind tunnel is built for the Aeronautical Society of Great Britain by Francis H. Wenham (1824-1908) and John Browning of England. This highly useful experimental device simulates a moving aircraft by passing air over a stationary model. Wind tunnels are still used today to study the aerodynamics of cars and other vehicles as well as planes.

c. 1873   First design of a safety bicycle using chain-drive to the rear wheel is made by H. J. Lawson of England. It has two medium-size wheels of equal diameter, with the rear being driven by means of a chain and sprockets.

MAY 24, 1874   First steel arch bridge, the St. Louis Bridge, opens to traffic. Designed and built by American engineer James Buchanan Eads (1820-1887) to span the Mississippi River, it is the first bridge project to use pneumatic caissons (enclosure with no bottoms in which compressed air keeps out the water), and also the first to build arches by the cantilever method without falsework or staging. Now called the Eads Bridge, it is also the first major construction of any sort with steel.

1875   Samuel Plimsoll (1824-1898), English politician and social reformer, invents the Plimsoll Line which is established by the British Parliament this year. This line is a fixed mark placed on the hull of every cargo ship indicating the maximum depth to which the ship can be safely loaded. Plimsoll offers the idea in order to save the lives of crew members aboard "coffin ships"—unseaworthy, overloaded vessels whose owners care little about crew safety.

1876   First practical, successful, lightweight four-cycle internal combustion engine is produced by German engineer Nikolaus August Otto (1832-1891). Utilizing the four-stroke cycle (four strokes of the piston for each explosion), it is this smooth-running, reliable engine that

offers the first practical alternative to the steam engine as a power source. It is the direct ancestor of all modern reciprocating engines.

APRIL 20, 1877   First cantilever bridge in the United States, the Kentucky River Bridge is completed. It is built by the self-taught American engineer C. Shaler Smith (1836-1886). A cantilever bridge is essentially two connected beams, each supported only at one end.

1878   Forerunner of the modern towed ship's log is first patented by Englishman Thomas F. Walker, Jr. It consists of a towed rotator connected to a log-line which is connected to a register or dial mounted on the rail of a ship. Reading this log, which tells a ship's speed and distance, involves only a walk to the stern (rear) to read a dial.

MAY 31, 1879   First practical electric railroad begins operations at the Berlin Trades Exhibition in Germany. This passenger line, built by German engineer Werner von Siemens (1816-1892), operates on the exhibition grounds and can pull 30 passengers at 4 mph (6.4 kph). It is the first practical electric train that gets its current from a station generator.

1884   Charles Algernon Parsons (1854-1931), English engineer, invents the steam turbine engine. This new engine does not use pistons since it has a jet of steam that turns a multibladed shaft, which is then connected directly to a propeller. Once perfected and applied to ships, it will revolutionize marine propulsion.

1884   Horatio F. Phillips (1845-1926), English inventor, first patents and publishes in England his design for what will become the true foundation of the modern airfoil (wing shape). He also demonstrates why a curved wing provides lift. This is a major breakthrough in understanding mechanical flight and will influence future pioneers.

1885   First true ancestor of the modern bicycle is the Stanley "Rover" Safety Bicycle. It has similarly sized wheels and is much lower and more stable than its predecessors. This rear-wheel driven bicycle sets the new bicycle style internationally and sends the high-wheelers and tricycles into irreversible decline. It is designed by John Kemp Starley (1854-1901) of England.

1885 Karl Friedrich Benz (1844-1929), German engineer, builds a two-seater tricycle powered by a four-cycle gas engine which drives well in tests. This is the first vehicle in which the engine and chassis form a single unit and can be considered the first true automobile. Benz did not attempt to convert a carriage, but rather made designs for an entirely new kind of self-propelled vehicle.

1885 Gottlieb Wilhelm Daimler (1834-1900) and Wilhelm Maybach (1846-1929), German engineers, build a much-improved high-speed, internal combustion engine and install a one-cylinder, four-cycle gas version on a bicycle frame and create the first true motorcycle.

1886 First four-wheel, gas-driven car with a four-stroke engine is built by German engineers Gottlieb Wilhelm Daimler (1834-1900) and Wilhelm Maybach (1846-1929). Their landmark vehicle is known today as the "Cannstatt-Daimler."

1886 First practical tandem bicycle (two or more persons sitting one behind the other) is designed and built by D. Albone and A. J. Wilson of England. As the first "bicycle built for two," it spawns versions that can transport up to six people.

SEPTEMBER 20, 1887 A patent is issued to John Charter of the United States for what becomes the first practical farm tractor. He then forms a company and produces in 1889 the first successful gasoline-powered farm tractor.

1887 First successful electric trolley line is built by Frank Julian Sprague (1857-1934), American engineer, in Richmond, Virginia.

(See also 1897)

1888 Edward Butler of England invents a powered tricycle which is the first vehicle to have a float-feed carburetor. This is basically a spray carburetor in which the fuel is sprayed from an opening into the center of an air stream. It precedes Wilhelm Maybach's (1846-1929) spray carburetor and has all the basic elements of a modern carburetor.

(See also 1894)

1888 John Boyd Dunlop (1840-1921), English inventor, first conceives the idea of pneumatic (air-filled) tires and applies it to his son's bicycle. Although invented for the bicycle, whose ride they transform, pneumatic tires come along just in time for the emerging automobile. Although Dunlop was preceded in this idea by Scottish engineer Robert William Thomson (1822-1873), who patented the concept in 1846, it is Dunlop who pursues it thoroughly and eventually creates an entirely new industry out of it.

1890 Clincher rim is first patented by William E. Barlett of the United States. This allows rubber, pneumatic (air-filled) tires to be firmly attached to the metal wheel rim, which has a channel or groove into which the tire's edge fits.

1890 First modern form of the diamond frame bicycle is produced in England. Made by the Humber firm, it has all the essential features of a modern bicycle: chain-drive to the rear wheel, a moveable rear wheel set in slotted fork ends to permit chain-tension adjustment, ball bearing raked and set steering head and fork, wheels of the same diameter, and a spring seat on an adjustable seat tube.

1891 Escalator is invented by Jesse W. Reno of the United States. His moving, inclined belt provides transportation to passengers riding on cleats attached to the belt. At first the handrail is stationary, but he builds an improved version with a moving handrail in this same year.

FEBRUARY 27, 1892 Rudolf Diesel (1858-1913),

German engineer, first files a patent for a new heat engine he has not yet built. He then begins to develop his new type of internal combustion engine that will be more efficient than Otto's four-cycle engine (which loses heat or energy because of temperature changes).

MARCH 6, 1893 First elevated electric city rail line, the Liverpool Overhead Railway opens in England.

1893 First successful gas-powered car in the United States is demonstrated in Springfield, Massachusetts, by American inventor Charles Edgar Duryea (1862-1938) and his brother, Frank (1870-1967). It is propelled by a 4hp, two-stroke motor and looks exactly like a buggy without a horse. The brothers set up the first American company to manufacture cars for sale in 1894.

1894 Wilhelm Maybach (1846-1929), German engineer, invents the atomizing carburetor which directs the fuel spray into the center of an air stream. His version is an improvement of the one invented by Edward Butler in England in 1888.

1894 Emile Levassor (1844-1897) and René Panhard (1841-1908), both French inventors, first design the Panhard auto which is the forerunner of the modern car. With a front-mounted engine driving the rear wheels via a clutch and gearbox, this vehicle is the archetype of nearly every auto for the next 50 years.

1895 First practical subway line in the United States is begun in Boston, Massachusetts. Completed in 1897, it is 1.5 mi (2.4 km) long and uses trolley streetcars or tramcars. It later switches to conventional subway trains.

AUGUST 9, 1896 Otto Lilienthal (1848-1896), German aeronautical engineer, crashes from a height of 50 ft (15 m) when gliding in one of his monoplane hang-gliders. He dies of a broken spine the next day in Berlin. Not only was he the first true aviator, but his life and work

directly inspired all those experimenters who followed him, especially the Wright brothers.

1896 Konstantin Eduardovich Tsiolkovsky (1859-1935), Russian physicist, first begins his groundbreaking scientific work on the theoretical aspects of liquid-propellant space rockets. Within two years he designs several rockets and solves the theoretical problem of how reaction (rocket) engines can escape from and reenter the earth's atmosphere.

1896 Frederick William Lanchester (1868-1946), English aeronautical and automobile pioneer, completes his first horizontally opposed ("flat") four-cylinder engine which he places in the center of his 1897 vehicle. This car has many new features and has a design that proves to contain a forerunner of the modern automatic transmission.

JUNE 1897 First ship to be powered by a Parsons turbine engine, the British *Turbinia* puts on a spectacular performance as it dashes through lines of assembled warships at a speed of 34.5 knots. Steam turbines are lighter, provide more power and higher speeds, and take up less space than other engines.

1897 First Stanley Steamer takes to the road. Built by two enterprising American twins, Francis E. Stanley (1849-1918) and Freeling O. Stanley (1849-1940), this simple, carriage-looking car with a tiller for steering sells very well.

1897 Frank Julian Sprague (1857-1934), American engineer, invents the multiple-unit, electric train in which a single driver can control a number of motor coaches from either end of the train. It is first used on the Chicago South Side Elevated Railway. This invention becomes the key component of all modern short-haul passenger railways.

1898 First truly practical submarine and the first underwater vessel accepted by the U. S. Navy, the *Holland* is built by American inven-

Electric elevated railroad in Paris.

tor John Philip Holland (1840-1914). His 50-ft (15 m) long, five-man submarine is powered by a 45 hp gasoline engine and achieves the first unqualified success in submarine history.

1898 Louis Renault (1877-1944) of France brings out his first automobile in which he disposes of the old chain drive and invents shaft drive. This system directly connects the engine and the power axle, revolutionizing the auto industry.

1899 First automobile to exceed one-mile-per-minute (60 mph [97 kph]) is the electric race car called *La Jamais Contente*. This bullet-shaped car is powered by large, expensive batteries. Electric cars will never hold another land speed record.

JULY 2, 1900 First Zeppelin airship makes its initial trial flight in Germany. The first of the rigid, monster airships, it is 420 ft (128 m) long and is housed in the first floating hangar.

NOVEMBER 3, 1901 World's longest railroad, the Trans-Siberian Railway, first opens to

Vladivostock, Russia, and uses a ferry to cross Lake Baikal. The full line is completed on September 25, 1904, and covers the 5,801 mi (9,333 km) from Moscow to Vladivostock.

1901 First American car to be manufactured in quantity is the "Curved Dash" Oldsmobile, designed by Ransom Eli Olds (1864-1950). This popular car has a two-speed gearbox with reverse, is steered by a tiller, and sells for about $650. This marks the real beginning of the American auto industry.

1902 Louis Renault (1877-1944), French automobile pioneer, invents the drum brake. Designed to operate by pressing two semi-circular brake shoes against a circular brake drum, they are soon adopted by nearly all vehicles.

JUNE 16, 1903 Henry Ford (1863-1947), American inventor, organizes the Ford Motor Company in Detroit, Michigan. The entire property of the newly opened factory consists of $28,000 capital put up by 12 shareholders. Ford builds the Model T in 1908, a car that many regard as the first really practical American automobile. He also establishes the moving assembly line to produce cars in volume, making them affordable to greater numbers of people.

DECEMBER 17, 1903 First sustained, controlled, powered flight is made by Orville Wright (1871-1948) in the *Flyer* at Kill Devil Hills near Kitty Hawk, North Carolina. He and Wilbur (1867-1912) make four flights this day, all from level ground and without any take-off assistance. The historic first flight lasts 12 seconds and covers 120 ft (36 m). The fourth and longest lasts 59 seconds and covers 852 ft (260 m). The age of flight has begun.

1903 Harley-Davidson Company of Milwaukee, Wisconsin, produces its first motorcycle. William S. Harley and Arthur Davidson are joined within a year by Davidson's brothers, William and Walter. Their first motorcycle is a single-cylinder machine with 2 hp.

1904 First person to exceed 100 mph (161 kph) in an automobile is Frenchman Louis E. Rigolly, whose large Gobron-Brille is timed at 103.56 mph (166 kph) in Belgium.

1905 First removable automobile tire rims are introduced by the Michelin Company of France. Before these, punctured or damaged tires had to be fixed on the spot with the wheels remaining attached to the car.

1905 First true hydrofoil is operated by its Italian inventor, Enrico Forlanini (1848-1930). The hydrofoil principle eliminates drag on the hull of a boat by going fast enough so that its foils (fins or skis attached to the hull) come to the surface and the vessel skids along the top of the water. This vehicle is powered by an airplane propeller.

OCTOBER 23, 1906 First successful powered airplane flight in Europe is made by Alberto Santos-Dumont (1873-1932), Brazilian aviation pioneer, at Bagatelle, France. His airplane flies some 200 ft (60 m).

1906 First Mack trucks are built in Allentown, Pennsylvania, by the Mack Brothers, John, William, and Augustus. Their 10-ton (9-metric ton) trucks have great hauling power and strength.

NOVEMBER 10, 1907 Louis Blériot (1872-1936), French aviator, first introduces in France what will become the modern configuration of the airplane. His plane has an enclosed or covered fuselage (body), a single set of wings (monoplane), and a propeller in front of the engine.

NOVEMBER 13, 1907 First piloted helicopter rises vertically in free flight in Liseux, France. Built by Paul Cornu (1881-1914) of France, it is powered by a 243-hp Antoinette engine driving two rotors. Although a major achievement, the practical helicopter will not appear until the 1930s.

1907 Automobile taxicabs first appear in New

York City. The standard fare for a short ride is a "jitney"—a common term for a nickel. The term soon becomes synonymous with the service itself.

SEPTEMBER 17, 1908   First fatality in a powered airplane occurs when Lt. Thomas E. Selfridge is killed when flying with Orville Wright (1871-1948) at Fort Meyer, Virginia. After a propeller blade breaks and severs control wires, the airplane crashes from a height of about 75 ft (23 m). Orville is severely injured and lives with back pain the rest of his life, but he will fly again.

OCTOBER 1908   Legendary Model T Ford "Tin Lizzie," the first of the "People's Cars," is introduced in the United States. It is the first Ford with left-handed steering and has a four-cylinder engine that gives 20 hp. Easy to drive and easy to repair, its production will continue until 1927, when total production reaches 16,536,075 vehicles—a record that stands until 1972. At the end of its run, it sells for an all-time low of $260.

1908   First workable gyrocompass is installed and used on the German battleship *Deutschland*. Different from the traditional magnetic compass, it employs the principle of the gyroscope—a device whose spinning wheel, if mounted correctly, maintains its orientation despite any movements about it. Since it is not influenced by any iron or steel around it as a magnetic compass is, it is the first major improvement on the compass in a thousand years.

1908   Tire treads are invented by Frank Seiberling who perfects a machine that cuts grooves in the tire surfaces. Up to now, tires have smooth surfaces and can give little traction when roads are bad.

JUNE 26, 1909   First commercial sale of an airplane in the United States is made as Glenn Hammond Curtis (1878-1930), American inventor, sells one of his planes to the Aeronautic Society of New York for $7,500. This action

spurs the Wright brothers to begin a patent suit to prevent him from selling airplanes without a license.

JULY 25, 1909   First airplane crossing of the English Channel is made by French aviator Louis Blériot (1872-1936), who flies his monoplane from Les Baraques near Calais to Dover, England, in 37 minutes. This event increases public and governmental awareness of the possible military aspects of the airplane.

1909   Derailleur-type gear for bicycles is first made in France. This still-popular system consists of two or more rear wheel sprockets of different diameters, a chain looped over a mechanism which both lifts and transfers it from one sprocket to another to provide a change of gear ratios, and also a chain-tensioning device to take up the slack. This extremely simple system also allows for direct drive on all ratios.

JANUARY 26, 1910   First practical seaplane flies. Built and flown by American inventor Glenn Hammond Curtiss (1878-1930), it lands and takes off in the waters off San Diego, California.

MARCH 8, 1910   First woman to become a qualified pilot is the Baroness de Laroche who receives her "brevet de pilote d'aeroplane," or pilot's license, in France. She dies in 1919 in an airplane accident.

SEPTEMBER 2, 1910   First American woman to fly an aircraft solo is Blanche Scott.

OCTOBER 16, 1910   First airship to cross the English Channel, flying from Europe to England, is the non-rigid *Clement-Bayard II*.

NOVEMBER 14, 1910   First airplane takeoff from a ship is accomplished as American aviator Eugene B. Ely (1886-1911) successfully gets airborne off the cruiser *USS Birmingham*.

1910   First modern bolt-on wheels are introduced by the Sankey Company of England. These eventually replace the old wooden-rim/cast-iron car wheel.

1911 Elmer Ambrose Sperry (1860-1930), American electrical engineer, first markets his gyrocompass. After working on his device since 1896, he perfects this electrically driven gyrocompass which always points north. During this year, his device is installed in an American battleship and stabilizes ships as well as aircraft.

1911 First electric starter for an automobile is invented by Charles Franklin Kettering (1876-1958), American engineer, and installed in a Cadillac. This finally does away with the difficult and sometimes dangerous business of handcranking and makes it possible for a person to start a car from behind the wheel with the touch of a toe. It also makes gasoline the fuel of choice over steam, since electric starters make gasoline-powered automobiles start up at once. Owners of steam vehicles must wait for them to heat to the required pressure.

FEBRUARY 1911 First amphibian airplane (able to takeoff and land on both water and land) is built by American inventor Glenn Hammond Curtiss (1878-1930), who fits wheels to his seaplane. Curtiss will become the world's leading pioneer and promoter of seaplanes.

APRIL 16, 1912 First woman to fly across the English Channel is Harriet Quimby (1884-1912) of the United States.

JULY 1, 1912 First female airplane pilot to be killed in flight is Harriet Quimby (1884-1912) of the United States, who crashes in Boston, Massachusetts.

1912 First diesel-powered locomotive is built by the Swiss firm Sulzer. It weighs 85 tons (77 metric tons) and provides 1,200 hp. Diesel-powered trains do not really become practical until the Germans develop a two-car, streamlined, diesel-electric in 1932. Diesel-electrics

then begin to appear in the United States and elsewhere.

1912 First use of the monocoque fuselage is made in France. Taken from the French word meaning "eggshell," this new design is a hollow, shell-like structure in which the shell itself carries most of the load and stress. Conceived by the Swiss designer Ruchonnet and applied by L. Bechereau on a Deperdussin monoplane, this invention is one of the milestones in aviation since it is the fuselage design of the future.

JANUARY 15, 1914 First regularly scheduled airplane passenger service in the United States begins operations. The Benoist Company runs a line between St. Petersburg and Tampa, Florida.

AUGUST 15, 1914 Panama Canal, linking the Atlantic and Pacific Oceans, officially opens. Built across the Isthmus of Panama under the direction of American engineer George Washington Goethals (1858-1928), it is really three major engineering projects in one. First an enormous dam is built to control the Chagras River and create a lake; next a deep cut is made through the hill whose top is the Continental Divide; and then giant locks are built at each end of the canal to move ships up and down the 85-ft (26-m) difference. It takes about eight years to complete this massive, 51.2-mi (85-km) canal. There are six sets of double locks, each 1,000 ft (305 m) long. Like the Suez Canal, the Panama Canal saves tremendous distances for ships—nearly 6,000 mi (9,654 km) are cut off a journey from England to California; nearly 8,000 mi (12,872 km) from a trip from New York to San Francisco; and over 10,000 mi (16,090 km) from San Francisco to the Strait of Gibraltar.

1914 First use of airplane flaps are made by the Royal Aircraft Factory's Farnborough *S.E.4* biplane in England. The flaps are movable, often hinged, surfaces that can be altered to provide more lift or more drag.

DECEMBER 1915 First successful, all-metal, fully cantilever-wing airplane, the *J-1* monoplane flies in Dessau, Germany. Built by German aircraft designer Hugo Junkers (1859-1935), the cantilever wings of this revolutionary, iron-and-steel airplane are supported at the base only and have no external bracing.

JULY 11, 1916 First comprehensive act of the United States federal government aimed at the establishment of a nationwide system of interstate highways becomes law. The commonly experienced difficulties of driving from one state with good roads into another with unimproved roads forces a national solution to this problem.

1917 Paul Langevin (1872-1946), French physicist, first succeeds in using sound waves or an acoustical echo as an underwater detector. His system uses piezoelectricity to create ultrasonic waves. This employs the principle that certain sound vibrations can cause an electrical effect. By World War II, the system called "sonar" (SOund NAvigation and Ranging) is perfected. Modern sonar, which employs a transducer (a device that converts energy from one form to another), is used for mapping ocean bottoms and fish or wreck locations as well as submarine detection.

1917 Steel discs for automobile tires first appear. Built in the shape of a disc, their drop center rim design allows lower pressure tires to be used. Today's auto wheels still use this rim concept for keeping the tires on the wheel.

1918 Ethyl gasoline is first sold by General Motors Laboratories in the United States. This is a fluid additive put into gasoline that reduces engine knocking by slowing down the air/fuel burn rate. As a lead compound, however, it is later found to be a harmful toxic contaminant and is no longer allowed in gasoline.

MAY 16, 1919 First transatlantic airplane flight begins in stages by the United States Navy's Curtiss *NC-4* seaplane flown by Lt. Cdr. A. C. Read (1887-1967) and his crew. Leaving Trepassey Bay, Newfoundland, it flies to Plymouth, England, via two stops in the Azores and Lisbon, Spain. It completes its trip on May 27.

JUNE 14, 1919 First direct, non-stop crossing of the Atlantic by airplane is made by a British two-man team. John Alcock (1882-1919) and Arthur Whitten-Brown (1886-1948) fly a Vickers *Vimy* bomber for 16 hours 27 minutes from St. Johns, Newfoundland to Clifden, Ireland.

(See also May 20, 1927)

JULY 2, 1919 First crossing of the Atlantic by airship is made by the British rigid airship *R-34*. This giant dirigible flies nonstop for 11 days from Scotland to New York. It later returns to Europe, marking the first two-way crossing of the Atlantic by any type of aircraft.

NOVEMBER 20, 1919 First municipal airport in the United States opens in Tucson, Arizona. It is still in use.

1919 Glenn Hammond Curtiss (1878-1930), American inventor, invents the first modern mobile home trailer. He designs and builds his own custom trailer, called the Aerocar, which is twice as long as his car and has four Pullman berths, a galley, closets, running water, and a telephone to the car pulling it.

FEBRUARY 21, 1921 First solo, coast-to-coast flight across the United States is begun by American aviator William D. Coney. He flies from San Diego, California, to Jacksonville, Florida, in 22 hours, 27 minutes.

1921 Prototype of the modern bus, the Safety Bus, is first introduced by the brothers Frank and William Fageol of the United States. This 22-passenger, single-deck bus has an enclosed but very wide body and a low-slung chassis. It has a totally new appearance and becomes very popular. In 1927, the brothers introduce another revolutionary bus design. Called the Twin Coach, it looks almost identical at the front and rear, seats 43, and is powered by two 4-cylinder engines mounted on each side of the chassis. Its radical design has a flat front and back, and the entrance door is placed ahead of the front axle. It also has driver-controlled pneumatic doors. This becomes the accepted design for modern buses.

1922   First airship to be filled with helium gas is the United States Army's *C-7* non-rigid dirigible. Helium is an inert gas that weighs twice as much as hydrogen and has slightly less lifting capacity. It is nonflammable but expensive.

January 9, 1923   First flight of a practical gyroplane or rotorcraft is made by Spanish designer Juan de la Cierva's (1886-1936) *C-3 Autogiro*, which is flown by Spenser Gomes in Spain. Unlike an airplane, a gyroplane derives its lift and/or thrust from rotating airfoils or wings. Although this version is not a true helicopter since it has wings, it does advance its development.

May 2, 1923   First non-stop flight across the continental United States is made by J. A. Macready and Oakley Kelly of the United States Air Service. They fly 2,520 mi (4,055 km) from New York to San Diego, California, in 26 hours, 50 minutes in a T-2 Fokker transport.

April 6, 1924   First successful flight around the world begins as four Douglas *World Cruisers* leave Seattle, Washington. Of the four, only two complete the circumnavigation as they each fly 27,553 mi (44,430 km) in 175 days and return to Seattle on September 28. The actual flying time is 371 hours, 11 minutes.

September 1924   First modern highway opens in Italy between Milan and Varese. Called the "autostrada" or automobile road, it has three undivided lanes on a 33-ft (10-m) roadway with 3-ft (91-cm) shoulders. It is the forerunner of the modern, high-volume, high-speed highway.

1924   Oil filters are first introduced for automobile engines. As filtering devices that strain and remove foreign particles from an engine's circulating oil, they must also be designed to retain oil after an engine stops.

March 16, 1926   First free flight of a liquid-fueled rocket takes place as American physicist Robert Hutchings Goddard (1882-1945) designs and builds a rocket that burns liquid oxygen and gasoline and successfully launches it at a farm in Auburn, Massachusetts. Launched from a 6.5-ft (2-m)-tall A-frame, it accelerates to a height of 41 ft (12.5 m) with an average speed of 60 mph (96 kph) in its 2.5-second flight. It lands 184 ft (56 m) away. This brief flight is to rocketry what the Wright Brothers' 12-second flight is to aviation.

1926   First frameless tractor/trailer truck in which a large cylindrical tank provides its own frame is made in England. These rear-wheel only cylinders are still used today to transport milk, chemicals, and other liquids and gases.

May 20, 1927   Charles Augustus Lindbergh (1902-1974), American aviator, leaves Roosevelt Field, Long Island, New York, and achieves the first solo, non-stop flight across the Atlantic Ocean. Navigating his Ryan monoplane *Spirit of St. Louis* by reckoning at times, he lands at Le Bourget Airfield, Paris, on May 21. He covers 3,600 mi (5,792 km) in 33 hours, 29 minutes. His daring and romantic solo flight has a tremendous impact and serves to make the United States and the world more aware of the true potential of aviation.

September 18, 1928   First flight of the Zeppelin *LZ-127 Graf Zeppelin* occurs. It becomes the most successful rigid airship ever built and flies over a million miles, carrying some 13,100 passengers.

Three Goodyear airships, the *Columbia, America,* and *Mayflower II.*

SEPTEMBER 18, 1928 First rotating-wing aircraft to fly the English Channel is the Cierva *C-8L Autogyro,* flown by its Spanish designer, Juan de la Cierva (1886-1936).

OCTOBER 11, 1928 First transatlantic crossing by an airship carrying paying passengers begins as the *Graf Zeppelin* leaves Germany for New Jersey.

NOVEMBER 1928 Ancestor of the modern container ship, the *Seatrain New Orleans* first be-

gins a new transport service. Its major innovation is its use of sealed containers which the shipper places on railroad flats. They are not unloaded until they reach their destination. This prepackaging of goods will become an efficient way of transporting cargo.

AUGUST 8, 1929 First flight around the world by an airship is begun as the *Graf Zeppelin* leaves New Jersey. Captained by Hugo Eckener of Germany, it makes its 21,500-mi (34,593 km) circumnavigation in 21 days, 7 hours, 34 minutes.

OCTOBER 25, 1930 First coast-to-coast air service across the United States is started by Transcontinental and Western Air (later TWA) between New York and Los Angeles.

1930 Frank Whittle, English inventor, takes out his first patent for the jet engine. Published in 1932, it is here that Whittle offers the conception for a simple, light-weight engine for propelling aircraft at high speeds by reaction propulsion.

1930 Bathysphere, the first deep sea exploration vessel, is developed by American naturalist Charles William Beebee (1877-1962) and American engineer Otis Barton. This steel spherical underwater vessel is capable of maintaining an interior environment of ordinary pressure when lowered beneath the sea. It has thick quartz windows and is suspended by a cable from a boat. In 1934, they descend to a depth of 3,028 ft (923 m). It proves difficult to operate and since it is not maneuverable, it is eventually replaced by the safer, more maneuverable bathyscaphe.

(See also 1948)

1931 First practical delta-wing aircraft flies in Germany. Built by Alexander Lippisch (1894-1976), the chief designer of the delta-wing airplane, this modern design has wings shaped to resemble an isosceles triangle. The true delta does not flourish until the development of the practical jet engine.

MAY 20, 1932 First solo flight by a woman across the Atlantic Ocean is made by American aviator Amelia Earhart (1898-1937). She flies from Newfoundland to Northern Ireland in a Lockheed *Vega* monoplane in 13 hours, 30 minutes.

1932 First fully modern highway system, the German autobahn network, opens to traffic. Conceived in 1926, this high-speed, limited access highway consists of dual roadways separated by a substantial median area.

FEBRUARY 25, 1933 First United States aircraft carrier, *Ranger*, is launched at Newport News, Virginia. It is essentially an airfield at sea, equipped with catapults for takeoffs and hooks and wires for landings.

FEBRUARY 1933 Modern airliner is born in the United States as the Boeing 247 flies for the first time. Along with the Douglas *DC-1* and the Lockheed *Electra*, the Boeing aircraft revolutionizes all aspects of air transport and defines the future shape and features of airliners. Its dominating features are: all-metal, low-wing monoplane; two powerful, supercharged, air-cooled engines mounted on the wings; variable-pitch propellers; and retractable undercarriage.

JULY 15, 1933 First solo flight around the world is made by one-eyed American aviator Wiley Post (1899-1935). Flying a Lockheed *Vega*, he covers 15,596 mi (25,094 km) in 7 days, 18 hours, 49 minutes. He also pioneers the early development of a pressure suit and proves the value of navigating instruments as well as the automatic pilot.

MAY 26, 1934 First diesel-powered, streamlined train in the United States runs the 1,015 mi (1,633 km) between Denver and Chicago at an average (nonstop) speed of 77.6 mph (125 kph).

JUNE 26, 1936 Premiere flight of the first practical helicopter with two side-by-side rotors is made in Germany. Designed by Heinrich Focke of Germany, its longest flight is well over one hour. It is a stable, controllable aircraft.

NOVEMBER 12, 1936 San Francisco-Oakland Bay Bridge is first opened to automobile traffic. Before this six-lane bridge is built, 35 million ferry boat commuters annually pass across the bay in trips that average one hour. The physical linking of San Francisco and Oakland is accomplished by what is actually two suspension bridges joined end to end at a common anchorage point, a 540 ft (165 m) tunnel, and a 1,400 ft (427 m) cantilever bridge to the mainland. The whole adds up to a 4.5 mi (7.3 km) crossing on two decks.

1937 First submarine turbine propulsion system using oxygen generated by hydrogen peroxide is invented by the German engineer Hellmuth Walter. Prior to this system, submarines have to surface and run their diesel engines to recharge the underwater batteries. This new engine uses chemical catalysts to break down hydrogen peroxide into water and oxygen. The freed oxygen is fed into a combustion chamber with diesel fuel to produce steam, which runs a turbine. Submarines can now carry their own oxygen for power consumption.

DECEMBER 31, 1938 First flight of Boeing's *307 Stratoliner*, the first passenger plane to have a pressurized cabin, is made. This four-engine aircraft carries 33 passengers over long distances. Now that the cabin is pressurized with warm air for passenger comfort, the new generation of airliners can take advantage of high-altitude operation where maximum speed can be obtained with minimum power.

1938 First version of what becomes known as the Volkswagen "Beetle" appears in Germany. Designed by Austro-Hungarian automotive designer Ferdinand Porsche (1875-1952), it is

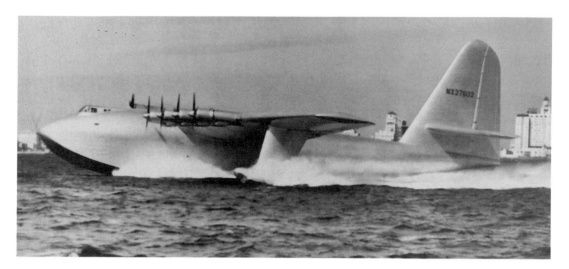

Howard Hughes's *Spruce Goose*, the largest seaplane ever flown.

sponsored by the German government as an ideal "people's car."

AUGUST 27, 1939   First fully jet-propelled aircraft to fly is Germany's *Heinkel 178*. It is powered by a centrifugal flow turbojet engine designed by Han von Ohain. Jet propulsion will revolutionize flying, but at this time, neither the Germans nor the British see its full potential, and operational jet aircraft do not appear until very late in World War II. When they do emerge, however, they outclass everything else.

1939   Longest underwater automobile tunnel in Japan and the world's first under-ocean tunnel, the Kanmon Tunnel, is begun. Connecting Honshu and Kyushu Islands, it is actually dug under the ocean (the Kanmon Straits).

1940   Jeep is first introduced by the Bantam Car Company of the United States, which makes a functional, four-wheel drive, light military scout car for the United States Army. These four-cylinder vehicles can climb a 69% grade fully loaded and, with special canvas sides and a snorkel, even cross streams up to 6 ft (1.8 m) deep. They eventually become known as Jeeps, a variation on the abbreviation for "General

Purpose." In 1941, the Ford and Willys-Overland begin producing them, since Bantam cannot meet the demand.

1940   First modern-type automatic transmission is offered as an option by General Motors on their Oldsmobile models. Called "Hydra-Matic," it makes gear changes according to varying road and load conditions without the driver having to do anything, since each gear is controlled through fluid pressure which regulates brake band clamping and engages or disengages a clutch.

JANUARY 13, 1942   First fully practical, single-rotor helicopter, built by Russian-American aeronautical engineer Igor Ivanovich Sikorsky (1889-1972), makes a successful flight. This is the first classic helicopter to use a single powered rotor to accomplish all the things that a helicopter should do: vertical take-off and landing, hovering stationary in the air, progressing through the air, and flying backwards, sideways, and forward.

1943   Jacques Yves Cousteau, French ocean explorer, and Emile Gagnan of France first develop the aqualung or scuba gear. This SCUBA

(Self-Contained Underwater Breathing Apparatus) system allows a diver to swim freely down to about 180 ft (55 m). It consists of three small bottles, containing highly compressed air, that the diver wears on his or her back. The bottles are connected to a demand regulator which automatically supplies air at the correct pressure according to the diver's depth. Later, this system will be improved and divers will use a mixture of oxygen and helium rather than normal air (oxygen and nitrogen), since this will allow them to operate as deep as 1,640 ft (500 m).

1944 First rocket-powered airplane enters service with the German Luftwaffe. The Messerschmitt *Me-163* has swept-back wings, a liquid-fuel rocket motor, and can climb to 30,000 ft (9,144 m) in two-and-one-half minutes. Because it is rocket and not jet-powered, it can only fly at maximum speed (590 mph) for 8-10 minutes. It enters the war too late to have any effect.

APRIL 24, 1946 Initial flights of the first Soviet-designed and built jet aircraft, *MiG-9* and *Yak-15*, are made. Olga Yamschikova, a member of the company test team, is probably the first woman to fly a jet aircraft when she pilots a *Yak-15* in 1947.

DECEMBER 8, 1946 Bell *X-1* rocket-powered test plane makes its first successful flight and reaches a speed of 550 mph (885 kph). This is the first American aircraft designed for supersonic speeds.

OCTOBER 14, 1947 First piloted aircraft to exceed the speed of sound in level flight is the Bell *X-1* flown by American aviator Charles Elwood Yeager. This rocket-powered aircraft is launched or released from a parent aircraft and reaches a speed of 670 mph (1,078 kph) or Mach 1.015. Mach speed is relative to an airplane's altitude, so that Mach 1.0 at sea level at 59°F (15°C) is 760.98 mph (1,224 kph).

In the stratosphere (above 36,089 ft [11,000 m]), Mach 1 is 659.78 mph (1,062 kph).

1948 First modern bathyscaphe, the *FNRS 2*, makes underwater trials. Built by Auguste Piccard (1884-1963) of Switzerland, this navigable diving vessel is designed to reach great depths in the ocean, and an improved version eventually descends to 13,000 ft (3962 m) in 1954. It consists of two main components: a heavy, steel crew cabin resistant to sea pressure and a light, gasoline filled container called a float. This float is lighter than water and thus provides the necessary lifting power. On January 23, 1960, Piccard and Dons Wals of the United States make a record manned underwater descent in the bathyscaphe, *Trieste*. They descend 35,810 ft (10,916 m) in the Mariana Trench in the Pacific Ocean.

1948 Reliable tubeless tires are first made available in the United States by B. F. Goodrich. These new tires require no inner tubes and are built to maintain air pressure by forming an airtight seal with a tire bead or edge and the wheel rim. Nearly all auto tires are tubeless today.

1948 First radial-ply automobile tire is marketed by Michelin of France. Called the Michelin X, it is first fitted on a Citroen. Radials have their plies or inner cord material running at right angles to the beads or sides of a tire. They prove to be strong tires that give good fuel economy and better road traction than traditional-ply tires.

FEBRUARY 26, 1949 First non-stop refuelled flight around the world is made by a United States Boeing *B-50*. Capt. James Gallagher and crew cover 23,452 mi (7,148 km) in 94 hours, 1 minute, and are refuelled in flight four times.

NOVEMBER 10, 1949 First flight of the practical tandem helicopter. The Piasecki *HRP-2*, a passenger-transport helicopter, is the heaviest

World's first supersonic jet in flight.

helicopter built to date. It has tandem (two) rotors at each end, and its long, slightly upcurved profile inspires the name "Flying Banana."

APRIL 12, 1950   First English Channel crossing in a sailplane is made by Lorne Welch of England. Flying a Weilhe sailplane, he catches the thermals (rising bodies of warm air) which form inland, and soars for a total of 210 mi (338 km).

OCTOBER 22, 1950   First non-stop jet flight across the Atlantic is made by one of two Republic *F-84E* fighters. Flown by David C. Schilling and William Ritchie of the United States, the planes are refuelled three times in flight. Schilling completes the London to Maine flight, but Ritchie is forced to bail out safely over Newfoundland.

APRIL 21, 1952   First jet airliner to enter commercial passenger service is the De Havilland *D.H. 106 Comet*. The British Overseas Airways Corporation (BOAC) offers service between London and Rome.

JULY 13, 1952   First Atlantic crossing by heli-

copter is begun by two Sikorsky *S-55*s. The flight is made in stages and is completed on July 31.

1952   Modern-type disc brakes, of Dunlop design, first are fitted to Jaguar entries in a European auto race. As distinct from drum brakes, they are designed to operate by pressing friction pads against a rotating steel disc. One piston brake cylinder is mounted on each side of a disc to press the pads against it. Disc brakes become common on a car's front wheels because they are not prone to brake fading.

MAY 16, 1953   First scheduled hydrofoil boat service begins passenger service between the Swiss and Italian parts of Lake Maggiore. Today, over 100 hydrofoil passenger vessels are operating throughout the world.

MAY 18, 1953   Jacqueline Cochrane of the United States becomes the first woman to fly faster than the speed of sound while flying a *F-86 Sabre*.

MAY 1953   First underwater crossing of the Atlantic Ocean is made by the British submarine

First automatic helicopter landing at a predetermined spot.

Andrew. It "snorkels" from Bermuda to the English Channel.

1953   First true American sports car is the Chevrolet Corvette. It is also the first series-production car with a reinforced fiberglass body fitted on a conventional steel frame. Like a true sports car, it seats only two, and also has a six-cylinder engine with three carburetors and an automatic transmission. Only 300 are built this year.

JANUARY 17, 1955   First nuclear-powered submarine, the U.S.S. *Nautilus*, is launched. Re-

sembling a conventional submarine, this 320-ft (98 m) long vessel has an underwater range that is nearly unlimited, being measured in years rather than miles. This can be considered the first true submarine, since all of its predecessors were really surface ships capable of brief underwater operations. The steam to run its two turbine engines is provided by a nuclear reactor.

NOVEMBER 28, 1956   First transition from horizontal flight to vertical flight and back again by a jet-powered aircraft is made in the United

A hydrofoil that rides on three metal plates, the Boeing Jetfoil's hull will lift clear of the water as its speed increases.

States by a Ryan *X-13 Vertijet*, a VTOL (vertical take-off and landing) aircraft.

1956   United States enacts the Federal-Aid Highway Act of 1956 and first begins its interstate highway program known formally as the National System of Interstate and Defense Highways. It is intended to be a 41,000-mi (65,969 km) network of modern freeways spanning America and linking together and serving more than 90% of all the cities with a population over 50,000.

JUNE 2, 1957   First solo balloon flight into the stratosphere (the upper portion of the atmosphere above 7 mi [11 km]) is made by Joseph W. Kittering, Jr., of the United States. In his plastic balloon, *Manhigh 1*, he stays aloft for six hours, 34 minutes and reaches an altitude of 96,000 ft (29,260 m).

OCTOBER 4, 1957   First man-made earth satellite is launched by the Soviet Union. Named *Sputnik 1*, it weighs 184 lbs (84 kg), contains a battery-powered transmitter, and resembles an aluminum globe. The easily-identifiable "beep" signals are monitored and replayed around the world. The United States is forced to recognize

that the U.S.S.R. has attained an unrealized degree of technological sophistication and possesses launch vehicles with large lifting power. The space age has begun.

1957   First true container ship, the American *Gateway City*, begins operations. Its platforms are designed to carry 35-ft (11 m) trailer vans which can be removed from the chassis of a truck and loaded directly onto the ship.

1957   First monorail passenger train opens for service in Japan in Ueno, Tokyo.

JANUARY 31, 1958   First United States satellite is launched into earth orbit by an Army Jupiter C launch vehicle. Named *Explorer 1*, it weighs 30.8 lbs (14 kg) and carries a scientific instrument designed to measure and count electrically charged atoms in space. Unexpectedly, it discovers an unknown thick belt of radiation around the earth that is later named the Van Allen Belt after the scientist conducting the experiment.

1958   Longest automobile tunnel in the world, either underwater or through rock, is first begun. The Mont Blanc Tunnel is cut through granite 8,000 ft (2,438 m) below snow-capped Mont Blanc, the highest mountain in the Alps. One opening is begun in Italy and the other in France, with the two crews meeting in August 1962. This 7-mile (11-km) long tunnel provides a year-round auto and truck route, cutting nearly 200 mi (322 km) off the distance between Paris and Milan.

1958   Harry Bertrand, American inventor, is granted the first patent for an automobile air bag system that uses a crash sensor. With this system, the bag inflates nearly simultaneously with the auto's impact, cushioning the driver's collision with the steering wheel and dashboard. In 1972, Ford Motor Company produces about 1,000 cars with air bags, but soon drops this demonstration project.

(See also 1973)

JANUARY 2, 1959   First human-made object to escape the earth's influence is launched. The Soviet Union's *Lunik 1* speeds within about 3,500 mi (5,632 km) of the moon and goes into orbit around the sun.

JUNE 7, 1959   First practical air-cushion vehicle, *SR.N1* hovercraft, makes its first successful trip over water. Designed by English engineer Christopher Cockerell, it is an oval craft with a diameter of 20 ft (6.1 m) and weighing 4 tons (3.6 metric tons). It features a flexible skirt that holds the cushion in place as it rides smoothly on a layer of air despite the unevenness of the surface.

JULY 25, 1959   First crossing of the English Channel by hovercraft is made by the English engineer Christopher Cockerell in his *SR.N1* vessel.

SEPTEMBER 15, 1959   First nuclear-powered surface ship, the Soviet ice-breaker *Lenin*, makes its maiden voyage. Designed to work in ice up to 6 ft (1.8 m) thick, it carries enough fuel to cruise for a year.

1959   First American compact car, the Chevro-

let Corvair, is introduced. Built to compete with the increasing number of small imports, it is the first United States rear-engine, air-cooled car. It later becomes a controversial vehicle when criticized by American consumer advocate Ralph Nader as being "unsafe at any speed," since it flips over easily. It goes out of production in 1970. Nader's campaign ultimately results in the introduction of federal safety standards for American cars.

APRIL 13, 1960   First navigational satellite, the experimental *Transit I-B*, is launched by the United States.

MAY 10, 1960   First undersea, around-the-world voyage is completed by the United States submarine *Trident*. Following underwater the route taken over 400 years earlier by Portuguese explorer Ferdinand Magellan (1480-1521), it travels 41,519 mi (66,804 km) in 84 days.

SEPTEMBER 24, 1960   First nuclear-powered aircraft carrier, the 1,123 ft (342 m) long *Enterprise*, is launched by the United States. Its 4.5 acres of flight deck can carry about 200 aircraft. It can reach 35 knots.

APRIL 12, 1961   First man in space and the first to orbit the earth is Soviet cosmonaut Yuri Gagarin (1934-1968). He circles the earth once in his *Vostok 1* spacecraft which reaches 187 mi (301 km) above the earth. During the entire flight, the spacecraft is controlled from the ground. On reentry, Gagarin ejects at 22,000 ft (6,706 m) and parachutes down safely after his one hour, 48 minute flight. The Soviets stun and amaze the world and the era of manned space flight begins.

MAY 5, 1961   First American in space is United States astronaut Alan Bartlett Shepard, Jr., who is launched on a suborbital trajectory by a Redstone booster rocket. He rides his *Freedom 7* space capsule to a height of 116 mi (187 km) and lands 297 mi (478 km) downrange in the Atlantic. He is weightless for 5 minutes of his 15 minute, 22 second flight.

1961 Dulles International Airport opens in Virginia. It is the first civil airport designed to handle jet aircraft and features innovative design and services like mobile passenger lounges which take passengers to their waiting planes instead of having the planes taxi in close to the hub.

FEBRUARY 20, 1962 First American to orbit the earth is John Herschel Glenn, Jr., in his Mercury capsule *Friendship 7*. Launched by an Atlas booster, he completes three orbits spanning 4 hours, 55 minutes before splashing down safely in the Atlantic.

AUGUST 20, 1962 First nuclear-powered cargo ship, the United States Maritime Administration's *Savannah*, makes its maiden voyage. It is able to carry 60 passengers and 10,000 tons (9,091 metric tons) of cargo at 20 knots for 3.5 years without having to refuel.

1962 First person to swim the English Channel underwater is Fred Baldasare of the United States. During his crossing, he remains in a submerged cage that is monitored by closed circuit TV cameras.

JUNE 16, 1963 First woman in space is the Soviet Union's Valentina Vladimirovna Tereshkova who flies in *Vostok 6*. As a cotton-mill worker and amateur parachutist, she is essentially a passenger in the ground-controlled spacecraft. She has some difficult moments but reenters safely after nearly three full days in orbit.

1963 First company to market a car with a Wankel rotary engine is NSU of West Germany. The Wankel is a radical departure from traditional piston gas engines, for instead of pistons and cylinders, it has a pair of triangular-shaped rotors that rotate inside an elliptical chamber, drawing in fuel, squeezing it until ignites, and expelling the burned gases, all in one continuous movement. Although rotary engines run very smoothly, they are less fuel efficient and have reliability problems.

MAY 20, 1964 First nuclear-powered lighthouse, the Baltimore Light, goes into operation in Chesapeake Bay, Baltimore Harbor, Maryland. This 60-watt radioisotope nuclear generator can produce a continuous supply of electricity for ten years without refueling.

OCTOBER 1, 1964 Modern high-speed railroad line opens in Japan between Tokyo and Osaka. This is the first of Japan's "Shinkansen" ("New Railways"), which become widely known in the West as "bullet trains." These electric passenger trains will hit top speeds of 132 mph (212 kph).

1964 Seat belts are first made standard equipment on cars produced by the Studebaker-Packard Corporation. Although this decision marks a radical break with other major American automakers, the company still goes out of business by year's end.

MARCH 18, 1965 First man to walk in space is Soviet cosmonaut Aleksey Leonov in *Voskhod 2*. During the second orbit, Leonov floats outside the spacecraft, tethered by a cord providing him with oxygen and suit pressurization.

JUNE 3, 1965 First American walk in space is conducted during *Gemini 4*. United States astronaut Edward H. White II (1930-1967) spends 20 minutes floating outside the capsule while tethered by a 25-ft (7.6 m) long cord that provides air and ventilation.

MARCH 1, 1966 First spacecraft to land on another planet is the Soviet Union's *Venus 3* which reaches the surface of Venus. Launched on November 16, 1965, it carries instrumentation which relays data back to earth until it crashes on the surface.

1966 United States Department of Transportation is first established at the Cabinet level. Its mission is to provide leadership in the identification of transportation problems and solutions, stimulate new technological advances, encourage cooperation among all interested parties, and recommend national policies and programs to accomplish these objectives.

APRIL 23, 1967 First Soviet cosmonaut to die in space is veteran Vladimir M. Komarov (1927-1967). After a successful launch, his spacecraft develops several severe system malfunctions and he attempts an emergency reentry. The Soviets announce that spacecraft tumbling causes its parachute to fail, resulting in a high-speed, fatal impact.

DECEMBER 21, 1968 First manned mission to orbit the moon is successfully conducted by the United States crew of *Apollo 8*. American astronauts Frank Borman, James A. Lovell, Jr., and William A. Anders enter lunar orbit on Christmas Eve. Their readings from the Book of Genesis as they circle the moon on Christmas Day makes this one of the most memorable space flights in history.

DECEMBER 31, 1968 First supersonic jet transport, the Tupolev *Tu-144*, makes its first flight. This Soviet aircraft soon begins regular service and later becomes the first commercial transport to exceed Mach 2 (twice the speed of sound).

FEBRUARY 9, 1969 First flight of the Boeing *747* "Jumbo Jet" airliner is made. This first wide-bodied, long-range transport is capable of carrying 347 passengers, and is, at the time, the largest aircraft in commercial airline service in the world.

MARCH 2, 1969 First flight of the British Aircraft Corporation/Aerospatiale *Concorde 001* supersonic transport is made in Toulouse, France. The product of joint British-French collaboration, it is assembled in France. In 1976, it becomes the first supersonic commercial transport to operate regularly scheduled passenger service. Both British Airways and Air France offer service to New York.

APRIL 30, 1969 First female airline pilot in the West, Turi Widerose of Norway, makes her first scheduled flight as a first officer for Scandinavian Airlines.

JULY 20, 1969 Man first walks on the moon. The United States *Apollo 11* crew of Neil Alden Armstrong, Edwin Eugene Aldrin, Jr, and Michael Collins achieves the national goal set by President John Fitzgerald Kennedy (1917-1963) on May 25, 1961. Armstrong first steps on the moon announcing, "That's one small step for a man, one giant leap for mankind." Aldrin joins him minutes later and they both gather lunar samples and conduct and set up experiments during their two hours outside the lunar module. The landing and moon walk are watched live on television by millions of people worldwide.

1969 First LASH ship, the Japanese-built *Acadia Forest*, is launched. This specialized LASH container ship (Lighter Aboard SHip) carries very large floating containers called "lighters." It is equipped with a massive gantry crane that straddles the deck and is used to load and unload lighters. Each lighter has a 400-ton (364-metric ton) capacity and is stowed in the holds on deck. Loading and discharging is rapid (15 minutes per lighter), and no port or dock facilities are necessary, since a group of lighters can be towed onto inland waterways.

APRIL 19, 1971 First space station, *Salyut 1*, is launched by the Soviet Union into earth orbit. It is occupied on June 6 of this year by cosmonauts Georgy T. Dobrovolsky and Vladislav N. Volkov and becomes the first manned space station. After a successful three-week mission, the cosmonauts undock and attempt to return to earth. During separation, however, an exhaust valve in their return craft is opened and the cabin depressurizes, killing both men within 45 seconds. After a routine, remote-controlled landing, the cosmonauts are found dead in their seats. Had they been wearing spacesuits inside the cabin, they would have survived.

1973 First automobile air bag is offered by General Motors. This is an inflatable cushion re-

The "Jumbo Jet," a Boeing 747 airplane.

straint installed in a car's steering wheel or dashboard that inflates with nitrogen upon collision and protects the driver or front passengers. The auto industry resists adopting this safety system, and it takes over 20 years to become standard equipment.

SEPTEMBER 1, 1975 First aircraft to make two return transatlantic flights (or four transatlantic crossings in one day) is the Aerospatiale/BAC *Concorde 001*. This supersonic transport flies from London, England, to Gander, Newfoundland, Canada, twice.

JANUARY 21, 1976 First passenger services by a supersonic airliner are made as British Airways and Air France *Concorde* supersonic transports take off simultaneously from their countries for Bahrain and Rio de Janeiro.

MAY 9, 1978 First crossing of the English Channel by a powered hang-glider is made by David Cook. He installs a piston engine on a Volmer *VJ-23 Swingwing* and crosses in one hour, 15 minutes.

AUGUST 12, 1978 First transatlantic crossing by a gas balloon begins as the *Double Eagle II*

establishes new endurance and distance records for gas balloons. Crewed by Americans Ben L. Abruzzo, Maxie L. Anderson, and Larry M. Newman, it flies 3,107 mi (4,999 km) in 137 hours and 5 minutes.

DECEMBER 19, 1978 First solar-powered aircraft, *Solar One*, makes a successful but brief hop-flight in England. David Williams and Fred To of England later fly the aircraft for almost three-quarters of a mile on June 13, 1979. Since it uses batteries to store the electricity generated by its 750 solar cells, some argue that it is an electric and not a solar aircraft.

JUNE 12, 1979 First human-powered aircraft to cross the English Channel is the *Gossamer Albatross*, designed and built by Paul MacCready of the United States. Flown by bicyclist Bryan Allen, it crosses from England in two hours, 49 minutes. Its wire bracing has high aerodynamic drag, and the tail-first design makes it difficult to fly, especially at higher speeds.

1980 First sail-assisted commercial ship to be built in the past 50 years, Japan's *Shin-Aitoku-Maru* is launched. This 1,750-ton (1,591-

Apollo II astronaut Edwin E. Aldrin, Jr. descending from the lunar module to the surface of the moon.

metric ton) tanker uses new materials and computer technology to build a sail system that supplements its engines. The entire mast and sail can be rotated mechanically on their own axes from a central control room. The sails are not hoisted or lowered but rather rolled out mechanically from the mast, and the system is able to respond to weather changes and obtain the most favorable angle to the wind.

APRIL 12, 1981 First winged, reusable spacecraft is launched into space as the United States space shuttle *Columbia* enters earth orbit. Flown by American astronauts John W. Young and Robert L. Crippen, this revolutionary spacecraft is launched on the 20th anniversary of Soviet cosmonaut Yuri Gagarin's (1934-1968) first flight in space. The shuttle uses solid-fuel booster rockets which it jettisons after they burn out, marking the first time solid fuel is used in a manned spaceflight. After two days of in-orbit tests, the crew pilots the shuttle to a perfect landing on the lake bed at Edwards Air Force Base, California.

JULY 7, 1981 First solar-powered aircraft flight

across the English Channel is made by *Solar Challenger* designed by Paul MacCready of the United States. Piloted by Steve Ptacek, it takes off from France and flies 180 mi (290 km) in five hours, 23 minutes. It reaches an altitude of 11,000 ft (3,353 m) and maintains an average speed of 35 mph (56 kph). Power is provided by 16,128 solar cells on the upper surfaces of the wing and tailplane which provide maximum power of 3 hp to its electric motor.

NOVEMBER 9, 1981 First manned crossing of the Pacific in a gas balloon is begun by Ben L. Abruzzo of the United States and a three-man crew. The helium-filled *Double Eagle V* covers 5,209 miles (8,381 km) during its three-day flight from Japan to California.

JUNE 18, 1983 First American woman astronaut, Sally K. Ride, is launched into space aboard the space shuttle *Challenger* on mission STS-7. She is accompanied by four male astronauts who, with her, comprise the first five-person crew to be launched into space. During their six-day mission, they deploy three satellites and demonstrate the shuttle's remote manipulator system by retrieving one

Footprint left by the first man on the moon, Neil A. Armstrong.

satellite from orbit. This is the first time a satellite is retrieved by a manned spacecraft.

FEBRUARY 3, 1984 First untethered space walk is made by American astronaut Bruce McCandless II during mission 41-B of the *Challenger* space shuttle. During the seven-day mission, McCandless uses the manned maneuvering unit (MMU), a self-contained, propulsive backpack, that enables him to move freely in space without having to be tethered to a spacecraft. The shuttle returns to Cape Canaveral, Florida, marking the first time a spacecraft lands at the place it was launched.

FEBRUARY 19, 1986 Soviet Union launches the core of the first permanent space station called *Mir* or "Peace." It is about 42.5 ft (13 m) long, 13.5 ft (4 m) maximum diameter, and weighs about 21 tons (19 metric tons). It contains several ports which will accept docking units that can be added on. It can accommodate a crew of twelve.

DECEMBER 14, 1986 First aircraft to fly around the world non-stop and unrefuelled is the specially-built *Voyager* aircraft. Designed by American Burt Rutan and flown by Dick Rutan and Jeana Yeager, it is a trimaran monoplane constructed of composite materials that include Magnamite graphite and Hexcel honeycomb. It has two British-built, high-efficiency engines, one of which is water-cooled. Its takeoff weight is 9,750 lbs (4,427 kg). It covers nearly 24,987 mi (40,212 km) flying westbound from California in 9 days, 3 minutes, 44 seconds. At landing, it has only 14 gal (53 l) of fuel left. This flight more than doubles all unrefueled, nonstop records and demonstrates the possibilities of all-composite structures.

JULY 2, 1987 First hot-air balloon to cross the Atlantic Ocean, the *Atlantic Flyer* is launched from Sugarloaf, Maine. Piloted by Richard Branson and Per Lindstrand, the balloon crashes just short of the coast of England after a flight of 31 hours, 41 minutes.

The first passenger service by a supersonic airliner.

JANUARY 15, 1991 First hot-air balloon to cross the Pacific Ocean takes off from Miyakonojo, Japan. Piloted by Richard Barnson and Per Lindstrand, it touches down two days later on a frozen lake in the North West Territories of Canada.

MARCH 3, 1991 First jet crash in United States history to be judged as "no known cause" by the National Transportation Safety Board (NTSB) occurs at Denver, Colorado's Stapleton International Airport. United Airlines Flight 505, with a crew of six and 20 passengers, unaccountably pitches up, yaws to the right, and nose-dives into the ground, killing everyone aboard. The NTSB rules out any mechanical failure on the Boeing 737-200, as well as human error. Weather is not considered a factor. This is the first unsolved case in the 25-year history of the NTSB, which keeps its files open and continues to reevaluate the evidence.

NOVEMBER 21, 1991 First person to row a boat across the North Pacific Ocean lands in Ilwaco, Washington. Forty-six-year-old Frenchman Gerard d'Aboville ends his 133-day solo voyage which covers 5,500 mi (8,850 km). Rowing an average speed of 2 knots and making 41 nautical miles a day in his kayak-like vessel, he claims to have capsized "35 or 36 times in incredibly violent storms."

JUNE 1992 First voyage of a full-sized MHD (magnetohydrodynamic propulsion) vessel is made in Japan. The 185-ton (168-metric ton) research ship *Yamato I* uses a revolutionary MHD drive system that is virtually silent and has no moving parts. The principle behind the ship's propulsion system is one found in all electric motors. When an electrical charge moves through a magnetic field, it is subject to a force (called the Lorentz force) that acts at right angles to the motion of the charge. The new drive system does not use propellers to push the ship through the water but rather employs the Lorentz force (via powerful superconducting electromagnets) to propel sea water out a nozzle at the stern of the ship, using a form of jet propulsion. This experimental drive system eliminates problems related to propeller systems as well as speed limitations, mechanical vibrations, and even detection by sonar.

FEBRUARY 1, 1993 Amtrak makes the first paying-passenger run of a three-month test of its new high-speed, X-2000 "tilt train." Running

This high-speed Japanese train can reach 155 mph (249 kph).

between Washington, D.C., and New York, this new train can take curves as much as 40% faster than current equipment. It does not have to slow down for a curve since each car tilts into the curve, and passengers feel little or no discomfort. Initial tests show the ride to be extremely quiet and smooth.

NOVEMBER 14, 1994 Eurotunnel, the underground "chunnel" connecting France and Britain by rail, opens officially as the Eurostar bullet train makes its first passenger run. Built by the Anglo-French consortium Eurotunnel at a cost of $16 billion, this undersea bridge is hailed as the greatest engineering feat of the century. The trip between Paris and London takes three hours and is offered twice a day in each direction. Ticket prices are close to those of airfares, and the tunnel is expected to herald a new age in continental travel.

FEBRUARY 3, 1995 First woman to pilot the United States Space Shuttle is American astronaut Eileen Collins. She conducts a successful *STS-63* shuttle mission.

MARCH 24, 1996 First American woman to board the orbiting Russian space station *Mir* is astronaut Shannon Lucid. She transfers from the United States space shuttle which is docked with *Mir*. Three days later, two Americans, Linda Godwin and Michael Clifford, make the first space "walk" from the shuttle to *Mir*.

SEPTEMBER 19, 1996 American astronaut Shannon Lucid transfers back to the United States space shuttle *Atlantis* from Russian space station *Mir*, six weeks later than scheduled. Within a week, she will return home after a record-breaking six months in orbit.

BIBLIOGRAPHY

A

Asimov, Isaac. *Asimov's Chronology of Science and Discovery.* New York: Harper & Row, 1989.

Augarten, Stan. *Bit by Bit: An Illustrated History of Computers.* New York: Ticknor & Fields, 1984.

B

Bendiner, Jessica and Elmer Bendiner. *Biographical Dictionary of Medicine.* New York: Facts On File, 1990.

Bloom, Alan. *250 Years of Steam.* Tadworth, Surrey: World's Work Ltd., 1988.

Bordley, James, III and A. McGehee Harvey. *Two Centuries of American Medicine: 1776-1976.* Philadelphia: W.B. Saunders Co., 1976.

Boyer, Carl B. and Uta C. Merzbach. *A History of Mathematics.* New York: John Wiley & Sons, 1989.

Brock, William Hodson. *The Norton History of Chemistry.* New York: W. W. Norton, 1993.

Bruno, Leonard C. *On The Move: A Chronology of Advances in Transportation.* Detroit: Gale Research Inc., 1993.

Bunch, Bryan H. and Alexander Hellemans. *The Timetables of Technology.* New York: Simon and Schuster, 1993.

C

Calder, Ritchie. *The Evolution of the Machine.* Toronto: McClelland and Stewart Ltd., 1968.

Castiglioni, Arturo. *A History of Medicine.* New York: J. Aronson, 1975.

Cipolla, Carlo M. and Derek Birdsall. *The Technology of Man.* New York: Holt, Rinehart and Winston, 1979.

Clair, Colin. *A Chronology of Printing.* New York: Frederick A. Praeger, 1969.

D

De Bono, Edward, ed. *Eureka! How and When Great Inventions Were Made.* London: Thames and Hudson, 1974.

Du Vall, Nell. *Domestic Technology: A Chronology of Developments.* Boston: G.K. Hall, 1988.

Dummer, G.W.A. *Electronic Inventions: 1745-1976.* New York: Pergamon Press, 1977.

Dunlap, Orrin E. *Radio & Television Almanac.* New York: Harper & Brothers, 1951.

E

Eves, Howard W. *An Introduction to the History of Mathematics.* Philadelphia: The Saunders Series, 1990.

F

Faul, Henry. *It Began with a Stone: A History of Geology from the Stone Age to the Age of Plate Tectonics.* New York: J. Wiley, 1983.

Finniston, Monty, Trevor Williams, and Christopher Bissell, eds. *Oxford Illustrated Encyclopedia of Invention and Technology.* Oxford: Oxford University Press, 1985.

G

Gardner, Eldon J. *History of Biology.* Minneapolis: Burgess Publishing Co., 1972.

Garrison, Fielding H. *An Introduction to the History of Medicine.* Philadelphia: W.B. Saunders Co., 1929.

Gascoigne, Robert Mortimer. *A Chronology of the History of Science, 1450-1900.* New York: Garland Publishing, Inc., 1987.

Gassan, Arnold. *A Chronology of Photography.* Athens, OH: Handbook Co., 1972.

Giscard d'Estaing, Valerie-Anne. *The Second World Almanac Book of Inventions.* New York: World Almanac, 1986.

Gohau, George. *A History of Geology.* New Brunswick: Rutgers University Press, 1990.

Great Scientific Achievements: The Twentieth Century. Pasadena: Salem Press, 1994.

Green, Joseph Reynolds. *A History of Botany, 1860-1900. Being a Continuation of Sachs' History of Botany.* New York: Russell & Russell, 1967.

H

Hellemans, Alexander and Bryan H. Bunch. *The Timetables of Science.* New York: Simon and Schuster: 1988.

Herrmann, Dieter B. *The History of Astronomy from Herschel to Hertzsprung.* Cambridge: Cambridge University Press, 1984.

History of Energy. Washington: Department of Energy, 1980.

A History of Paper. New York: Fraser Paper, 1964.

Hooper, Tony. *Genetics.* Austin: Raintree Steck-Vaughn Publishers, 1994.

Hunter, Dard. *Papermaking: The History and Technique of an Ancient Craft.* New York: Dover Publications, Inc., 1978.

K

Kaufman, Morris. *The First Century of Plastics.* London: Plastics Institute, 1963.

L

Lay, M.G. *Ways of the World: A History of the World's Roads and of the Vehicles That Used Them.* New Brunswick, NJ: Rutgers University Press, 1992.

Lemagny, Jean-Claude and Andre Rouille, eds. *A History of Photography.* Cambridge: Cambridge University Press, 1987.

M

Magill, Frank N. *Great Events From History II: Science and Technology Series.* Pasadena: Salem Press, 1991.

Magner, Lois N. *A History of Medicine.* New York: M. Dekker, 1992.

———. *A History of the Life Sciences.* New York: M. Dekker, 1994.

Marcorini, Edgardo, ed. *The History of Science and Technology:* A Narrative Chronology. New York: Facts On File, 1988.

Mayr, Ernst. *The Growth of Biological Thought.* Cambridge, MA: Belknap Press, 1982.

McGraw-Hill Encyclopedia of Astronomy. New York: McGraw-Hill, 1993.

McGrew, Roderick E. *Encyclopedia of Medical History*. New York: McGraw-Hill, 1985.

Mez-Mangold, Lydia. *A History of Drugs*. Basle: F. Hoffman-La Roche & Co., 1971.

Moore, Patrick. *History of Astronomy*. London: Macdonald & Co., 1983.

Morton, Alan G. *History of Botanical Science: An Account of Botany from Ancient Times to the Present Day*. New York: Academic Press, 1981.

Motz, Lloyd. *The Story of Astronomy*. New York: Plenum Press, 1995.

Mould, Richard F. *A Century of X-Rays and Radioactivity in Medicine*. Bristol: Institute of Physics Publishing, 1993.

Mount, Ellis and Barbara A. List. *Milestones in Science and Technology*. Phoenix: Oryx Press, 1994.

Muller, Heinrich. *Guns, Pistols, Revolvers*. New York: St. Martin's Press, 1980.

N

Nordenskiold, Erik. *The History of Biology*. St. Clair Shores, MI: Scholarly Press, 1976.

O

Ochoa, George and Melinda Corey. *The Timeline Book of Science*. New York: The Stonesong Press, 1995.

P

Pannekoek, Anton. *A History of Astronomy*. New York: Dover Publications, 1989.

Partington, James R. *A History of Chemistry*. London: Macmillan, 1961-70.

Presence, Peter, ed. *Purnell's Encyclopedia of Inventions*. Bristol: Purnell & Sons Ltd., 1976.

R

Ralston, Anthony and Edwin D. Reilly, eds. *Encyclopedia of Computer Science*. New York: Van Nostrand Reinhold, 1993.

Ronan, Colin A. *Science: Its History and Development Among the World's Cultures*. New York: Facts On File, 1982.

S

Sachs, Julius. *History of Botany, 1530-1860*. New York: Russell & Russell, 1967.

Sambursky, Samuel, ed. *Physical Thought from the Presocratics to the Quantum Physicists*. London: Hutchinson, 1974.

Schapsmeier, Edward L. *Encyclopedia of American Agricultural History*. Westport, CT: Greenwood Press, 1975.

Smith, Maryanna S. *Chronological Landmarks in American Agriculture*. Washington: Department of Agriculture, 1980.

Swetz, Frank J. *From Five Fingers to Infinity: A Journey Through the History of Mathematics*. Chicago: Open Court, 1994.

T

Thompson, Susan J. *A Chronology of Geological Thinking from Antiquity to 1899*. Metuchen, NJ: Scarecrow Press, 1988.

Travers, Bridget, ed. *World of Invention*. Detroit: Gale Research Inc., 1994.

————. *World of Scientific Discovery*. Detroit: Gale Research Inc., 1994.

Trow-Smith, Robert. *Man the Farmer*. London: Priory Press Ltd., 1973.

V

Veit, Stan. *Stan Veit's History of the Personal Computer*. Asheville, NC: WorldComm, 1993.

Von Braun, Wernher and Frederick I. Orway III. *History of Rocketry and Space Travel*. New York: Thomas Y. Crowell Co., 1966.

W

Weaver, Jefferson Hane, ed. *The World of Physics*. New York: Simon and Schuster, 1987.

Westcott, Gerald F. *Mechanical and Electrical Engineering*. London: Her Majesty's Stationery Office, 1955.

Williams, Brian. *Inventions and Discoveries*. New York: Warwick Press, 1979.

Williams, Michael R. *A History of Computing Technology*. Englewood Cliffs, NJ: Prentice-Hall, Inc., 1985.

Williams, Trevor I. *The Triumph of Invention: A History of Man's Technological Genius*. London: Macdonald Orbis, 1987.

————. *Science: A History of Discovery in the Twentieth Century*. Oxford: Oxford University Press, 1990.

Wolf, Abraham. *A History of Science, Technology, and Philosophy in the Eighteenth Century*. New York: Macmillan Co., 1939.

————. *A History of Science, Technology, and Philosophy in the 16th & 17th Centuries*. London: George Allen & Unwin Ltd., 1950.

S